Natural and artificial metalloenzymes

Online virtual event

31 January – 3 February 2022

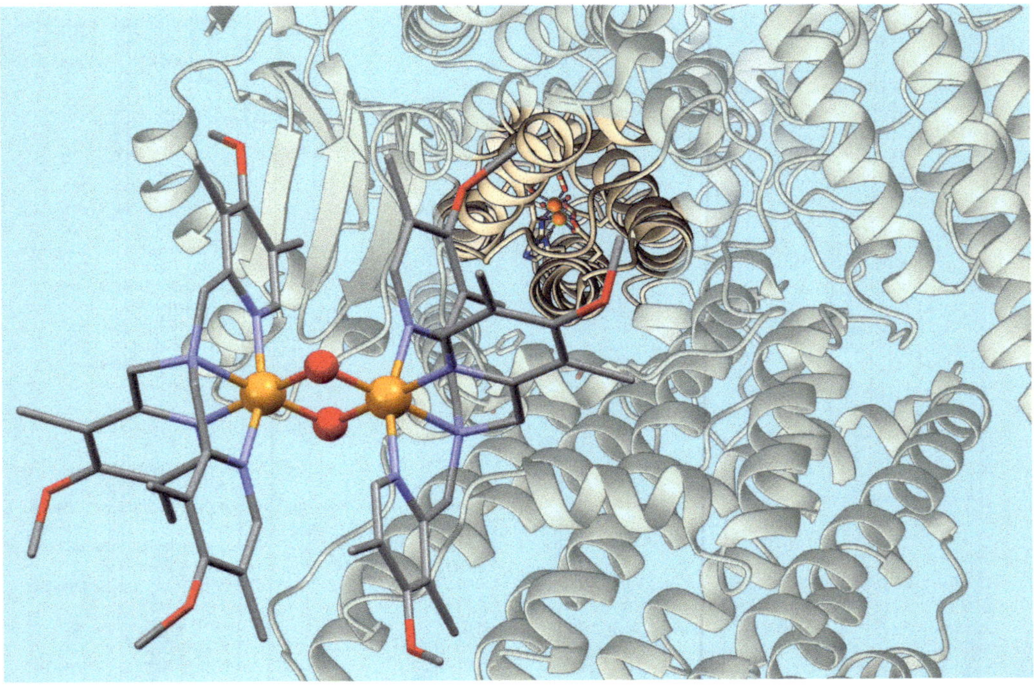

FARADAY DISCUSSIONS

Volume 234, 2022

ROYAL SOCIETY OF CHEMISTRY

The **Faraday Division** of the Royal Society of Chemistry, previously the Faraday Society, was founded in 1903 to promote the study of sciences lying between chemistry, physics and biology.

Faraday Discussions (Print ISSN 1359-6640, Electronic ISSN 1364-5498) is published 8 times a year by the Royal Society of Chemistry, Thomas Graham House, Science Park, Milton Road, Cambridge, UK CB4 0WF.

Volume 234 ISBN 978-1-83916-566-5

2022 annual subscription price: print+electronic £1133 US $1994; electronic only £1079, US $1899. Customers in Canada will be subject to a surcharge to cover GST. Customers in the EU subscribing to the electronic version only will be charged VAT.

All orders, with cheques made payable to the Royal Society of Chemistry, should be sent to the Royal Society of Chemistry Order Department, Royal Society of Chemistry, Thomas Graham House, Science Park, Milton Road, Cambridge, CB4 0WF, UK Tel +44 (0)1223 432398; E-mail **orders@rsc.org**

If you take an institutional subscription to any Royal Society of Chemistry journal you are entitled to free, site-wide web access to that journal. You can arrange access via Internet Protocol (IP) address at **www.rsc.org/ip**

Customers should make payments by cheque in sterling payable on a UK clearing bank or in US dollars payable on a US clearing bank.

Printed in the UK

MIX Paper from responsible sources FSC® C013604

ROYAL SOCIETY OF CHEMISTRY

Faraday Discussions

Faraday Discussions are unique international discussion meetings that focus on rapidly developing areas of chemistry and its interfaces with other scientific disciplines.

Information for Authors

♾ The paper used in this publication meets the requirements of ANSI/NISO Z39.48–1992 (Permanence of Paper).

Registered charity number: 207890

Natural and Artificial Metalloenzymes

Faraday Discussions

www.rsc.org/faraday_d

A General Discussion on Natural and Artificial Metalloenzymes was held virtually on the 31st of January, 1st of February, 2nd of February and 3rd of February 2022.

The Royal Society of Chemistry is the world's leading chemistry community. Through our high impact journals and publications we connect the world with the chemical sciences and invest the profits back into the chemistry community.

CONTENTS

ISSN 1359-6640; ISBN 978-1-83916-566-5

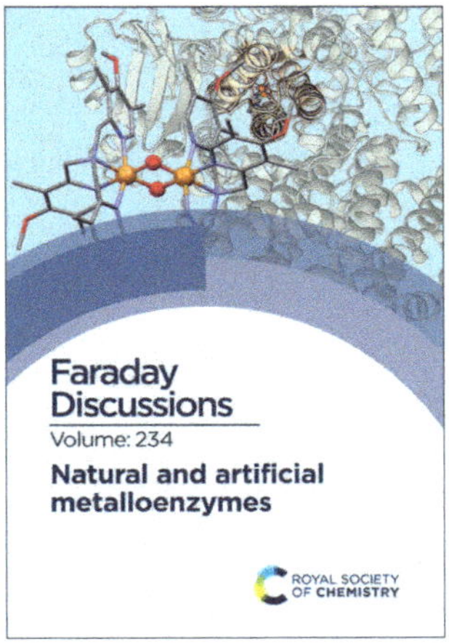

Cover
See L. Que, Jr *et al.*, *Faraday Discuss.*, 2022, **234**, 109–128.

A crystallised bis-μ-oxo-diiron(IV) complex has emerged as one of the competing structures for the oxidant generated by the methane-oxidising soluble methane monooxygenase.

The structure of sMMO hydroxylase was described by RCSB PDB: 6YD0 (*J. Am. Chem. Soc.*, 2020, **142**, 14249). The Mercury program (*J. Appl. Cryst.*, 2020, **53**, 226–235) and the ChimeraX program (*Protein Sci.*, 2021, **30**, 70–82) were used to prepare the molecular and protein images, respectively. Image reproduced by permission of Gregory Rohde from Faraday Discuss., 2022, **234**, 109–128.

INTRODUCTORY LECTURE

PAPERS AND DISCUSSIONS

POSTER SPONSORSHIP

ROYAL SOCIETY OF CHEMISTRY | FARADAY DIVISION

Faraday Discussions

ROYAL SOCIETY OF CHEMISTRY

PAPER

Spiers Memorial Lecture: activating metal sites for biological electron transfer

Edward I. Solomon *[ab] and Anex Jose [a]

Received 6th January 2022, Accepted 17th January 2022
DOI: 10.1039/d2fd00001f

Metal sites in biology often exhibit unique spectroscopic features that reflect novel geometric and electronic structures imposed by the protein that are key to reactivity. The blue copper active site involved in long range, rapid biological electron transfer is a classic example. This review presents an overview of both traditional and synchrotron based spectroscopic methods and their coupling to electronic structure calculations to understand the unique features of the blue copper active site, their contributions to function and the role of the protein in determining the geometric and electronic structure of the active site (called the "entatic state"). The relation of this active site to other biological electron transfer sites is further developed. In particular, ultrafast XFEL spectroscopy is used to evaluate the methionine–S–Fe bond in cytochrome *c*, and its entatic control by the protein in determining function (electron transfer *vs.* apoptosis).

In Bioinorganic Chemistry, we mostly focus on a small region of the large biomolecule that is the metal and its ligands, as this is the active site for catalysis (Fig. 1).[1] The wide range of spectroscopic methods of physical-inorganic chemistry that are used to probe these active sites are summarized below.[2]

1. Spectroscopy and electronic structure calculations

The left side of Fig. 2 shows the complete electronic structure diagram for square planar D_{4h}-$[CuCl_4]^{2-}$ (d^9) that will be a useful reference in this presentation.[4] All energy levels are occupied up to the highest energy half-occupied $d_{x^2-y^2}$ orbital that is the GS. On the right in Fig. 2 are the main spectroscopic methods of physical inorganic chemistry ordered by increasing energy. At low resolution, the GS is studied by magnetic susceptibility and, for ^{57}Fe, Mössbauer spectroscopy. These mostly give the redox and spin states of the metal site. At higher resolution, the GS is studied by electron paramagnetic resonance (EPR) spectroscopy, which

[a]*Department of Chemistry, Stanford University, Stanford, CA 94305, USA. E-mail: solomone@stanford.edu*
[b]*Stanford Synchrotron Radiation Lightsource, SLAC National Accelerator Laboratory, Stanford University, Menlo Park, CA 94025, USA*

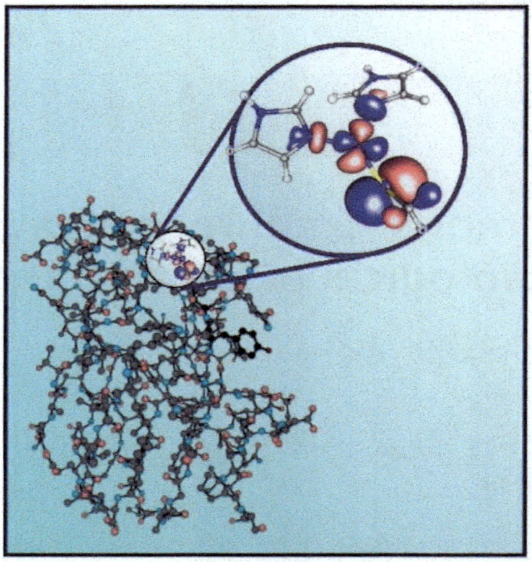

Fig. 1 The protein as a whole with its active site and ground state (GS) wave function in the expanded scale. Reprinted with permission from ref. 3. Copyright (2006) American Chemical Society.

gives the g_i values that define the nature of the half-occupied orbital (*i.e.* whether the electron is in the $d_{x^2-y^2}$ or d_{z^2} orbital) and the hyperfine couplings of the electron spin with the metal and the ligand nuclear spins that define its delocalization. This delocalization is probed at a still higher resolution using double

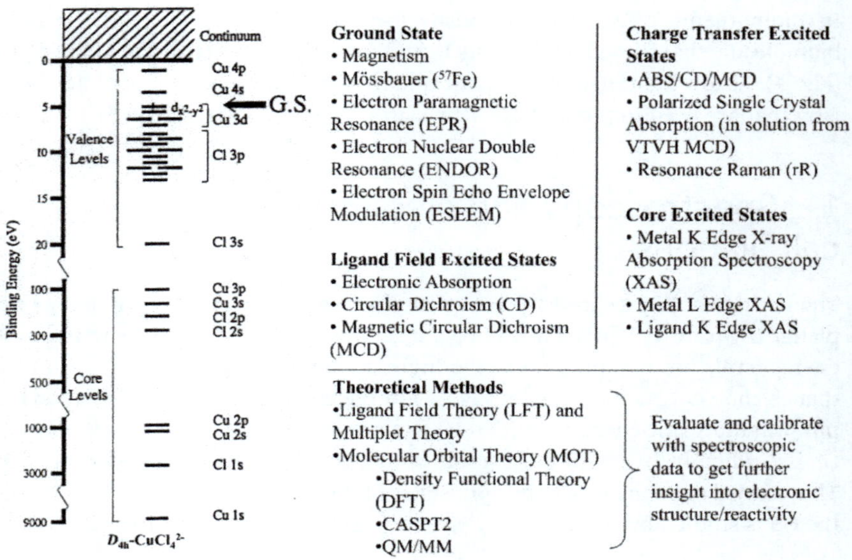

Fig. 2 Electronic structure diagram of D_{4h}-$[CuCl_4]^{2-}$ (left) and spectroscopic and electronic structure methods used in Bioinorganic Chemistry (right). Reprinted with permission from ref. 3. Copyright (2006) American Chemical Society.

resonance and pulsed EPR methods. These GS studies are generally done in a magnetic field using a microwave source at an energy in the order of 1 cm^{-1}.

Going up into the near-infrared (near-IR)/visible (Vis) energy range, electrons from the filled d orbitals in Fig. 2 (left) are excited into the half-occupied $d_{x^2-y^2}$ orbital, producing ligand field (LF) transitions. These d → d transitions are parity forbidden and, hence, weak in absorption, but can be studied by circular dichroism (CD) (in an optically active site as is generally the case in metalloproteins) and in magnetic CD (MCD) in a longitudinal magnetic field at low temperatures, as their different selection rules enhance the LF transition intensities. It is important to note that the energy splittings of the LF states are sensitive to the ligand environment of the metal active site and, thus, probe geometry and bonding.

At higher energies [Vis/ultra-violet (UV)], the photon can excite electrons from the filled ligand-based valence molecular orbitals into the hole in the $d_{x^2-y^2}$ orbital, shown in the left hand side of Fig. 2 . These are the ligand to metal charge transfer (LMCT) transitions and, since they involve transfer of the electron from a formally ligand-centered molecular orbital to a formally metal-centered molecular orbital, they are electric dipole allowed, hence intense in the absorption spectrum (and relatively weak in CD and MCD) and polarized along the L–M bond. These highly intense transitions enable tuning a laser into the absorption band to resonance enhance the Raman vibrational modes of the active site specific to the CT state in resonance. These CT transitions directly probe the σ- and π-bonding interactions of the specific ligand with the metal center.

Proceeding to still higher energies, now using a synchrotron for intense, polarized, continuous X-ray photons, electrons are excited from filled metal and ligand core levels into the half-occupied d orbital. For D_{4h}-$[CuCl_4]^{2-}$, these are the Cu K-edge at 8979 eV, the Cu L-edge at 930 eV and the Cl K-edge at 2820 eV. As will be presented below, these directly define the GS wave function of a metal active site.

These methods define the electronic structure experimentally. It is important then to correlate these results to electronic structure calculations. These are generally of two types. Ligand field theory (LFT) focuses on the ground and low-lying LF excited states. It is quite powerful if the metal site is not very covalent. For more covalent sites and to correlate to the complete energy level diagram in Fig. 2, molecular orbital theory is required, and this mostly means density functional theory (DFT). However, there are a range of different functionals and hybrids and these calculations can give significantly different electronic structure descriptions. Thus, it is important to correlate calculations to experiments and, when supported by spectroscopy, these can provide detailed insight into frontier molecular orbitals (FMO) and reaction coordinates in catalysis.

There are two general goals of spectroscopy in Bioinorganic Chemistry: (1) to determine the active site geometric and electronic structure and understand how structure leads to function and (2) to define the reaction mechanism on a molecular level. The latter involves using the different spectroscopic methods of Fig. 2 to define the reaction intermediates and calculations supported by these spectroscopic data to evaluate the reaction coordinates. These general goals have led to important spectroscopic goals. One has been the development of new spectroscopic methods to address important problems in Bioinorganic Chemistry that were not accessible through existing methods. The second has been to understand the unique spectroscopic features observed for many major classes of active sites in biology (relative to small molecule inorganic complexes of the same metal ion). These reflect novel geometric and electronic structures that are imposed on the metal site by the protein (as in Fig. 1) that activate the metal site for function. This presentation will emphasize these goals, where a range of spectroscopies (including new spectroscopic methods) coupled with electronic structure calculations are used to understand the unique spectral features of the blue Cu proteins that function in rapid, long range electron transfer (ET) in biology. Since the focus of this presentation is on unique spectroscopic features, we first summarize the "normal" structural and spectral features of Cu(II) complexes using D_{4h}-$[CuCl_4]^{2-}$ as a reference.

2. Geometric and electronic structure of "normal" Cu(II) complexes

From Fig. 3A, placing a Cu(II) ion in an octahedral field splits the five d-orbitals into an e_g set above the t_{2g} set by 10 D_q (Fig. 3B left). Cu(II) has 9 d-electrons, resulting in an electron hole in the e_g set, giving a ^2E_g ground state. This is orbitally degenerate and, thus, undergoes a Jahn–Teller distortion of the site that splits the orbital degeneracy and lowers the energy of the system. Cu(II) complexes generally show a tetragonal elongation along the z-axis and contraction in the

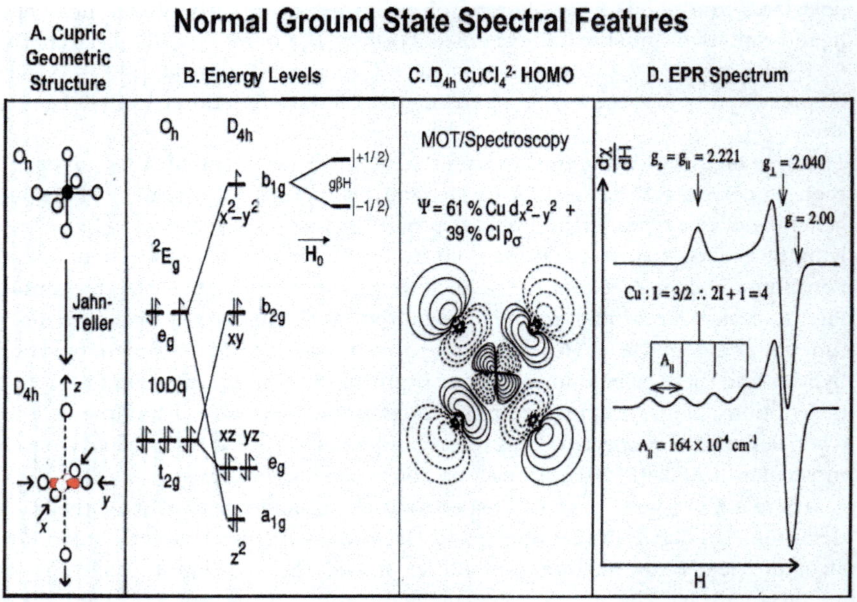

Fig. 3 Normal Cu(II) complexes. (A) Cupric geometric structure in O_h symmetry and upon Jahn–Teller distortion to D_{4h}; the highest energy half-occupied $d_{x^2-y^2}$ orbital is highlighted in red, (B) Energy levels of Cu(II) d-orbitals in O_h and D_{4h}, and the Zeeman splitting of the GS in a magnetic field (H_o), (C) D_{4h}-$[CuCl_4]^{2-}$ HOMO (from methods in Fig. 2) and (D) EPR spectrum without (top) and with (bottom) metal hyperfine coupling. Reprinted with permission from ref. 3. Copyright (2006) American Chemical Society.

equatorial plane and, in the limit of a large Jahn–Teller effect, this results in the square planar D_{4h}-$[CuCl_4]^{2-}$ reference complex. As shown in the middle panel of Fig. 3B, the energy splitting of the d orbitals is sensitive to the ligand environment and this is very different for the square planar $[CuCl_4]^{2-}$ complex, where the highest energy d orbital is $d_{x^2-y^2}$ because its lobes are oriented along the metal–ligand bonds and, thus, have the largest antibonding interaction with the ligands (shown in red in the bottom of Fig. 3A). This d orbital is σ antibonding with its four Cl^- ligands and, as shown in Fig. 3C, based on the spectroscopic methods in Fig. 2 covering ten orders of magnitude in photon energy, it has 61% $d_{x^2-y^2}$ character with the rest of the wave function equally delocalized over the four Cl^- σ-antibonding orbitals.[5] The $d_{x^2-y^2}$ orbital is half-occupied and this $^2B_{1g}$ ground state will Zeeman split in a magnetic field (Fig. 3B right, β: Bohr magneton) by an amount defined by its g_i values. As shown in the top of Fig. 3D, D_{4h}-$[CuCl_4]^{2-}$ has a $g_{\|}$ of 2.221 (Zeeman splitting with the magnetic field (H) along the z-axis of D_{4h}-$[CuCl_4]^{2-}$, bottom of Fig. 3A) and a $g_{\perp}$ of 2.040 (corresponds to the magnetic field along x- and y-axes in the bottom of Fig. 3A).[6] Cu also has a nuclear spin of $I = 3/2$ that hyperfine couples to the electron spin and splits each g_i value into $2I + 1 = 4$ lines. From the EPR spectrum of D_{4h}-$[CuCl_4]^{2-}$ at the bottom of Fig. 3D the hyperfine splitting of the $g_{\|}$ region, $|A|_{\|}$, is large and much larger than that in the $g_{\perp}$ region. This EPR signal is observed at a microwave energy corresponding to 0.3 cm^{-1}.

Proceeding up in energy into the near-IR/Vis region, the photon excites electrons from the filled d orbitals in Fig. 3B into the half-occupied $d_{x^2-y^2}$ orbital. These d–d transitions are parity forbidden and weak in the absorption spectrum of D_{4h}-$[CuCl_4]^{2-}$ (Fig. 4B red). These are the bands between 10 000 and 17 000 cm^{-1} with $\varepsilon \sim 40\ M^{-1}\ cm^{-1}$. Proceeding up in energy are the intense LMCT transitions (blue) with ε values of > 2000 $M^{-1}\ cm^{-1}$ that reflect the bonding interactions of the Cl^- ligands with the Cu(II). From Fig. 4C, Cl^- has three valence 3p orbitals (note a second pπ is perpendicular to the plane) that are split in energy, with the σ (along the Cl–Cu bonds) more stabilized in energy than the π due to better overlap. This leads to the Cl π and σ energy order in Fig. 4A. This

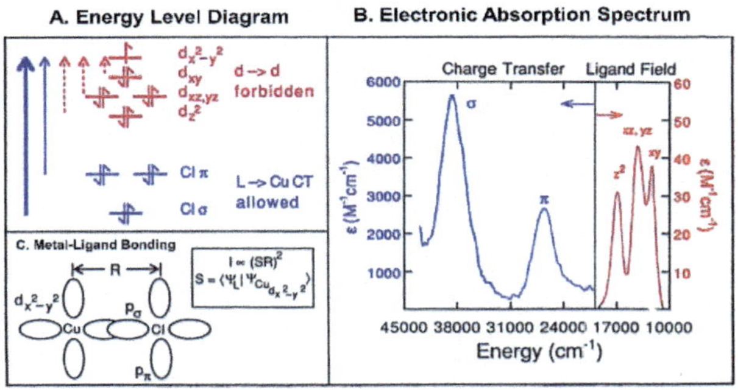

Fig. 4 Valence excited state spectral features of D_{4h}-$[CuCl_4]^{2-}$. (A) Energy level diagram and (B) electronic absorption spectrum, d-orbitals and LF transitions in red; ligand bonding orbitals and CT transitions in blue. (C) Metal–ligand bonding interactions; inset: relation of CT intensity (I) to donor (ψ_L)–acceptor ($\psi_{Cu,d_{x^2-y^2}}$) orbital overlaps.[7] Reprinted with permission from ref. 3. Copyright (2006) American Chemical Society.

bonding also controls the intensity of the CT transition (see inset in Fig. 4C), which is proportional to the overlap of the donor and acceptor orbitals involved in the LMCT transition. From Fig. 4C the pσ has a significant overlap with the $d_{x^2-y^2}$ orbital and produces an intense CT transition at higher energy than the π CT (which has no overlap with $d_{x^2-y^2}$ but gains some intensity due to their configuration interaction).[8] Thus, the Vis/UV absorption spectrum of D_{4h}-$[CuCl_4]^{2-}$ shown in Fig. 4B (blue) has low energy weak π and high energy intense σ CT transitions. The EPR spectrum, shown in the bottom of Fig. 3D, and the absorption spectrum in Fig. 4B are characteristic of "normal" tetragonal Cu(II) complexes and serve as reference spectra for the unique spectral features of the blue Cu active sites and their relation to its ET function.

3. Electron transfer sites in Bioinorganic Chemistry

There are three classes of metal sites in biology that function in ET (Fig. 5). For copper proteins, there is the blue Cu site that is mononuclear with thiolate and thioether ligands, and Cu_A that is binuclear, also having thiolate (bridging) and thioether ligation. In iron–sulfur chemistry, there are the rubredoxins, ferrodoxins and HiPIP (high potential iron protein) that have different nuclearities but all with tetrahedral sulfur ligand fields that are weak and function as high spin sites in both redox states. In heme chemistry, there are the cytochromes (cyt), all having strong equatorial porphyrin ligation and two axial ligands: cyt *c* has thioether and histidine as axial ligands. The cytochromes are low spin in both redox states.

There are quite a variety of structures and they cover almost a 2 V potential range, but there are two common features among all the sites in Fig. 5 enabling their efficient ET: (1) there is minimal geometry change with redox (*i.e.* a low reorganization energy (λ) in Marcus theory of ET)[9] and (2) there is efficient electronic coupling between the electron donor and acceptor (H_{DA}), often over long distances through the protein. This presentation mostly focuses on how the

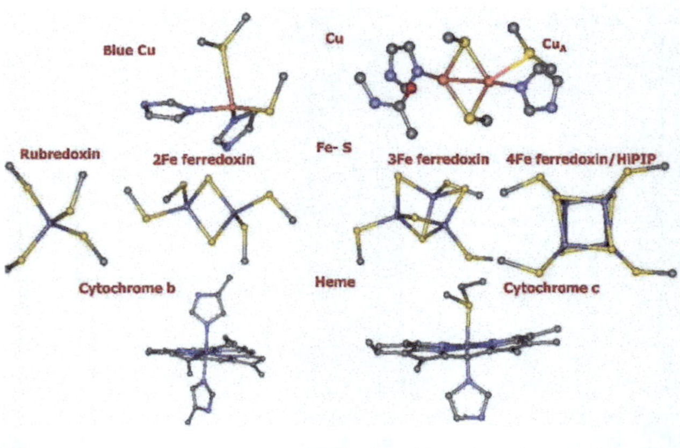

Fig. 5 Electron transfer sites in Bioinorganic Chemistry. Reprinted with permission from ref. 3. Copyright (2006) American Chemical Society.

unique spectral features of the blue Cu site reflect a geometric and electronic structure imposed by the protein that enables this ET function. These concepts are then extended to cyt *c* in the later part of this presentation.

4. The unique geometric and electronic structure of the blue copper active site

The first crystal structure of a blue Cu protein was that of plastocyanin, solved by Hans Freeman in 1978.[10] As shown in Fig. 6A the Cu(II) site had a tetrahedral structure, not the tetragonal structure generally observed for normal Cu(II) complexes. This led to the idea that the protein must oppose the Jahn–Teller distortion of the oxidized site, thus activating it for rapid ET. This concept of the protein constraining the ligand environment of a metal site activating it for function is known as the "entatic state" in Bioinorganic Chemistry.[11,12] The entatic nature of the blue Cu site will be elucidated later in this presentation. As mentioned above, the blue Cu site has two special ligands: a short cysteine (Cys) thiolate S–Cu(II) bond at 2.1 Å and a long methionine (Met) thioether S–Cu(II) bond at 2.9 Å. The remaining ligands are His N's with fairly normal bond lengths.

Associated with the unique tetrahedral structure and Cys/Met ligation are the unique spectral features of the blue Cu proteins. Instead of weak LF transitions with $\varepsilon \sim 40\ M^{-1}\ cm^{-1}$ in the 16 000 cm^{-1} region, the blue Cu site has an intense absorption band ($\varepsilon \sim 5000\ M^{-1}\ cm^{-1}$) in this region (in Fig. 6B) and the magnitude of the parallel hyperfine splitting of the blue Cu site in Fig. 6C is reduced by more than a factor of two relative to that of normal Cu(II). It will now be elucidated that these unique spectral features reflect a novel GS wave function that is the redox active molecular orbital (RAMO) key to its biological function of long range rapid ET.

From Fig. 6C, the EPR spectrum of the blue Cu site in plastocyanin has $g_{\parallel} > g_{\perp} > 2.00$. From the LF spectroscopy in section 2, this requires that the half-occupied

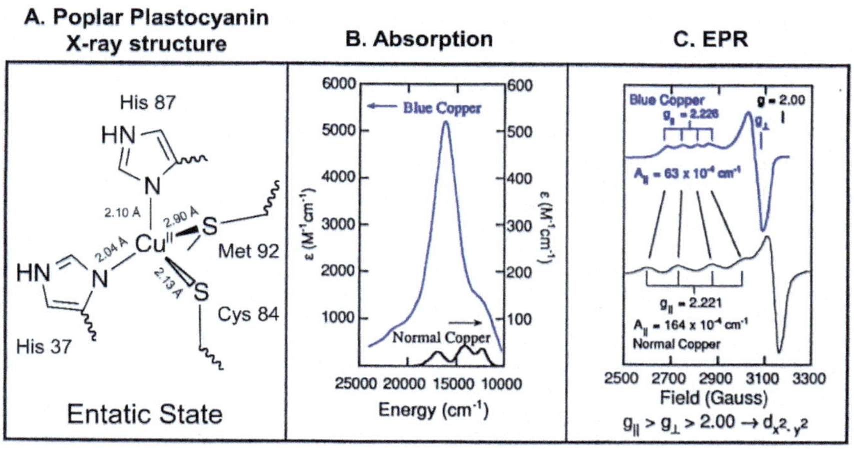

Fig. 6 Oxidized blue Cu site: (A) schematic of the X-ray structure of the blue Cu site in plastocyanin, (B) absorption and (C) EPR spectra of normal Cu(II) (black) and blue Cu (blue) sites. Reprinted with permission from ref. 3. Copyright (2006) American Chemical Society.

orbital is $d_{x^2-y^2}$ and this had to be correlated with the distorted tetrahedral geometric structure of the blue Cu site in Fig. 6A. This was accomplished by single crystal EPR spectroscopic studies of plastocyanin.[13]

Plastocyanin crystallizes in an orthorhombic space group with four symmetry related molecules in the unit cell. An expanded view of the active sites of these proteins is shown in Fig. 7A. Rotating the crystal around the *a*-axis with the field along different orientations in the *b*/*c*-plane produced the single crystal EPR spectra shown in the left hand side of Fig. 7B. It can be seen that the $g_{\parallel}$ spectrum with the four observable hyperfine lines is present when the field is along the *c*-axis, while rotation by ~90° leads to a $g_{\perp}$ EPR signal (Fig. 7B left bottom). Thus, $g_{\parallel}$ is approximately along the *c*-axis of the unit cell which, from the expanded view in Fig. 7A, is the approximate orientation of the long Met thioether S–Cu bond. Three single crystal rotations were performed and simulated with the four molecules in the unit cell (Fig. 7B right), which locked in the $g_{\parallel}$ at an angle of just 5° off the long (2.9 Å) Cu(II)–S_{Met} bond. Thus, the $d_{x^2-y^2}$ orbital, which is perpendicular to this direction, is in the plane of the strong Cu(II)–S_{Cys} and two Cu(II)–N_{His} bonds. This changed how the active site is visualized relative to the crystallography (Fig. 6A) and the focus is now on this $x^2 - y^2$ plane.

We now consider one of the unique spectral features of the blue Cu active site, the origin of the small metal parallel hyperfine splitting ($A_{\parallel}$ in the top panel of Fig. 6C). As presented in Fig. 8, there are three contributions to the hyperfine coupling of the nuclear spin *I* of the Cu to the electron in the Cu $d_{x^2-y^2}$ orbital. These are: (1) the Fermi contact interaction of the electron spin with the nuclear spin at the nucleus, which is indirect and occurs through the spin polarization of core s-electrons; (2) the spin dipolar coupling of the electron spin averaged over the $d_{x^2-y^2}$ orbital with the nuclear spin on the Cu and (3) the orbital dipolar coupling of the electron with the nuclear spin that has the same origin as the g_i

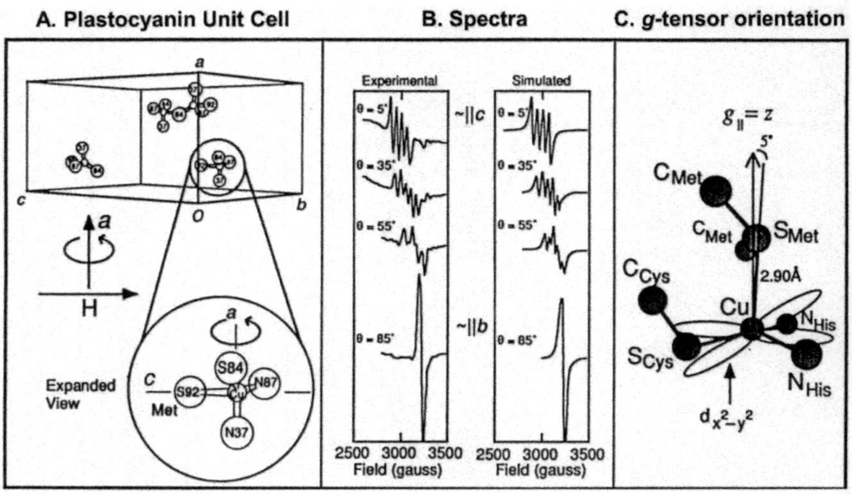

Fig. 7 Orientation of the $d_{x^2-y^2}$ orbital in plastocyanin from single crystal EPR spectroscopy. (A) Plastocyanin unit cell, (B) experimental (left) and simulated (right) single crystal EPR spectra, and (C) resultant *g*-tensor orientation. Reprinted with permission from ref. 3. Copyright (2006) American Chemical Society.

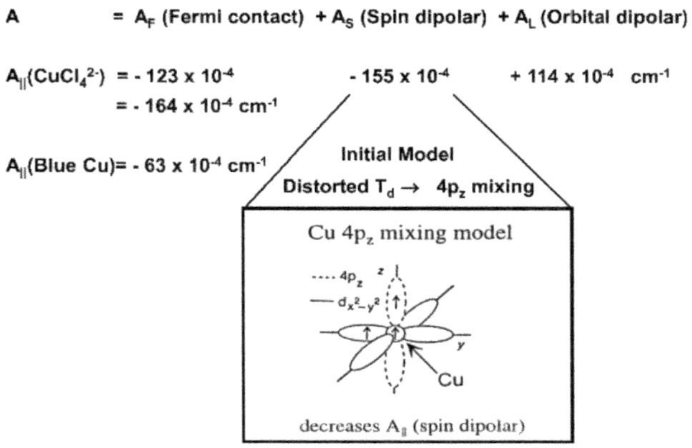

Fig. 8 The three contributions to the metal $A_{\parallel}$ in the Cu(II) EPR spectrum and the initial model for the origin of the small $|A|_{\parallel}$ of the blue Cu site (inset). Reprinted with permission from ref. 3. Copyright (2006) American Chemical Society.

values deviating from 2.00 (*vide infra*) and can be estimated from these. As shown in Fig. 8, all three contributions are comparable in magnitude but have different signs and sum to the large negative $A_{\parallel}$ of -164×10^{-4} cm^{-1} observed for D_{4h}-$[CuCl_4]^{2-}$ in Fig. 3D. For the blue Cu site in plastocyanin, the magnitude of $A_{\parallel}$ is greatly reduced to -63×10^{-4} cm^{-1}. Initially, this decreased magnitude of $A_{\parallel}$ was assumed to derive from the distorted tetrahedral structure of the blue Cu site in Fig. 6A.[14] This was reasonably thought to mix Cu $4p_z$ character into the $d_{x^2-y^2}$ orbital (dashed orbital in the inset of Fig. 8). Electron spin in this p_z orbital has the opposite sign for spin dipolar coupling with the nuclear spin on the Cu and ~12% $4p_z$ mixing would reduce the magnitude of $A_{\parallel}$ to the experimental value for plastocyanin.

While this possibility was reasonable, it was important to evaluate it experimentally. This was accomplished by going up by 10^{10} in photon energy, from the 10^{-4} cm^{-1} region of hyperfine coupling to the ~9000 eV region of X-ray absorption spectroscopy at the Cu K-edge.[15].

As shown in Fig. 9A the Cu K-pre-edge 1s → 3d transition occurs at 8979 eV. A pure 1s → 3d transition is electric dipole forbidden and, thus, is very weak. However, the distorted T_d geometry of the blue Cu site enables 4p mixing into this RAMO and the 1s → 4p transition is electric dipole allowed, so even a low percent mixing (α^2 in Fig. 9A) would give rise to dominant pre-edge intensity. As shown in Fig. 9B (arrow), the pre-edge intensity of the blue Cu XAS spectrum at 8979 eV is indeed much higher than that of D_{4h}-$[CuCl_4]^{2-}$, where 4p mixing is forbidden by group theory. The important experiment here was to determine the nature of the 4p mixing in plastocyanin. This was accomplished by the analysis of its polarized single crystal XAS data. The single crystal EPR studies in Fig. 7 showed that the electronic *z*-axis was approximately along the Cu–S_{Met} bond, so the polarized single crystal spectra obtained with the ***E*** vector of the synchrotron oriented along this bond would reveal the Cu $4p_z$ character. From the dashed $\boldsymbol{E}\|z$ spectrum in Fig. 9C, there is no pre-edge intensity, thus no Cu $4p_z$ mixing. Alternatively, when the ***E*** of the synchrotron is oriented in the (x,y) plane (the solid spectrum in

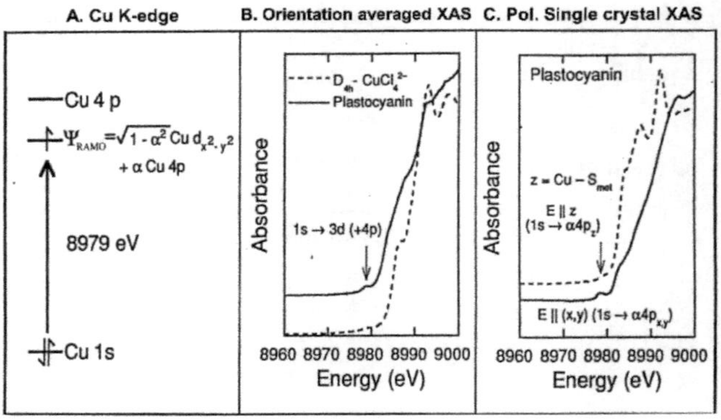

Fig. 9 X-ray absorption spectroscopy to evaluate the Cu 4p mixing into the $3d_{x^2-y^2}$ orbital in plastocyanin. (A) Cu K-pre-edge 1s → 3d transition at 8979 eV, (B) orientation averaged XAS spectra of D_{4h}-$[CuCl_4]^{2-}$ and plastocyanin and (C) Polarized single crystal XAS spectrum of plastocyanin with $E||z$ and $E||(x,y)$ (z-axis defined by the single crystal EPR in Fig. 7). Reprinted with permission from ref. 16. Copyright (2004) American Chemical Society.

Fig. 9C), dominant pre-edge intensity is observed. This Cu $4p_{x,y}$ mixing in plastocyanin demonstrated by these polarized XAS data would increase, not decrease the magnitude of the parallel hyperfine splitting. This eliminated the generally accepted explanation for the small $A_{||}$ of blue Cu sites and focused our attention on the alternative possibility for the small $A_{||}$ of blue Cu sites: a highly covalent GS which shifts the electron spin onto the ligand(s) and reduces its interaction with the nuclear spin on the Cu.

This possibility of a highly covalent blue Cu active site was initially addressed through a quantitative consideration of its g_i values.[17] The table at the top of Fig. 10, left column, gives the experimental g_i values of plastocyanin. Moving from left to right in this table, for an unpaired electron in a pure $d_{x^2-y^2}$ orbital, thus having only spin angular momentum, the g_i values would be isotropic and equal to the spin only value of 2.0023. This is similar to what is observed in organic radicals where spin–orbit coupling (SOC) is small. However, transition metal sites have large values of their SOC parameter ($\lambda[\mathrm{Cu(II)}] = -830\ \mathrm{cm}^{-1}$) and this mixes excited LF states into the GS, giving them an orbital angular momentum contribution to their g_i values. A complete LFT calculation of the blue Cu site in plastocyanin including SOC gives the g_i values in the third column of the table in Fig. 10, with $g_{||} > g_{\perp} > 2.00$. However, these deviate too much from the experimental values in the first column. Thus, there is too much orbital angular momentum mixed into the GS. This is due to the nature of LFT that treats the orbitals as having pure metal d character. Covalency delocalizes the electron onto the ligand(s) and this decreases the orbital angular momentum in the GS. The fourth column in the table uses one of the earliest DFT methods (SCF-Xα scattered wave) with SOC included to calculate the effects of covalency on the GS g_i values. Comparison to the third column demonstrates that covalency does indeed decrease the deviation from 2.00 relative to LFT, however comparison to the first column shows that these are decreased by too much, indicating that this "state of the art" calculation at the time (1985) overestimated the covalency of the blue Cu

	experimental Q-band EPR	$(x^2\text{-}y^2)^1$ spin only	L.F.T d orbitals $+ \lambda\, L{\cdot}S$	SCF-Xα-SW LF + CT levels $\lambda_{Cu}\, L{\cdot}S + \lambda_L\, L{\cdot}S$ Norman Radii	Adjustec Radii
g_x	2.047	2.00	2.125	2.046	2.059
g_y	2.059	2.00	2.196	2.067	2.076
g_z	2.226	2.00	2.479	2.159	2.226

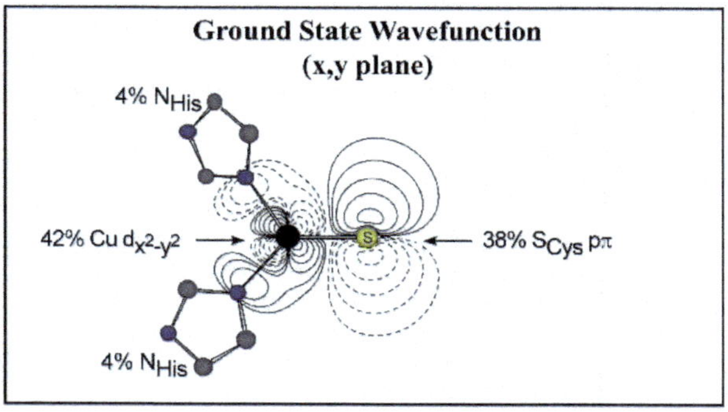

Fig. 10 Evaluation of the GS wave function of the blue Cu site in plastocyanin. The top table gives the g_i values observed experimentally (from Q-band EPR) and the corresponding values predicted by various electronic structure methods. The panel below shows the experimentally adjusted GS wave function in the (x,y) plane with % character on Cu and S_{Cys} and two N_{His}. Reprinted with permission from ref. 3. Copyright (2006) American Chemical Society.

site. Thus, parameters in this DFT calculation were adjusted to calculate a GS wave function that agrees with the experimental g_i values. This is shown below the table [in the (x,y) plane]. This was a major insight into the electronic structure of the blue Cu sites, as it showed a highly covalent GS which has 42% $d_{x^2-y^2}$ character, and this character is anisotropically delocalized into the thiolate ligand having 38% character and this delocalization involved the S_{Cys} π-, not σ-orbital, of the thiolate.

While this experimentally adjusted DFT calculation was performed >35 years ago using a relatively simple code, modern DFT calculations give parallel results. From the model on the top half of the left hand side of Fig. 11 and the first two rows in the table on the right, both a pure functional (BP86) and a hybrid functional (B3LYP) containing 20% Hartree–Fock exchange gave equivalent GS wavefunctions, a $d_{x^2-y^2}$ orbital highly delocalized into the S pπ-orbital of the Cys sulfur ligand, but again too covalent with 30% Cu and 60% S_{Cys} compared to experiment (blue row, *vide infra*) with ∼40% character on each.

However, the calculations using the model on the top left of Fig. 11 only included the first coordination sphere of the active site. Inclusion of the extended protein environment (modelled by QM/MM or using a large QM region, combined with a hybrid DFT method) now gives reasonable agreement with experiment, with ∼43% of the wave function on the Cu and the S_{Cys}.[3,18–20] Thus, the environment decreases the covalent delocalization onto the thiolate, and this reflects the combined effects of an H-bond of the backbone to the thiolate and nearby carbonyl dipoles. This change in the covalency of the Cu–S_{Cys} bond provides

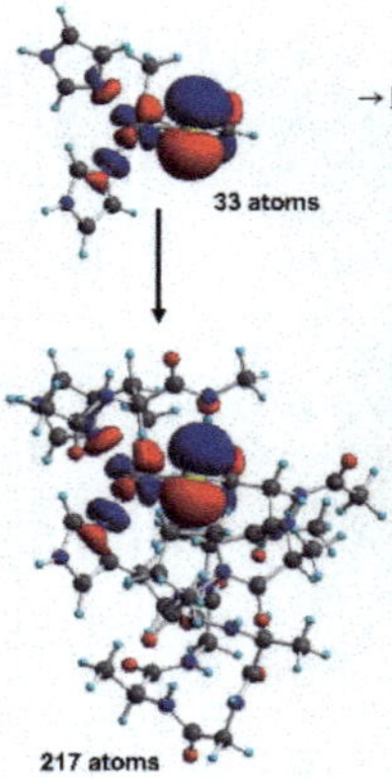

Model	Functional	Atomic Spin Density	
		Cu	S_{Cys}
33-atom QM	BP86	0.27	0.61
	B3LYP	0.33	0.57
Experiment (XAS)		0.41 d	0.38 p
217-atom QM (includes protein dipoles and H-bond to S_{Cys})	B3LYP	0.43	0.43
QM/MM, all protein	B3LYP	0.44	0.42

Fig. 11 GS wave function of the blue Cu active site by modern DFT. The RAMO contours for a small model (33 atoms) and a larger model (217 atoms) are shown on the left. The molecular orbital characters obtained using various functionals for both models, along with those obtained experimentally (from XAS in blue), are given in the table on the right. Reprinted with permission from ref. 3. Copyright (2006) American Chemical Society.

a second sphere mechanism for the protein to tune the redox potential of the active site over a range of ~200 mV for its ET function.[21]

We now focus on the experimental results in the middle row (blue) in Fig. 11 and the thiolate S π-orbital contribution in the RAMO contour of the blue Cu site (on the left), which were derived from new spectroscopic methods that we developed over the years to quantify covalency and define the nature of the metal–ligand bonding.

5. New spectroscopic methods in Bioinorganic Chemistry

Cu L-edge XAS enabled us to experimentally determine the amount of metal d character in the RAMO in Fig. 11. For Cu, the L-edge absorption is at ~930 eV and involves the Cu 2p transition to this RAMO (Fig. 12 right).[22] The 2p is localized on the Cu and 2p → 3d is electric dipole allowed, thus the intensity of this transition quantifies the amount of 3d character in the GS wave function of the blue Cu site [$\sqrt{(1-\alpha^2)}$ in the right hand side of Fig. 12]. From the left hand side of Fig. 12, the L-edge absorption peak of the blue Cu site in plastocyanin (specifically the L_3-edge) is significantly lower in intensity than that of D_{4h}-$[CuCl_4]^{2-}$, indicating that it has less d character and, thus, is more covalent. Since from the wide range of spectroscopies summarized in section 1, D_{4h}-$[CuCl_4]^{2-}$ has 61 ± 2% $d_{x^2-y^2}$ character in its GS (Fig. 3C), the intensity ratio in the left hand side of Fig. 12 quantifies the blue Cu site as having 41% $d_{x^2-y^2}$ character, a very covalent active site, but with its covalency tuned by the second sphere effects of the protein, as shown in the bottom of Fig. 11.

The ligand responsible for this highly covalent blue Cu active site was defined by S K-edge XAS.[15] We developed ligand K-edge XAS as a direct experimental method to quantify ligand character in GS wavefunctions.[23] The S K-edge is at ~2470 eV and, as shown in the right hand side of Fig. 13, it involves a transition

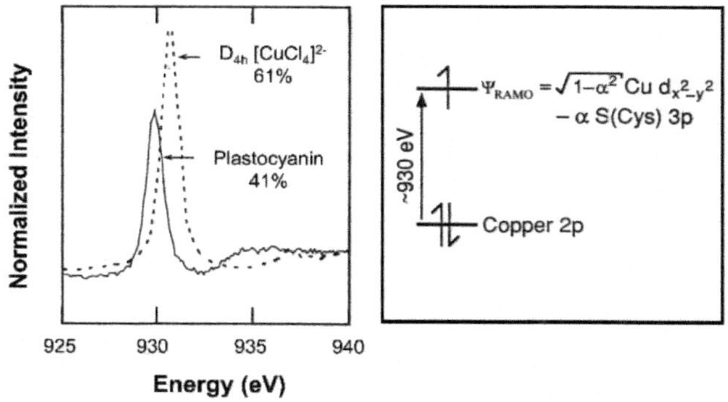

Fig. 12 Cu L-edge spectroscopy. L_3-edge XAS spectrum of D_{4h}-$[CuCl_4]^{2-}$ (dashed) and plastocyanin (solid) showing %Cu 3d character in RAMO on the left and the 2p → 3d transition on the right. Reprinted with permission from ref. 16. Copyright (2004) American Chemical Society.

from the S 1s orbital into the RAMO of the blue Cu center. Since the 1s orbital is localized on the sulfur and s → p is electric dipole allowed, the intensity of this S K-pre-edge transition directly reflects the Cys S_p character in the GS of the blue Cu site (α^2 in the right hand side of Fig. 13). From the left hand side of Fig. 13, the S K-pre-edge intensity of plastocyanin is 2.5 times as that of the tet b thiolate–Cu(II) model complex[24] that has 14% S_p character in its GS wave function. This quantified the blue Cu site in plastocyanin as having 38% Cys S_p character[15] that will be proven in the next section to be key to its function in long range, rapid ET.

The third spectroscopic method that provided a key insight was the development of low temperature MCD to assign the unique absorption features of the blue Cu site (Fig. 6B) and from this determine the π- *vs.* σ-bonding interaction of the S_{Cys} with the Cu site.[25] From Fig. 14A (top two panels), at low temperature the absorption bands of the blue Cu absorption spectrum sharpen and, by correlating low temperature absorption with the low temperature CD (not shown) and low temperature MCD (Fig. 14A bottom), each method having a different selection

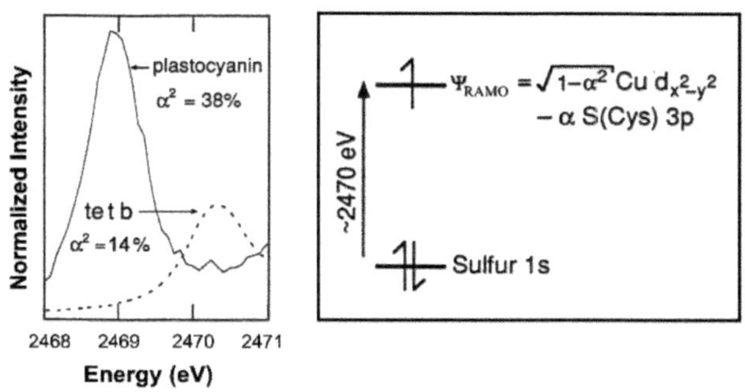

Fig. 13 Sulfur K-edge XAS. Left: K-edge XAS spectrum of the thiolate–Cu(II) model complex (dashed) and plastocyanin (solid), giving % S character in the GS. Right: 1s (S) → RAMO transition. Reprinted with permission from ref. 16. Copyright (2004) American Chemical Society.

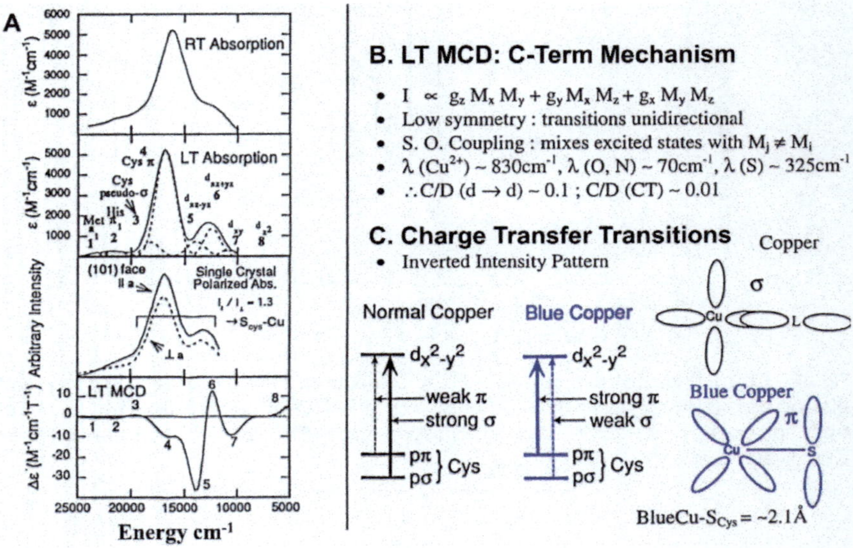

Fig. 14 Correlation of (A) room temperature, low temperature and single crystal polarized absorption and low temperature MCD spectra of plastocyanin, (B) low temperature MCD C-term mechanism and (C) charge transfer transitions in normal Cu (black) and blue Cu (blue) sites. Reprinted with permission from ref. 3. Copyright (2006) American Chemical Society.

rule, there are eight transitions required to fit the low temperature absorption spectrum of the blue Cu site in plastocyanin. Band 4 in the second panel of Fig. 14A is the characteristic intense absorption feature of the blue Cu site, while the lower energy band 6 has ~20% of this intensity. From polarized single crystal spectroscopy on plastocyanin (third panel of Fig. 14A), both bands have the same polarization ratio and, from correlation to the crystal structure (Fig. 7A), these are polarized along the Cu–S_{Cys} bond.

From section 2 and as illustrated in Fig. 14C, in "normal" Cu(II) complexes, the lobes of the $d_{x^2-y^2}$ orbital are oriented along the Cu–ligand bonds and the charge transfer spectrum exhibits a low energy weak π and higher energy intense σ transition. However, this is not the correct assignment of bands 6 and 4 of the blue Cu absorption spectrum, as revealed by the low temperature MCD spectrum at the bottom of Fig. 14A. Bands 5–8 are the most intense in the low temperature MCD spectrum relative to the absorption spectrum. As shown in Fig. 14B, MCD intensity requires two perpendicular transition moments (M_i), while in a low symmetry protein active site, all energy levels are non-degenerate. Thus, the electronic transitions are unidirectional. SOC is required to mix excited states with different polarizations to induce MCD intensity. Since the SOC parameter of Cu is large relative to that in the ligands, transitions that are intense in MCD relative to absorption (the C/D ratio) are assigned as d → d transitions. This is the case for bands 5–8. Alternatively, band 4, that is intense in absorption but weak in MCD, is the lowest energy, hence S_{Cys}-π CT, while a higher energy transition that is weak in absorption (band 3) is the S_{Cys}-σ to Cu $d_{x^2-y^2}$ CT. Thus, the blue Cu site has an inverted, lower energy intense π and higher energy weak σ CT and, since

CT intensity results from orbital overlap (right hand side of Fig. 14C), the $d_{x^2-y^2}$ orbital is rotated by 45° with its lobes bisected by the Cu–S_{Cys} bond. This orientation reflects the strong π-antibonding interaction of the thiolate with the Cu due to its short 2.1 Å Cu–S bond length.

Importantly the unique spectral features of the blue Cu site reflect a highly covalent thiolate S–Cu π-bond.

6. Electronic structure contributions to electron transfer

We now consider how this highly covalent thiolate S–Cu π-bond enables the function of the blue Cu site in long range, rapid ET. The focus is on H_{DA}, the electronic coupling between the electron donor and acceptor through protein pathways. A particularly interesting correlation is to the blue Cu site in multicopper oxidases, as the pathway for ET is well defined. As shown in Fig. 15A, the blue Cu site is near the surface of the enzyme, with its Cys ligand flanked on either side by His residues that are ligands of a buried trinuclear Cu cluster (TNC) site.[26] Electrons from the substrate are transferred to the blue Cu site and rapidly transferred over 13 Å through the Cis–His pathway to the TNC, where O_2 is reduced to water. As qualitatively visualized in Fig. 15A (green and yellow contours), the high covalency of the S_{Cys} in the GS wave function activates the Cis–His pathway for ET to the TNC. Quantitatively,[18] as presented in Fig. 15B, H_{DA} increases linearly with the covalency of the Cu–S_{Cys} bond, with the rate of ET (k_{ET}) proportional to $(H_{DA})^2$ (see the Marcus equation for non-adiabatic ET, the bottom of Fig. 15A). From these calculations, there is one dominant superexchange pathway between the blue Cu center and Cu of the TNC, and this requires the π donation of the thiolate to the copper to activate a π electron–hole pathway through the protein backbone to go through a π to σ crossover to the catalytic site (Fig. 16C inset). Thus, the unique spectroscopic features of the blue Cu site reflect an electronic structure that is maximized for its ET function. We now consider the role of the protein in determining this electronic structure.

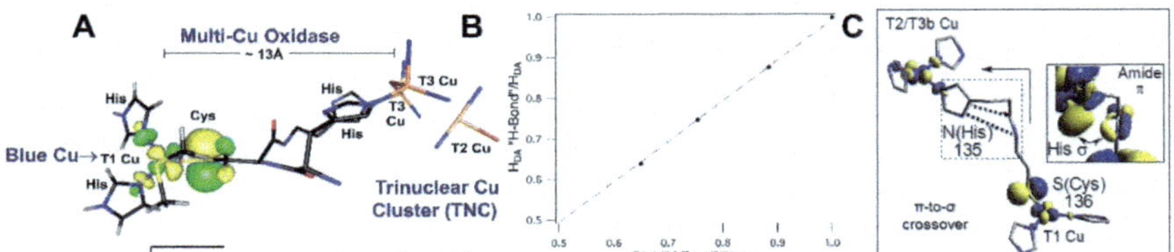

$$k_{ET} = \sqrt{\frac{4\pi}{h^2\lambda k_bT}}|H_{DA}|^2\exp\left(-\frac{(\Delta G^0+\lambda)^2}{4\lambda k_bT}\right)$$

Fig. 15 (A) Active site of multicopper oxidases, showing blue Cu and TNC separated by ~13 Å and connected by a Cis–His pathway. The RAMO of the blue Cu site is shown in yellow and green contours. Given below is the Marcus equation for non-adiabatic ET; reprinted with permission from ref. 3. Copyright (2006) American Chemical Society. (B) Linear correlation between Cu–S covalency and H_{DA} and (C) molecular orbital involved in the Cis–His superexchange pathway from the blue Cu site to TNC. The inset shows the π to σ crossover. Reprinted with permission from ref. 18. Copyright (2014) American Chemical Society.

7. Role of the protein in determining the geometric and electronic structure of the blue copper site: the entatic state

Fig. 16A shows the absorption spectra of a series of blue Cu related proteins with the same Cys, Met and two His ligand set that vary from the blue color of plastocyanin to green in nitrite reductase (NiR). From Fig. 16A this color change reflects the fact that the CT transitions go from the inverted low energy intense π, high energy weak σ pattern of the blue Cu site to the "normal" low energy weak π and higher energy intense σ CT of the green site.[27]

The structural changes associated with these spectral changes are given in Fig. 16B and C. In going from the blue Cu site to the green site, the thioether S_{Met} comes in from 2.9 to 2.5 Å, the thiolate goes out by 0.1 Å and the S–Cu–S plane rotates into the N–Cu–N plane, reflecting a Jahn–Teller distortion to a more tetragonal structure.

Thus, the protein is tuning the geometric and electronic structure of the Cu site and, from the correlation in Fig. 16B and C, this appeared to be controlled by the thioether Cu–S bond. This was tested by mutating the strong thioether Cu–S bond of the green site to a non-coordinating threonine.[28] As shown in Fig. 17A, this variant (M182T) expressed as a blue Cu site. Thus, the thioether ligation to the Cu is controlled by the protein. Its contribution to the function of the blue Cu site was revealed by additional studies on the green copper site in wild type NiR.[29] From Fig. 17B, this green site is in a thermodynamic equilibrium with a blue Cu site. The green Cu site is dominant at low temperatures, indicating that it is enthalpically favored and, from resonance Raman (rR), it has the strong Cu–S_{Met} bond (Fig. 16B).

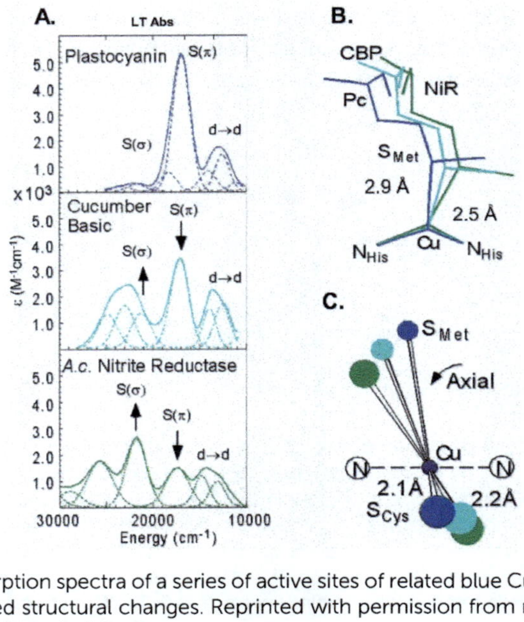

Fig. 16 (A) Absorption spectra of a series of active sites of related blue Cu proteins and (B) and (C) associated structural changes. Reprinted with permission from ref. 27. Copyright (1998) American Chemical Society.

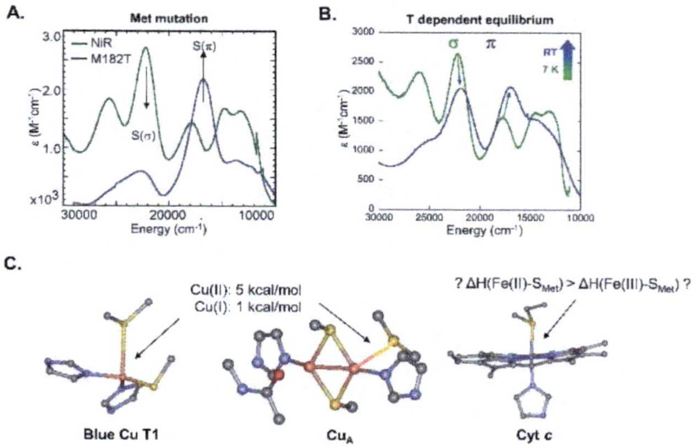

Fig. 17 (A) Absorption spectra of wild type NiR (green) and its M182T variant (blue); reprinted with permission from ref. 28. Copyright (2003) American Chemical Society. (B) Temperature dependent equilibrium between green and blue Cu sites; reprinted with permission from ref. 29. (C) Estimates of the Cu–S_{Met} bond enthalpies in blue Cu and Cu_A sites; presumed Fe–S_{Met} enthalpies in cyt *c*.

Increasing temperature results in its conversion to a blue Cu site which is, therefore, entropically favored and, from rR, there is no longer a Cu–S_{Met} bond. From a Van't Hoff analysis of this blue/green equilibrium, the enthalpic strength of the thioether Cu(II)–S bond is 5 kcal mol^{-1}. This is reproduced by DFT calculations which further give the strength of the thioether Cu(I)–S bond of the reduced site to be 1 kcal mol^{-1} (Fig. 17C left). This study thus defined the entatic state of the blue Cu active site in proteins like plastocyanin (Fig. 6A). The thioether–Cu bond is weak, and the protein opposes the entropic contribution to the Gibbs free energy in keeping the thioether bound at physiological temperature. From the left hand side of Fig. 17C, this stabilizes the oxidized more than the reduced state of blue Cu sites and provides a first coordination sphere mechanism of lowering the reduction potential by ~200 mV. Parallel studies on the binuclear Cu_A site showed the same function of the thioether S–Cu bond, stabilizing the oxidized over the reduced state and lowering the potential (middle of Fig. 17C).[30]

Alternatively, in cyt *c* (right hand side of Fig. 17C), it had been generally thought that the thioether Fe–S bond stabilized the reduced state more than the oxidized state and, thus, in contrast to the Cu protein ET active sites, the thioether Fe–S bond would raise the potential of cyt *c* by several hundred mV (relative to the partially unfolded protein).[31–34] This potential difference in thioether Fe–S *vs.* Cu–S bonding led us to the experimental evaluation of the covalency of this Fe–S_{Met} bond in ferric and ferrous cyt *c*.[35]

8. The Fe–S_{Met} bond in cyt *c*: contributions to function

A direct spectroscopic method to determine the covalency of the Fe–S_{Met} bond is L-edge XAS. From Fig. 18C for iron, this can probe both the dπ- and dσ-orbitals

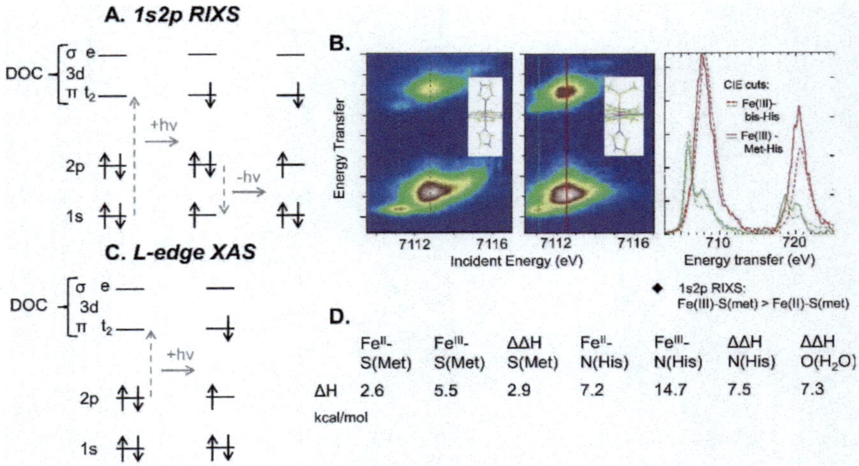

Fig. 18 (A) The photon absorption in the 1s → 3d transition and photon energy emission in the subsequent 2p → 1s transition in 1s2p RIXS, (B) 1s2p RIXS planes for a low spin bis-imidazole ferric compound (left), low spin ferric cyt *c* (middle) and constant incident energy cuts (L-edge like spectra) from RIXS planes (indicated by vertical lines in the RIXS spectra; green = π and red = σ contribution). (C) The 2p → 3d transition in L-edge XAS. (D) Table showing DFT calculated $Fe–S_{Met}$, $Fe–N_{His}$ and $Fe–OH_2$ enthalpic bond strengths for ferrous and ferric cyt *c*. Reprinted with permission from ref. 35. Copyright (2014) American Chemical Society.

and, thus, the differential orbital covalency (DOC) of cyt *c*.[36] However, there are issues with directly experimentally measuring L-edges on frozen protein solutions, so the K-edge was used to obtain the L-edge data though resonant inelastic X-ray scattering (1s2p RIXS). As shown in Fig. 18A, this involves measuring both the 1s → 3d photon absorption (along the *x*-axis in Fig. 18B) and the photon emitted by the 2p electron filling the 1s core hole (*y*-axis in Fig. 18B). From the right side of Fig. 18A, this RIXS process results in the same final states as in L-edge XAS (Fig. 18C) but involves different selection rules.[37]

The 1s2p RIXS spectrum of ferric cyt *c* is shown in the middle of Fig. 18B and vertical cuts through specific K-pre-edge features (on the *x*-axis) produced the L-edge like spectra, enhancing the π (green in the right hand side of Fig. 18B) and σ (red in the right hand side of Fig. 18B) contributions to the L-edge.[35] By modelling the L-edge like RIXS spectra of cyt *c*, it was experimentally determined that the Fe(III)–S_{Met} bond is in fact more covalent than the Fe(II)–S_{Met} bond, as determined above for Cu(II) *vs.* Cu(I) bonds. This was reproduced using DFT calculations, which further gave the enthalpic strength of the Fe(III)–S_{Met} bond as 2.9 kcal mol^{-1} higher than that of Fe(II)–S_{Met} bond (Fig. 18D).

1s2p RIXS data were also obtained for a porphyrin model complex where the axial Met is replaced with an imidazole ligand (left hand side of Fig. 18B). Parallel studies on this complex show that the Fe(III)–N_{His} bond is, in fact, even stronger than that of Fe(II)–N_{His} bond with a $\Delta\Delta H$ of 7.5 kcal mol^{-1} (Fig. 18D), which from calculations is also the case for an axial water ligand (Fig. 18D right). Thus, the Fe–S_{Met} bond does indeed increase the reduction potential, but this is relative to the His or water ligand that replaces it in a partially unfolded protein solution. The results in Fig. 18D further showing that, as with the Cu active sites, the Fe–S_{Met} bond in cyt *c* is also very weak, in an entatic state imposed by the protein. For ferrous cyt *c*, this unconstrained Fe(II)–S_{Met} bond strength is only 2.6 kcal mol^{-1}.

The entatic nature of the cyt *c* active site was evaluated and quantified by ultrafast X-ray free electron laser (XFEL) experiments at the Linac Coherent Light Source (LCLS) at the SLAC National Accelerator Laboratory.[38] The experimental design is given in Fig. 19A. A 100 μ jet of ferrous cyt *c* is excited by a 50 fs 520 nm optical laser pulse. This excites the β-band of the porphyrin and heats the heme by ~100 °C. It takes ~6 ps for this heat to dissipate through the protein, thus the X-ray laser pulse was used to excite the Fe from 100 fs to ~20 ps. Two types of spectroscopies were utilized. The Kβ X-ray emission (XES) spectra in Fig. 19B are sensitive to the spin state of the Fe. The GS of ferrous cyt *c* in blue is a singlet, while the hot heme at 600 fs in red exhibits a lower energy multiplet feature at 7045 eV, showing that the heme is now in a quintet state. The K-edge XAS near edge spectra shown in Fig. 19C are sensitive to the local structure of the Fe. In going from the GS of ferrous cyt *c* in blue to the hot heme at 600 fs in red, there are shifts in the near edge features. These are shape resonances modelled using multiple scattering to determine that the thioether S is no longer bound to Fe when the heme is in the $S = 2$ state. Thus, the Kβ XES data could be used to monitor the binding of the thioether S to the Fe(II) as the heat dissipated from the heme (inset in Fig. 19D). This enabled an estimate of the strength of the thioether

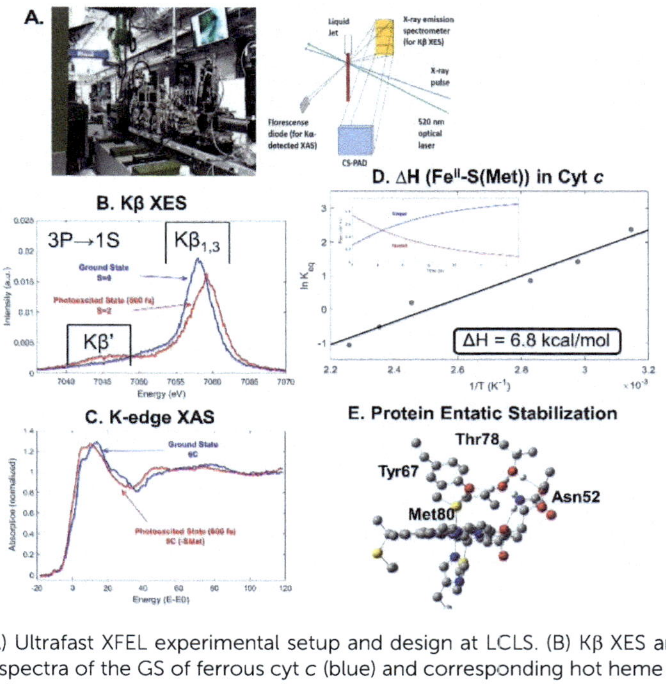

Fig. 19 (A) Ultrafast XFEL experimental setup and design at LCLS. (B) Kβ XES and (C) K-edge XAS spectra of the GS of ferrous cyt *c* (blue) and corresponding hot heme at 600 fs (red). (D) Plot used to estimate the strength of the Fe(II)–S_{Met} bond. In inset: populations of singlet (blue) and quintet (red) species obtained from XES modelling. (E) Active site of cyt *c*, showing an H-bond network to the Met in the protein pocket which stabilizes its Fe–S bond. From ref. 38. Reprinted with permission from AAAS.

Fe(II)–S bond in the protein as 6.8 kcal mol^{-1}. Relative to the ΔH of the free thioether Fe(II)–S bond in cyt *c* (Fig. 18D), this shows that the protein entatic state contribution stabilized this bond by ~4 kcal mol^{-1}. As shown in Fig. 19E, this stabilization derives from an H-bond network to the Met in the protein pocket. This 4 kcal mol^{-1} is in a "sweet spot", enabling the two physiological functions of cyt *c*. The presence of this bond enables its normal function of rapid ET to cytochrome *c* oxidase at the appropriate reduction potential to enable proton pumping for ATP synthesis. However, the Fe–S_{Met} bond is weak enough that in the membrane, this bond is ruptured by cardiolipin binding to cyt *c*, turning on its peroxidase function that permeabilizes the membrane, triggering apoptotic cell death.

9. Summary

This presentation emphasized three general themes of Bioinorganic Spectroscopy:

1. The wide range of spectroscopic methods in Fig. 2 provide complementary insights into the geometric and electronic structures of active sites, enabling structure–function correlations.

2. The correlation of spectroscopic results to electronic structure calculations provides deep insight into frontier molecular orbitals and reaction coordinates in catalysis.

3. The unique spectroscopic features that have dominated many areas of Bioinorganic Chemistry reflect novel geometric and electronic structures imposed on the active site by protein constraints that activate the metal center for its specific, efficient biological function.

Conflicts of interest

There are no conflicts to declare.

Acknowledgements

This work was supported by the National Institutes of Health grant R01DK031450 to E. I. S., who thanks his past students and collaborators for their contributions to this research.

References

1 R. H. Holm, P. Kennepohl and E. I. Solomon, *Chem. Rev.*, 1996, **96**, 2239–2314.
2 E. I. Solomon and M. A. Hanson, in *Inorganic Electronic Structure and Spectroscopy Vol. II*, eds. E. I. Solomon and A. P. B. Lever, John Wiley & Sons, Inc., New York, 1999, p. 1.
3 E. I. Solomon, *Inorg. Chem.*, 2006, **45**, 8012–8025.
4 E. I. Solomon, in *Comments on Inorganic Chemistry*, ed. N. Sutin, Gordon and Breach, New York, 1984, p. 225.
5 A. A. Gewirth, S. L. Cohen, H. J. Schugar and E. I. Solomon, *Inorg. Chem.*, 1987, **26**, 1133–1146.
6 P. Cassidy and M. A. Hitchman, *Inorg. Chem.*, 1977, **16**, 1568–1570.

7 R. S. Mulliken, C. A. Rieke, D. Orloff and H. Orloff, *J. Chem. Phys.*, 1949, **17**, 1248–1267.

8 S. R. Desjardins, K. W. Penfield, S. L. Cohen, R. L. Musselman and E. I. Solomon, *J. Am. Chem. Soc.*, 1983, **105**, 4590–4603.

9 R. A. Marcus and N. Sutin, *Biochim. Biophys. Acta, Rev. Bioenerg.*, 1985, **811**, 265–322.

10 P. M. Colman, H. C. Freeman, J. M. Guss, M. M urata, V. A. Norris, J. A. M. Ramshaw and M. P. Venkatappa, *Nature*, 1978, **272**, 319–324.

11 B. L. Vallee and R. J. Williams, *Proc. Natl. Acad. Sci. U. S. A.*, 1968, **59**, 498–505.

12 H. B. Gray, B. G. Malmström and R. J. P. Williams, *J. Biol. Inorg. Chem.*, 2000, **5**, 551–559.

13 K. W. Penfield, R. R. Gay, R. S. Himmelwright, N. C. Eickman, V. A. Norris, H. C. Freeman and E. I. Solomon, *J. Am. Chem. Soc.*, 1981, **103**, 4382–4388.

14 J. E. Roberts, T. G. Brown, B. M. Hoffman and J. Peisach, *J. Am. Chem. Soc.*, 1980, **102**, 825–829.

15 S. E. Shadle, K. O. Hodgson, E. I. Solomon, B. Hedman and H. J. Schugar, *J. Am. Chem. Soc.*, 1993, **115**, 767–776.

16 E. I. Solomon, R. K. Szilagyi, S. DeBeer George and L. Basumallick, *Chem. Rev.*, 2004, **104**, 419–458.

17 K. W. Penfield, A. A. Gewirth and E. I. Solomon, *J. Am. Chem. Soc.*, 1985, **107**, 4519–4529.

18 R. G. Hadt, S. I. Gorelsky and E. I. Solomon, *J. Am. Chem. Soc.*, 2014, **136**, 15034–15045, DOI: 10.1021/ja508361h.

19 S. Moon, S. Patchkovskii and D. R. Salahub, *J. Mol. Struct.*, 2003, **632**, 287–295.

20 S. Sinnecker and F. Neese, *J. Comput. Chem.*, 2006, **27**, 1463–1475.

21 R. G. Hadt, N. Sun, N. M. Marshall, K. O. Hodgson, B. Hedman, Y. Lu and E. I. Solomon, *J. Am. Chem. Soc.*, 2012, **134**, 16701–16716.

22 S. J. George, M. D. Lowery, E. I. Solomon and S. P. Cramer, *J. Am. Chem. Soc.*, 1993, **115**, 2968–2969.

23 B. Hedman, K. O. Hodgson and E. I. Solomon, *J. Am. Chem. Soc.*, 1990, **112**, 1643–1645.

24 J. L. Hughey, T. G. Fawcett, S. M. Rudich, R. A. Lalancette, J. A. Potenza and H. J. Schugar, *J. Am. Chem. Soc.*, 1979, **101**, 2617–2623.

25 A. A. Gewirth and E. I. Solomon, *J. Am. Chem. Soc.*, 1988, **110**, 3811–3819.

26 A. Messerschmidt and R. Huber, *Eur. J. Biochem.*, 1990, **187**, 341–352.

27 L. B. LaCroix, D. W. Randall, A. M. Nersissian, C. W. G. Hoitink, G. W. Canters, J. S. Valentine and E. I. Solomon, *J. Am. Chem. Soc.*, 1998, **120**, 9621–9631.

28 L. Basumallick, R. K. Szilagyi, Y. Zhao, J. P. Shapleigh, C. P. Scholes and E. I. Solomon, *J. Am. Chem. Soc.*, 2003, **125**, 14784–14792.

29 S. Ghosh, X. Xie, A. Dey, Y. Sun, C. P. Scholes and E. I. Solomon, *Proc. Natl. Acad. Sci. U. S. A.*, 2009, **106**, 4969–4974.

30 M.-L. Tsai, R. G. Hadt, N. M. Marshall, T. D. Wilson, Y. Lu and E. I. Solomon, *Proc. Natl. Acad. Sci. U. S. A.*, 2013, **110**, 14658–14663.

31 A. L. Raphael and H. B. Gray, *Proteins: Struct., Funct., Bioinf.*, 1989, **6**, 338–340.

32 T. Ye, R. Kaur, F. T. Senguen, L. V Michel, K. L. Bren and S. J. Elliott, *J. Am. Chem. Soc.*, 2008, **130**, 6682–6683.

33 F. A. Tezcan, J. R. Winkler and H. B. Gray, *J. Am. Chem. Soc.*, 1998, **120**, 13383–13388.

34 H. Senn and K. Wuthrich, *Q. Rev. Biophys.*, 1985, **18**, 111–134.

35 T. Kroll, R. G. Hadt, S. A. Wilson, M. Lundberg, J. J. Yan, T.-C. Weng, D. Sokaras, R. Alonso-Mori, D. Casa, M. H. Upton, B. Hedman, K. O. Hodgson and E. I. Solomon, *J. Am. Chem. Soc.*, 2014, **136**, 18087–18099, DOI: 10.1021/ja5100367.

36 E. C. Wasinger, F. M. F. de Groot, B. Hedman, K. O. Hodgson and E. I. Solomon, *J. Am. Chem. Soc.*, 2003, **125**, 12894–12906.

37 M. Lundberg, T. Kroll, S. DeBeer, U. Bergmann, S. A. Wilson, P. Glatzel, D. Nordlund, B. Hedman, K. O. Hodgson and E. I. Solomon, *J. Am. Chem. Soc.*, 2013, **135**, 17121–17134.

38 M. W. Mara, R. G. Hadt, M. E. Reinhard, T. Kroll, H. Lim, R. W. Hartsock, R. Alonso-Mori, M. Chollet, J. M. Glownia, S. Nelson, D. Sokaras, K. Kunnus, K. O. Hodgson, B. Hedman, U. Bergmann, K. J. Gaffney and E. I. Solomon, *Science*, 2017, **356**, 1276–1280.

Faraday Discussions

ROYAL SOCIETY OF CHEMISTRY

DISCUSSIONS

Introductory lecture: general discussion

Abhishek Dey, Debabrata Goswami, Kenneth Karlin, Edward I. Solomon and T. Daniel P. Stack

DOI: 10.1039/d2fd90011d

Abhishek Dey opened a general discussion of the introductory lecture by Edward Solomon: Could you comment on the potential benefits of introducing entatic states in artificial catalysis?

Edward I. Solomon answered: We believe this can be an important component in enhancing reactivity. We have found key examples of entatic states in metallozeolite catalysis. Both Cu and Fe active sites in zeolite lattices take CH_4 to CH_3OH at low temperature. In the Cu zeolites, an entatic state provides two Cu(I) in close proximity to enable conversion with N_2O or O_2 to the active $[Cu_2O]^{2+}$ form and the relative orientation of the zeolite bidentate ligation on each Cu(II) activates the $[Cu_2O]^{2+}$ site for H-atom abstraction (HAA). In the Fe zeolite systems, an entatic state activates a mononuclear Fe(IV)=O $S = 2$ site by constraining it in a square pyramidal structure without a trans axial ligand. This destabilizes the Fe(IV)=O increasing the energy of the O–H bond formed in HAA and the lack of a trans axial ligand leads to a high oxyl-Fe(III) character in the transition state for HAA.

Abhishek Dey remarked: When one thinks of making artificial catalyst/lattices, there are of course two ways of doing it, control the thermodynamic effect or kinetics of it. Did you see entatic states contribute to both? The role of it in lowering reorganization energy in blue Cu during ET was very clear from your lecture. How about other reactions like HAT or oxo transfer?

Edward I. Solomon responded: In non-heme Fe(IV)=O HAT and oxo transfer reactions we have observed entatic contributions to both the thermodynamics and the kinetics. In a number of enzyme and metallozeolite sites we have studied the environment can destabilize Fe(IV)=O therefore increasing the strength of the Fe(III)O–H formed in HAT, a thermodynamic contribution. Marcus theory can be used to correct for the thermodynamic contribution to give the intrinsic barrier. This then relates to the frontier molecular orbital (FMO), the promotion energy to reach this orbital, its polarization (*i.e.* amount of oxyl-Fe(III) character) and its sigma *vs.* pi nature relative to the Fe–O bond which can affect selectivity.

Abhishek Dey commented: Regarding the reactivity of peroxidases *vs.* oxo transfer – we are seeing complementary reactivity. High-valent iron porphyrins that can do electron transfer can not do oxo transfer and *vice versa*.

Edward I. Solomon responded: Interesting. At least in part, this could relate to the requirement of substrate access to the Fe═O for the oxo transfer.

Kenneth Karlin said: You talked about the Cu–thiolate bond in 'blue' copper as being 42% $d_{x^2-y^2}$ and 38% S π-orbital. What about the "other" 20%? Can you answer with respect to a spectroscopist/computational chemist, and also answer with respect to someone like myself, please.

Edward I. Solomon replied: There is some delocalization into the rest of the Cys ligand associated with the superexchange pathway for electron transfer (*i.e.* the Cys–His pathway to the trinuclear Cu cluster in the multicopper oxidase example I use in my presentation) and some into the two His ligands in the $x^2 - y^2$ plane. These are quantified by the superhyperfine couplings of their nuclear spins to the electron spin as provided by ENDOR and ESEEM.

Kenneth Karlin remarked: You talked about the copper–thioether (Cu–S(met)) or even iron–histidine bonds being 1–14 kcal per mole from various different analyses. How is this, when C–H bonds are 75–105 kcal mmol^{-1}? Must it be we're talking NOT about bond dissociation energies (BDEs)? What is this really talking about, what do mean by an iron–histidine bond that is only 14 kcal per mole?

Edward I. Solomon answered: Yes, the Cu–thioether bonds are weak at 5 kcal mol^{-1} for the Cu(II) and 1 kcal mol^{-1} for the Cu(I) in the blue copper site. This is due to the strong donor interaction of the Cys with the Cu. For the His–Fe bond in cyt *c* it is 2.6 kcal mol^{-1} for the Fe(II) and 5.5 kcal mol^{-1} for the Fe(III). Note that for the Fe in addition to the covalent and ionic contributions to the bond there is the difference in exchange stabilization in going from low spin to high spin upon ligand loss. Importantly, the entropic contribution to the Gibbs free energy ($-T\Delta S$) at room temperature (8–10 kcal mol^{-1}) is greater than all of these enthalpies and without the protein entatic constraints these ligands would no longer be bound to the metal sites.

Kenneth Karlin commented: I never knew that metal–ligand bonds were typically as low as 30 or 40 in their BDEs.

Debabrata Goswami asked: Given that the ligand bond is so weak compared to the regular bonds, would it also mean that they would be much farther off in terms of distance? Is there a structure–function correlation as a result of this?

Edward I. Solomon answered: The weak thioether–Cu bond is indeed long at 2.9 A in the blue Cu site of plastocyanin, but constrained to be present by the protein. This is the entatic state that overcomes the entropic contribution to the free energy maintaining this bond at RT.

Dan Stack enquired: Does a correlation exist between the redox potential of a blue copper site and its charge transfer band in the Cu(II) form?

Edward I. Solomon responded: In the blue Cu proteins you don't observe MLCT in the Cu(I) state as these are higher in energy than the 280 nm intense absorption of the protein aromatic residues. However, in the oxidized blue Cu site the energy and intensity of the LMCT of the thiolate correlates with its strong covalent pi donor bonding and also provides a hole superexchange pathway for ET through the valence orbital of the protein.

Dan Stack asked: Is the redox potential of a blue copper site attenuated by adjusting the hydrogen bonding to the thiolate ligand to reduce the covalency of the copper–sulfur bond?

Edward I. Solomon answered: The H-bond to the thiolate does indeed reduce its covalent donor interaction with the Cu that destabilizes the oxidized more than the reduced state contributing to an increase in redox potential.

Dan Stack commented: Does the imidazole ligation to the copper center in a blue site compensate when the hydrogen bonding from the protein backbone to the thiolate sulfur is strong?

Edward I. Solomon replied: Only to a limited extent in blue Cu sites like plastocyanin as the imidazole ligands are weak donors (a few percent covalency) while the thiolate is a very strong donor ($\sim$40% covalency).

Conflicts of interest

There are no conflicts to declare.

Faraday Discussions

ROYAL SOCIETY OF CHEMISTRY

PAPER

A bio-inspired heterodinuclear hydrogenase CoFe complex†

Lili Sun,‡[a] Suzanne M. Adam,‡[ab] Walaa Mokdad,[a] Rolf David,[a] Anne Milet,[a] Vincent Artero*[b] and Carole Duboc*[a]

Received 8th November 2021, Accepted 2nd December 2021

DOI: 10.1039/d1fd00085c

Herein, a new heterobimetallic CoFe complex is reported with the aim of comparing its performance in terms of H_2 production within a series of related MFe complexes (M = Ni, Fe). The fully oxidized $[(L^{N2S2})Co^{II}(CO)Fe^{II}Cp]^+$ complex (**Co^{II}Fe^{II}**, $L^{N2S2\ 2-}$ = 2,2′-(2,2′-bipyridine-6,6′-diyl)bis(1,1′-diphenylethanethiolate), Cp^- = cyclopentadienyl anion) can be (electro)chemically reduced to its **Co^{I}Fe^{II}** form, and both complexes have been isolated and fully characterized by means of classic spectroscopic techniques and theoretical calculations. The redox properties of **Co^{II}Fe^{II}** have been investigated in DMF, revealing that this complex is the easiest to reduce by one-electron among the analogous MFe complexes (M = Ni, Fe, Co). Nevertheless, it displays no electrocatalytic activity for H_2 production, contrary to the FeFe and NiFe analogs, which have proven remarkable performance.

Introduction

Hydrogenases are metalloenzymes present in certain organisms – bacteria or microalgae – which have the capacity to produce or oxidize hydrogen.[1–4] Remarkably, they reversibly catalyze the conversion of protons into molecular hydrogen at a potential close to the thermodynamic potential ($E^0_{app}(H^+/H_2) = -413$ mV in water at pH 7 and 25 °C under 0.1 bar of H_2). Two main classes of hydrogenases are known, differing by the nature of their active sites. The active site of the [NiFe]-hydrogenases consists of a heterodinuclear NiFe complex, where the two metal ions are bridged by thiolate moieties from the cysteine residues of the polypeptide chain. In the case of the [FeFe]-hydrogenases, the active site is a dinuclear iron complex with the two Fe ions also connected by two thiolate functionalities, but arising from an unusual ligand in biology, *i.e.*, dithiomethylamine. These two organometallic MFe clusters (M = Ni or Fe) present a unique structure in biology with the presence of carbon monoxide (CO) and

[a]*Univ. Grenoble Alpes, CNRS UMR 5250, DCM, F-38000 Grenoble, France. E-mail: carole.duboc@univ-grenoble-alpes.fr*

[b]*Univ. Grenoble Alpes, CNRS, CEA, Laboratoire de Chimie et Biologie des Métaux, F-38000 Grenoble, France*

† Electronic supplementary information (ESI) available. See DOI: 10.1039/d1fd00085c

‡ Equal contribution of the co-authors.

cyanide (CN^-) as iron ligands. These structures contribute to these systems' unique electronic properties, which facilitate the protonation of the metal sites under mild conditions, thereby making hydrogenases great catalysts for H_2 production. Thus, mimicking the activity of these enzymes with bio-inspired complexes to develop new catalysts for H_2 production has quickly emerged as an attractive strategy for chemists.[5–8]

Recently, we described the H_2 production reactivity of two dinuclear $M^{II}Fe^{II}$ (M = Fe or Ni) complexes with similar structures. We have demonstrated that both $[(L^{N2S2})Ni^{II}Fe^{II}Cp(CO)]^+$ (**Ni^{II}Fe^{II}**)[9,10] and $[(L^{N2S2})(MeCN)Fe^{II}(CO)Fe^{II}Cp]^+$ (**Fe^{II}Fe^{II}**)[11,12] complexes (L^{N2S2}: 2,2′-(2,2′-bipyridine-6,6′-diyl)bis(1,1′-diphenylethanethiolate), Cp = cyclopentadienyl, Scheme 1) display comparable electrocatalytic performance for H_2 production, both following an *E*[ECEC] catalytic mechanism (E = electron transfer, C = chemical reaction, *e.g.*, a protonation step in the present case). This absence of notable changes in the reactivity was rationalized by the non-innocent redox activity of the L^{N2S2} ligand that is proposed to dominate the catalytic process.[13] Since it has been shown that Co complexes can display high performance for H_2 production,[14–17] we describe herein the investigation of the **Co^{II}Fe^{II}** parent complex, with the aim of comparing its reactivity with its NiFe and FeFe analogs. The target $[(L^{N2S2})Co^{II}(CO)Fe^{II}Cp]^+$ complex, **Co^{II}Fe^{II}**, has been synthesized and characterized, and its reduction chemistry as well as its H_2 production reactivity have been investigated.

Results and discussion

Synthesis and characterization of Co^{II}Fe^{II}

An equimolar mixture of $[Co^{II}(L^{N2S2})]$ and $[Fe^{II}(CO)(Cp)(CH_3CN)_2]BF_4$ stirred overnight in dichloromethane (DCM) at room temperature under an inert atmosphere affords the dark brown $[(L^{N2S2})Co^{II}(CO)Fe^{II}(Cp)]BF_4$ target complex

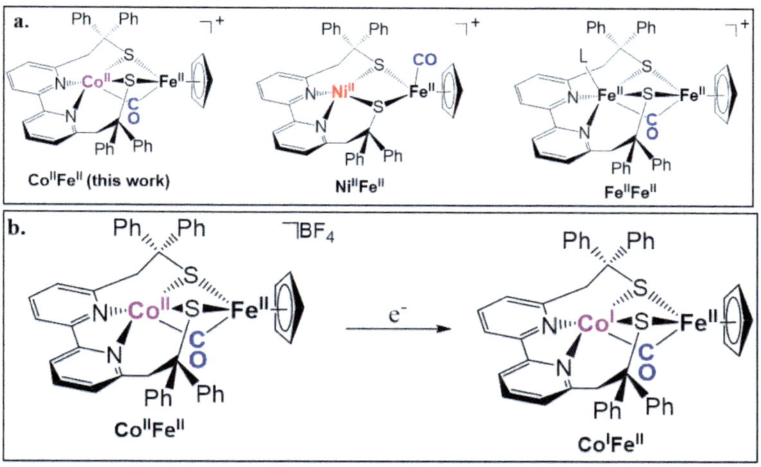

Scheme 1 (a) Family of dinuclear model complexes (CoFe (this work), NiFe,[11] and FeFe,[10] with L = CO or solvent). (b) Reduction reactivity of the **Co^{II}Fe^{II}** complex described in this work.

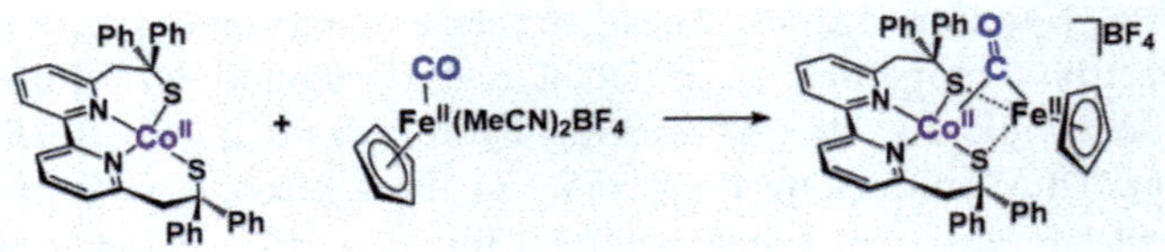

Scheme 2 Synthetic pathway to afford $Co^{II}Fe^{II}$.

(**$Co^{II}Fe^{II}$**), which could be isolated with high yield (Scheme 2). The optical spectrum of a solution of **$Co^{II}Fe^{II}$** in dimethylformamide (DMF) displays two features in the visible region of the spectrum with $\lambda_{max} = 468$ nm ($\varepsilon = 4900$ M^{-1} cm^{-1}) and 584 nm ($\varepsilon = 3100$ M^{-1} cm^{-1}), distinct from the spectra of the mononuclear reagents (see Fig. S1†). The ESI mass spectrum supports the overall formulation, $[(L^{N2S2})Co^{II}Fe^{II}(CO)(Cp)]^{+}$, with the major mass peak being observed at m/z = 786.04 (calc. 786.09).

Infrared spectroscopy was employed to monitor the binding mode of the CO moiety in **$Co^{II}Fe^{II}$**. A stretching frequency is observed at 1856 cm^{-1} in the range

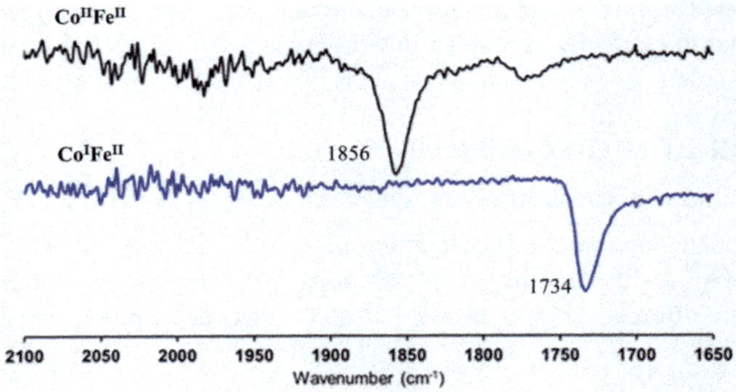

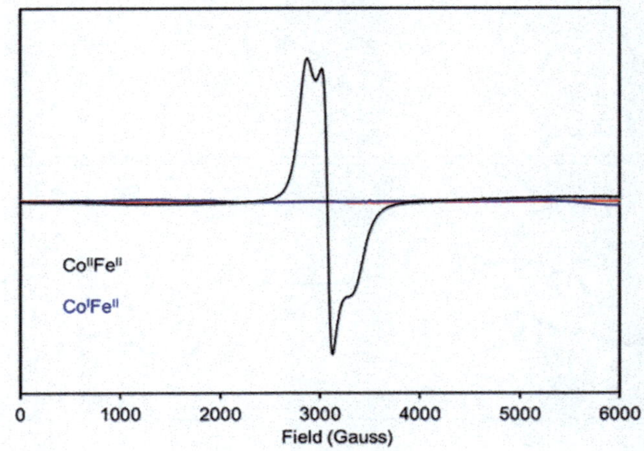

Fig. 1 Attenuated total reflectance infrared (top) and 100 K solid-state EPR (bottom) spectra of $Co^{II}Fe^{II}$ (black) and $Co^{I}Fe^{II}$ (blue) complexes.

for metal-bridging CO moieties (Fig. 1), at an energy close to that found in the **Fe^{II}Fe^{II}** derivative (1822 cm^{-1}), and at a significantly lower energy than the terminal M–CO stretching frequencies present in the [Fe^{II}(CH$_3$CN)$_2$(CO)(Cp)]BF$_4$ reagent[18] and the analogous **Ni^{II}Fe^{II}** complex[9] (1984 and 1929 cm^{-1}, respectively). Based on these data, it can be concluded that the CO bridges the Co^{II} and Fe^{II} ions in **Co^{II}Fe^{II}**, both in solution (see Fig. S2†) and in the solid state (Fig. 1, bottom).

In order to gain insight into the electronic structure of this heterodinuclear complex, EPR spectra were recorded both as a powder sample (Fig. 1) and as a frozen DMF solution (Fig. S2†) sample at 100 K. The solid-state sample clearly shows a rhombic spectrum ($g_x = 2.24$; $g_y = 2.09$; $g_z = 1.98$) consistent with a low-spin d^7 Co^{II} species. In this case, the d^6 Fe^{II} center does not contribute to the spectrum indicating that the single unpaired electron of the overall $S = \frac{1}{2}$ complex is localized on the Co center. A hyperfine interaction ($I = 7/2$ for ^{59}Co) is resolved in the frozen DMF solution sample spectrum, supporting the Co^{II} assignment (see ESI, Fig. S2†).

To further investigate the structural and spectroscopic properties of **Co^{II}Fe^{II}**, DFT calculations were carried out. Its optimized structure, **Co^{II}Fe^{II}***, displays predicted spectroscopic parameters consistent with the experimental ones, especially regarding the calculated ν_{CO} vibration at 1887 cm^{-1} (*vs.* ν_{CO} exp. at 1856 cm^{-1}). The overall spin state of the complex is confirmed to be a doublet (the quartet is predicted to be at a much higher energy, see Table S1†). The Mulliken spin population of 1.1 on the Co ion is fully consistent with a localized $S = \frac{1}{2}$ **Co^{II}Fe^{II}** complex, as proposed by the experimental EPR data (see Fig. S3† for the spin density).

Redox properties of Co^{II}Fe^{II}

The cyclic voltammogram (CV) of a DMF solution of **Co^{II}Fe^{II}** reveals two diffused-controlled (quasi-)reversible one-electron processes, centered at $E_{1/2} = -1.00$ V ($\Delta E_p = 100$ mV) and $E_{1/2} = -2.04$ V ($\Delta E_p = 80$ mV) *vs.* Fc$^{+/0}$ (Fig. 2). The redox system at -1.00 V can be attributed to the Co$^{II/I}$ reduction, and that at more negative potential, to the reduction of the redox-non-innocent bipyridine ligand. The Co$^{II/I}$ reduction potential against ferrocene, obtained *via* DFT (see ESI for details and Table S2†) is in perfect agreement with a value of -1.02 V *vs.* Fc$^{+/0}$.

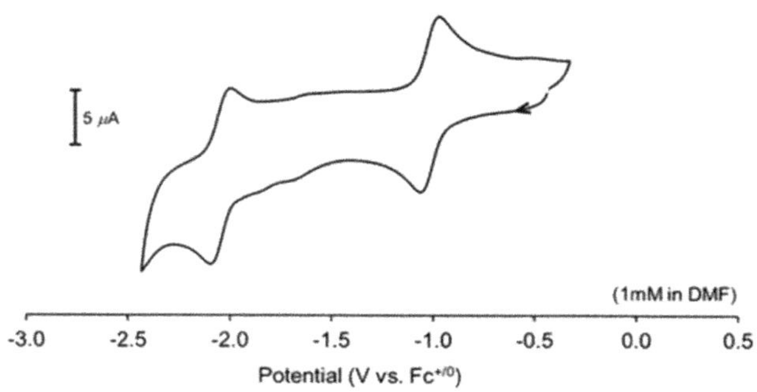

Fig. 2 Cyclic voltammogram of a DMF solution of **Co^{II}Fe^{II}** (1 mM); electrolyte: 0.1 M (Bu)$_4$NClO$_4$; reference: Ag/AgNO$_3$; counter: Pt wire; scan rate: 100 mV s^{-1}.

Such assignments can be proposed based on the comparison of its redox properties to those of the analogous $\mathbf{Fe^{II}Fe^{II}}$ and $\mathbf{Ni^{II}Fe^{II}}$ complexes.[9,11] The CV of $\mathbf{Ni^{II}Fe^{II}}$ reported in MeCN shows two reversible redox systems at −1.29 V and −1.90 V *vs.* $Fc^{+/0}$, attributed to the $Ni^{II/I}$ and L^{N2S2} reduction processes, respectively. The reduction of the Ni^{II} ion in $\mathbf{Ni^{II}Fe^{II}}$ occurs at a significantly more positive potential than in the mononuclear $[Ni^{II}(L^{N2S2})]$ complex under the same conditions ($E_{1/2} = -1.82$ V *vs.* $Fc^{+/0}$ corresponding to a potential shift of 530 mV), highlighting the strong electronic effect of the Lewis acidic Fe component. The $Co^{II/I}$ reduction potential in $\mathbf{Co^{II}Fe^{II}}$ is shifted to an even greater extent (700 mV) as compared to the mononuclear $[Co^{II}(L^{N2S2})]$, which displays a quasi-reversible $Co^{II/I}$ wave at −1.7 V *vs.* $Fc^{+/0}$ in DMF.[14] Regarding $\mathbf{Fe^{II}Fe^{II}}$, the diiron analog exists as the mono-CO form in acetonitrile at a concentration of 0.2 mM, with a bridging CO ligand similar to that found in $\mathbf{Co^{II}Fe^{II}}$ (see Scheme 1). Under these conditions, the first one-electron reduction process occurs at −1.21 V *vs.* $Fc^{+/0}$ resulting in the delocalized, mixed-valent $Fe^{1.5}Fe^{1.5}$ species. In this series of $M^{II}Fe^{II}$ complexes, $\mathbf{Co^{II}Fe^{II}}$ is, therefore, the easiest to reduce.

In the analogous $\mathbf{Fe^{II}Fe^{II}}$ and $\mathbf{Ni^{II}Fe^{II}}$ complexes, it was shown that the second one-electron reduction process occurs at the bipyridine moiety of the L^{N2S2} ligand, at $E_{1/2} = -1.65$ V and $E_{1/2} = -1.90$ V *vs.* $Fc^{+/0}$, respectively, in MeCN. In the case of $\mathbf{Co^{II}Fe^{II}}$, the redox system associated with the bipyridine ligand occurs at a less negative potential, $E_{1/2} = -2.04$ V *vs.* $Fc^{+/0}$ in DMF. The small difference with $\mathbf{Ni^{II}Fe^{II}}$ can be explained by the difference in the solvent (DMF *vs.* acetonitrile), while that with $\mathbf{Fe^{II}Fe^{II}}$ likely arises from differences in the structural and electronic properties of the one-electron reduced species (see below).

Synthesis and characterization of the one-electron reduced $Co^{I}Fe^{II}$ species

The $\mathbf{Co^{II}Fe^{II}}$ complex can be chemically reduced using one molar equiv. of cobaltocene ($CoCp_2$), to generate the one-electron reduced species $\mathbf{Co^{I}Fe^{II}}$ (see ESI‡ for details). Based on the electrochemical analysis, it has been proposed that the reduction occurs at the cobalt center. This conclusion is experimentally supported by the disappearance of the X-band EPR feature present in the

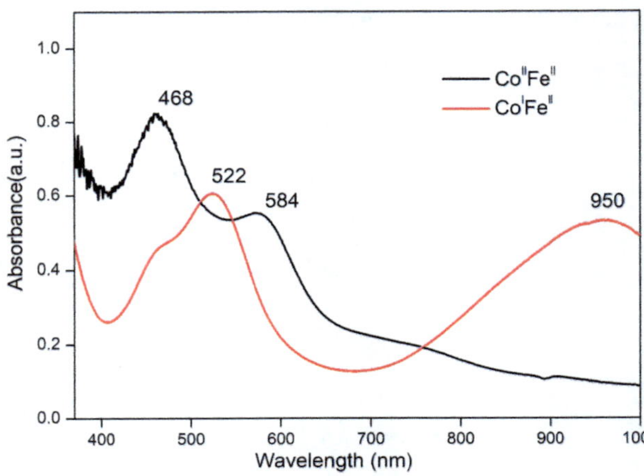

Fig. 3 UV-visible spectra of the $\mathbf{Co^{II}Fe^{II}}$ (black), $\mathbf{Co^{I}Fe^{II}}$ (red) complexes (0.1 mM, DMF).

spectrum of $\mathbf{Co^{II}Fe^{II}}$ (Fig. 1). The EPR-silent spectrum of $\mathbf{Co^{I}Fe^{II}}$ thus confirms an overall $S = 0$ complex (Co^{I}, d^8; Fe^{II}, d^6). The IR spectrum displays a CO vibrational frequency at 1734 cm^{-1} consistent with a bridging mode (Fig. 1). With respect to $\mathbf{Co^{II}Fe^{II}}$ (1856 cm^{-1}) the energy of this vibration is notably weaker, indicating an elongation of the bridging C=O bond. Distinct changes in the optical profile are also observed upon reduction (Fig. 3), including the growth of a broad, low-energy d–d transition feature (950 nm; $\varepsilon = 1500$ M^{-1} cm^{-1}).

The structural and spectroscopic properties of $\mathbf{Co^{I}Fe^{II}}$ have been further investigated by DFT calculations. The predicted CO vibration calculated from its optimized structure, $\mathbf{Co^{I}Fe^{II}*}$, is consistent with the experimental ones, *i.e.*, ν_{CO}

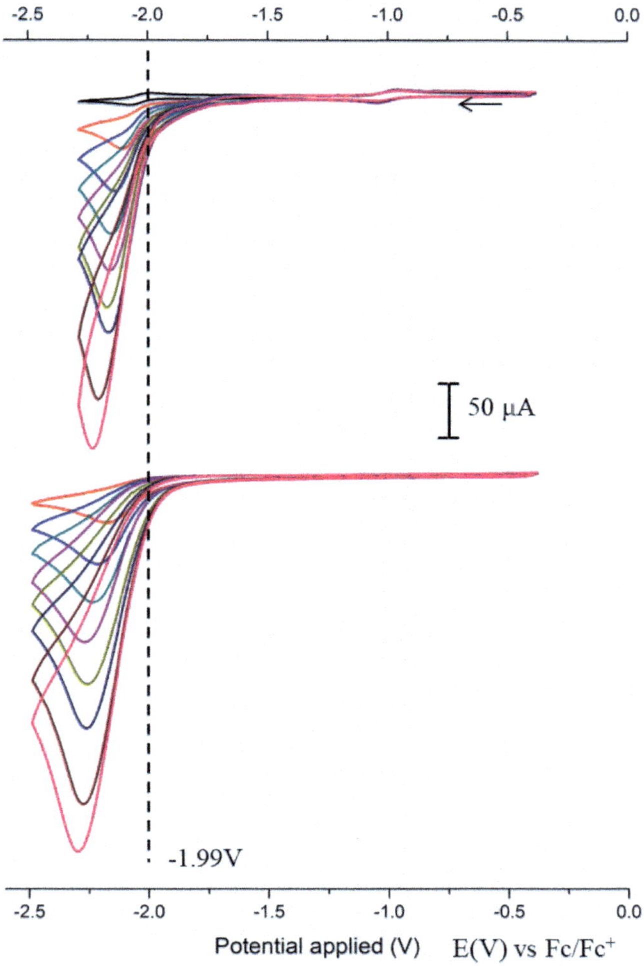

Fig. 4 Top: CVs of $\mathbf{Co^{II}Fe^{II}}$ (0.73 mM, top, black) in the absence or presence of various amounts of Et_3NHBF_4 in DMF solution, 0.1 M n-Bu_4NClO_4, on a glassy carbon electrode at 100 mV s^{-1}: 5 equiv. (red); 10 equiv. (blue); 15 equiv. (cyan); 20 equiv. (pink); 25 equiv. (dark yellow); 30 equiv. (dark blue); 40 equiv. (deep red); 50 equiv. (magenta). Bottom: CVs of the corresponding blank samples (no catalyst).

vibration at 1779 cm^{-1} *vs.* ν_{CO} exp. at 1734 cm^{-1}. The overall spin state of the complex is confirmed to be a singlet in agreement with the EPR-silent spectrum.

Reactivity of $Co^{II}Fe^{II}$ in the presence of H^+

Following the addition of Et_3NHBF_4 as a proton source, a catalytic process (E_{cat} = −1.99 V *vs.* Fc/Fc^+, measured at the mid wave at a concentration of 10.95 mM of Et_3NHBF_4) develops on top of the second wave of the $\mathbf{Co^{II}Fe^{II}}$ complex (Fig. 4). Nevertheless, when compared with the CVs recorded under the same conditions, but in the absence of $\mathbf{Co^{II}Fe^{II}}$, the reduction appears to occur at a similar potential evidencing that the presence of $\mathbf{Co^{II}Fe^{II}}$ has a marginal impact, if any, on the H_2 production catalyzed by the glassy carbon surface of the electrode. This electrochemical behavior was completely unexpected since the investigation of both $\mathbf{Ni^{II}Fe^{II}}$ and $\mathbf{Fe^{II}Fe^{II}}$ supported the fact that the nature of the metal bound by the L^{N2S2} ligand has no notable impact on the performance of these electrocatalysts, and that the bipyridine unit was the key factor in controlling their reactivity.[11]

Conclusions

The target $\mathbf{Co^{II}Fe^{II}}$ complex was synthesized with the aim of comparing its reactivity towards H_2 production with its NiFe and FeFe analogs. We were also able to isolate and characterize the corresponding one-electron reduced species. Unexpectedly, this new complex was not active to produce H_2 electrocatalytically, while the analogous NiFe and FeFe complexes have demonstrated remarkable performance. Current work in our laboratories focuses on understanding the origin of this lack of reactivity, since $\mathbf{Co^{II}Fe^{II}}$ displays the most promising redox properties, *i.e.*, it is the easiest to reduce within the series of $M^{II}Fe^{II}$ complexes.

Conflicts of interest

There are no conflicts to declare.

Acknowledgements

Financial support for this work was provided by the Agence National de la Recherche in the framework of the "Investissements d'avenir" Program (ANR-15-IDEX-02), the Labex ARCANE (ANR-11-LABX-003), and the CBH-EUR-GS (ANR-17-EURE-0003), and ANR-DFG (ANR-16-CE92_0012_01 NiFeMim) and by the China Scholarship Council.

References

1 W. Lubitz, H. Ogata, O. Rüdiger and E. Reijerse, *Chem. Rev.*, 2014, **114**, 4081–4148.

2 D. Schilter, J. M. Camara, M. T. Huynh, S. Hammes-Schiffer and T. B. Rauchfuss, *Chem. Rev.*, 2016, **116**, 8693–8749.

3 Y. Nicolet, C. Piras, P. Legrand, C. E. Hatchikian and J. C. Fontecilla-Camps, *Structure*, 1999, **7**, 13–23.

4 A. Volbeda, M.-H. Charon, C. Piras, E. C. Hatchikian, M. Frey and J. C. Fontecilla-Camps, *Nature*, 1995, **373**, 580–587.

5 S. Kaur-Ghumaan and M. Stein, *Dalton Trans.*, 2014, **43**, 9392–9405.

6 A. C. Ghosh, C. Duboc and M. Gennari, *Coord. Chem. Rev.*, 2021, **428**, 213606.

7 N. Coutard, N. Kaeffer and V. Artero, *Chem. Commun.*, 2016, **52**, 13728–13748.

8 J. T. Kleinhaus, F. Wittkamp, S. Yadav, D. Siegmund and U.-P. Apfel, *Chem. Soc. Rev.*, 2021, **50**, 1668–1784.

9 D. Brazzolotto, M. Gennari, N. Queyriaux, T. R. Simmons, J. Pécaut, S. Demeshko, F. Meyer, M. Orio, V. Artero and C. Duboc, *Nat. Chem.*, 2016, **8**, 1054–1060.

10 M. E. Ahmed, S. Chattopadhyay, L. K. Wang, D. Brazzolotto, D. Pramanik, D. Aldakov, J. Fize, A. Morozan, M. Gennari, C. Duboc, A. Dey and V. Artero, *Angew. Chem., Int. Ed.*, 2018, **57**, 16001–16004.

11 L. Wang, M. Gennari, A. Barrozo, J. Fize, C. Philouze, S. Demeshko, F. Meyer, M. Orio, V. Artero and C. Duboc, *ACS Catal.*, 2020, **10**, 177–186.

12 M. E. Ahmed, D. Saha, L. Wang, M. Gennari, S. Ghosh Dey, V. Artero, A. Dey and C. Duboc, *ChemElectroChem*, 2021, **8**, 1674–1677.

13 M. Gennari and C. Duboc, *Acc. Chem. Res.*, 2020, **53**, 2753–2761.

14 V. Artero and J.-M. Saveant, *Energy Environ. Sci.*, 2014, **7**, 3808–3814.

15 G.-G. Luo, H.-L. Zhang, Y.-W. Tao, Q.-Y. Wu, D. Tian and Q. Zhang, *Inorg. Chem. Front.*, 2019, **6**, 343–354.

16 N. Queyriaux, R. T. Jane, J. Massin, V. Artero and M. Chavarot-Kerlidou, *Coord. Chem. Rev.*, 2015, **304–305**, 3–19.

17 D. Dolui, S. Khandelwal, P. Majumder and A. Dutta, *Chem. Commun.*, 2020, **56**, 8166–8181.

18 A. R. Cutler and A. B. Todaro, *Organometallics*, 1988, **7**, 1782–1787.

Faraday Discussions

PAPER

Electrocatalytic alcohol oxidation by a molecular iron complex†

Himangshu Kuiry,[a] Debasree Das,[b] Soumadip Das,[a] Soham Chakraborty,[a] Bittu Chandra[a] and Sayam Sen Gupta*[a]

Received 18th October 2021, Accepted 14th December 2021
DOI: 10.1039/d1fd00074h

An efficient electrochemical method for the selective oxidation of alcohols to their corresponding aldehydes/ketones using a biomimetic iron complex, $[(bTAML)Fe^{III}-OH_2]^-$, as the redox mediator in an undivided electrochemical cell with inexpensive carbon and nickel electrodes using water as an oxygen source is reported. The substrate scope also includes alcohols that contain O and N heteroatoms in the scaffold, which are well tolerated under these reaction conditions. Mechanistic studies show the involvement of a high-valent $Fe^V(O)$ species, $[(bTAML)Fe^V(O)]^-$, formed *via* PCET (overall $2H^+/2e^-$) from $[(bTAML)Fe^{III}-OH_2]^-$ at 0.77 V (*vs.* Fc^+/Fc). Moreover, electrokinetic studies of the oxidation of C–H bonds indicate a second-order reaction, with the C–H abstraction by $Fe^V(O)$ being the rate-determining step. The overall mechanism, studied using linear free energy relationships and radical clocks, indicates a "net hydride" transfer, leading to the oxidation of the alcohol to the corresponding aldehyde or ketone. When the reaction was carried out at pH > 11, the reaction could be carried out at a ∼500 mV lower potential than that at pH 8, albeit with reduced reaction rates. The reactive intermediate involved at pH > 11 is the corresponding one-electron oxidized $[(bTAML)Fe^{IV}(O)]^{2-}$ species.

Introduction

Oxidation reactions are ubiquitous, and they are one of the most significant transformations in organic chemistry, whereby fossil fuels with high H/C ratios are converted to a variety of oxygenated products with reduced H/C ratios. Oxidations are also widely used to increase molecular complexity to synthesize several chemical entities, including pharmaceuticals. Common oxidants, including metal oxyanions (*e.g.*, $Cr_2O_7^{2-}$, MnO_4^-) and hypervalent iodine reagents, are stoichiometric and generate waste products, which is against the

[a]*Department of Chemical Sciences, Indian Institute of Science Education and Research Kolkata, Mohanpur, West Bengal, India-741246. E-mail: sayam.sengupta@iiserkol.ac.in*

[b]*Chemical Engineering and Process Development Division, CSIR-National Chemical Laboratory, Pune, Maharashtra-411008, India*

† Electronic supplementary information (ESI) available. See DOI: 10.1039/d1fd00074h

principles of green chemistry.[1] On the other hand, nature uses O_2 as the terminal oxidant and enzymes as catalysts that activate O_2 to carry out a wide range of oxidations with very high selectivity. The formation of H_2O from O_2 constitutes the only byproduct during such oxidations. In the last two decades, there has been significant progress in developing oxidation processes that mimic enzyme-based oxidations, but further advances are still required to deploy them commercially. In this field, electrochemistry provides a unique opportunity in which the driving force for the redox reaction is provided by electricity, which is expected to be produced from renewable resources like solar or wind. Many electrochemical oxidation reactions employ H^+ as the terminal electron acceptor, generating H_2 as the sole byproduct. The ability to control the anodic potential also affords great possibilities of controlling the rate and selectivity of reactions, especially for those which involve single-electron transfer (SET).[2–5] However, a fundamental limitation of direct electrochemical oxidation is the high overpotentials required for SET reactions, since they generally lead to the formation of high-energy radical cation intermediates upon oxidation of the organic molecule at the anode. This limits their application, since other functional groups in the molecule could also be oxidized at such high potentials, leading to its decomposition. This problem is being addressed using "electron–proton transfer mediators" (EPTMs) which have several advantages, including prospects for improved reaction selectivity and the ability to operate with electrode potentials which are several hundred mV lower than those of reactions conducted by direct electrolysis.[6]

The electrocatalytic oxidation of alcohols has attained great importance from the perspective of energy and chemical conversion. The attractiveness of alcohols as chemical fuels, ranging from fuel cells to their utilization in biomass, originates from their high energy densities, established manufacturing infrastructure, and ability to be extracted from renewable resources. Recent advances in the electrochemical oxidation of alcohols by Ni, Ru and Cu/nitroxyl based catalysts have shown promising results. For example, the chemical oxidation of alcohols using the stable aminoxyl radical TEMPO (2,2,6,6-tetramethylpiperidine *N*-oxyl) mediator and NaOCl as the terminal oxidant can be performed electrochemically. For the electrochemical oxidation of alcohols, the anodic oxidation features a two-electron redox cycle between the hydroxylamine and oxoammonium form of the catalyst (*i.e.*, TEMPOH/TEMPO$^+$), with the oxoammonium form acting as the catalyst for the dehydrogenative oxidation of alcohols.[7–9] This has been employed for several applications, including pharmaceutical synthesis and biomass conversion. Stahl *et al.* developed a complementary electrochemical alcohol oxidation method that uses a cooperative Cu/TEMPO mediator system that operates at a 0.5 V lower electrode potential than that of the TEMPO-only mediated process.[10] This difference is because the new process is mediated by the Cu^I/Cu^{II} oxidation cycle, which allows C–H abstraction from the alkoxide bound to Cu^{II} by TEMPO. This addition of a Cu-complex allows the exploitation of the low potential of the TEMPO/TEMPOH (rather than TEMPO/TEMPO$^+$) redox process, thus reducing the overpotential by 0.5 V.

Several bio-inspired methodologies inspired by metalloenzymes have been attempted for alcohol oxidation. For example, nickel phosphine complexes with pendant amines have been used to mimic hydrogenase to function as an electrocatalyst for the oxidation of primary and secondary alcohols, with turnover frequencies as high as 3.3 s^{-1}.[11] The reaction involved the initial base-assisted binding of the alcohol to the Ni complex, followed by dehydrogenation, leading to a Ni^{II}–H intermediate which then undergoes 2-electron oxidation, leading to the regeneration of the starting complex. The authors claimed that these complexes were the first reported electrocatalysts for alcohol oxidation based on non-precious metals. The other approach used HAT-mediators, such as the electrochemically generated high-valent oxometal intermediates that would catalyze alcohol oxidation. High valent oxometal complexes are crucial reactive intermediates because of their vital roles in the selective transfer of oxygen atoms into a target substrate, like an alkane, alkene, or alcohol, during a large number of biological and chemical reactions. The reactivities of high-valent oxometal complexes and their roles in the inorganic reaction mechanisms have been thoroughly investigated for years. The strategy for electrochemical oxidation involves the generation of high-valent oxometal species from the starting M–OH_2 complex *via* electrochemical proton-coupled oxidation, which is expected to react with the substrate of interest subsequently.[12] Such reactivity has been extensively investigated in electrocatalytic water oxidation, but reports of the preparative-scale electrochemical oxidation of organic molecules involving this approach have been much rarer. The generation of the oxometal species electrochemically bypasses the need for stoichiometric chemical oxidants, and the only byproduct formed is H_2 due to proton reduction at the cathode. The earliest examples reported involved the electrochemical generation of Ru^{V}(O) from Ru^{II}–OH_2 complexes and their subsequent oxidation of benzyl alcohol *via* a pathway involving concerted hydride proton transfer (HPT).[13,14] Subsequently, Meyer *et al.* synthesized Ru^{IV}(O)/Ru^{V}(O) species from their corresponding Ru^{II}–OH_2 complexes *via* multiple PCET processes and showed significant rate enhancements for the one-electron oxidized Ru^{V}(O) intermediate over the corresponding Ru^{IV}(O).[13] Recently, Stahl *et al.* reported the oxidation of activated benzylic C–H bond and alcohols *via* an electrochemically generated dimeric μ-oxo-(Fe^{IV}-TAML)$_2$ (TAML = tetra-amido macrocyclic ligand) intermediate in a divided cell.[15] Our group had previously reported the 5th generation biuret modified TAML, $[Et_4N]_2$[Fe-bTAML(Cl)],[16] which catalyzes the oxidation of 3° C–H bonds with unprecedented selectivity over 2° C–H bonds using NaOCl or mCPBA as the terminal oxidant.[17,18] The epoxidation of C=C bonds with NaOCl and O_2 has also been demonstrated.[19,20] The active intermediate, characterized by several spectroscopic techniques, was identified as the Fe^{V}(O) intermediate, [(bTAML)$Fe^{V}(O)]^{-}$.[17,21] This intermediate is remarkably stable at RT with a half-life of ~30 min.[17,22] The mechanism of the C–H oxidation of alkanes and oxidation of alkenes and alcohols by the Fe^{V}(O) intermediate has been reported earlier.[17,20] This Fe^{V}(O) intermediate can also be generated electrochemically in mixed $CH_3CN : H_2O$ solvent from the starting aqua-bound Fe-bTAML complex *via* two subsequent PCET reactions.[23] The electrochemical generation of the Fe^{V}(O) intermediate allowed us to perform selective electrochemical alkane hydroxylation and epoxidation.[24] The electrochemical oxidations were carried out in a simple undivided cell using inexpensive electrodes made from carbon and nickel. In this procedure, we have effectively oxidized sp^3 C–H bonds with BDE >97 kcal mol^{-1} using water as the oxygen source. Electrochemical and spectroelectrochemical methods also studied the detailed mechanism of the 3° C–H bond oxidation using this Fe^{V}(O) intermediate. More than 80% yield was obtained during the bulk electrolysis process for a broad range of substrates.

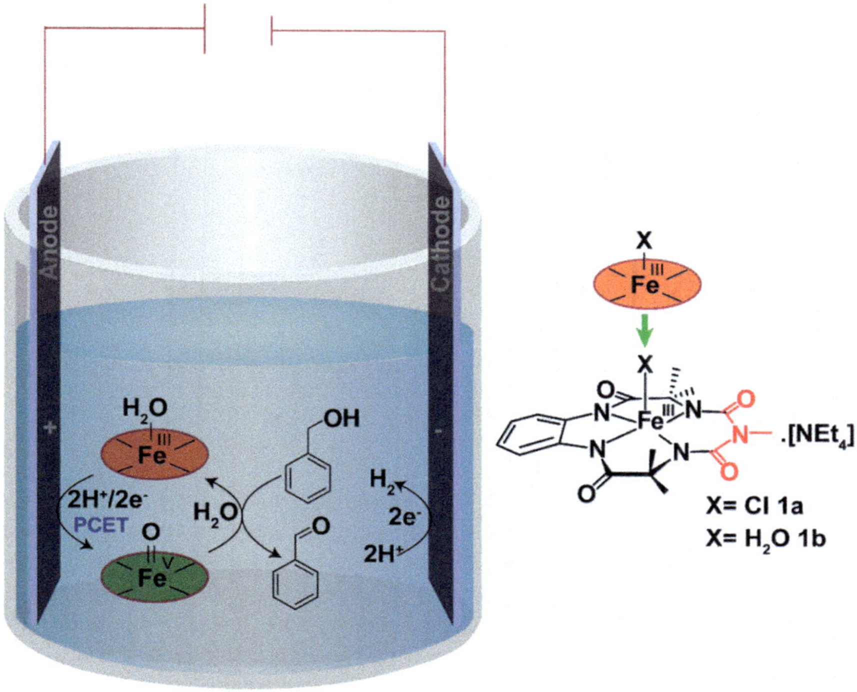

Scheme 1 Electrochemical oxidation of benzyl alcohol to benzaldehyde.

Herein, we report the electrochemical oxidation of alcohols with Fe^{III}-bTAML as an electrocatalyst. We show that some benzylic and aliphatic secondary alcohols can be oxidized using our methodology with high yields.[25,26] We demonstrate that the electrochemically generated $Fe^{V}(O)$ intermediate is likely responsible for this oxidation. Additionally, we have determined the rate and the order of the reaction using cyclic voltammetry. The electrochemical oxidation mechanism was also studied using methods including kinetic isotope effect values, linear free energy relationships, and radical clocks (Scheme 1).

Results and discussions

Cyclic voltammetry of 1a

The electrochemical properties of the bTAML complexes **1a** and **1b** in 0.1 M tetrabutylammonium hexafluorophosphate ($TBAPF_6$) acetonitrile solution using glassy carbon (GC) as the working electrode, a Pt wire as the counter electrode, and $Ag/AgNO_3$ (0.01 M) as the reference electrode were previously reported by us.[23,24] In short, for complex **1a**, a reversible one-electron wave at $E_{1/2} = 0.13$ V (0.06 V *vs.* Fc^+/Fc) and a second quasi-reversible one-electron wave at 0.72 V (0.65 V *vs.* Fc^+/Fc) with a near Nernstian peak-to-peak separation of $\Delta E_p = 66$ mV was observed by cyclic voltammetry. The peak-to-peak separation of 66 mV of the first and second oxidation processes indicates that the processes at 0.13 V and 0.72 V are both $1e^-$ redox processes and were assigned to the Fe^{IV}/Fe^{III} and Fe^{V}/Fe^{IV}

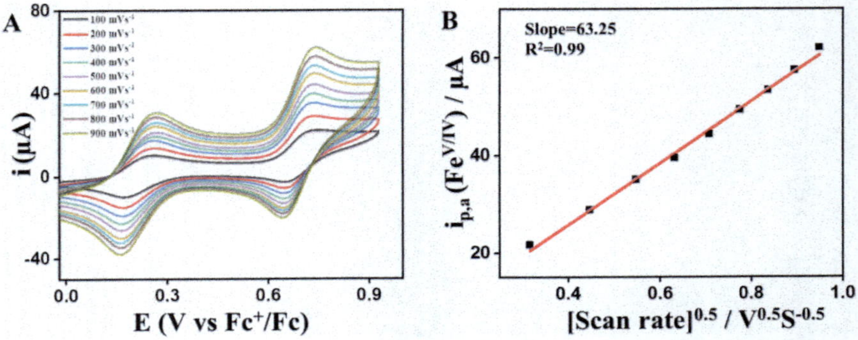

Fig. 1 (A) Cyclic voltammogram of 1 mM complex **1a** in (19 : 1) CH_3CN–H_2O at different scan rates varying from 100 mV s^{-1} to 900 mV s^{-1}. (B) The plot of the catalytic current (i_p) at 0.73 V vs. square root of the scan rate ($v^{1/2}$); linear fit (slope = 63.25, $R^2 = 0.99$).

redox couples, respectively (Fig. S1†). Upon the addition of 5% H_2O to acetonitrile, a 0.2 V shift of the anodic wave of the Fe^{III}/Fe^{IV} redox couple was observed due to the substitution of the axial anionic chloro ligand by the neutral water molecule. For the $Fe^{V/IV}$ couple, the peak current (I_p) at 0.80 V *versus* Ag/$AgNO_3$ (0.73 V *vs.* Fc^+/Fc) varied linearly with $v^{1/2}$ (Fig. 1). This linearity suggested that diffusion-limited electron transfer was taking place at the electrode surface. The diffusion coefficient D was calculated from the peak current *versus* square root of the scan rate variation plot, as shown in Fig. 1B, using eqn (1), where n, v, R and T are the number of electrons transferred, the scan rate of voltammetry, the universal gas constant and temperature, respectively. Here, $n = 1$, which is the number of electrons transferred per catalyst in the redox wave. From the equation,

$$i_p = 0.446nFA[\mathrm{Fe}]\sqrt{\frac{nFvD}{RT}} \tag{1}$$

D was obtained as 1.1×10^{-5} cm^2 s^{-1}.

In our earlier report, we showed that the controlled potential electrolysis (CPE) of a dilute solution (<0.2 mM) of **1a** in CH_3CN–H_2O at 0.80 V *vs.* Ag/$AgNO_3$ led to the immediate formation of the $[(\mathrm{bTAML})Fe^V(O)]^-$ intermediate, while at higher concentrations, the formation of the μ-O-Fe^{IV}_2 dimer was observed due to the diffusion of $[(\mathrm{bTAML})Fe^V(O)]^-$ into the bulk and a subsequent comproportionation reaction with **1b** (Fig. S3†). The electrochemical generation of $Fe^V(O)$ at 0.80 V *vs.* Ag/$AgNO_3$ on the CV time scale prompted us to explore the possibility of the electrochemical oxidation of alcohols. Several alcohols have been shown to react with the chemically synthesized $Fe^V(O)$ with high selectivity and fast reaction rates. The CVs of **1a** recorded with 0.10 M $TBAPF_6$ as an electrolyte in (19 : 1) CH_3CN–H_2O medium in the presence of benzyl alcohol displayed an almost unperturbed first oxidation wave at $E_{p,a} = 0.28$ V (*vs.* Fc^+/Fc), and a slight increase in the current at $E_{p,a}$ of 0.73 V (*vs.* Fc^+/Fc) (Fig. 2). The current enhancement in the second oxidation wave indicated a chemical reaction coupled to the oxidation of **1b**.

Furthermore, the peak currents (i_p) did not vary linearly with the square root of the scan rate. The possibility of the chemical reaction was ascertained by

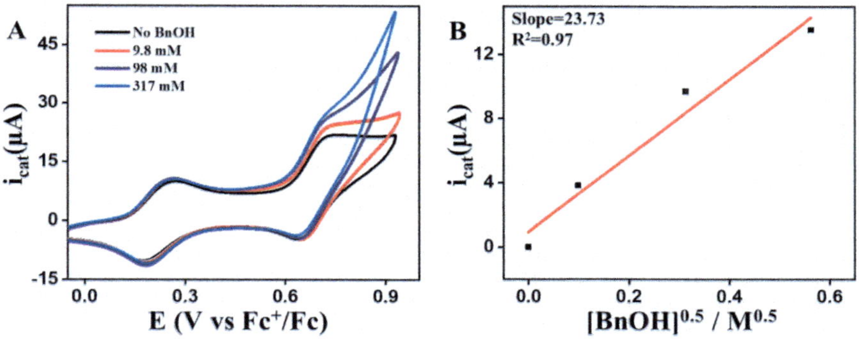

Fig. 2 (A) Cyclic voltammogram of catalyst **1a** in (19 : 1) CH_3CN-H_2O showing the steadily increasing catalytic current with increasing the benzyl alcohol concentration from 9.8 mM to 317 mM at a scan rate of 100 mV s^{-1}. (B) Plot of background-corrected i_{cat} *vs.* the square root of the conc. of benzyl alcohol in (19 : 1) CH_3CN-H_2O at a scan rate of 100 mV s^{-1}.

performing CPE at 0.77 V (*vs.* Fc^+/Fc) in the presence of benzyl alcohol for 3.5 hours. GC-MS analysis of the reaction medium showed the complete disappearance of the starting material with the concomitant formation of the product benzaldehyde. In the CV trace, the cathodic waveform displayed a hybrid peak–plateau shape, similar to that observed for the adamantane oxidation previously reported by us.[24] The formation of benzaldehyde during the CPE experiments prompted us to perform bulk electrochemical oxidation reactions with various alcohol-containing substrates.

Bulk electrolysis of a variety of alcohols

The oxidation of various benzylic alcohols was explored with the optimized electrochemical reaction conditions (acetonitrile aqueous phosphate buffer (4 : 1 v/v, 5 mM pH 8) at a constant potential of 0.77 V (*vs.* Fc^+/Fc) at room temperature, using a reticulated vitreous carbon (RVC) electrode as the anode and a nickel foam electrode as the cathode for 3–6 h). For example, for benzyl alcohol oxidation, analysis of the reaction mixture after the completion of the reaction (3.5 h) indicated that >97% of the benzyl alcohol was consumed and ~93% of benzaldehyde was formed with ~29% faradaic efficiency. In the control experiments, where no catalyst was added, <5% of benzyl alcohol was consumed, and <1% of benzaldehyde was formed. When different electron-donating or electron-withdrawing *para*-substituted benzyl alcohols were used as the substrates, we obtained greater than 90% yields of the corresponding benzaldehydes. In some cases, such as 4-phenyl benzyl alcohol and 4-chloro benzyl alcohol, very small amounts (<5%) of the overoxidized corresponding benzoic acid product were obtained. Benzylic alcohols were also oxidized efficiently to their corresponding aldehydes irrespective of heteroatoms such as O and N in the substrate. Pyridine and furan-based moieties were well tolerated under the reaction conditions, and no *N*-oxide formation was observed for the oxidation of 2-pyridinemethanol. Earlier attempt to oxidize 2-ethylpyridine electrochemically by the first-generation Fe-TAML was unsuccessful,[15] while we obtained 98% conversion of 2-pyridinemethanol, along with 96% yield of the corresponding aldehyde (Table 1).

Table 1 Electrochemical oxidation of a wide range of alcohols using Fe-bTAML[b]

Entry	Substrate (% conversion)	Product[a] (% yield)
1	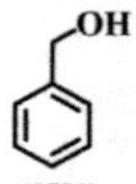 (97%)	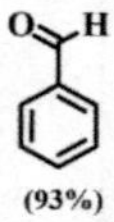 (93%)
2	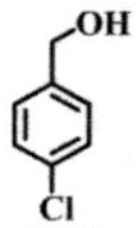 (99%)	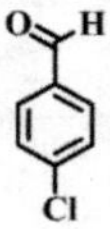 (96%)
3	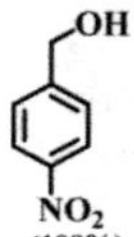 (100%)	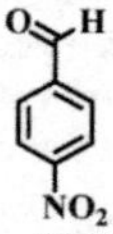 (99%)
4	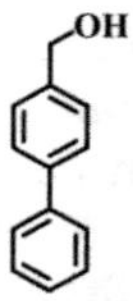 (99%)	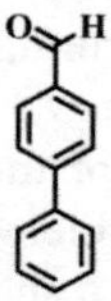 (96%)
5	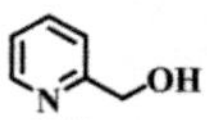 (98%)	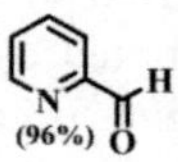 (96%)
6	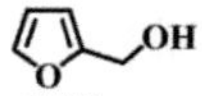 (97%)	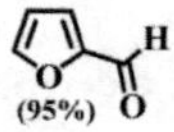 (95%)
7	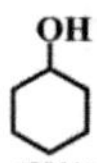 (80%)	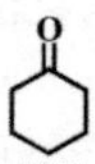 (75%)

[a] Based on GC yield. [b] Reaction conditions: **1a** (0.75 mM) and substrate (15 mM) in acetonitrile aqueous phosphate buffer (4 : 1 v/v, 5 mM, pH 8) at a constant potential of 0.77 V (*vs.* Fc^+/Fc) at room temperature, with an RVC electrode as the anode and a nickel foam electrode as the cathode using 0.10 M $TBAPF_6$ as an electrolyte for 3–6 h.

Evaluation of the kinetic parameters for the electrochemical oxidation of benzyl alcohols

We used cyclic voltammetry (CV), as a common electroanalytical tool, for evaluating the kinetic parameters (*e.g.*, reaction order, rate constant and KIE) of the oxidation of benzyl alcohols. Our previous studies on electrochemical water oxidation[23] and electrochemical alkane and alkene oxidations[24] revealed that a $Fe^V(O)$ intermediate is formed from the $[(bTAML)Fe^{III}–OH_2]^-$ (**1b**) complex *via* a PCET process (overall $2e^-/2H^+$) during the electrochemical reactions. Herein, we speculated the formation of the same $Fe^V(O)$ species *via* PCET during the electrochemical alcohol oxidations and subsequent reaction of $Fe^V(O)$ with benzyl

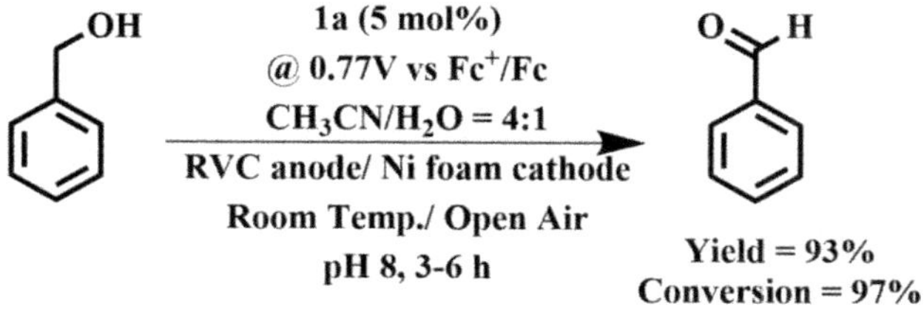

Scheme 2 Reaction conditions for electrochemical oxidation of benzyl alcohol.

alcohols to give benzaldehydes as the sole products (confirmed by GC-MS) and the subsequent regeneration of Fe^{III} (**1b**) (Scheme 2).

An enhancement in the current was observed at the second wave (*i.e.*, the Fe^{V}/Fe^{IV} redox peak at 0.73 V) in the presence of benzyl alcohol as substrates (Fig. 2). The substrate consumption is negligible in the reaction layer, and the plateau current is governed by the slow chemical reaction. The benzyl alcohol oxidation is believed to be followed by an E_rC_i' mechanism,[29] where a reversible electrode electron transfer occurs, followed by an irreversible homogeneous chemical reaction. This current enhancement (catalytic current, i_{cat}) could be used to extract the kinetic parameters using eqn (2) and (3). When we varied the concentration of the benzyl alcohol (BnOH) and the catalyst, the catalytic current, i_{cat}, varied according to eqn (2) and (3), where [BnOH] is the concentration of the benzyl alcohol and n is the number of electrons transferred in the catalytic reaction.

$$i_{cat} = nFAC_{cat}(Dk_{cat})^{1/2} \tag{2}$$

$$i_{cat} = nFAC_{cat}(Dk_2[\mathrm{BnOH}])^{1/2} \tag{3}$$

The equations used to analyze the catalytic currents obtained for catalyst **1a** were based on the theory formulated by Savéant and coworkers[27,28] and are relevant to reactions where the solution-phase chemical reaction is rate-limiting (*i.e.*, electron transfer to the electrode does not contribute to the reaction rate).

Accordingly, the background-corrected i_{cat} increases with increasing catalyst concentrations and shows a linear dependence on the catalyst concentration (Fig. S2†). Hence, the catalytic activity has a first-order relationship with the concentration of **1a.** The catalytic current also varies linearly with the square root of [BnOH], suggesting a first-order kinetic dependence on the concentration of the benzyl alcohol (eqn (3), Fig. 2). So, the rate of benzyl alcohol oxidation can be expressed as rate = k_2[Fe][BnOH]. Similar second-order dependence on the substrate has been found for the electrochemical oxidation of alcohols, which involves high-valent oxoruthenium intermediates.[13] Finally, the same rate expression was also obtained previously for benzyl alcohol oxidation with chemically synthesized Fe^{V}(O) under single turnover conditions.[25]

It is worth noting that the CV shapes observed when varying the catalyst and benzyl alcohol concentrations were not precisely the S-shaped curves obtained in zone KS and KD,[29] which can be achieved by increasing the substrate concentration or increasing the scan rate to limit substrate consumption.[29a] As the CVs show a hybrid peak and plateau at varying benzyl alcohol concentrations (Fig. 2A),

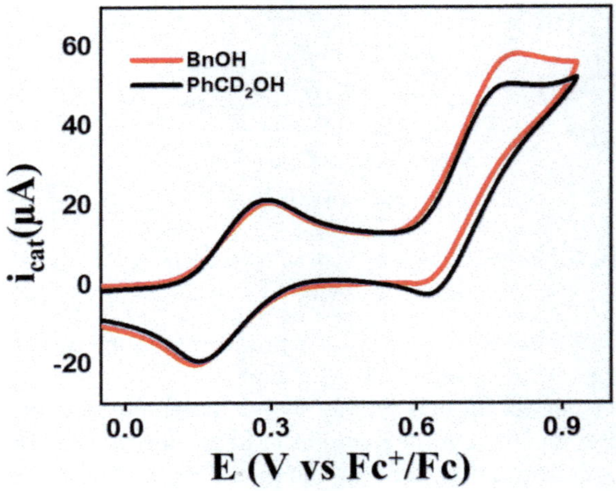

Fig. 3 CVs of 1 mM **1a** in (19 : 1) CH_3CN-H_2O containing 0.48 M benzyl alcohol (black) and deuterated 0.48 M benzyl alcohol (red) (scan rate: 100 mV s^{-1}).

the plateau current analysis according to eqn (3) to determine the kinetic parameters would result in an approximated maximum rate value (k_2). The k_2 value of 0.28 M^{-1} s^{-1} was determined from the slope (Fig. 2B) using eqn (3). The second-order rate constant (k_2) was determined from the catalytic response obtained at various substrate concentrations at a fixed scan rate (Fig. 2A), using values of D estimated from the Randles–Sevcik equation. Moreover, an approximate kinetic isotope effect (KIE) value of 1.8 was obtained by comparing the catalytic plateau current for benzyl alcohol and benzyl alcohol-d_2 (Fig. 3). A similarly low KIE of 3.2 was also observed for TEMPO mediated electrochemical benzyl alcohol oxidation.[10]

Linear free energy relationship (LFER) with *para*-substituted benzyl alcohols

The oxidation of several *para*-substituted benzyl alcohols by $Fe^{V}(O)$ was explored to understand if the abstraction of a hydrogen atom from the α-CH of the benzyl alcohol was the rate-determining step. From eqn (2), the catalytic rate constant values (k) for different substituted alcohols were calculated (Fig. 4). This predicted the potential effect of the *para* substituents on the benzyl alcohols on the equilibrium process. The k_2 value of the benzyl alcohol decreased seven times for the oxidation of the electron-withdrawing nitro group of 4-nitrobenzyl alcohol. The k_2 values for the oxidation of 4-fluoro and 4-chloro benzyl alcohols did not show substantial differences. Hammett plots were made for the four different *p*-substituted benzyl alcohols (Fig. 5). The independent rate measurements revealed a dependence on the rate of oxidation. For various *p*-substituted alcohols, the Hammett plots show preferential oxidation of electron-rich alcohols (with $\rho = -0.98$). The relatively small value of ρ indicates that the cleavage of the α-CH of benzyl alcohol is the rate-determining step in alcohol oxidation. It also suggests that the reaction rates were not significantly influenced by the electronic environment of the *para* substituents. The Hammett value obtained by us was higher than those of the (bpy)Cu/TEMPO mediated electrochemical oxidations of alcohols ($\rho = -0.24$)[30]

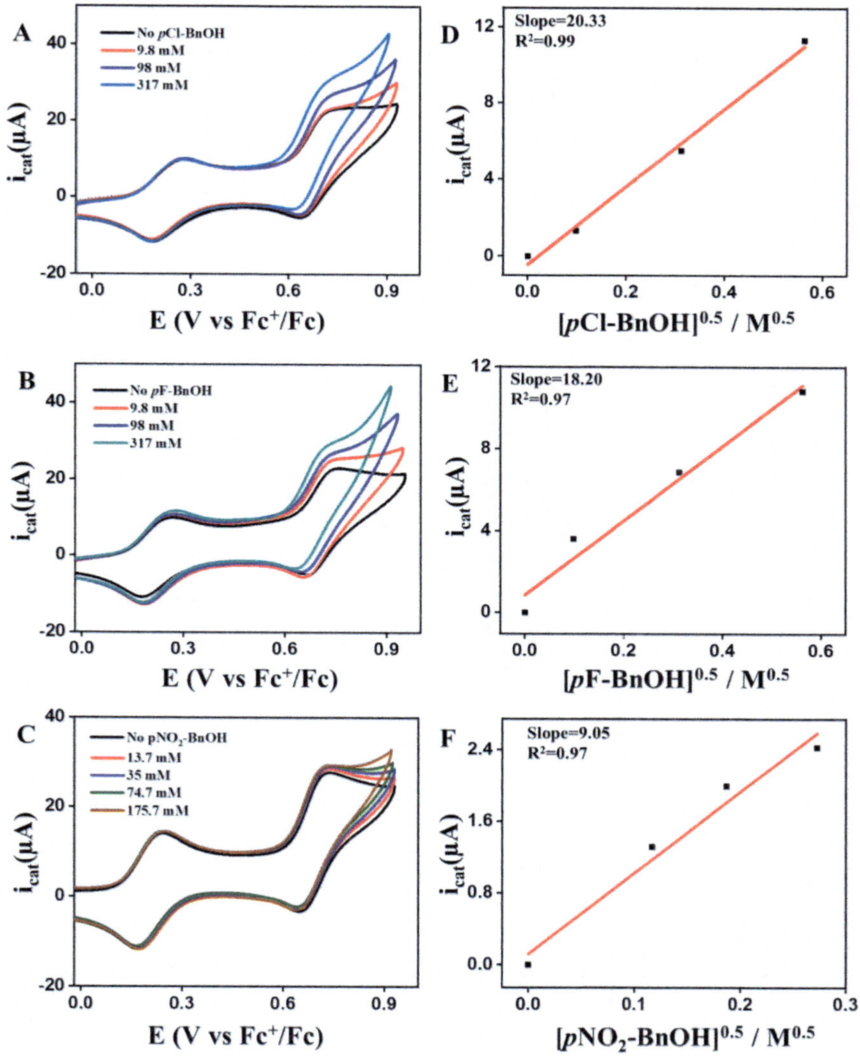

Fig. 4 Cyclic voltammograms of catalyst **1a** in (19 : 1) CH_3CN-H_2O, *i.e.* i_{cat} *vs.* *E* (V *vs.* Fc^+/Fc) plots, to show catalytic current enhancement with different *para*-substituted benzyl alcohols: (A) *p*Cl-BnOH, (B) *p*F-BnOH and (C) pNO_2-BnOH, and plots of the background-corrected i_{cat} *vs.* $[sub]^{0.5}$ for (D) *p*Cl-BnOH, (E) *p*F-BnOH and (F) pNO_2-BnOH.

and synthetic Fe^{IV}(O)–porphyrin$^{\bullet+}$[31] mediated chemical oxidations of alcohols ($\rho = -0.39$), but was lower than those obtained for the Cyt P450 enzymes ($\rho = -1.19$).[32]

Oxidation with cyclobutanol

For benzyl alcohol oxidation, the carbon radical of the benzyl alcohol formed after α-CH abstraction can undergo further oxidation, leading to aldehyde formation *via* two possible pathways. The first involves the classical "rebound mechanism", in which the Fe^{IV}–OH formed upon C–H abstraction rebounds the –OH group to form a gem-diol, which is subsequently dehydrated to form the aldehyde. In an

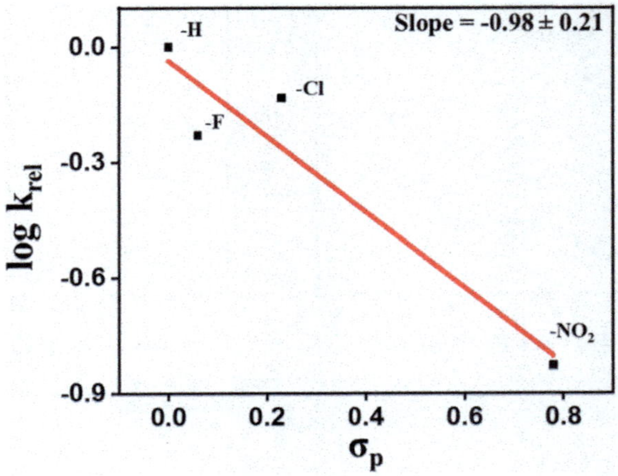

Fig. 5 Hammett plots showing $\log(k_X/k_H)$ *vs. para* substituent constants of benzyl alcohols.

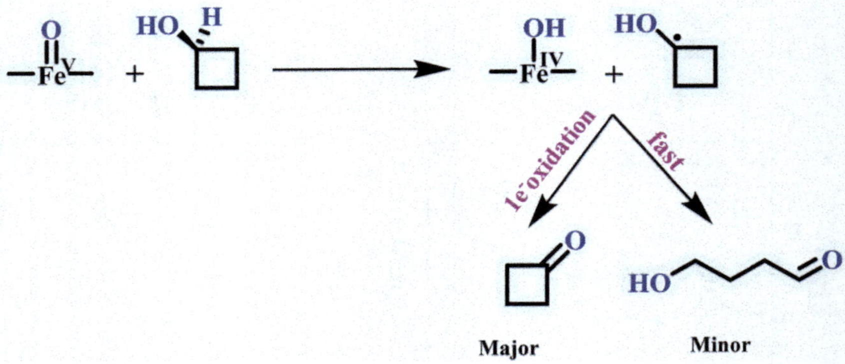

Fig. 6 The oxidation of cyclobutanol by $Fe^V(O)$ shows cyclobutanone as the primary product.

alternative scenario, the carbon radical of the benzyl alcohol can be oxidized, which then transfers an H^+ to the solvent cage, leading to aldehyde formation. Cyclobutanol has been taken as a substrate to distinguish between the one-electron and two-electron processes (Fig. 6).[33,34] In this experiment, if ketone is the sole product upon oxidation, the reaction occurs by hydrogen atom abstraction followed by electron transfer. Conversely, if a carbon radical is formed by the abstraction of the H atom of cyclobutanol and then escapes from the solvent cage, an open ring product, 4-hydroxybutyraldehyde, can be formed. Our experiment with $Fe^V(O)$ shows the predominant formation of cyclobutanone, thus indicating that the oxidation by $Fe^V(O)$ is a two-electron process.

Mechanistic interpretation

The first step of electrocatalytic alcohol oxidation involves the formation of the bTAML $Fe^V(O)$ from the starting bTAML $Fe^{III}(OH_2)$ *via* PCET reactions. This is followed by H-atom abstraction, evidenced by the KIE values and the relatively small value of ρ obtained from the Hammett plot. Following the H-atom abstraction, which leads to the formation of bTAML Fe^{IV}(OH) and the cyclobutanol radical cage, the following possibilities exist, leading to product formation: (i) diffusion of the cyclobutanol radical from the cage, followed by further oxidation, or (ii) ET/PT from the carbon radical to the Fe^{IV}–OH, leading to aldehyde formation. The possibility of (i) is negated by the radical clock experiment using cyclobutanol, where cyclobutanone is obtained as the major product (only a minor ring-opened product was observed). We believe that Fe^{IV}–OH is a strong enough oxidant (the $Fe^{IV/III}$ redox potential is ~0.6 V) to oxidize the benzyl alcohol radical *via* ET/PT to form the aldehyde product. Hence, the likely mechanism for alcohol oxidation by bTAML Fe^{V}(O) involves a 2-electron process leading to a "net hydride transfer" (Fig. 7).

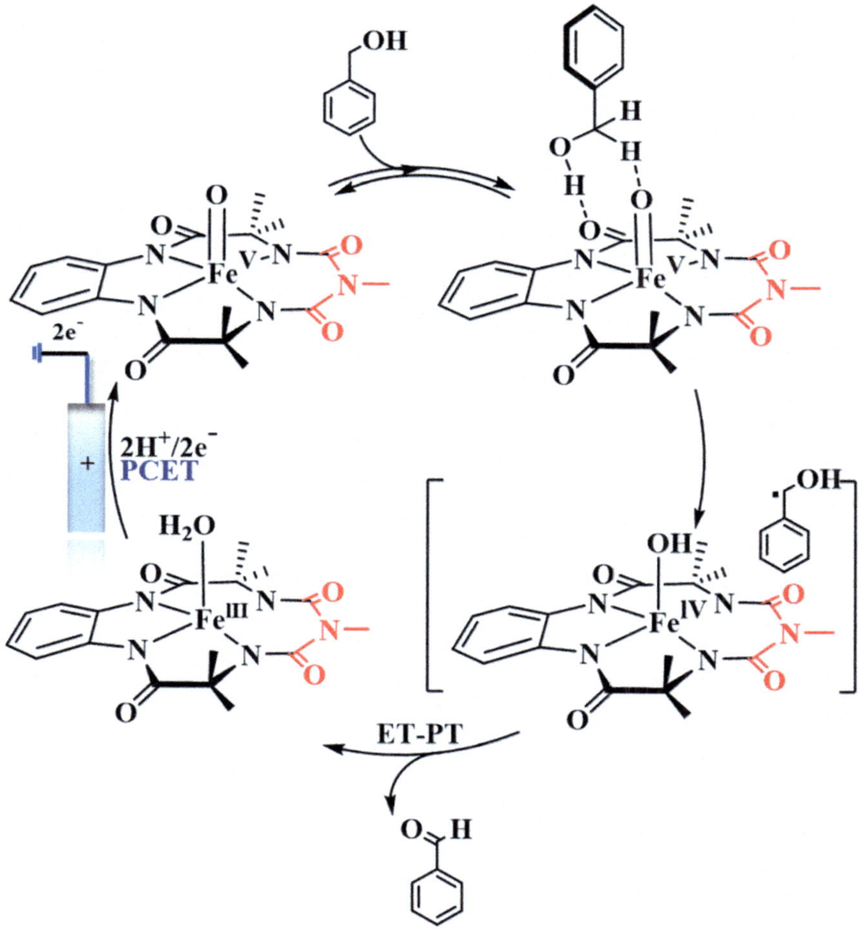

Fig. 7 Mechanistic cycle of the oxidation of benzyl alcohols by the Fe^{V}(O) bTAML framework.

Before we conclude, it is necessary to comment on a few crucial observations that help us better understand the reaction mechanism. The first is on the role of the μ-oxo-Fe^{IV}_2 dimer, which is the resting state of the complex during bulk electrolysis. Based on prior work from our group,[25] we have shown that the μ-oxo-Fe^{IV}_2 dimer is able to oxidize benzyl alcohol; however, the rates are significantly slower (~400 folds) than those of the Fe^{V}(O) complex under single-turnover experiments. We have also demonstrated that the μ-oxo-Fe^{IV}_2 dimer exists in equilibrium with Fe^{V}(O) and Fe^{III} species, and that Fe^{V}(O) was the species responsible for the oxidation of benzyl alcohol. To understand the role of the μ-oxo-Fe^{IV}_2 dimer towards the oxidation of benzyl alcohol, we attempted the oxidation of benzyl alcohol with the Fe-bTAML complex at 0.47 V (*vs.* Fc^+/Fc) at pH 8. At this potential, the μ-oxo-Fe^{IV}_2 dimer is exclusively formed and does not get overoxidized at the anode to the corresponding Fe^{V}(O) complex. Analysis of the reaction mixture after 6 hours of electrolysis indicated minimal product formation. This showed that the μ-oxo-Fe^{IV}_2 dimer is not the active intermediate for alcohol oxidation under these conditions. Instead, the anodic oxidation of the μ-oxo-Fe^{IV}_2 dimer into the Fe^{V}(O) intermediate is responsible for alcohol oxidation. Secondly, the faradaic efficiency for benzyl alcohol oxidation was found to be only 29%. We believe that such a low value is due to the competing water oxidation at that potential.

Reducing the overpotential for alcohol oxidation

One of the major limitations in using oxometal intermediates for the electrocatalytic oxidation of alcohols is the high overpotential (η) required for oxidation. For example, the oxidation of benzyl alcohols performed using Fe^{V}(O) intermediates was carried out at 0.77 V *vs.* Fc^+/Fc, a potential that is significantly higher than those of the widely studied Cu(I)/TEMPO systems[10] (−0.18 V *vs.* Fc^+/Fc). Therefore, we were interested in exploring if the oxidation of alcohols could be performed at the Fe^{IV}/Fe^{III} couple, thereby reducing the overpotential. We have previously shown that the corresponding Fe^{IV}(O) intermediate, which is generated at the Fe^{IV}/Fe^{III} couple at pH > 11, is also able to oxidize benzylic alcohols, albeit with lower rates (~2500 times)[21] than that of the corresponding one-electron oxidized Fe^{V}(O) complex under single-turnover experiments. Since Fe^{IV}(O) can be electrochemically formed at the potential of the $Fe^{IV/III}$ couple at pH > 11, we performed CPE for $Fe^{IV/III}$ at 0.26 V (*vs.* Fc^+/Fc) at pH 12 in (1 : 1) CH_3CN–H_2O. Spectroelectrochemistry studies clearly indicate the formation of the Fe^{IV}(O) intermediate (Fig. 8A). Upon adding benzyl alcohol during the CV experiments, an enhancement of the current for the Fe^{IV}/Fe^{III} couple was observed (Fig. 8B). We, therefore, performed bulk electrolysis for the oxidation of different alcohols (*p*-nitro benzyl alcohol, benzyl alcohol and 2-pyridinemethanol) at 0.26 V *vs.* Fc^+/Fc (pH 12, 0.10 M KPF_6 as a supporting electrolyte, in (1 : 1) CH_3CN–H_2O using a RVC electrode as the anode and a nickel foam electrode as the cathode). Analysis of the reaction mixture after 16–20 hours indicated 50–55% yields of their corresponding aldehydes. But for cyclohexanol (2° alcohol), less than 5% of the product was observed, which was consistent with our previous studies which indicated that the Fe^{IV}(O) intermediate is only able to oxidize alcohols with weak C–H bonds. Hence, by increasing the pH to >11, the overpotential for alcohol oxidation is reduced by ~500 mV. This overpotential is similar to that of the

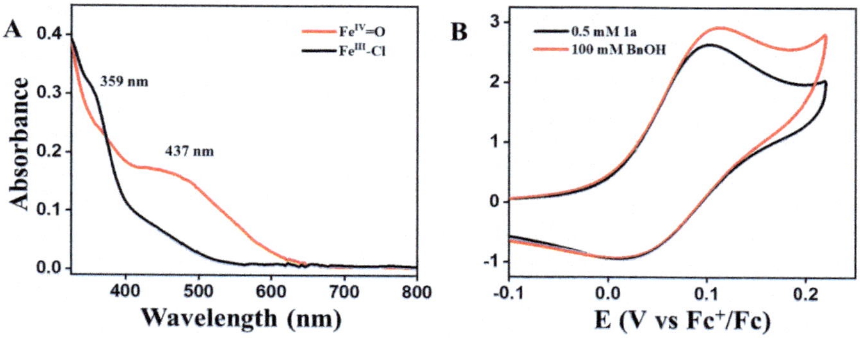

Fig. 8 (A) UV–vis spectral changes of **1a** (black) during the spectrochemical analysis of $Fe^{IV}(O)$ (red) species from **1a** (0.4 mM) in (1 : 1) CH_3CN–H_2O and 0.10 M KPF_6 as a supporting electrolyte at pH 12 using Pt mesh as the working electrode and a Pt wire as the counter electrode during CPE at 0.26 V *vs.* Fc^+/Fc. (B) CV of catalyst **1a** (0.5 mM) in (1 : 1) CH_3CN–H_2O and 0.10 M KPF_6 as a supporting electrolyte at pH 12, showing an increasing catalytic current in the presence of benzyl alcohol (100 mM) at a scan rate of 50 mV s^{-1}.

electrochemical TEMPO-catalyzed oxidation of benzyl alcohol. However, the reduced rates of the reaction significantly affected the yields of the products formed. Before we conclude, it is important to discuss the interplay between pH and overpotential for electrochemical alcohol oxidation. When electrochemical benzyl alcohol oxidation was performed at 0.47 V *vs.* Fc^+/Fc at pH 8, benzaldehyde formation was not observed. In contrast, significant benzaldehyde formation was observed when the reaction was carried out at pH 12 at lower potentials (0.26 V *vs.* Fc^+/Fc). The reactive intermediate formed at the anode at pH 8, μ-O-Fe^{IV}_2, was unable to oxidize benzyl alcohol, in contrast to the $Fe^{IV}(O)$ formed at pH > 11. It is possible that under these electrochemical conditions, the base might have a role in assisting the "net hydride" transfer, thus promoting the reaction. More studies need to be performed to understand the role of pH.

Conclusion

In conclusion, we have demonstrated an electrochemical method for the selective oxidation of alcohols to their corresponding aldehydes. Spectroelectrochemical studies indicated the involvement of a high-valent $Fe^V(O)$ species, $[(bTAML)Fe^V(O)]^-$, formed *via* PCET from the iron(III) complex $[(bTAML)Fe^{III}–OH_2]^-$. Determination of the rate equation from cyclic voltammetry experiments and product analysis allowed us to conclude that the mechanism of hydroxylation involves abstraction of the H-atom by $Fe^V(O)$, followed by another electron transfer to $[(bTAML)Fe^{IV}–OH]^-$, resulting in a "net-hydride" transfer. This electrochemical oxidation was performed with a series of benzylic and secondary alcohols. The substrate scope also included alcohols that contained O and N heteroatoms in the scaffold, which were well tolerated under these reaction conditions. For example, 2-pyridinemethanol was oxidized to the corresponding aldehyde with a yield of ~98% oxidation using the bTAML–Fe^{III} complex. In contrast, Fe–TAML complexes were unreactive towards *N*-heteroatom-containing substrates like 2-ethylpyridine due to their coordinating nature towards the Fe center of the catalyst. The main attractions of this method include: (i) the use of inexpensive reticulated vitreous carbon (RVC) electrodes in a simple undivided cell (unlike previously reported alcohol oxidations with Fe-TAMLs where a divided electrochemical cell was used), (ii) an earth-abundant iron complex as a redox mediator, (iii) a greener source for O-atoms, (iv) it is under an open atmosphere. We also show that by increasing the pH to >11, the overpotential could be reduced by nearly ~500 mV, albeit with reduced reaction yields. The increase in pH allows the formation of the one-electron oxidized $Fe^{IV}(O)$ intermediate, which is able to oxidize benzylic alcohols, but with reduced reaction rates.

Conflicts of interest

There are no conflicts to declare.

Acknowledgements

S. S. G. acknowledges SERB, New Delhi (grant no. EMR/2017/003258), and IISER Kolkata ("Start-up Grant") for funding. H. K. and S. D. acknowledge CSIR-New Delhi for their fellowship. S. C. acknowledges KVPY for the scholarship. B. C. acknowledges UGC-New Delhi for his fellowship. We acknowledge IISER Kolkata and the CSIR-National Chemical Laboratory for the research facilities and infrastructure. We acknowledge Debasmita Chatterjee for her constant help and support. We acknowledge Dr Rahul Banerjee and Dr Biplab Maji for their instrumental support.

References

1 S. Caron, R. W. Dugger, S. G. Ruggeri, J. A. Ragan and D. H. B. Ripin, *Chem. Rev.*, 2006, **106**, 2943–2989.
2 B. A. Frontana-Uribe, R. D. Little, J. G. Ibanez, A. Palma and R. Vasquez-Medrano, *Green Chem.*, 2010, **12**, 2099–2119.
3 J. B. Sperry and D. L. Wright, *Chem. Soc. Rev.*, 2006, **35**, 605–617.
4 J.-I. Yoshida, K. Kataoka, R. Horcajada and A. Nagaki, *Chem. Rev.*, 2008, **108**, 2265–2299.
5 M. Yan, Y. Kawamata and P. S. Baran, *Chem. Rev.*, 2017, **117**, 13230–13319.
6 F. Wang and S. S. Stahl, *Acc. Chem. Res.*, 2020, **53**, 561–574.
7 J. M. Hoover and S. S. Stahl, *J. Am. Chem. Soc.*, 2011, **133**, 16901–16910.
8 J. E. Steves and S. S. Stahl, *J. Am. Chem. Soc.*, 2013, **135**, 15742–15745.
9 B. L. Ryland and S. S. Stahl, *Angew. Chem., Int. Ed.*, 2014, **53**, 8824–8838.
10 A. Badalyan and S. S. Stahl, *Nature*, 2016, **535**, 406–410.
11 C. J. Weiss, E. S. Wiedner, J. A. S. Roberts and A. M. Appel, *Chem. Commun.*, 2015, **51**, 6172–6174.
12 S. Ohzu, T. Ishizuka, Y. Hirai, H. Jiang, M. Sakaguchi, T. Ogura, S. Fukuzumi and T. Kojima, *Chem. Sci.*, 2012, **3**, 3421–3431.
13 A. Paul, J. F. Hull, M. R. Norris, Z. Chen, D. H. Ess, J. J. Concepcion and T. J. Meyer, *Inorg. Chem.*, 2011, **50**, 1167–1169.
14 S. G. Robinson, J. B. C. Mack, S. N. Alektiar, J. Du Bois and M. S. Sigman, *Org. Lett.*, 2020, **22**, 7060–7063.
15 A. Das, J. E. Nutting and S. S. Stahl, *Chem. Sci.*, 2019, **10**, 7542–7548.

16 C. Panda, M. Ghosh, T. Panda, R. Banerjee and S. Sen Gupta, *Chem. Commun.*, 2011, **47**, 8016–8018.
17 M. Ghosh, K. K. Singh, C. Panda, A. Weitz, M. P. Hendrich, T. J. Collins, B. B. Dhar and S. Sen Gupta, *J. Am. Chem. Soc.*, 2014, **136**, 9524–9527.
18 S. Jana, M. Ghosh, M. Ambule and S. Sen Gupta, *Org. Lett.*, 2017, **19**, 746–749.
19 K. K. Singh, M. K. Tiwari, B. B. Dhar, K. Vanka and S. Sen Gupta, *Inorg. Chem.*, 2015, **54**, 6112–6121.
20 K. K. Singh and S. Sen Gupta, *Chem. Commun.*, 2017, **53**, 5914–5917.
21 S. Pattanayak, A. J. Jasniewski, A. Rana, A. Draksharapu, K. K. Singh, A. Weitz, M. Hendrich, L. Que, A. Dey and S. Sen Gupta, *Inorg. Chem.*, 2017, **56**, 6352–6361.
22 M. Ghosh, S. Pattanayak, B. B. Dhar, K. K. Singh, C. Panda and S. Sen Gupta, *Inorg. Chem.*, 2017, **56**, 10852–10860.
23 S. Pattanayak, D. R. Chowdhury, B. Garai, K. K. Singh, A. Paul, B. B. Dhar and S. Sen Gupta, *Chem.–Eur. J.*, 2017, **23**, 3414–3424.
24 B. Chandra, K. M. Hellan, S. Pattanayak and S. Sen Gupta, *Chem. Sci.*, 2020, **11**, 11877–11885.
25 M. Ghosh, Y. L. K. Nikhil, B. B. Dhar and S. Sen Gupta, *Inorg. Chem.*, 2015, **54**, 11792–11798.
26 S. Jana, J. Thomas and S. Sen Gupta, *Inorg. Chim. Acta*, 2019, **486**, 476–482.
27 C. Costentin and J.-M. Savéant, *ChemElectroChem*, 2014, **1**, 1226–1236.
28 J. M. Savéant and E. Vianello, *Electrochim. Acta*, 1965, **10**, 905–920.
29 (*a*) E. S. Rountree, B. D. McCarthy, T. T. Eisenhart and J. L. Dempsey, *Inorg. Chem.*, 2014, **53**, 9983–10002; (*b*) K. J. Lee, N. Elgrishi, B. Kandemir and J. L. Dempsey, *Nat. Rev. Chem.*, 2017, **1**, 0039; (*c*) J.-M. Savéant, *Elements of Molecular and Biomolecular Electrochemistry*, John Wiley & Sons, Inc., Hoboken NJ, 2006; (*d*) A. J. Bard and L. R. Faulkner, *Electrochemical Methods*, John Wiley & Sons, Inc., Hoboken NJ, 2001.
30 J. M. Hoover, B. L. Ryland and S. S. Stahl, *J. Am. Chem. Soc.*, 2013, **135**, 2357–2367.
31 N. Y. Oh, Y. Suh, M. J. Park, M. S. Seo, J. Kim and W. Nam, *Angew. Chem., Int. Ed.*, 2005, **44**, 4235–4239.
32 Q. Wang, X. Sheng, J. H. Horner and M. Newcomb, *J. Am. Chem. Soc.*, 2009, **131**, 10629–10636.
33 J. Rocek and A. E. Radkowsky, *J. Am. Chem. Soc.*, 1968, **90**, 2986–2988.
34 O. Pestovsky and A. Bakac, *J. Am. Chem. Soc.*, 2004, **126**, 13757–13764.

Faraday Discussions

ROYAL SOCIETY OF CHEMISTRY

PAPER

Oxidative dehalogenation of halophenols by high-valent nonheme iron(IV)-oxo intermediates†

Umesh Kumar Bagha,[a] Jagnyesh Kumar Satpathy,[a] Gourab Mukherjee,[ab] Prasenjit Barman,[ac] Devesh Kumar,*[d] Sam P. de Visser *[ae] and Chivukula V. Sastri *[a]

Received 22nd September 2021, Accepted 21st October 2021

DOI: 10.1039/d1fd00064k

Mononuclear high-valent iron(IV)-oxo intermediates are excellent oxidants towards oxygenation reactions by heme and nonheme metalloenzymes and their model systems. One of the most important functions of these intermediates in nature is to detoxify various environmental pollutants. Organic substrates, such as halogenated phenols, are known to be water pollutants which can be degraded to their less hazardous forms through an oxidation reaction by iron(IV)-oxo complexes. Metalloproteins in nature utilize various types of second-coordination sphere interactions to anchor the substrate in the vicinity of the active site. This concept of substrate-binding is well-known for natural enzymes, but is elusive for the relevant biomimetic model systems. Herein, we report the oxidative reactivity patterns of an iron(IV)-oxo intermediate, $[Fe^{IV}(O)(2PyN2Q)]^{2+}$, (2PyN2Q = 1,1-di(pyridin-2yl)-*N*,*N*-bis(quinolin-2-ylmethyl)methanamine) with a series of mono-, di- and tri-halophenols. A detailed experimental study shows that the dehalogenation reactions of the halophenols by such iron(IV)-oxo intermediates proceed *via* an initial hydrogen atom abstraction from the phenolic O–H group. Furthermore, based on the size and nucleophilicity of the halophenol, an intermediate substrate-bound species forms that is a phenolate adduct to the ferric species, which thereafter leads to the formation of the corresponding products.

[a]*Department of Chemistry, Indian Institute of Technology Guwahati, Assam, 781039, India. E-mail: sastricv@iitg.ernet.in; Fax: +91-361-258-2349*

[b]*Department of Chemical Science, Tata Institute of Fundamental Research, Dr. Homi Bhabha Road, Colaba, Mumbai, 400005, India*

[c]*Department of Chemistry, Kaliyaganj College, West Bengal, 733129, India*

[d]*Department of Applied Physics, School for Physical Sciences, Babasaheb Bhimrao Ambedkar University, Lucknow, 226025, India. E-mail: dkclcre@yahoo.com*

[e]*The Manchester Institute of Biotechnology, Department of Chemical Engineering and Analytical Science, The University of Manchester, 131 Princess Street, Manchester, M1 7DN, UK. E-mail: sam.devisser@manchester.ac.uk*

† Electronic supplementary information (ESI) available. See DOI: 10.1039/d1fd00064k

Introduction

High-valent iron(IV)-oxo complexes have been invoked as the key reactive intermediates in the catalytic cycles of nonheme iron enzymes and their model systems.[1,2] The nonheme iron(IV)-oxo species is reportedly the active oxidant involved in enzymatic and biomimetic reactions.[3,4] These iron(IV)-oxo intermediates have been shown to catalyse a wide variety of biological oxidation processes, such as alkane hydroxylation, olefin epoxidation, alcohol oxidation, sulfoxidation, *N*-dealkylation, *etc.*[5] Mononuclear nonheme iron enzymes perform the oxidative transformations of various organic substrates in biological systems.[6–11] High valent iron(IV)-oxo intermediates have been identified as important species in such oxidative transformations[12–17] and several nonheme dioxygenases have been isolated and characterized.[18,19] These are versatile intermediates towards hydrogen atom transfer (HAT) and oxygen atom transfer (OAT) reactions.[20–25] In addition, mononuclear high-valent iron(IV)-oxo complexes have been proposed to be reactive intermediates in the degradation of toxic organic pollutants,[26] which are potential threats to the continuous problem of environmental pollution. Among these organic pollutants, halophenols are a major class of toxic organic pollutants, which includes 4-fluorophenols, 4-chlorophenols, 4-bromophenols, 2,6-difluorophenols, 2,6-dichlorophenols, 2,6-dibromophenols, 2,4,6-trifluorophenols, 2,4,6-trichlorophenols and 2,4,6-tribromophenols. The main sources of these compounds are industrial waste, pesticides, herbicides, disinfectants and dyes.[27,28] Their extensive usage and resistance to biodegradation cause their accumulation in the soil and groundwater.[29–31] As a consequence, these pollutants enter the human body through the food chain, causing serious health concerns as they are potentially carcinogenic.[27,32,33] In addition, many chlorophenols, such as 2,4,6-trichlorophenols, are included in the US Environmental Protection Agency's priority list of dangerous pollutants because of their high persistence in the environment.[34] Therefore, it is necessary to implement different techniques that can help in the detoxification of these pollutants and convert them to non-hazardous chemicals.

A variety of techniques have been developed for the degradation of halophenols that include physical, chemical and biological processes.[35–38] Despite the fact that a variety of physical and chemical processes are available for the degradation of halophenols, such as adsorption, ion exchange, liquid–liquid extraction, chemical oxidation and advanced oxidation processes, all of those are costly and energy-intensive. Moreover, many of these processes produce by-products or more toxic intermediates. For example, the incineration of chlorophenol wastes produces even more toxic compounds, such as polychlorinated dibenzofurans or dibenzo-*p*-dioxins.[37] On the other hand, bacterial degradation is one of the most efficient methods to detoxify halophenols. Chlorophenols and their derivatives are removed from the environment by several bacteria, such as *Pseudomonas knackmussii* B-13, *Rhodococcus opacus* 1G, *Ralstonia pickettii etc.*[27] This is considered to be a cost-effective and eco-friendly method to remove chlorophenols.

The dehalogenation mechanism by dehaloperoxidase (DHP) has been investigated by several groups but as yet remains unresolved.[39–43] Enzymatic assays suggest that the mechanism is similar to that for the same reaction catalysed by

Caldariomyces fumago chloroperoxidase (CCPO),[44] horseradish peroxidase (HRP)[45] and lignin peroxidases.[46,47] The proposed mechanism for the conversion of halophenols to quinones by these enzymes involves two successive one-electron transfer steps mediated by iron(IV)-oxo intermediates and it is seen that the oxygen atom in the quinone product is derived from a solvent water molecule.[44] Interestingly, a different dehalogenation mechanism has been proposed for dehaloperoxidase-hemoglobin A (DHP A), which involves substrate binding in the internal active site of an enzyme where the substrate gets oxidized to quinone in two sequential one-electron steps, followed by the addition of a water molecule, as shown in Scheme 1.[48]

The first step is initiated with the oxidative electron transfer from the substrate and loss of a proton, resulting in the formation of a phenoxy radical, 2,4,6-TXPO (X = Cl or Br) (Scheme 1a). In the second step, oxidation of the 2,4,6-TXPO radical takes place by two probable alternative mechanisms: disproportionation and internal oxidation. In the disproportionation mechanism, which occurs in solution, one equivalent of each of the reactant and phenoxy cation are formed from two equivalents of phenoxy radicals, whereas in internal oxidation, the substrate remains immobilized inside the enzyme for a second electron transfer to produce a phenoxy cation (Scheme 1b). Thereafter, addition of a water molecule to the phenoxy cation in the *para*-position forms the 2,6-dihaloquinone product with the loss of HX (Scheme 1c).

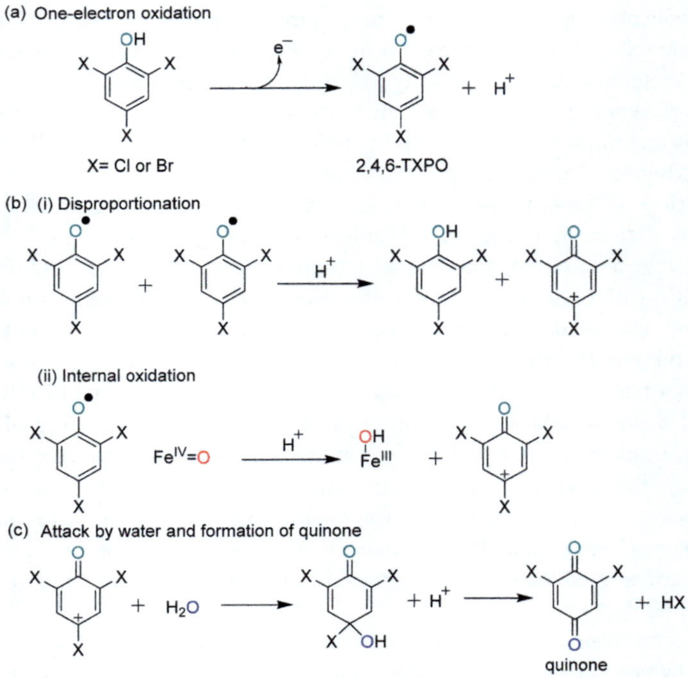

Scheme 1 Schematic representation of the mechanisms for the oxidation of substrates by DHP A.[48]

However, limited literature is available on the detoxification of organic pollutants by the nonheme iron(IV)-oxo species. Recently, Paine and co-workers synthesized a series of iron(IV)-oxo complexes supported by polydentate nitrogen-donor ligands which performed the oxidative degradation of halophenols in an aqueous medium.[49] It was observed that water plays a vital role in most of the processes used for the degradation of halophenols. However, all the intermediates might not be stable in aqueous medium. We therefore, envisaged that the solvent system has a huge role to play in the catalytic dehalogenation of the phenolic systems and, in particular, in stabilizing the much-elusive substrate-bound species involved in the mechanism. In this work, we employed an iron(IV)-oxo species supported by a pentadentate ligand, 2PyN2Q (1,1-di(pyridin-2yl)-*N*,*N*-bis(quinolin-2-ylmethyl)methanamine),[50] in an acetonitrile solvent, which degrades halophenols by an oxidative pathway. The proposed mechanism of the oxidative dehalogenation of halophenols by this iron(IV)-oxo intermediate proceeds *via* an initial hydrogen atom abstraction from the phenolic O–H, resulting in the formation of a new intermediate, a phenolate adduct of the ferric species, which leads to the formation of corresponding quinone product.

Results and discussion

We started our work by synthesizing the ferrous complex $[Fe^{II}(2PyN2Q)(OTf)_2]$, **1**, supported by a pentadentate ligand, 2PyN2Q.[50] The asymmetric geometry of the ligand framework has been taken into account to study the substrate approach parameters. The dark yellow compound (**1**) was synthesized, following previously reported procedures,[51–53] by the dropwise addition of an acetonitrile solution of $Fe^{II}(OTf)_2{\cdot}2CH_3CN$ to the ligand. UV–vis spectroscopy and electrospray ionization-mass spectrometry (ESI-MS) confirmed the ligation of the iron(II) species in the complex. Upon the addition of 1.5 equivalents of $PhI(OAc)_2$ to **1** in CH_3CN under ambient conditions, a pale green iron(IV)-oxo complex $[Fe^{IV}(O)(2\text{-}PyN2Q)]^{2+}$, **2**, was generated with distinctive absorption features ($\lambda_{max} = 770$ nm, $\varepsilon = 340$ L M^{-1} cm^{-1}, $t_{1/2} = 50$ min at RT) (see Fig. S1, ESI†), as reported previously.[50] This absorption band corresponds to the d–d transition features of the iron(IV)-oxo species generated *in situ*. Furthermore, the formation of the oxo-species was confirmed by ESI-MS, which showed a major peak at *m*/*z* 269.57

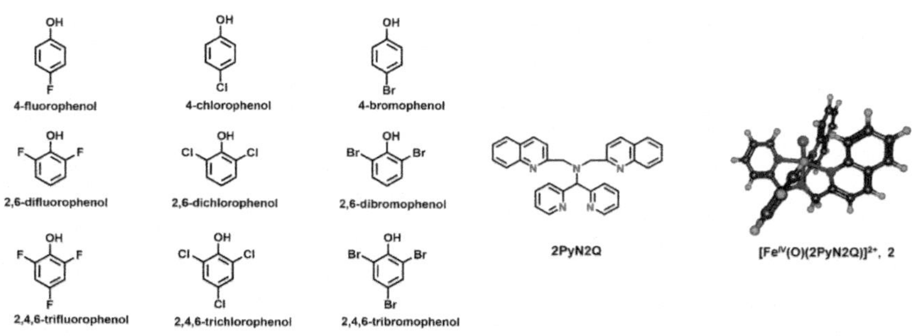

Scheme 2 Substrates, ligand 2PyN2Q and oxidant **2** used in this work.

and a minor peak at *m*/*z* 688.08 corresponding to $[Fe^{IV}(O)(2PyN2Q)]^{2+}$ and $[Fe^{IV}(O)(2PyN2Q)(OTf)]^{1+}$, respectively, as shown in Fig. S2 (ESI†).

The complex $[Fe^{IV}(O)(2PyN2Q)]^{2+}$, as shown in Scheme 2, has an asymmetric skeleton with two quinolines on one side of it – guarding the iron-oxo core and directing the approach of the substrates from the other side of the complex. We, therefore, decided to investigate several halophenols with various steric demands and different electronic impulses toward the iron–oxo active site to see how the backbone affects the substrate binding and approach. As exhibited in Scheme 2, several mono-, di- and tri-halophenols were employed as substrates to study the oxidative dehalogenation reactions of complex **2**. However, the use of iodophenols was precluded due to solubility issues in our generalized conditions. The reactions were monitored by following the changes in the UV–vis absorption spectra as a function of time upon the addition of the substrates. Initially, tri-halophenols, such as 2,4,6-trifluorophenol (2,4,6-TFP), 2,4,6-trichlorophenol (2,4,6-TCP) and 2,4,6-tribromophenol (2,4,6-TBP) were reacted with **2**. Complex **2** is known to be a better catalyst for sulfoxidation and C–H abstraction reactions.[50] Similar trends were observed in the reactivity with halophenols.

At −40 °C, the addition of tri-halophenols to **2** led to the gradual pseudo first-order decay of the iron(IV)-oxo characteristic band at 770 nm in the UV–vis spectrum, with the concomitant formation of the corresponding iron(II) precursor with an isosbestic point at 620 nm, as shown in Fig. 1. The pseudo first-order rate constants increased linearly with an increase in the concentration of the substrates, enabling us to procure the second-order rate constants (k_2). The second-order rate constants (k_2) for **2** with different tri-halophenols were evaluated, as shown in Table 1, and the reactivity order was found to be 2,4,6-TBP > 2,4,6-TCP > 2,4,6-TFP. This trend in the reactivity established the electrophilicity of the iron(IV)-oxo species involved in the reactions. A similar order was also observed for the reactivity of heme metalloenzymes with tri-halophenols.[54]

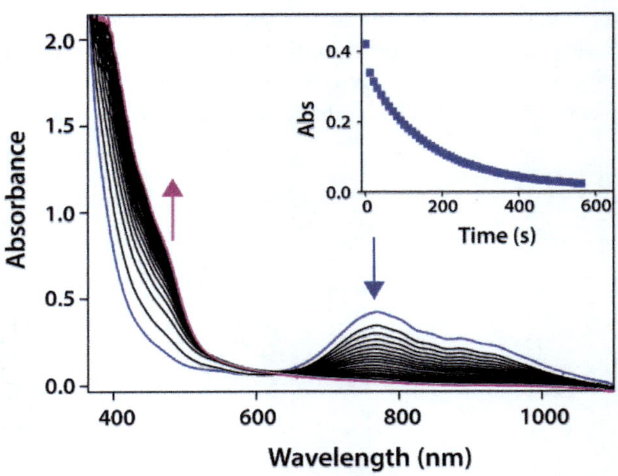

Fig. 1 UV–vis spectral changes of **2** upon the addition of 10 equiv. 2,4,6-TCP in CH_3CN at −40 °C. The inset shows the decay profile of the 770 nm band.

Table 1 Second-order rate constants (k_2) and product distributions for **2** for the reaction of 1 mM of **2** in CH_3CN with mono-, di-, and tri-halophenols

Substrate	$10^{-1} \times k_2$ (in $M^{-1}\ s^{-1}$)	Temperature (°C)	Product distribution, % (*ortho* : *para*) quinone
2,4,6-TFP	4.98	−40	35 : 21
2,4,6-TCP	9.3	−40	30 : 15
2,4,6-TBP	18.3	−40	50 : 10
2,6-DFP	3.55	−40	30 : 0
2,6-DCP	8.54	−40	33 : 0
2,6-DBP	14.6	−40	21 : 0
4-FP	20.8	0	30 : 25
4-CP	19.8	0	40 : 31
4-BP	16	0	38 : 10

When we substituted the phenolic hydrogen with deuterium and repeated the experiments with 2,4,6-TCP-d_1 (2,4,6-trichlorophenol-d_1), a second-order rate constant of $3.54 \times 10^{-2}\ M^{-1}\ s^{-1}$ was obtained at −40 °C. This gave a non-classical kinetic isotope effect (KIE) value of 26 that confirmed the abstraction of the phenolic hydrogen and the corresponding formation of the phenolate radical in the mechanism of the dehalogenation process (see Fig. 2b).[21,55] With that established, it was also understood that the size of the substrate, however, had no role to play in the H-abstraction step, since 2,4,6-tribromophenol was found to be the most reactive among the three. This size *vs.* nucleophilicity balance among the substrates investigated led us to further check the dehalogenation of **2** with di- and mono-halophenols. The addition of di-halophenols to the intermediate **2** in CH_3CN under the same conditions resulted in the decay of the absorption band at 770 nm in a pseudo first-order manner as a function of time (Fig. S3, ESI†). A similar pattern was obtained for the di-halophenols as well *i.e.* 2,6-DBP > 2,6-DCP > 2,6-DFP. By contrast, for *para*-halophenols, the reactions were sluggish at −40 °C. Clearly, the phenolate radicals resulting from the H-atom abstraction were less stabilized for the mono-halo substrates compared to those of the di- and tri-halophenols. Interestingly, the addition of the *p*-halophenols to the intermediate **2** at 0 °C resulted in the decay of the d–d transition band of **2**. However, the decay profile was not clean at 770 nm. A new feature was observed at about 600 nm (see Fig. 3a).

Thus, second-order rate constants for the oxidation of *p*-halophenols were evaluated as a function of the change in the absorption band at 885 nm with time (Table 1). These rates were found to be comparable for all the three *p*-halophenols, indicating that the formation of this new species was independent of the nucleophilicity of the employed substrate. However, there was a definite effect of the size of the substrate utilized for dehalogenation, as this feature was not observed in the cases of di- and tri-halophenols under identical conditions. When the reaction of 4-FP with **2** was carried out at room temperature, a new absorption band at 614 nm was observed, which is typical of the characteristic LMCT band, usually seen for oxygen adducts of iron such as iron(III)–OOH and/or iron(III)–OOR.[56–61] Although, unlike such intermediates, this new species was comparatively stable, having a half-life time of about 10 h at room temperature. To further characterize this species, high-resolution mass spectrometry (HRMS) was

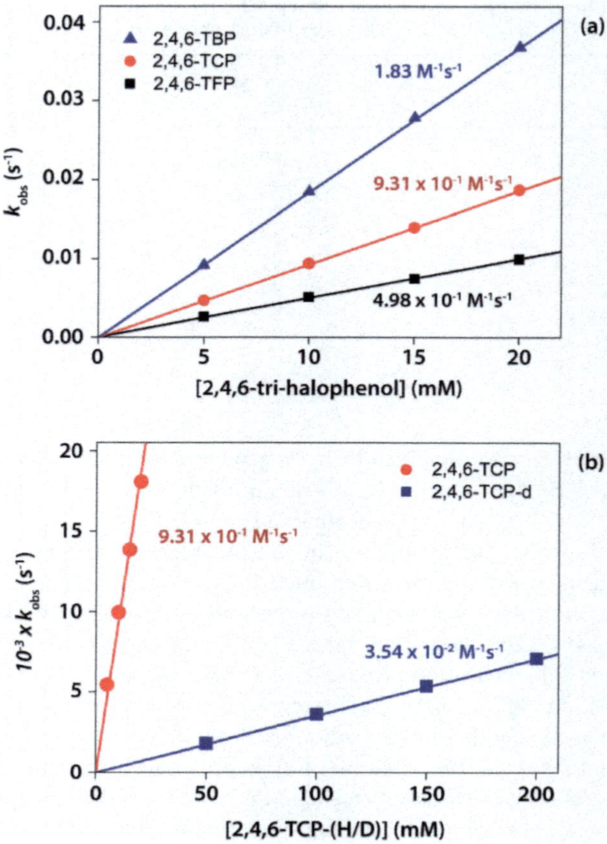

Fig. 2 Second-order rate constants determined for **2** (1 mM) with (a) 2,4,6-TFP (■), 2,4,6-TCP (●) and 2,4,6-TBP (▲) at −40 °C; (b) plot of k_{obs} against the concentration of 2,4,6-TCP (●) and 2,4,6-TCP-d (■) with **2** in CH_3CN at −40 °C.

conducted, but unfortunately, it resulted in excessive fragmentation, thereby precluding the correct assessment of the molecular ion fragment. The X-band EPR spectrum of the adduct was recorded at 77 K (Fig. 3b) and showed a broad signal at $g = 5.04$, along with additional high-end g values at 6.55, 8.39 and 13.74. This indicated the presence of a Fe(III)-species with dominant high-spin characteristics.[59,62]

We speculate that the new species is a substrate-bound intermediate, where the phenolate radical, deprived of its hydrogen, binds with the iron(III)–OH species, thereby forming a hepta-coordinated adduct of iron. Arguably, the active site structure is crowded at this point of time. The steric demand of the ligand framework is hereby brought into effect as the new substrate-bound adduct should be mostly stable for *p*-halophenols, and less stable for di- and tri-halophenols that have bulky halogen atoms in the *ortho*-position. This was indeed observed in the UV–vis spectrum, where the formation of the LMCT band was not observed for phenols which had chloro and/or bromo at the *ortho*-positions of the phenol. The bulky size of the bromo component in the *ortho*-position

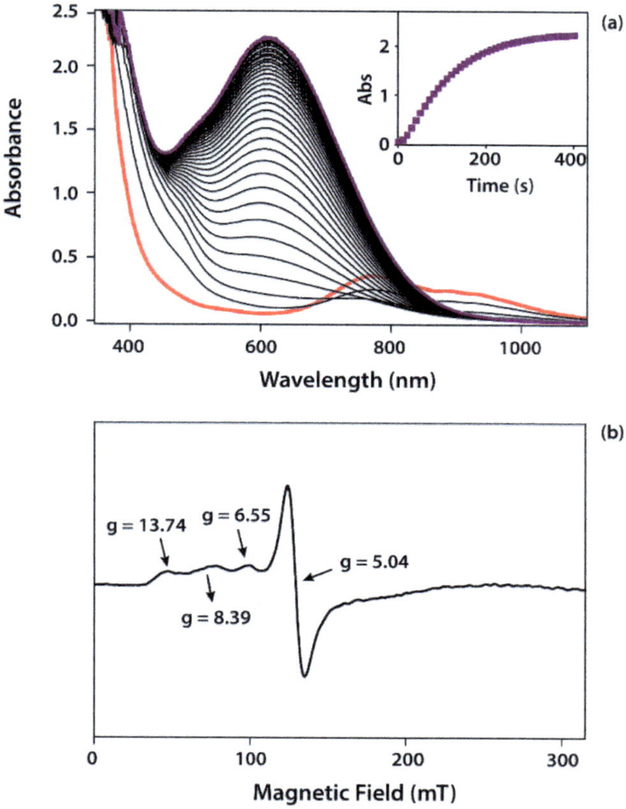

Fig. 3 (a) UV–vis spectral changes of **2** upon the addition of 10 equiv. 4-CP in CH_3CN at room temperature. The inset shows the formation profile of the absorption band at 614 nm; (b) X-band EPR spectrum of the adduct at 77 K in CH_3CN. EPR parameters: microwave power: 0.995 mW, modulation frequency: 100 kHz, modulation amplitude: 1 mT, microwave frequency: 9136 MHz.

restricts the ease of approach of the phenolate radical to bind with the iron(III)-species. On the other hand, fluorine, which has a comparatively much smaller atomic size, was tolerated by the hepta-coordinated iron(III)-adduct and the formation of new bands was observed at 558 nm and 583 nm for 2,4,6-TFP and 2,6-DFP, respectively (see Fig. S5 and S6, ESI†). Also, a shift of the λ_{max} from 614 nm to 558 nm for 4-FP and 2,4,6-TFP, respectively, further reinforces the argument that this is a substrate-bound iron(III)-species.

To understand the details of the reaction mechanism, analyses of the oxidized products were carried out. In the reaction of **2** with 2,4,6-TCP, we found mixtures of 2,4-dichloro-*ortho*-benzoquinone (*o*-quinone) and 2,6-dichloro-*para*-benzoquinone (*p*-quinone), with the former being the major product. To gain further insights, the *ortho-* : *para*-quinone ratios for all the halophenols were evaluated and are listed in Table 1. Interestingly, *o*-quinone was found to be the major product in every case. In fact, for all three 2,6-dihalophenols discussed here, *o*-quinone was found to be the sole product.

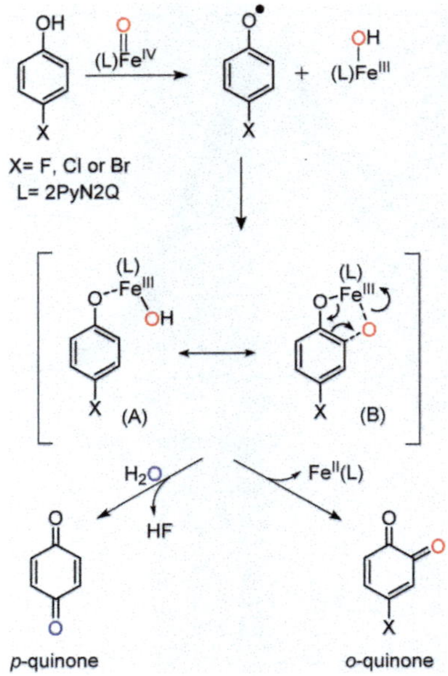

Scheme 3 A plausible mechanism for the oxidative dehalogenation of *p*-halophenols by the intermediate **2**.

The product distribution hereby obtained guided us to propose a plausible mechanism for the dehalogenation reaction. The reaction mechanism involves two consecutive one-electron reductions of the iron(IV)-oxo species.[44–48] The reaction commences with the abstraction of the phenolic hydrogen atom, thereby forming iron(III)–OH and a phenolate radial species. Thereafter, the phenolate radical overcomes the steric fencing around the iron centre to coordinate to the ferric species as a substrate-bound adduct (**A**). In the adduct (**A**), the hydroxyl group (–OH) is in the vicinity of the *ortho*-position of the phenolate cyclize to form the adduct (**B**). Finally, the adduct (**B**) undergoes a second electron transfer step to produce *o*-quinone in a concerted fashion, as shown in Scheme 3. The formation of *p*-quinone could be attributed to the presence of a trace amount of water in the solvent.

Conclusions

In summary, we have synthesized an iron(IV)-oxo complex which acts as an excellent oxidant for the oxidative degradation of halophenolic compounds that are potential environmental pollutants. Previously, we have seen that this complex is a highly active catalyst for heteroatom oxidation. Likewise, in oxidative dehalogenation, a detailed comparative analysis of several mono-, di and tri-halophenols has been undertaken. We have observed a substrate-bound species being generated as an intermediate during the course of the reaction. Although the reaction rate of the iron(IV)-oxo compound is guided by its inherent electrophilicity, interestingly, the formation of the substrate-bound adduct is dependent upon the steric factor imposed by the bulkiness of the halogen atom in its *ortho*-

position. As such, the ligand backbone, therefore, also plays a crucial role in directing the approach of the halophenols towards the active site. Furthermore, the tentative formation of a hepta-coordinated species along the reaction coordinate brings forth the selective formation of the *ortho*-quinone product in a major amount. Thus, oxidative dehalogenation by iron(IV)-oxo complexes is substrate-dependent of course, but the generality of this mechanism with other biomimetic catalysts is yet to be tested. Also, what governing factors and non-covalent interactions are operational behind the extraordinary stability of such a species is yet to be discovered. Further detailed investigations on the identification and mechanistic aspects of these reactions are currently under way in our laboratory.

Author contributions

U. K. B., D. K., S. P. d. V. and C. V. S. designed the research; U. K. B. and J. K. S. performed the research; U. K. B., G. M., P. B., D. K., S. P. d. V. and C. V. S. analysed the data; and U. K. B., G. M., S. P. d. V., D. K. and C. V. S. wrote the manuscript.

Conflicts of interest

There are no conflicts to declare.

Acknowledgements

Research support was provided to CVS by the Department of Science and Technology (SERB), India through the grant code CRG/2019/000387. S. P. d. V. thanks IIT Guwahati for a visiting professorship.

References

1 M. Costas, M. P. Mehn, M. P. Jensen and L. Que Jr, *Chem. Rev.*, 2004, **104**, 939.
2 M. M. Abu-Omar, A. Loaiza and N. Hontzeas, *Chem. Rev.*, 2005, **105**, 2227.
3 J. M. Bollinger Jr, J. C. Price, L. M. Hoffart, E. W. Barr and C. Krebs, *Eur. J. Inorg. Chem.*, 2005, 4245.
4 C. V. Sastri, M. S. Seo, M. J. Park, K. M. Kim and W. Nam, *Chem. Commun.*, 2005, 1405.
5 W. Nam, *Acc. Chem. Res.*, 2007, **40**, 522.
6 E. I. Solomon, T. C. Brunold, M. I. Davis, J. N. Kemsley, S. K. Lee, N. Lehnert, F. Neese, A. J. Skulan, Y. S. Yang and J. Zhou, *Chem. Rev.*, 2000, **100**, 235.
7 M. Costas, M. P. Mehn, M. P. Jensen and L. Que Jr, *Chem. Rev.*, 2004, **104**, 939.
8 M. M. Abu-Omar, A. Loaiza and N. Hontzeas, *Chem. Rev.*, 2005, **105**, 2227.
9 P. C. A. Bruijnincx, G. van Koten and R. J. M. Klein Gebbink, *Chem. Soc. Rev.*, 2008, **37**, 2716.
10 A. R. McDonald and L. Que Jr, *Coord. Chem. Rev.*, 2013, **257**, 414.
11 W. Nam, Y.-M. Lee and S. Fukuzumi, *Acc. Chem. Res.*, 2014, **47**, 1146.
12 M. Sono, M. P. Roach, E. D. Coulter and J. H. Dawson, *Chem. Rev.*, 1996, **96**, 2841.
13 J. T. Groves, *Proc. Natl. Acad. Sci. U. S. A.*, 2003, **100**, 3569.
14 B. Meunier, S. P. de Visser and S. Shaik, *Chem. Rev.*, 2004, **104**, 3947.

15 I. G. Denisov, T. M. Makris, S. G. Sligar and I. Schlichting, *Chem. Rev.*, 2005, **105**, 2253.
16 *Iron-containing Enzymes: Versatile Catalysts of Hydroxylation Reactions in Nature*, ed. S. P. de Visser and D. Kumar, RSC Publishing, Cambridge, 2011.
17 S. P. de Visser, Y. T. Lin, H. S. Ali, U. K. Bagha, G. Mukherjee and C. V. Sastri, *Coord. Chem. Rev.*, 2021, **439**, 213914.
18 D. A. Proshlyakov, T. F. Henshaw, G. R. Monterosso, M. J. Ryle and R. P. Hausinger, *J. Am. Chem. Soc.*, 2004, **126**, 1022.
19 J. M. Bollinger Jr, J. C. Price, L. M. Hoffart, E. W. Barr and C. Krebs, *Eur. J. Inorg. Chem.*, 2005, 4245.
20 S. T. Kleespies, W. N. Oloo, A. Mukherjee and L. Que Jr, *Inorg. Chem.*, 2015, **54**, 5053.
21 C. V. Sastri, J. Lee, K. Oh, Y. J. Lee, J. Lee, T. A. Jackson, K. Ray, H. Hirao, W. Shin, J. A. Halfen, J. Kim, L. Que Jr, S. Shaik and W. Nam, *Proc. Natl. Acad. Sci. U. S. A.*, 2007, **104**, 19181.
22 D. Wang, K. Ray, M. J. Collins, E. R. Farquhar, J. R. Frisch, L. Gómez, T. A. Jackson, M. Kerscher, A. Waleska, P. Comba, M. Costas and L. Que Jr, *Chem. Sci.*, 2013, **4**, 282.
23 G. Mukherjee, J. K. Satpathy, U. K. Bagha, M. Q. E. Mubarak, C. V. Sastri and S. P. de Visser, *ACS Catal.*, 2021, **11**, 9761.
24 G. Mukherjee, A. Alili, P. Barman, D. Kumar, C. V. Sastri and S. P. de Visser, *Chem.–Eur. J.*, 2019, **25**, 5086.
25 G. Mukherjee, F. G. Cantú Reinhard, U. K. Bagha, C. V. Sastri and S. P. de Visser, *Dalton Trans.*, 2020, **49**, 5921.
26 S. C. Peck and W. A. van der Donk, *JBIC, J. Biol. Inorg. Chem.*, 2017, **22**, 381.
27 A. O. Olaniran and E. O. Igbinosa, *Chemosphere*, 2011, **83**, 1297.
28 P. K. Arora and H. Bae, *Microb. Cell Fact.*, 2014, **13**, 31.
29 S. L. Simonich and R. A. Hites, *Science*, 1995, **269**, 1851.
30 J. Gao, L. Liu, X. Liu, H. Zhou, S. Huang and Z. Wang, *Chemosphere*, 2008, **71**, 1181.
31 W. J. Sim, S. H. Lee, I. S. Lee, S. D. Choi and J. E. Oh, *Chemosphere*, 2009, **77**, 552.
32 E. O. Igbinosa, E. E. Odjadjare, V. N. Chigor, I. H. Igbinosa, A. O. Emoghene, F. O. Ekhaise, N. O. Igiehon and O. G. Idemudia, *Sci. World J.*, 2013, 1.
33 J. Pumarega, M. Gasull, D. H. Lee, T. López and M. Porta, *PLoS One*, 2016, **11**, e0160432.
34 Agency for Toxic Substances and Disease Registry (ATSDR), *Toxicological Profile for Chlorophenols*, U.S. Department of Health and Human Services, Public Health Service, Atlanta, GA, 1999.
35 M. Anbia and S. Amirmahmoodi, *Sci. Iran.*, 2011, **18**, 446.
36 M. Pera-Titus, V. Garcia-Molina, M. A. Banos, J. Gimenez and S. Esplugas, *Appl. Catal., B*, 2004, **47**, 219.
37 A. Karci, *Chemosphere*, 2014, **99**, 1.
38 S. Y. Oh and Y. D. Seo, *Environ. Sci. Pollut. Res.*, 2016, **23**, 951.
39 R. L. Osborne, L. O. Taylor, K. P. Han, B. Ely and J. H. Dawson, *Biochem. Biophys. Res. Commun.*, 2004, **324**, 1194.
40 J. Belyea, L. B. Gilvey, M. F. Davis, M. Godek, T. L. Sit, S. A. Lommel and S. Franzen, *Biochemistry*, 2005, **44**, 15637.

41 C. Colomban, A. H. Tobing, G. Mukherjee, C. V. Sastri, A. B. Sorokin and S. P. de Visser, *Chem.–Eur. J.*, 2019, **25**, 14320.

42 M. W. LaCount, E. L. Zhang, Y.-P. Chen, K. Han, M. M. Whitton, D. E. Lincoln, S. A. Woodin and L. Lebioda, *J. Biol. Chem.*, 2000, **275**, 536.

43 S. Franzen, M. P. Roach, Y.-P. Chen, R. B. Dyer, W. H. Woodruff and J. H. Dawson, *J. Am. Chem. Soc.*, 1998, **120**, 4658.

44 R. L. Osborne, G. M. Raner, L. P. Hager and J. H. Dawson, *J. Am. Chem. Soc.*, 2006, **128**, 1036.

45 R. P. Ferrari, E. Laurenti and F. Trotta, *JBIC, J. Biol. Inorg. Chem.*, 1999, **4**, 232.

46 K. E. Hammel and P. J. Tardone, *Biochemistry*, 1988, **27**, 6563.

47 N. B. J. Vollaard, J. P. Shearman and C. E. Cooper, *Sports Med.*, 2005, **35**, 1045.

48 J. Zhao, V. de Serrano, J. Zhao, P. Le and S. Franzen, *Biochemistry*, 2013, **52**, 2427.

49 S. Munshi, R. D. Jana and T. K. Paine, *Dalton Trans.*, 2021, **50**, 5590.

50 G. Mukherjee, C. W. Z. Lee, S. S. Nag, A. Alili, F. G. Cantú Reinhard, D. Kumar, C. V. Sastri and S. P. de Visser, *Dalton Trans.*, 2018, **47**, 14945.

51 W. K. C. Lo, C. J. McAdam, A. G. Blackman, J. D. Crowley and D. A. McMorran, *Inorg. Chim. Acta*, 2015, **426**, 183.

52 A. A. Massie, M. C. Denler, L. T. Cardoso, A. N. Walker, M. K. Hossain, V. W. Day, E. Nordlander and T. A. Jackson, *Angew. Chem., Int. Ed.*, 2017, **56**, 4178.

53 S. Rana, J. P. Biswas, A. Sen, M. Clémancey, G. Blondin, J.-M. Latour, G. Rajaraman and D. Maiti, *Chem. Sci.*, 2018, **9**, 7843.

54 Y. Wang and A. Liu, *Chem. Soc. Rev.*, 2020, **49**, 4906.

55 P. Barman, A. K. Vardhaman, B. Martin, S. J. Wörner, C. V. Sastri and P. Comba, *Angew. Chem., Int. Ed.*, 2015, **54**, 2095.

56 Y. Sheng, C. S. Abelson, J. Prakash, A. Draksharapu, V. G. Young Jr and L. Que Jr, *Angew. Chem., Int. Ed.*, 2021, **60**, 20991.

57 A. K. Vardhaman, P. Barman, S. Kumar, C. V. Sastri, D. Kumar and S. P. de Visser, *Chem. Commun.*, 2013, **49**, 10926.

58 A. K. Vardhaman, C. V. Sastri, D. Kumar and S. P. de Visser, *Chem. Commun.*, 2011, **47**, 11044.

59 F. Namuswe, T. Hayashi, Y. Jiang, G. D. Kasper, A. A. N. Sarjeant, P. M. Loccoz and D. P. Goldberg, *J. Am. Chem. Soc.*, 2010, **132**, 157.

60 U. K. Bagha, J. K. Satpathy, G. Mukherjee, C. V. Sastri and S. P. de Visser, *Org. Biomol. Chem.*, 2021, **19**, 1879.

61 G. Mukherjee and C. V. Sastri, *Isr. J. Chem.*, 2020, **60**, 1032.

62 S. Sahu, M. G. Quesne, C. G. Davies, M. Dürr, I. I. Burmazović, M. A. Siegler, G. N. L. Jameson, S. P. de Visser and D. P. Goldberg, *J. Am. Chem. Soc.*, 2014, **136**, 13542.

Faraday Discussions

PAPER

Syntheses and investigation of metal complexes with macrocyclic polythioether ligands†

Thomas Rotärmel, Jonathan Becker and Siegfried Schindler *

Received 6th September 2021, Accepted 14th October 2021
DOI: 10.1039/d1fd00062d

Copper(I) complexes with the macrocyclic thioether ligands 1,4,8,11-tetrathiacyclotetradecane (tetrathiacyclam, 14-S_4) and 1,8-dithia-4,11-diazacyclotetradecane (dithiacyclam, 14-N_2S_2) were synthesised and structurally characterised. While the copper(I) complexes showed no reactivity towards dioxygen, the formation of "dioxygen adduct complexes" could be spectroscopically detected with ozone using low temperature stopped-flow techniques. Furthermore, it was possible to synthesise and characterise iron(II) and cobalt(II) complexes with the tetrathiacyclam ligand. No "dioxygen adduct" intermediates were observed when these complexes were reacted with dioxygen or ozone. Depending on the reaction conditions, the coordination of the metal ions could be controlled (*endo*- vs. *exo*-coordination and *cis*- vs. *trans*-coordination) and in addition to mononuclear complexes, also coordination polymers were obtained.

Introduction

With regard to gaining a better understanding of important selective oxidation/oxygenation reactions there has been a large number of investigations of copper and iron complexes that can be regarded as model compounds for corresponding metalloenzymes.[1] In particular, more recently interest increased in the analysis of the active site of the copper monooxygenases peptidylglycine-α-hydroxylating mono-oxygenase (PHM) and dopamine β-monooxygenase (DβM), enzymes that are important for the synthesis of neuropeptides and hormones.[2] Only in 2004, structural characterisation of a copper dioxygen complex in PHM could show unambiguously a mononuclear *end-on*-superoxido copper complex in the active site.[3] While up to then model compounds had supported the formation of such a species spectroscopically it finally was possible to structurally characterise an *end-on* copper superoxido complex with the tripodal ligand

Institut für Anorganische und Analytische Chemie, Justus-Liebig-Universität Gießen, Heinrich-Buff-Ring 17, 35392 Gießen, Germany. E-mail: siegfried.schindler@ac.jlug.de

† Electronic supplementary information (ESI) available. CCDC 2107318–2107328. For ESI and crystallographic data in CIF or other electronic format see DOI: 10.1039/d1fd00062d

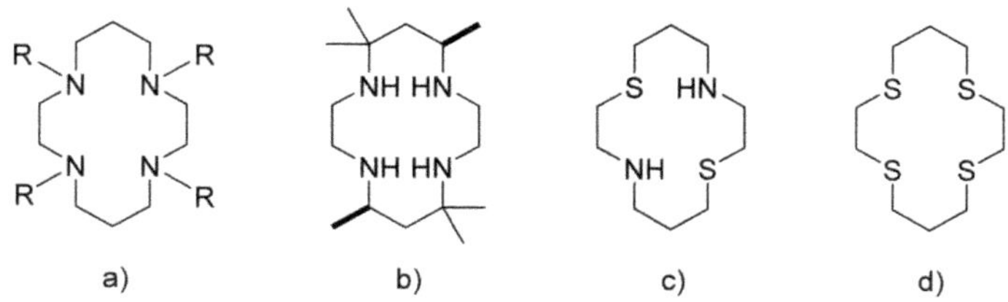

Scheme 1 Macrocyclic ligands (a) cyclam (R = H), TMC (R = CH_3), (b) tet b, (c) 14-N_2S_2 and (d) 14-S_4.

tris(tetramethylguanidino)tren (TMG_3tren).[4] Alternatively, dioxygen can be coordinated as a *side-on* superoxide which has been thoroughly described previously by Kitajima and Tolman;[5] however, this is not a binding mode so far observed in nature. In an effort to get closer to modelling the active site of PHM, Itoh and co-workers[6] used tridentate ligands instead of tetradentate tripodal ligands. However, while most of these compounds were based on all nitrogen ligands,[7] PHM and DβM contain thioether ligand residues, such as cysteine (Cys) and methionine (Met).[3,7,8] The coordination of Met plays an important role in copper dioxygen chemistry and is of high interest for understanding how thioether-ligation influences the structure, spectroscopy and reactivity.[3,9] Therefore, in previous work, a number of different thioether ligands already have been studied.[10] Bhadra *et al.* recently described the synthesis of an *end-on* superoxido copper complex system with a $^{TMG}N_3S$ thioether ligand (a derivative of TMG_3tren). It is the first reported *end-on* superoxido copper complex with an experimentally proven Cu–S bond which also possesses demonstrated hydrogen atom abstraction (HAA) reactivity.[11]

Based on earlier work by Valentine and co-workers, we have been successful in spectroscopically observing the formation of an *end-on* superoxido copper complex with the ligand tet b (*rac*-5,5,7,12,12,14-hexamethyl-1,4,8,11-tetraazacyclotetradecane), a derivative of the well-known macrocyclic ligand cyclam (Scheme 1).[12,13] With regard to these results we decided to investigate the reactivity of copper complexes with the known sulfur containing derivatives of cyclam (Scheme 1) 14-N_2S_2 (1,8-dithia-4,11-diazacyclotetradecane, dithiacyclam) and 14-S_4 (1,4,8,11-tetrathiacyclotetradecane, tetrathiacyclam) towards dioxygen and ozone.[14,15]

Results and discussion

As observed by us[16] and reported previously by Nam, Karlin and co-workers, the copper(I) complex with TMC as ligand does not react with dioxygen.[17] In contrast, as described above, the complex $[Cu(tet\ b)]^+$ reacts with dioxygen to form an *end-on* superoxido complex.[12]

More recently we started to use ozone as an oxidant to trap an oxygen adduct complex, *e.g.* to obtain an iron(IV) oxido complex with the ligand TMC, an approach based on earlier work by Wieghardt and co-workers as well as by Pestovsky *et al.*[18] Furthermore, the oxygenation with ozone is well known for iron and manganese porphyrinoid complexes.[19] However, to the best of our knowledge, studies on the reactions of ozone and copper(I) complexes to trap an "oxygen adduct" have not been reported so far. Therefore, we thought it might be useful to extend our investigations on the reactions of copper(I) complexes with ozone.

Interestingly, formation of the superoxido complex $[(\text{tet b})Cu(O_2)]^+$ is still observed (characteristic absorbance maxima at 410 nm, Fig. 1) if the reaction is performed with ozone, however, much faster decomposition takes place in comparison with the reaction with pure dioxygen. Low temperature stopped-flow measurements allowed obtaining the time resolved UV-vis spectra presented in Fig. 1.

Because ozone is always applied together with dioxygen, with our ozone generator a mixture of 6–14% ozone and 86–94% dioxygen is obtained (depending on the gas pressure, time and current settings), the direct reaction with dioxygen is still occurring. However, the presence of ozone obviously does lead to a fast decomposition of the superoxido complex. Interestingly, even in the presence of ozone, we did not observe a reaction of $[Cu(TMC)]^+$ with either dioxygen or ozone (under the applied stopped-flow conditions).

Synthesis and characterisation of copper(I) complexes with the ligands 14-S_4 and 14-N_2S_2

In order to better understand the effect of sulfur in the TMC ligand system, the two derivatives 14-S_4 and 14-N_2S_2 were synthesised. While 14-S_4 was commercially available, 14-N_2S_2 had to be prepared in a multi-step synthesis according to the literature.[15]

Previous work was reported on copper complexes with the 14-S_4 ligand in order to investigate the importance of the geometry of copper(I/II)-complexes in copper redox chemistry and revealed a polymeric crystal structure for the copper(I) complex where each copper(I)-ion is coordinated to four sulfur atoms, three of them originating from one ligand and the fourth one from an adjacent ligand.[20]

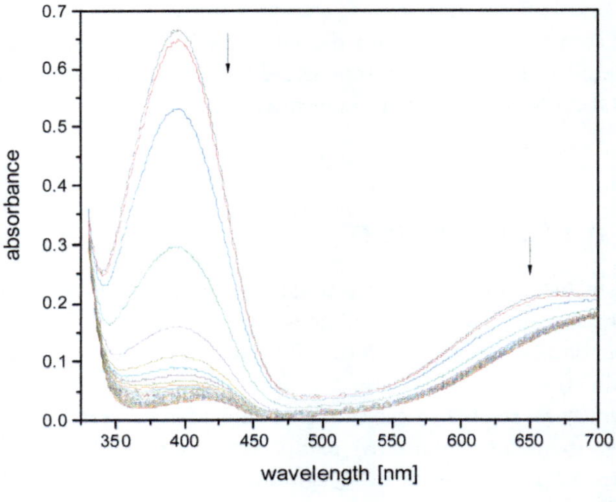

Fig. 1 Time-resolved UV/vis spectra of the reaction of ozone/dioxygen with $[Cu^I(\text{tet b})]PF_6$ in acetone ($c_{complex}$ = 0.4 mmol L^{-1} after mixing) at −90 °C, total time: 0.3 s.

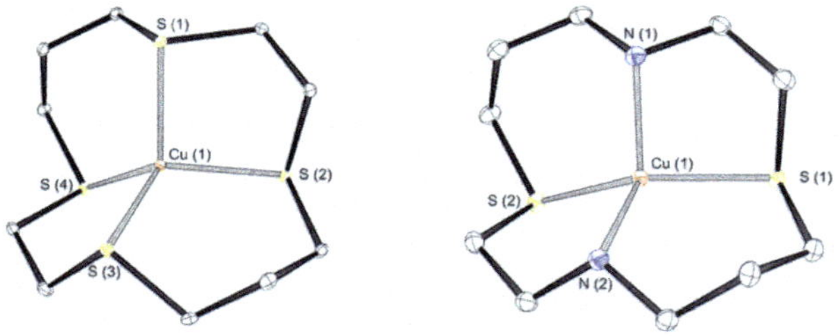

Fig. 2 Molecular structures of $[Cu^{I}(14\text{-}S_4)]^+$ and $[Cu^{I}(14\text{-}N_2S_2)]^+$ in ORTEP representations at 50% probability. H atoms and counterions were omitted for clarity.

The ability of the 14-S_4 ligand to maintain four coordinating bonds to a single copper(I)-ion was indicated in solution by NMR measurements.[21] In contrast to these previous results we were able to prepare and structurally characterise the monomeric copper species with different copper salts. The reaction of the 14-S_4 ligand with $[Cu(MeCN)_4]OTf$ in THF generated a colourless solid in high yield. The formation of $[Cu^{I}(14\text{-}S_4)]OTf$ was confirmed by electrospray-ionisation mass-spectrometry (ESI-MS), revealing the characteristic peak at $m/z = 330.975$ for the $[Cu^{I}(14\text{-}S_4)]^+$ cation. Moreover, we obtained suitable single crystals for X-ray structural characterisation. The crystal structure revealed that the copper(I)-ion is coordinated in a distorted tetrahedron by four sulfur-donor atoms. The molecular structure of $[Cu^{I}(14\text{-}S_4)]OTf$ is presented in Fig. 2 (crystallographic data are reported in ESI†) and selected bond lengths and angles are given in Table 1. Furthermore, it was possible to synthesise and structurally characterise $[Cu^{I}(14\text{-}S_4)]PF_6$ and $[Cu^{I}(14\text{-}S_4)]BPh_4$ (see ESI†) which only differ with regard to the anions.

Changing the solvent in the reaction with the 14-S_4 ligand and a copper salt from a non-coordinating solvent such as THF or dichloromethane to a well-coordinating solvent, *e.g.* MeCN, also led to the formation of a 1-dimensional polymeric structure. However, this polymer is different to the one reported previously and described above. In this new polymer, the copper(I)-ion is *exo*-coordinated in a distorted tetrahedral geometry by three sulfur-donor atoms and

Table 1 Selected bond lengths/Å and angles/° for $[Cu^{I}(14\text{-}S_4)]OTf$ and $[Cu^{I}(14\text{-}N_2S_2)]OTf$

$[Cu^{I}(14\text{-}S_4)]OTf$		$[Cu^{I}(14\text{-}N_2S_2)]OTf$	
Bond length		**Bond length**	
Cu(1)–S(1)	2.270(4)	Cu(1)–S(1)	2.257(6)
Cu(1)–S(2)	2.281(4)	Cu(1)–S(2)	2.266(5)
Cu(1)–S(3)	2.262(4)	Cu(1)–N(1)	2.091(17)
Cu(1)–S(4)	2.283(4)	Cu(1)–N(2)	2.096(17)
Bond angle		**Bond angle**	
S(1)–Cu(1)–S(2)	95.8°(16)	S(1)–Cu(1)–S(2)	146.4°(2)
S(1)–Cu(1)–S(3)	126.2°(16)	S(1)–Cu(1)–N(1)	90.8°(5)
S(1)–Cu(1)–S(4)	112.4°(16)	S(1)–Cu(1)–N(2)	106.4°(5)
S(2)–Cu(1)–S(3)	111.3°(16)	S(2)–Cu(1)–N(1)	106.3°(5)
S(2)–Cu(1)–S(4)	118.5°(16)	S(2)–Cu(1)–N(2)	91.2°(5)
S(3)–Cu(1)–S(4)	94.6°(15)	N(1)–Cu(1)–N(2)	117.7°(7)

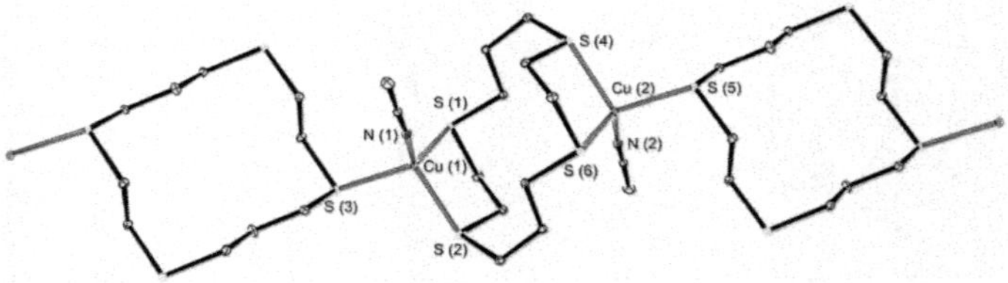

Fig. 3 Excerpt of the crystal structure of $\{[Cu^I(14\text{-}S_4)(MeCN)PF_6]\}_n$ in ORTEP representation at 50% probability. H atoms and counterions were omitted for clarity.

one nitrogen-donor atom from MeCN. The three S–Cu bonds arising from two different ligands coordinate alternately in a monodentate and bidentate fashion. A section of the polymer chain is presented in Fig. 3 (crystallographic data are reported in the ESI†).

In contrast to the reaction with the 14-S_4 ligand, no solvent effect was observed for the formation of copper(I) complexes with the 14-N_2S_2 ligand. It therefore seems that the coordination of the complex (*endo*- *vs.* *exo*-coordination) depends on the donor-atoms (two *vs.* four sulfur-atoms) and the type of solvent (non- *vs.* good-coordinating solvent) used in the reaction. In fact, sulfur-containing macrocycles often show metal coordination in an *exo*-fashion, caused by the torsion arrangements from the donor atoms.[22] However, with the "right" reaction conditions it is possible to force the ligand to coordinate the metal-ion inside the cavity (*endo*-coordination).

A copper(I) complex with the 14-N_2S_2 ligand, $[Cu^I(14\text{-}N_2S_2)]OTf$, was easily obtained and the molecular structure is presented in Fig. 2 (crystallographic data are reported in the ESI†). The structure of its cation is nearly identical to the molecular structure of the cation of the complex $[Cu^I(14\text{-}N_2S_2)]PF_6$ which had been reported previously.[23]

In both complexes, $[Cu^I(14\text{-}S_4)]OTf$ as well as $[Cu^I(14\text{-}N_2S_2)]OTf$, the copper ion is four-coordinated either by four sulfur atoms or by two sulfur atoms and two nitrogen atoms in a distorted tetrahedral geometry. However, the complexes differ in bond lengths and angles (Table 1). For example, comparing the 14-S_4 complex to the 14-N_2S_2 complex the S–Cu–S angle decreases by $\sim$28° based on the shorter Cu–N bond lengths, which are incorporated in the 14-N_2S_2 complex, while the Cu–S bond lengths in both complexes are almost identical.

Reactivity of the copper(I) complexes $[Cu^I(14\text{-}S_4)]OTf$ and $[Cu^I(14\text{-}N_2S_2)]OTf$ towards dioxygen and ozone

Similar to the reaction of the copper(I) complex with the ligand TMC described above, copper complexes $[Cu^I(14\text{-}S_4)]OTf$ and $[Cu^I(14\text{-}N_2S_2)]OTf$ did not react with dioxygen. The coordinated sulfur atoms might even lead to a higher stability towards oxidation due to the fact that sulfur containing ligand systems can form quite stable Cu(I) complexes that are inert to dioxygen.[7] Furthermore, no reaction was observed under stopped-flow conditions in dichloromethane at −80 °C. In contrast, the reaction of $[Cu^I(14\text{-}S_4)]OTf$ with ozone at −80 °C in dichloromethane immediately led to a lime green species. Warming the solution led to the formation of a light green coloured suspension. The light green solid could be separated, however, up to now we did not succeed in characterising this product. Time-resolved UV-vis spectra obtained in a low temperature stopped-flow experiment are shown in Fig. 4.

Absorbance maxima at 402 nm and 627 nm are consistent with the formation of an *end-on* superoxido copper complex (compare with Fig. 1).[7] However, as can be seen in Fig. 4 (inset), the complex was not stable enough to allow isolation/crystallisation of this transient intermediate.

Still, in an effort to obtain crystals, in one approach (sample stored for 6 months in a freezer at −80 °C) the colour of the solution turned from a lime green to a purple colour and a few red purple crystals were obtained, suitable for X-ray structural characterisation. The molecular structure revealed formation of the copper(II) complex $[Cu^{II}(14\text{-}S_4)]OTf_2$ (molecular structure and crystallographic data are presented in the ESI†). The copper(II) ion is *endo*-coordinated by the macrocyclic ligand system with additional two triflate ions coordinated in axial positions. The molecular structure is very similar to the perchlorate analogue reported previously.[24]

The overall finding is quite important because it shows that it is possible to obtain a transition metal superoxido complex by applying ozone as an oxidant.

In contrast to the reaction of $[Cu^{I}(14\text{-}S_4)]OTf$ with ozone, $[Cu^{I}(14\text{-}N_2S_2)]OTf$ produced immediately a dark green species with ozone in dichloromethane at −80 °C. Warming up the solution led to the formation of a light green suspension. Again, the light green solid was separated, but could not be characterised so far. Stopped-flow measurements of the reaction of the copper(I) complex with ozone at low temperature were performed and time resolved UV-vis spectra are shown in Fig. 5. These investigations confirmed the formation of a short-lived "oxygen adduct" complex. A characteristic UV absorbance band around 386 nm could be detected, which in comparison with literature data most likely indicates the formation of a *side-on* peroxido copper(II) complex.[7,17] Unfortunately, in a series of

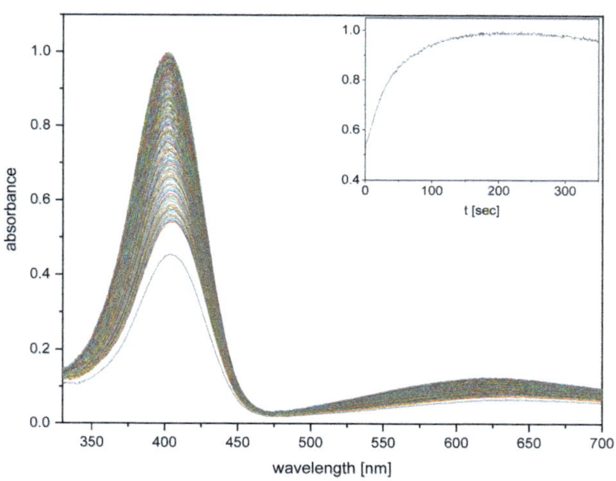

Fig. 4 Time-resolved UV/vis-spectra of the reaction of ozone with $[Cu^{I}(14\text{-}S_4)]OTf$ in dichloromethane ($c_{complex}$ = 0.2 mmol L^{-1} after mixing) at −80 °C, total time: 375 s. Inlay (timetrace): absorbance *vs.* time at 402 nm.

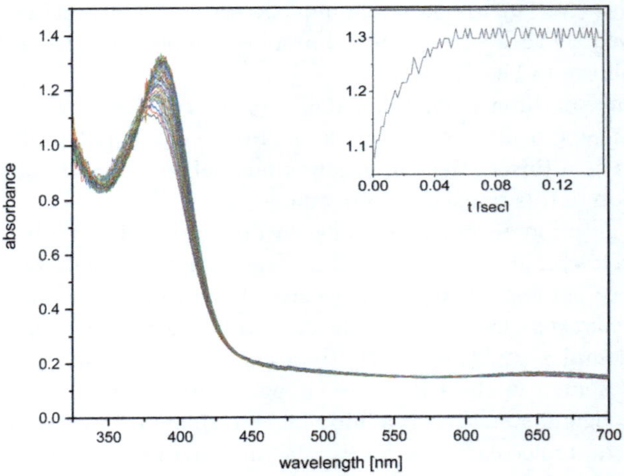

Fig. 5 Time-resolved UV/vis-spectra of the reaction of ozone with $[Cu^{I}(14\text{-}N_2S_2)]OTf$ in dichloromethane ($c_{complex} = 0.2$ mmol L^{-1} after mixing) at −80 °C, total time: 0.15 s. Inlay (timetrace): absorbance *vs.* time at 386 nm.

crystallisation approaches no crystals could be obtained for a further characterisation of this complex.

Unfortunately, it was not possible (due to the instability of the complexes and the special setup for ozone chemistry) to characterise the postulated complexes $[(14\text{-}S_4)Cu(O_2)]OTf$ and $[(14\text{-}N_2S_2)Cu(O_2)Cu[(14\text{-}N_2S_2)](OTf)_2$ further, however, the UV-vis spectra in combination with the results for the copper tet b complex system strongly support this assignment. In contrast to the iron complexes described above, we think that the formation of a copper(III) oxido or copper(II) oxygenyl complex is less likely during the reaction with ozone. However, we cannot completely exclude it so far. Instead, we propose the mechanism shown in Scheme 2 for the formation of the copper superoxido and copper peroxido complex.

The quite labile copper ozonido complex can react with dioxygen (that is present in excess) leading to the copper superoxido complex $[(14\text{-}S_4)Cu(O_2)]OTf$. The opposite reaction is well known for the synthesis of alkali metal ozonido complexes. Thus, potassium and rubidium ozonido salts are prepared from the corresponding superoxido compounds and ozone.[25] If there is no steric hindrance the copper superoxido complex can react further with another copper(I) complex, leading to the corresponding dinuclear copper peroxido complex. Further side

$$[Cu^{I}(L)]^{+} + O_3 \longrightarrow [Cu^{II}(L)O_3]^{+} \quad (1)$$

$$[Cu^{II}(L)O_3]^{+} + O_2 \longrightarrow [Cu^{II}(L)O_2]^{+} + O_3 \quad (2)$$

$$[Cu^{II}(L)O_2]^{+} + [Cu^{I}(L)]^{+} \longrightarrow [(L)Cu^{II}(O_2)Cu^{II}(L)]^{2+} \quad (3)$$

Scheme 2 Proposed reaction mechanism for the formation of a copper superoxido and copper peroxide complex.

reactions occur due to the fact that we observe decomposition of the formed complexes as well as precipitation of a so far not identified product in more concentrated solutions.

Synthesis and characterisation of iron(II) complexes with the 14-S_4 ligand

Hughes *et al.* reported the first investigations of iron(II) complexes with the 14-S_4 ligand.[26] The authors already were quite successful in describing the reaction between the 14-S_4 ligand and FeI_2, however, without reporting a crystal structure of the complex. They prepared $[Fe^{II}(14\text{-}S_4)(MeCN)_2]FeI_4$ and analysed the complex by Mössbauer spectroscopy, elemental analysis and IR spectroscopy. We now could crystallise $[Fe^{II}(14\text{-}S_4)(MeCN)_2]FeI_4$ by reacting FeI_2 and 14-S_4 in a ratio of 2 : 1. The molecular structure is presented in Fig. 6 (crystallographic data are reported in the ESI†).

The Fe^{II} center is *endo*-coordinated and octahedrally surrounded by the 14-S_4 ligand and two additional acetonitrile molecules, which are coordinated in *cis* positions. Furthermore, the successful complexation was supported by ESI-MS revealing the characteristic mass peak at $m/z = 161.989$ for the $[Fe(14\text{-}S_4)]^{2+}$ fragment.

Moreover, we succeeded in synthesising the complex $[Fe^{II}(14\text{-}S_4)(MeCN)_2][(FeI_3)_2(14\text{-}S_4)]$ in a reaction between FeI_2 and 14-S_4 in a ratio of 1 : 1. The molecular structure presented in Fig. 7 (crystallographic data are reported in the ESI†) is quite interesting, because of the fact that the ligand is involved both as a cation as well as an anion and on the one hand in *exo*- and on the other hand in *endo*-coordination. In comparison to $[Fe^{II}(14\text{-}S_4)(MeCN)_2]FeI_4$ the acetonitrile molecules are coordinated in *trans* positions. The formation of the expected Fe^{II} complex was also confirmed by ESI-MS revealing a mass peak at $m/z = 161.989$ for the $[Fe(14\text{-}S_4)]^{2+}$ fragment.

A comparison of the molecular structure between both iron(II) complexes shows almost the same bond lengths and angles (bond lengths and angles are presented in Table 2). In addition, an iron(II) complex could be prepared with $[Fe(MeCN)_4](PF_6)_2$ and 14-S_4 followed by an anion exchange with $AgBAr_4^F$. The

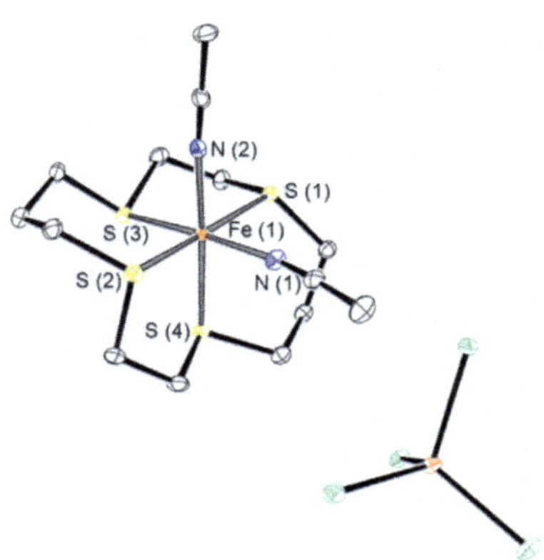

Fig. 6 Molecular structure of $[Fe^{II}(14\text{-}S_4)(MeCN)_2]FeI_4$ in ORTEP representation at 50% probability. H atoms were omitted for clarity.

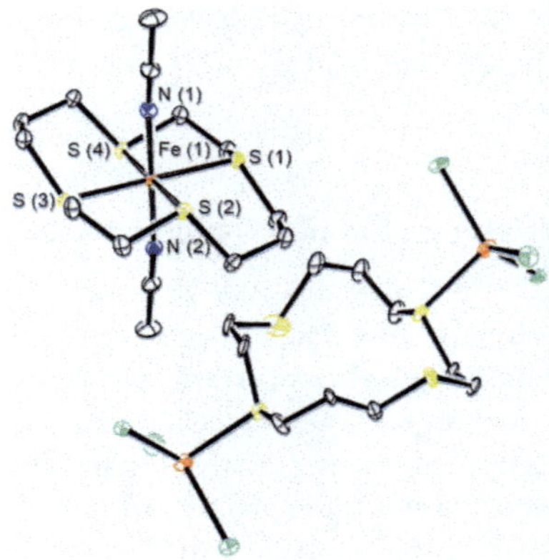

Fig. 7 Molecular structure of $[Fe^{II}(14\text{-}S_4)(MeCN)_2][(FeI_3)_2(14\text{-}S_4)]$ in ORTEP representation at 50% probability. H atoms were omitted for clarity.

molecular structure of $[Fe^{II}(14\text{-}S_4)(MeCN)_2](BAr_4^F)_2$ is presented in the ESI† and shows the same octahedral geometry as the complex $[Fe^{II}(14\text{-}S_4)(MeCN)_2][(FeI_3)_2(14\text{-}S_4)]$.

Bench-top experiments at different temperatures for the reaction of the iron complexes with dioxygen and ozone did not provide evidence for the formation of

Table 2 Selected bond lengths/Å and angles/° for *cis* Fe^{II} complex and *trans* Fe^{II} complex

Cis Fe^{II} complex		*Trans* Fe^{II} complex	
Bond length		**Bond length**	
Fe(1)–S(1)	2.252(9)	Fe(1)–S(1)	2.249(11)
Fe(1)–S(2)	2.249(10)	Fe(1)–S(2)	2.246(11)
Fe(1)–S(3)	2.233(9)	Fe(1)–S(3)	2.253(11)
Fe(1)–S(4)	2.240(9)	Fe(1)–S(4)	2.246(11)
Fe(1)–N(1)	1.945(3)	Fe(1)–N(1)	1.923(4)
Fe(1)–N(2)	1.955(3)	Fe(1)–N(2)	1.926(4)
Bond angle		**Bond angle**	
N(1)–Fe(1)–N(2)	89.7°(11)	N(1)–Fe(1)–N(2)	179.1°(15)
N(1)–Fe(1)–S(1)	91.2°(9)	N(1)–Fe(1)–S(1)	87.4°(11)
N(1)–Fe(1)–S(2)	86.2°(9)	N(1)–Fe(1)–S(2)	86.1°(11)
N(1)–Fe(1)–S(3)	175.4°(9)	N(1)–Fe(1)–S(3)	92.5°(11)
N(1)–Fe(1)–S(4)	90.8°(9)	N(1)–Fe(1)–S(4)	93.6°(11)
N(2)–Fe(1)–S(1)	87.9°(8)	N(2)–Fe(1)–S(1)	92.7°(11)
N(2)–Fe(1)–S(2)	88.4°(3)	N(2)–Fe(1)–S(2)	94.9°(11)
N(2)–Fe(1)–S(3)	94.8°(8)	N(2)–Fe(1)–S(3)	92.5°(11)
N(2)–Fe(1)–S(4)	176.5°(9)	N(2)–Fe(1)–S(4)	85.5°(11)
S(1)–Fe(1)–S(2)	175.5°(4)	S(1)–Fe(1)–S(2)	89.5°(4)
S(1)–Fe(1)–S(3)	88.4°(3)	S(1)–Fe(1)–S(3)	179.6°(5)
S(1)–Fe(1)–S(4)	95.5°(3)	S(1)–Fe(1)–S(4)	90.1°(4)
S(2)–Fe(1)–S(3)	94.6°(4)	S(2)–Fe(1)–S(3)	90.1°(4)
S(2)–Fe(1)–S(4)	88.2°(3)	S(2)–Fe(1)–S(4)	179.5°(5)
S(3)–Fe(1)–S(4)	84.7°(3)	S(3)–Fe(1)–S(4)	90.3°(4)

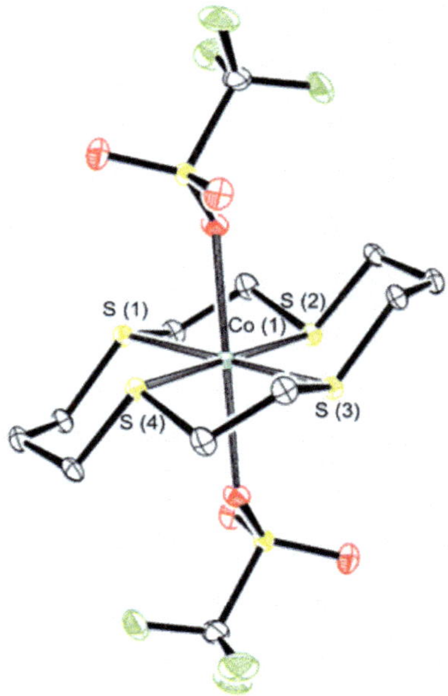

Fig. 8 Molecular structure of $[Co^{II}(14\text{-}S_4)(OTf)_2]$ in ORTEP representation at 50% probability. H atoms were omitted for clarity.

an "oxygen adduct" complex as an intermediate. This was furthermore supported by cyclic voltammetry, which showed irreversible redox reactions (ESI†).

Synthesis and characterisation of cobalt(II) complexes with the 14-S_4 ligand

Furthermore, a cobalt(II) complex could be prepared with $Co^{II}(OTf)_2$ and the 14-S_4 ligand in THF. In previous work cobalt(II) complexes have already been synthesised with the cyclam and 14-N_2S_2 ligand.[27,28] Nevertheless, to the best of our knowledge, this is the first synthesised and structurally characterised cobalt(II) complex with the 14-S_4 ligand. The successful complexation was supported by ESI-MS revealing the characteristic mass peaks at $m/z = 163.489$ and 475.924 for the $[Co(14\text{-}S_4)]^{2+}$ and $[Co(14\text{-}S_4)OTf]^+$ fragments. The molecular structure, presented in Fig. 8 (crystallographic data are reported in the ESI†), shows an octahedral geometry. Selected bond lengths and angles are reported in Table 3. A comparison of the cobalt(II) complexes with cyclam,[27] 14-N_2S_2 (ref. 28) and 14-S_4 ligand systems shows similarities in terms of bond angles, lengths and geometry.

However, it is interesting to note, that while the previous reported cobalt(II) complexes showed no solvent dependency with regard to the *exo*- *vs.* *endo*-coordination of the cobalt(II) ion, this is the case for the cobalt(II) 14-S_4 complex. Similar to the solvent dependency of the copper(I) complexes described above, the same effect could be observed with cobalt(II) ions and the 14-S_4 ligand.

If the reaction is carried out in MeCN, a polymeric structure could be obtained revealing an *exo*-coordination of the cobalt(II)-ions (molecular structures and crystallographic data are reported in the ESI†).

Table 3 Selected bond lengths/Å and angles/° for $[Co^{II}(14\text{-}S_4)](OTf)_2$

$[Co^{II}(14\text{-}S_4)](OTf)_2$			
Bond length		**Bond angle**	
Co(1)–S(1)	2.2262(5)	S(1)–Co(1)–S(2)	90.2°(18)
Co(1)–S(2)	2.2262(5)	S(1)–Co(1)–S(3)	180.0°
Co(1)–S(3)	2.2206(5)	S(1)–Co(1)–S(4)	89.8°(18)
Co(1)–S(4)	2.2206(5)	S(2)–Co(1)–S(3)	89.8°(18)
Co(1)–O(1)	2.357(14)	S(2)–Co(1)–S(4)	180.0°
Co(1)–O(2)	2.357(14)	S(3)–Co(1)–S(4)	90.2°(18)
		S(1)–Co(1)–O(2)	89.8°(4)
		S(3)–Co(1)–O(2)	90.2°(4)
		S(2)–Co(1)–O(2)	96.9°(4)
		S(4)–Co(1)–O(2)	83.1°(4)
		S(1)–Co(1)––O(1)	90.2°(4)
		S(3)–Co(1)–O(1)	89.8°(4)
		S(2)–Co(1)–O(1)	83.1°(4)
		S(4)–Co(1)–O(1)	96.9°(4)
		O(1)–Co(1)–O(2)	180.0°

Furthermore, as with the iron complexes, cyclic voltammetry of $[Co^{II}(14\text{-}S_4)](OTf)_2$ only showed irreversible redox behaviour (ESI†).

Experimental section

All reagents and solvents used were of p.a. quality and were purchased from Sigma Aldrich and AcrosOrganics. The solvents were purified by distillation with a drying agent by standard procedures. The ligand 14-N_2S_2,[15,29] $[Cu(TMC)]PF_6$ (ref. 17) and $[Cu(tet\ b)]PF_6$ (ref. 30) were synthesised according to literature protocols. The air sensitive syntheses of the metal complexes were performed under inert conditions in an MBraun glove box (equipped with water and dioxygen sensors; H_2O and O_2 concentrations less than 0.1 ppm) under an argon atmosphere. Stopped-flow measurements were carried out with a HI-Tech SF-61SX2 and CSF-61DX2 low-temperature stopped-flow unit equipped with a diode array spectrophotometer. NMR measurements were performed with a Bruker Avance II 400 MHz (AV II 400). For electrospray-ionisation MS (ESI-MS) measurements a Bruker microTOF mass spectrometer was used. UV/vis spectra were recorded with an Agilent 8453 spectrometer. Experimental data in single-crystal X-ray diffraction measurements were collected by using a Bruker D8 Venture system, equipped with dual IμS-sources (Mo(Cu)), a PHOTON 100 detector and an Oxford Cryostream 700 low temperature system (details are described in the ESI†). Cyclic voltammetry measurements were performed with a glassy carbon electrode as working electrode, a platinum/titanium electrode (counter electrode) and an Ag/AgCl reference electrode. MeCN was used as solvent and NBu_4BF_4 was applied as conducting salt. Ferrocene was used as standard and the electrochemical data were recorded with an e-corder 410 edaq (eDAQ, Colorado Springs, US) and the program eChem. The CHN elemental analysis was performed using the CHN-Analysator: Thermo FlashEA – 1112 Series.

Ozonolysis of copper complexes

Ozonolysis was performed with the OZONOSAN PM 83 K. For the reactions with ozone and the copper complexes absolute dry glassware was used. The complex solutions were prepared in a glove box. Approximately 100 mg of the copper complexes were dissolved in 2 mL DCM for the reaction at −80 °C. Ozone was bubbled for 5 min through the complex solution with a glass cannula, which allowed bubbling ozone through the complex solution without any contact with the metal of the syringe needle. After the reaction with ozone, the complex solutions were stored in a −80 °C freezer.

Synthesis of [Cu(14-S_4)]OTf

A solution of [Cu(MeCN)$_4$]OTf (69.3 mg, 0.184 mmol) in THF (2 mL) was added dropwise to a solution of 14-S_4 (50.0 mg, 0.186 mmol) in THF (2 mL). The colourless solution was stirred overnight at room temperature. Slow evaporation of the solvent led to colourless crystals that were suitable for X-ray structural characterisation. Yield: 73.5 mg (83%). HRMS (ESI): *m*/*z* calcd for $C_{10}H_{20}CuS_4^+$: 330.973; found: 330.975 $[M + L]^+$. EA: calcd for [$C_{11}H_{20}CuF_3O_3S_5$]: C, 27.46%; H, 4.19%; found: C, 27.81%; H, 4.21%.

Synthesis of [Cu(14-S_4)]PF_6

A solution of [Cu(MeCN)$_4$]PF_6 (65.9 mg, 0.177 mmol) in THF (2 mL) was added dropwise to a solution of 14-S_4 (50.0 mg, 0.186 mmol) in THF (2 mL). The colourless solution was stirred overnight at room temperature. Slow evaporation of the solvent led to colourless crystals that were suitable for X-ray structural characterisation. Yield: 67.5 mg (80%). HRMS (ESI): *m*/*z* calcd for $C_{10}H_{20}CuS_4^+$: 330.973; found: 330.972 $[M + L]^+$. EA: calcd for [$C_{10}H_{20}CuF_6PS_4$ + 0.25CH_3CN]: C, 25.88%; H, 4.29%; N, 0.72%; found: C, 26.19%; H, 4.14%; N, 0.57%.

Synthesis of [Cu(14-S_4)]BPh_4

A solution of [Cu(MeCN)$_4$$PF_6$ (67.5 mg, 0.181 mmol) in DCM/acetone (1/1, 2 mL) was added dropwise to a solution of 14-S_4 (50.0 mg, 0.186 mmol) in DCM/acetone (1/1, 2 mL). The colourless solution was stirred for 30 minutes at room temperature. An acetone solution (2 mL) of sodium tetraphenylborate (61.9 mg, 0.181 mmol) was added to the reaction mixture and stirred at room temperature overnight. On the next day, the formed colourless solid was filtrated and 2 mL diethyl ether was added to the solution. The precipitated solid was filtrated, washed with diethyl ether and dried. Colourless crystals suitable for X-ray structural characterisation were obtained by slow diffusion of diethyl ether into a DCM solution of the complex at −30 °C for one day. Yield: 38.2 mg (32%). HRMS (ESI): *m*/*z* calcd for $C_{10}H_{20}CuS_4^+$: 330.973; found: 330.970 $[M + L]^+$. EA: calcd for [$C_{34}H_{40}BCuS_4$]: C, 62.70%; H, 6.19%; found: C, 62.21%; H, 6.02%.

Synthesis of {[Cu(14-S_4)(MeCN)PF_6]}$_n$

A solution of [Cu(MeCN)$_4$]PF_6 (69.3 mg, 0.177 mmol) in a THF/MeCN mixture (1/1, 2 mL) was added dropwise to a solution of 14-S_4 (50.0 mg, 0.186 mmol) in THF (2 mL). The colourless solution changed to a milky suspension after the solution of the copper-salt was added. The colourless solid was filtrated and a small amount of the solid was dissolved again. Slow evaporation of the solvent led to colourless crystals that were suitable for X-ray structural characterisation. Yield: 44.8 mg.

Synthesis of [Cu(14-N_2S_2)]OTf

A solution of [Cu(MeCN)$_4$]OTf (63.7 mg, 0.169 mmol) in THF (2 mL) was added dropwise to a solution of 14-N_2S_2 (40.0 mg, 0.171 mmol) in THF (2 mL). The colourless solution was stirred overnight at room temperature. Slow evaporation of the solvent led to colourless crystals that were suitable for X-ray structural characterisation. Yield: 55.2 mg (73%). HRMS (ESI): *m*/*z* calcd for $C_{10}H_{22}CuN_2S_2^+$: 297.051; found: 297.051 $[M + L]^+$. EA: calcd for [$C_{11}H_{22}CuF_3N_2O_3S_3$ + 0.1C_4H_8O]: C, 30.14%; H, 5.06%; N, 6.17%; found: C, 30.49%; H, 4.90%; N, 6.11%.

Synthesis of [Cu(14-N_2S_2)]PF_6

A solution of [Cu(MeCN)$_4$]PF_6 (62.2 mg, 0.167 mmol) in DCM/THF (1/1, 2 mL) was added dropwise to a solution of 14-N_2S_2 (40.0 mg, 0.171 mmol) in DCM (2 mL). The colourless solution was stirred overnight at room temperature. Slow evaporation of the solvent led to colourless crystals that were suitable for X-ray structural characterisation. Yield: 62.1 mg (84%). HRMS (ESI): *m*/*z* calcd for $C_{10}H_{22}CuN_2S_2^+$: 297.051; found: 297.051 $[M + L]^+$. EA: calcd for [$C_{10}H_{20}CuF_6N_2PS_2$ + 0.1C_4H_8O]: C, 27.75%; H, 5.11%; N, 6.22%; found: C, 27.79%; H, 4.95%; N, 6.23%.

Synthesis of [Fe(14-S_4)(MeCN)$_2$]FeI_4 [26]

To a stirred solution of FeI_2 (116 mg, 0.373 mmol) in MeCN (2 mL), a solution of 14-S_4 (50.0 mg, 0.186 mmol) in DCM (2 mL) was added. The reaction mixture was heated at 50 °C for 4 h in order to start the reaction. Subsequently the reaction mixture was stirred for two days at room temperature. Slow evaporation of the solvent led to purple crystals that were suitable for X-ray structural characterisation. Yield: 20.2 mg (7%). HRMS (ESI): *m*/*z* calcd for $C_{10}H_{20}FeS_4^{2+}$: 161.989; found: 161.989 $[M + L]^{2+}$. EA: calcd for [$C_{14}H_{26}Fe_2I_4N_2S_4$ + CH_2Cl_2]: C, 17.08%; H, 2.68%; N, 2.66%; found: C, 16.80%; H, 2.45%; N, 2.26%.

Synthesis of [Fe(14-S_4)(MeCN)$_2$][(FeI_3)$_2$(14-S_4)]

To a stirred solution of 14-S_4 (50.0 mg, 0.186 mmol) in THF (2 mL), a solution of FeI_2 (57.6 mg, 0.186 mmol) in MeCN (2 mL) was added dropwise. The reaction mixture was heated at 50 °C for 4 h in order to start the reaction. Subsequently the reaction mixture was stirred for two days at room temperature. Slow evaporation of the solvent led to purple crystals that were suitable for X-ray structural characterisation. Yield: 34.5 mg (19%). HRMS (ESI): *m*/*z* calcd for $C_{10}H_{20}FeS_4^{2+}$: 161.989; found: 161.989 $[M + L]^{2+}$. EA: calcd for [$C_{24}H_{46}Fe_3I_6N_2S_8$ + 3CH_3CN + FeI_2]: C, 18.19%; H, 2.80%; N, 3.54%; found: C, 18.30%; H, 2.60%; N, 3.90%.

Synthesis of [Fe^{II}(14-S_4)(MeCN)$_2$](BAr_4^F)$_2$

To a stirred solution of 14-S_4 (50.0 mg, 0.186 mmol) in THF (2 mL), FeI_2 (57.6 mg, 0.186 mmol) in MeCN (2 mL) was added dropwise. The reaction mixture was heated at 50 °C and stirred for one day. After one day $AgBAr_4^F$ (404 mg, 0.372 mmol) was added and the suspension was stirred for 6 h. The colourless precipitate was filtered off. Slow evaporation of the solvent led to purple crystals that were suitable for X-ray structural characterisation. Yield: 19.8 mg (5%). HRMS (ESI): *m*/*z* calcd for $C_{10}H_{20}FeS_4^{2+}$: 161.989; found: 161.989 $[M + L]^{2+}$. EA: calcd for $[C_{78}H_{50}B_2F_{48}FeN_2S_4 + 3.5CH_3CN]$: C, 44.85%; H, 2.68%; N, 3.38%; found: C, 44.88%; H, 2.21%; N, 3.76%.

Synthesis of $\{[Co(14\text{-}S_4)(MeCN)(Cl)_2]\}_n$

A suspension of $CoCl_2$ (24.2 mg, 0.186 mmol) in MeCN/THF (1/1, 2 mL) was added dropwise to a solution of 14-S_4 (50.0 mg, 0.186 mmol) in THF (2 mL) and stirred overnight. Unreacted $CoCl_2$ was filtrated and slow evaporation of the blue solvent led to blue crystals that were suitable for X-ray structural characterisation. Yield: 39.3 mg.

Synthesis of $[Co(14\text{-}S_4)(OTf)_2]$

A solution of $Co(OTf)_2$ (66.4 mg, 0.186 mmol) in THF (2 mL) was added dropwise to a solution of 14-S_4 (50.0 mg, 0.186 mmol) in THF (2 mL) and stirred at 60 °C over 3 days. After 3 days the solvent was evaporated and the pink solid was redissolved in MeCN. The orange/pink suspension was filtrated and the orange filtrate-solution was retained for evaporation. Slow evaporation of the solvent led to orange crystals that were suitable for X-ray structural characterisation. Yield: 10.0 mg (9%). HRMS (ESI): *m*/*z* calcd for $C_{10}H_{20}CoS_4^{2+}$: 163.488; found: 163.489 $[M + L]^{2+}$, $C_{11}H_{20}CoF_3O_3S_5^{+}$: 475.929; found: 475.924 $[M + L + OTf]^{+}$. EA: calcd for $[C_{12}H_{20}CoF_6O_6S_6]$: C, 23.04%; H, 3.22%; found: C, 23.30%; H, 3.00%.

Conclusions

Within this report, we describe the synthesis and characterisation of a series of mononuclear metal complexes (M = Cu, Fe, Co) containing the thioether ligands 14-S_4 and 14-N_2S_2. The iron and the cobalt complexes have been prepared mainly to investigate the coordination behaviour and solvent dependency and did not form “dioxygen adduct” complexes when reacted with either dioxygen or ozone. All complexes with the 14-S_4 ligand showed interesting metal ion coordination behaviour depending on the conditions. The copper(I) complexes were synthesise especially to investigate their reactivity towards different oxidants (dioxygen and ozone). Both copper(I) complexes, $[Cu^I(14\text{-}S_4)]OTf$ and $[Cu^I(14\text{-}N_2S_2)]OTf$, showed no reactivity towards dioxygen, similar to the copper(I) TMC complex. In contrast, both complexes reacted with ozone to form either an *end-on* superoxido or a *side-on* peroxido complex, an important finding with regard to opening a new possibility to prepare these complexes. Low temperature stopped-flow measurements showed that these complexes were not stable over time and therefore it was not possible to isolate and to crystallise them. However, by applying a suitable ligand system it should be possible to obtain more inert/stable “dioxygen adduct” complexes this way.

Conflicts of interest

There are no conflicts to declare.

Acknowledgements

We gratefully acknowledge support by the Justus-Liebig-Universität Gießen and the Deutsche Forschungsgemeinschaft for financial support (SS, SCHI 377/18-1). Furthermore, we would like to thank Stefan Schaub for his help in the laboratory with ozone handling.

Notes and references

1 D. A. Quist, D. E. Diaz, J. J. Liu and K. D. Karlin, *J. Biol. Inorg. Chem.*, 2017, **22**, 253.

2 S. T. Prigge, A. S. Kolhekar, B. A. Eipper, R. E. Mains and L. M. Amzel, *Science*, 1997, **278**, 1300.

3 S. T. Prigge, B. A. Eipper, R. E. Mains and L. M. Amzel, *Science*, 2004, **304**, 864.

4 C. Würtele, E. Gaoutchenova, K. Harms, M. C. Holthausen, J. Sundermeyer and S. Schindler, *Angew. Chem.*, 2006, **118**, 3951.

5 (*a*) K. Fujisawa, M. Tanaka, Y. Moro-oka and N. Kitajima, *J. Am. Chem. Soc.*, 1994, **116**, 12079; (*b*) A. M. Reynolds, B. F. Gherman, C. J. Cramer and W. B. Tolman, *Inorg. Chem.*, 2005, **44**, 6989.

6 A. Kunishita, M. Kubo, H. Sugimoto, T. Ogura, K. Sato, T. Takui and S. Itoh, *J. Am. Chem. Soc.*, 2009, **131**, 2788–2789.

7 L. M. Mirica, X. Ottenwaelder and T. D. P. Stack, *Chem. Rev.*, 2004, **104**, 1013.

8 (*a*) E. I. Solomon, D. E. Heppner, E. M. Johnston, J. W. Ginsbach, J. Cirera, M. Qayyum, M. T. Kieber-Emmons, C. H. Kjaergaard, R. G. Hadt and L. Tian, *Chem. Rev.*, 2014, **114**, 3659; (*b*) R. E. Cowley, L. Tian and E. I. Solomon, *Proc. Natl. Acad. Sci. U. S. A.*, 2016, **113**, 12035.

9 G. Y. Park, Y. Lee, D.-H. Lee, J. S. Woertink, A. A. Narducci Sarjeant, E. I. Solomon and K. D. Karlin, *Chem. Commun.*, 2010, **46**, 91.

10 (*a*) S. Kim, J. W. Ginsbach, A. I. Billah, M. A. Siegler, C. D. Moore, E. I. Solomon and K. D. Karlin, *J. Am. Chem. Soc.*, 2014, **136**, 8063; (*b*) S. Kim, J. Y. Lee, R. E. Cowley, J. W. Ginsbach, M. A. Siegler, E. I. Solomon and K. D. Karlin, *J. Am. Chem. Soc.*, 2015, **137**, 2796; (*c*) J. Y. Lee and K. D. Karlin, *Curr. Opin. Chem. Biol.*, 2015, **25**, 184.

11 M. Bhadra, W. J. Transue, H. Lim, R. E. Cowley, J. Y. C. Lee, M. A. Siegler, P. Josephs, G. Henkel, M. Lerch, S. Schindler, *et al.*, *J. Am. Chem. Soc.*, 2021, **143**, 3707.

12 M. Nappa, J. S. Valentine, A. Miksztal, H. J. Schugar and S. S. Isied, *J. Am. Chem. Soc.*, 1979, **101**, 7744.

13 T. Hoppe, S. Schaub, J. Becker, C. Würtele and S. Schindler, *Angew. Chem., Int. Ed.*, 2013, **52**, 870.

14 (*a*) X. Liang and P. J. Sadler, *Chem. Soc. Rev.*, 2004, **33**, 246; (*b*) M. Studer and T. A. Kaden, *Helv. Chim. Acta*, 1986, **69**, 2081.

15 T. L. Walker, W. Malasi, S. Bhide, T. Parker, D. Zhang, A. Freedman, J. M. Modarelli, J. T. Engle, C. J. Ziegler, P. Custer, *et al.*, *Tetrahedron Lett.*, 2012, **53**, 6548.

16 S. Schindler, Unpublishedresults.

17 I. Garcia-Bosch, R. E. Cowley, D. E. Díaz, M. A. Siegler, W. Nam, E. I. Solomon and K. D. Karlin, *Chem.–Eur. J.*, 2016, **22**, 5133.

18 (*a*) C. A. Grapperhaus, B. Mienert, E. Bill, T. Weyhermüller and K. Wieghardt, *Inorg. Chem.*, 2000, **39**, 5306; (*b*) O. Pestovsky, S. Stoian, E. L. Bominaar, X. Shan, E. Münck, L. Que and A. Bakac, *Angew. Chem.*, 2005, **117**, 7031; (*c*) S. Schaub, A. Miska, J. Becker, S. Zahn, D. Mollenhauer, S. Sakshath, V. Schünemann and S. Schindler, *Angew. Chem.*, 2018, **130**, 5453.

19 R. A. Baglia, J. P. T. Zaragoza and D. P. Goldberg, *Chem. Rev.*, 2017, **117**, 13320.

20 E. R. Dockal, L. L. Diaddario, M. D. Glick and D. B. Rorabacher, *J. Am. Chem. Soc.*, 1977, **99**, 4530.

21 M. M. Bernardo, M. J. Heeg, R. R. Schroeder, L. A. Ochrymowycz and D. B. Rorabacher, *Inorg. Chem.*, 1992, **31**, 191.

22 (*a*) S. Park, S. Y. Lee, K.-M. Park and S. S. Lee, *Acc. Chem. Res.*, 2012, **45**, 391; (*b*) M. Heller, *Z. Anorg. Allg. Chem.*, 2006, **632**, 441.

23 T. L. Walker, S. Mula, W. Malasi, J. T. Engle, C. J. Ziegler, A. van der Est, J. Modarelli and M. J. Taschner, *Dalton Trans.*, 2015, **44**, 20200.

24 M. D. Glick, D. P. Gavel, L. L. Diaddario and D. B. Rorabacher, *Inorg. Chem.*, 1976, **15**, 1190.

25 W. Schnick and M. Jansen, *Angew. Chem., Int. Ed. Engl.*, 1985, **24**, 54.

26 D. L. Hughes, M. Jimenez-Tenorio, G. J. Leigh, A. Houlton and J. Silver, *J. Chem. Soc., Dalton Trans.*, 1992, 2033.

27 J. F. Endicott, J. Lilie, J. M. Kuszaj, B. S. Ramaswamy, W. G. Schmonsees, M. G. Simic, M. D. Glick and D. P. Rillema, *J. Am. Chem. Soc.*, 1977, **99**, 429.

28 L. Iffland, D. Siegmund and U.-P. Apfel, *Z. Anorg. Allg. Chem.*, 2020, **646**, 746.

29 P. Gerschel, K. Warm, E. R. Farquhar, U. Englert, M. L. Reback, D. Siegmund, K. Ray and U.-P. Apfel, *Dalton Trans.*, 2019, **48**, 5923.

30 M. Wern, T. Hoppe, J. Becker, S. Zahn, D. Mollenhauer and S. Schindler, *Eur. J. Inorg. Chem.*, 2016, **2016**, 3384.

Faraday Discussions

ROYAL SOCIETY OF CHEMISTRY

PAPER

Phenolate-bonded bis(μ-oxido)-bis-copper(III) intermediates: hydroxylation and dehalogenation reactivities†

Peng Kang,[a] Bo-Lin Lin,[a] Tao A. G. Large,[a] Jasper Ainsworth,[a] Erik C. Wasinger[b] and T. Daniel P. Stack*[a]

Received 16th October 2021, Accepted 2nd November 2021
DOI: 10.1039/d1fd00071c

Exogenous phenolate *ortho*-hydroxylation by copper oxidants formed from dioxygen is generally thought to occur through one of two limiting mechanisms defined by the structure of the active oxidant: an electrophilic μ-η^2:η^2-peroxo-bis-copper(II) species as found in the oxygenated form of the binuclear copper enzyme tyrosinase (oxyTyr), or an isomeric bis(μ-oxido)-bis-copper(III) species (**O**) with ligated phenolate(s) as evidenced by most synthetic systems. The characterization of the latter is limited due to their limited thermal stability. This study expands the scope of an **O** species with ligated phenolate(s) using *N,N′*-di-*tert*-butyl-1,3-propanediamine (**DBPD**), a flexible secondary diamine ligand. Oxygenation of the [(**DBPD**)Cu(I)]$^{1+}$ complex at low temperatures (*e.g.*, 153 K) forms a spectroscopically and structurally faithful model to oxyTyr, a side-on peroxide intermediate, which reacts with added phenolates to form a bis(μ-oxido)-bis-copper(III) species with ligated phenolates, designated as an **A** species. The proposed stoichiometry of **A** is best understood as possessing 2 rather than 1 bonded phenolate. Thermal decomposition of **A** results in regiospecific phenolate *ortho*-hydroxylation with the *ortho*-substituent as either a C–H or C–X (Cl, Br) group, though the halogen displacement is significantly slower. DFT and experimental studies support an electrophilic attack of an oxide ligand into the π-system of a ligated phenolate. This study supports a hydroxylation mechanism in which O–O bond cleavage of the initially formed peroxide by phenolate ligation, which precedes phenolate aromatic hydroxylation.

[a]*Department of Chemistry, Stanford University, Stanford, California 94305, USA*
[b]*Department of Chemistry and Biochemistry, California State University, Chico, California 95929, USA. E-mail: stack@stanford.edu*

† Electronic supplementary information (ESI) available: Full reference of Gaussian 03 and Gaussian 09, fit of the titration of $^{S}\mathbf{P}^{\mathbf{DBPD}}$ with phenolate **1**, UV-Vis spectra of $^{m}\mathbf{A}^{\mathbf{DBPD}}$ (**m** = **6**, **7**, **9**, **10**), X-ray absorption spectroscopy of $^{S}\mathbf{P}^{\mathbf{DBPD}}$ and $^{1}\mathbf{A}_2{}^{\mathbf{DBPD}}$, product analyses of selected catechols, NTO analyses and Cartesian coordinates of DFT optimized structures. See DOI: 10.1039/d1fd00071c

Introduction

Tyrosinases (Tyr) are a binuclear copper enzyme family capable of hydroxylating phenols to catechols and catechols to quinones using an oxidant created from dioxygen and Cu(I).[1,2] Oxygenated Tyr (oxyTyr) has a spectroscopically and structurally characterized μ-η^2:η^2-peroxo-bis-copper(II) active site (^{S}**P** or a side-on peroxide), which is generally accepted as the active electrophilic oxidant in the hydroxylation reaction.[3–6] Surprisingly, several synthetic ^{S}**P** species are known to exist in a facile equilibrium with their valence-tautomeric form, a bis(μ-oxido)-bis-copper(III) (**O**) species,[7–13] in line with small free energy differences between the two isomeric species under certain conditions.[14–16] The small kinetic and small thermodynamic differences suggest an alternative oxidation mechanism of Tyr in which the ^{S}**P** species may isomerize to its **O** isomer upon phenolate ligation, and the nascent **A** species serves as the active electrophilic oxidant in the phenolate hydroxylation. This mechanism differs fundamentally from the generally accepted mechanism of tyrosinases in which the O–O bond cleavage precedes the C–O bond formation in the hydroxylation reaction.[4,5]

Hydroxylation of exogenous phenolates by synthetic complexes has been attributed to both ^{S}**P** [17–20] and **O** isomers.[21–23] Since the ^{S}**P** $\rightleftharpoons$ **O** equilibrium position is set largely by the copper coordination environment, most analyses have focused solely on the influence of the nitrogen ligation, excluding the role of phenolate ligation on the reaction mechanism. One would anticipate that direct phenolate ligation to a copper center could influence a ^{S}**P** $\rightleftharpoons$ **O** equilibrium position, given the strong electron-donation properties of an anionic phenolate relative to a neutral nitrogen-based ligand. Experimentally such influences are documented.[5,21,22,24]

Characterization of most synthetic ^{S}**P** or **O** species that react with phenolate to form a characterizable intermediate is challenging due to their extreme thermal instability. Indeed, only four phenolate-bonded ^{S}**P** or **O** systems have been characterized spectroscopically, and all possess a bis(μ-oxido)-bis-copper(III) core.[5,25] In 2005, we reported that the Cu(I) complex of *N*,*N*′-di-*tert*-butyl-ethylenediamine (**DBED**) oxygenates at 193 K to form a spectroscopically and structurally faithful side-on peroxide species to oxyTyr, ^{S}**P**$^{\mathbf{DBED}}$, which isomerizes to a bis(μ-oxido)-bis-copper(III) core, **A**$^{\mathbf{DBED}}$, with equatorial ligation of phenolate(s); it is postulated that equatorial phenolate ligation triggers the Cu_2O_2 core isomerization.[21] This species is most simply understood as the active oxidant in the hydroxylation reaction. Company *et al.* also reported reaction of an **O** species, formed using a xylyl-linked *m*-XYL^{MeAN} ligand, with added phenolates through the formation of an intermediate, which thermally decays to hydroxylate the phenolates.[23,24] Two additional **A** species have been characterized most recently: one using a tridentate alkylamine ligand,[5] and one using imidazole nitrogenous ligation; key in the stabilization of this latter **A** species was the use of unhindered monodentate imidazoles and extreme low temperatures of 130 K.[26] In all these cases, the optical properties of the formed intermediate are similar, and more than 2 equivalents of phenolates to the dimeric copper complex maximized its optical formation. Whether this corresponds to a 1 : 1 phenolate : Cu_2O_2 core equilibrium that is shifted by excess phenolate, or to a 2 : 1 phenolate : Cu_2O_2 core composition is difficult to discern given the thermal instabilities as noted in the first publication.

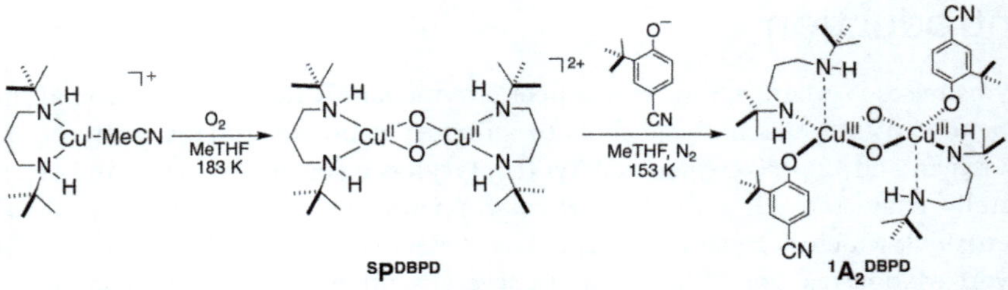

Scheme 1 Proposed formation of $^{S}P^{DBPD}$ and $^{1}A_2^{DBPD}$ species.

However, all subsequent computational papers have assumed a 1 : 1 composition.[22,27,28]

Herein, we report a new diamine system, *N*,*N*′-di-*tert*-butyl-1,3-propanediamine (**DBPD**), that creates a ^{S}**P** species upon oxygenation of its Cu(I) complex (Scheme 1). This ^{S}**P**^{DBPD} species reacts with a wide variety of phenolates to form spectroscopically congruent species, designated as ^{m}**A**$_n^{DBPD}$ where **m** represents the phenolate and ***n*** the number of phenolates ligated to the Cu_2O_2 core. The spectroscopic characterization of this family of intermediates provides insight not only into the composition of ^{m}**A**$_n^{DBPD}$ but also probes its structure through correlations to its hydroxylation reactivity.

Results

Optical characterization of ^{S}P^{DBPD}

Oxygenation of a solution of [(**DBPD**)Cu(I)(MeCN)](CF_3SO_3) in aprotic solvent (MeTHF) at 183 K (Scheme 1) forms an intensely blue species with characteristic

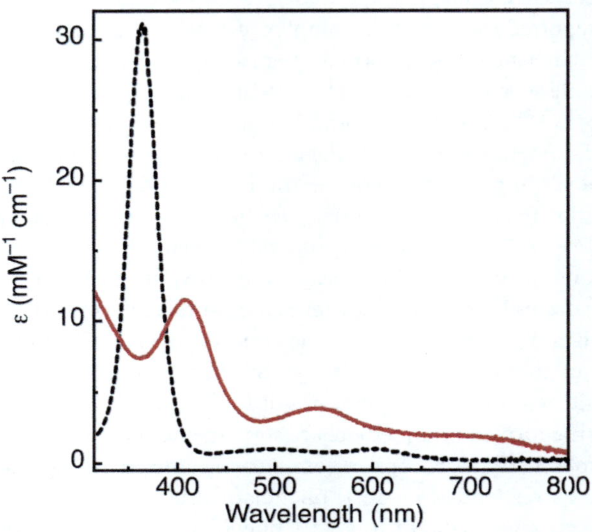

Fig. 1 UV-Vis spectra of $^{S}P^{DBPD}$ (dashed line) and $^{1}A_2^{DBPD}$ species (solid line, 3 equivalents of 2-*tert*-butyl-4-cyanophenolate (1) added; 1 mM Cu, $CF_3SO_3^-$, MeTHF, 153 K).

optical absorption bands of a side-on peroxide species, designated $^{\mathbf{S}}\mathbf{P}^{\mathbf{DBPD}}$. This complex shows an intense 365 nm band (ε = 31 mM^{-1} cm^{-1}) and two lesser intensity visible bands at 550 nm (ε = 1 mM^{-1} cm^{-1}) and 605 nm (ε = 0.95 mM^{-1} cm^{-1}) (Fig. 1). As the oxygenation reaction is rapid ($t_{1/2}$ < 2 s) under conditions of saturating O_2, the formation kinetics are inaccessible by conventional mixing methods at 183 K. $^{\mathbf{S}}\mathbf{P}^{\mathbf{DBPD}}$ does not decompose upon removal of the excess O_2 by purging with N_2 or evacuation under vacuum, indicating that the binding of O_2 is effectively irreversible at these temperatures.

Intense optical absorptions of nitrogenously-ligated $^{\mathbf{S}}\mathbf{P}$ species typically appear in the range of 335–365 nm with extinction coefficients ranging from 15 to 30 mM^{-1} cm^{-1}.[7] This band of $^{\mathbf{S}}\mathbf{P}^{\mathbf{DBPD}}$ at 365 nm is red-shifted compared to the 350 nm feature of the closely related, well-characterized $^{\mathbf{S}}\mathbf{P}^{\mathbf{DBED}}$ species, both of which ligate through secondary amines.[29] This feature is assigned as an O_2^{2-} π^*_σ → Cu d_{xy} ligand-to-metal charge transfer (LMCT) band.[30–32] Also characteristic of a $^{\mathbf{S}}\mathbf{P}$ species are the less intense visible 550 and 605 nm features assigned as O_2^{2-} π^*_v → Cu d_{xy} LMCT bands.[33] Cu K-edge X-ray absorption spectroscopy also supports a $^{\mathbf{S}}\mathbf{P}$ species with a Cu–Cu separation of 3.56 Å and a pre-edge feature consistent with a Cu(II) center (*vide infra*).

Thermal stability of $^{\mathbf{S}}\mathbf{P}^{\mathbf{DBPD}}$ and alcohol oxidation

UV-Vis spectroscopy indicates that the $^{\mathbf{S}}\mathbf{P}^{\mathbf{DBPD}}$ is stable at 183 K (1 mM Cu, THF, $CF_3SO_3^-$) for several days, but decomposes at higher temperatures through a single exponential process with a rate constant of 7.2×10^{-4} s^{-1} at 233 K. This decay rate is nearly 5-fold faster than $^{\mathbf{S}}\mathbf{P}^{\mathbf{DBED}}$ (1.6×10^{-4} s^{-1}, 231 K, THF).[34] An Eyring analysis affords the activation parameters of $\Delta H^\ddagger$ of 9.2(9) kcal mol^{-1} and $\Delta S^\ddagger$ of −33(6) cal K^{-1} mol^{-1} (Fig. 2). GC product analysis recovers *ca.* 80% of the intact **DBPD** ligand from the reaction without any other significant products. As ligand hydroxylation is

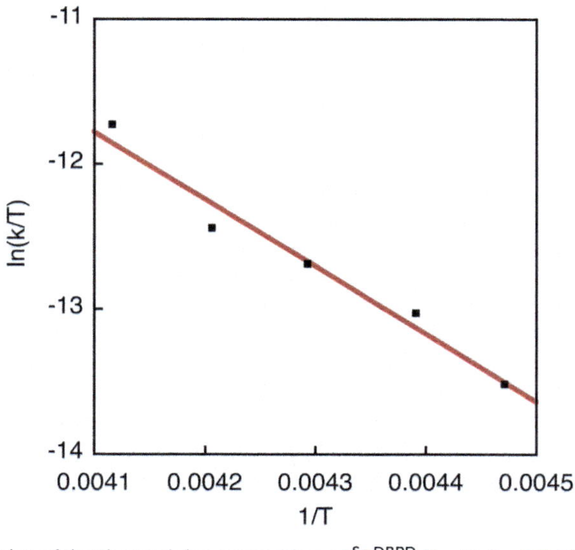

Fig. 2 Eyring plot of the thermal decomposition of $^{\mathrm{S}}\mathrm{P}^{\mathrm{DBPD}}$ (1 mM Cu, $CF_3SO_3^-$, THF, 223–243 K).

apparently not involved in the predominant decay pathway, the thermal decomposition pathway is currently unknown. However, **$^{S}P^{DBPD}$** also decays in the presence of two equivalents of benzylic or aliphatic alcohols over a 12 hour period (1 mM Cu, THF, 193 K). Benzyl alcohol and octanol are oxidized to their aldehydes in 80% and 40% yields, respectively, based on oxidant.

$^{S}P^{DBPD}$ to A^{DBPD} conversion

Addition of the sodium salt of 2-*tert*-butyl-4-cyanophenolate (**1**, Scheme 1) to **$^{S}P^{DBPD}$** (1 mM Cu, under N_2, MeTHF, 153 K) results in formation of a new species, **$^{1}A^{DBPD}$**, optically distinct from **$^{S}P^{DBPD}$** (Fig. 1). Titration of **$^{S}P^{DBPD}$** with phenolate **1** is followed readily through the disappearance of its 365 nm feature and the appearance of a 410 nm band (ε = 11.5 mM^{-1} cm^{-1}) along with more intense visible absorptions at 550 nm (ε = 3.9 mM^{-1} cm^{-1}) and 700 nm (ε = 1.8 mM^{-1} cm^{-1}) (Fig. 3). Each step in the **$^{S}P^{DBPD}$** to **$^{1}A^{DBPD}$** titration is rapid ($t_{1/2}$ < 5 seconds) and slightly more than 2 equivalents of the phenolate are required to complete the conversion; beyond two equivalents the overall spectrum in the 350–1000 nm range does not change significantly and the relative ratios of the UV-Vis bands are invariant to limited temperature changes.

Almost every phenolate examined (Scheme 2) reacts with **$^{S}P^{DBPD}$** to form an **A**-like species. In each case, the optical features of the corresponding **$^{m}A^{DBPD}$** (**m** = phenolate **1–14**) species are maximized with slightly more than 2 equivalents of phenolate. However, the extinction coefficients of **$^{m}A^{DBPD}$** species vary; more electron deficient phenolates result in greater absorbances at comparable copper concentrations. For example, the apparent optical yield of **$^{13}A^{DBPD}$** and **$^{14}A^{DBPD}$** are comparable, whereas that of **$^{5}A^{DBPD}$** is much less and decays at a much faster rate at 153 K.

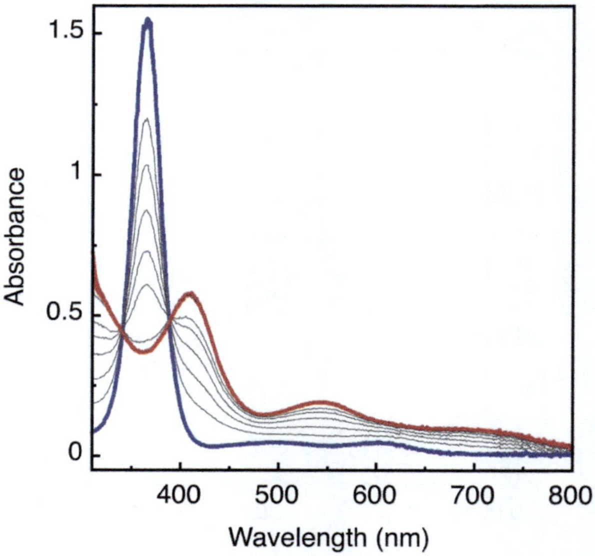

Fig. 3 Optical titration of $^{S}P^{DBPD}$ (blue) with three equivalents of phenolate 1 to form $^{1}A^{DBPD}$ (red) (1 mM Cu, $CF_3SO_3^-$, MeTHF, 153 K).

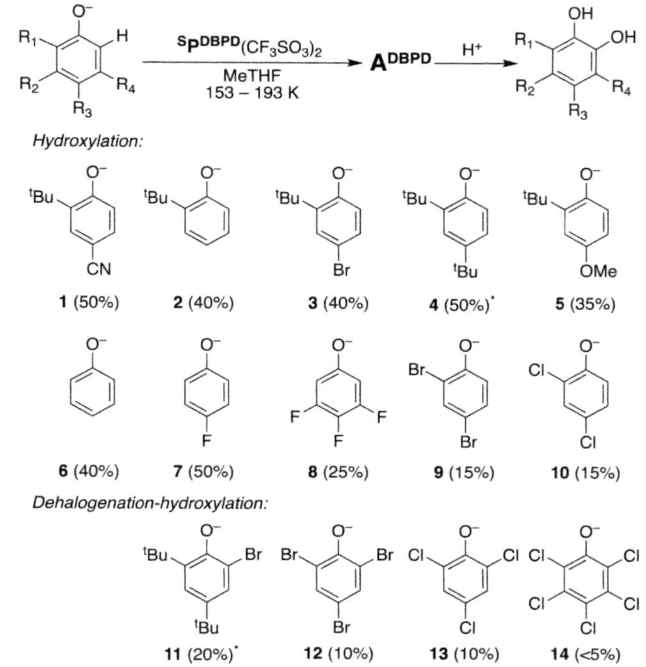

Scheme 2 Phenolates (two equivalents) for $^{m}A^{DBPD}$ (**m** = phenolate **1–14**) formation from $^{S}P^{DBPD}$, and the yield based on oxidant of hydroxylation or dehalogenation determined by ^{1}H-NMR and/or LC-MS relative to $^{S}P^{DBPD}$, assuming $^{m}A^{DBPD}$ as a 2e^{-} oxidant (*: a nearly equimolar mixture of corresponding catechol and quinone; this yield is based on the phenolate hydroxylated).

UV-vis spectra of $^{m}A^{DBPD}$ species

The UV-Vis optical features of the family of $^{m}A^{DBPD}$ species (**m** = phenolate **1–14**) possess similar absorption features: an intense band in the range 360 to 410 nm

Table 1 Selected UV-Vis bands (λ, nm; ε, mM^{-1} cm^{-1}) of $^{m}A^{DBPD}$[a]

Phenolate	UV	Vis-1	Vis-2
1	407 (11.5)	540 (3.9)	715 (1.8)
2	407 (9.3)	555 (4.0)	713 (2.8)
3	405 (10.6)	555 (5.1)	690 (3.9)
4	405 (9.0)	575 (4.3)	685 (5.1)
5[b]	409 (4.3)	540 (0.72)	730 (0.74)
6	390 (10.3)	545 (5.1)	670 (2.1)
7	385 (10.3)	545 (6.4)	673 (2.7)
8	379 (6.1)	514 (4.0)	663 (0.7)
9	360 (10.0)	440 (2.4)	555 (10.9)
10	367 (11.1)	441 (3.1)	553 (10.9)
11	387 (7.0)	595 (3.8)	685 (4.0)
12	358 (9.8)	432 (3.3)	555 (13.4)
13	355 (9.9)	432 (3.8)	548 (8.5)
14	368 (11.2)	435 (4.7)	540 (11.3)

[a] Conditions: 1 mM Cu, 3 equivalents of phenolate, $CF_3SO_3^{-}$, MeTHF, 153 K. [b] Incomplete formation due to thermal instability.

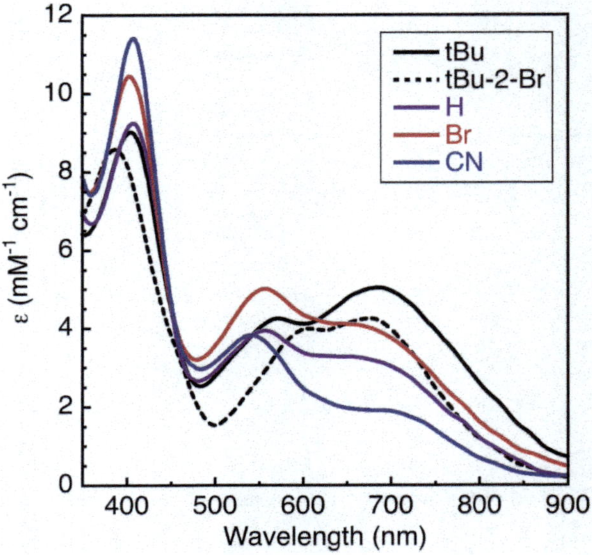

Fig. 4 UV-Vis spectra of ^{m}A^{DBPD} species: m = **1–3**, **4** and **11** (1 mM Cu, 3 equivalents of phenolate, $CF_3SO_3^-$, MeTHF, 153 K).

(ε = 7–11 mM^{-1} cm^{-1}), and additional features in the visible range from 500 to 800 nm (ε = 4–13 mM^{-1} cm^{-1}) (Table 1, Fig. 4). These features correlate well with other characterized phenolate-bonded species possessing a bis-Cu(III)-bis-oxide core.[21,23] The high energy features are analogous to the characteristic 400 nm transition of a classic bis-Cu(III)-bis-oxide species attributable to an oxide $\sigma^* \rightarrow$ Cu d_{xy} LMCT band.[35] The visible bands are attributable to phenolate out-of-plane $\pi_{op} \rightarrow$ Cu(III) d_{xy} LMCT by analogy to the previous assignment of the **A** species formed with the **DBED** ligand.[5,21,22]

The energy and intensity of the three major optical bands of **^{m}A^{DBPD}** are subject to phenolate variations, providing clear evidence of strong, direct phenolate ligation to the copper centers. Among phenolates **1–4** that share identical *ortho* substituents (^{t}Bu, H) but different *para*-substituents, each form an **^{m}A^{DBPD}** species with nearly identical *ca.* 400 nm features, but distinct visible features. Electron-deficient phenolates (*e.g.*, **9**, **10** and **12–14**) form **^{m}A^{DBPD}** species with UV and visible bands significantly blue-shifted as compared to more electron-rich phenolates (Table 1).

Titration of ^{S}P^{DBPD} to A^{DBPD}

An optical titration of **^{S}P^{DBPD}** with phenolate **1** at 153 K provides insight into the phenolate stoichiometry in **^{1}A^{DBPD}** species (Fig. 3). The clean isosbestic behavior at *ca.* 400 nm suggests only two dominant absorbing species. Indeed, a full wavelength analysis assuming only two colored species produces two significant eigenvectors; the assumption of three colored species leads to one physically unreasonable projected absorption spectrum. An alternative approach to this spectral analysis is to consider the variation of the concentration of **^{S}P^{DBPD}** ([P]) *versus* the total concentration of phenolate added in the titration (C_{PhO}) (Fig. 5).

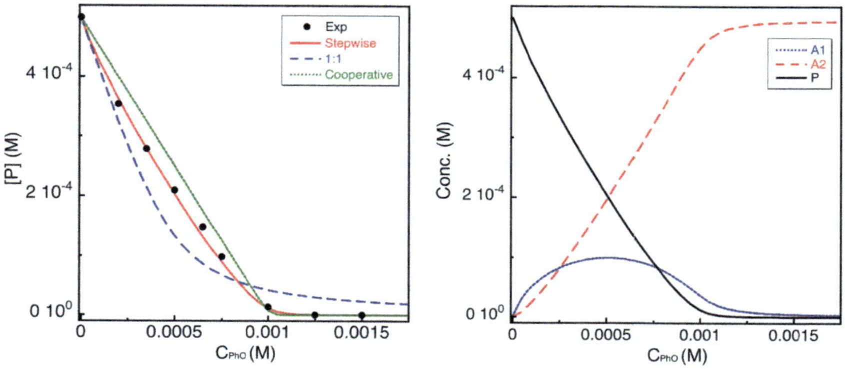

Fig. 5 (Left) Fit of the optical titration of $^{S}P^{DBPD}$ with a 1 : 1 model (dashed line), cooperative 2 : 1 model (dotted line), stepwise 2 : 1 model (solid line) and experimental data points (solid circles). (Right) Simulated speciation of $^{S}P^{DBPD}$ (solid line), $^{1}A_1^{DBPD}$ (dotted line) and $^{1}A_2^{DBPD}$ (dashed line) species in the stepwise model (one equivalent of phenolate = 0.0005 M).

[P] is determined readily by Gaussian decomposition of each spectrum along with the measured extinction coefficient of $^{\mathbf{S}}\mathbf{P}^{\mathbf{DBPD}}$. Three models were assessed to fit these data: a 1 : 1 phenolate : $^{\mathbf{S}}\mathbf{P}^{\mathbf{DBPD}}$ model forming $^{\mathbf{1}}\mathbf{A_1}^{\mathbf{DBPD}}$, a 2 : 1 phenolate : $^{\mathbf{S}}\mathbf{P}^{\mathbf{DBPD}}$ cooperative model forming $^{\mathbf{1}}\mathbf{A_2}^{\mathbf{DBPD}}$, and a stepwise model with both $^{\mathbf{1}}\mathbf{A_1}^{\mathbf{DBPD}}$ and $^{\mathbf{1}}\mathbf{A_2}^{\mathbf{DBPD}}$ species (Scheme 3).

As *ca.* two equivalents of phenolate are necessary to maximize the 408 nm band of the $^{\mathbf{1}}\mathbf{A}^{\mathbf{DBPD}}$ species, the 1:1 model with one phenolate ligated to a bis(μ-oxido)-bis-copper(III) core assumes that excess phenolate is required to bias an equilibrium towards $^{\mathbf{1}}\mathbf{A_1}^{\mathbf{DBPD}}$. Alternatively, strong binding of two phenolates in a cooperative manner to a bis(μ-oxido)-bis-copper(III) core to form $^{\mathbf{1}}\mathbf{A_2}^{\mathbf{DBPD}}$ would be consistent with the measured stoichiometry of the titration. Finally, a 2 : 1 stepwise model assumes that the ultimate formation of $^{\mathbf{1}}\mathbf{A_2}^{\mathbf{DBPD}}$ occurs through the intermediacy of $^{\mathbf{1}}\mathbf{A_1}^{\mathbf{DBPD}}$. The mathematical relationships between the total concentration of the copper dimeric species (C_O), the equilibrium concentration of $^{\mathbf{S}}\mathbf{P}^{\mathbf{DBPD}}$ ([P]), the total concentration of added phenolate (C_{PhO}), and the cumulative stability constants β_1 for $^{\mathbf{1}}\mathbf{A_1}^{\mathbf{DBPD}}$ and β_2 for $^{\mathbf{1}}\mathbf{A_2}^{\mathbf{DBPD}}$, are derived for each model. In the limit of $\beta_2 \rightarrow 0$, eqn (3) (Scheme 3) for the stepwise model, converges with eqn (1) for the 1 : 1 model and in the limit of $\beta_1 \rightarrow 0$, eqn (3) converges with eqn (2) for the cooperative model as anticipated.

Of the three models, the stepwise model agrees most closely with the experimental titration data (Fig. 5, left) albeit this fit has one additional degree of freedom compared to the cooperative 1 : 1 and 2 : 1 models. The fit of the 1 : 1 model deviates from the experimental data significantly throughout the titration ($\log \beta_1 = 3.8(2)$). The cooperative 2 : 1 model fits the rate of change in the concentration of $^{\mathbf{S}}\mathbf{P}^{\mathbf{DBPD}}$ comparably well, except that it systematically overestimates the value of [P] ($\log \beta_2 = 11.5(7)$). The stepwise model best fits the data in that the added equilibrium allows the over estimation of [P] in the cooperative 2 : 1 model to be attenuated. The converged cumulative stability constants of $^{\mathbf{1}}\mathbf{A_1}^{\mathbf{DBPD}}$ and $^{\mathbf{1}}\mathbf{A_2}^{\mathbf{DBPD}}$ in the stepwise model are $\log \beta_1 = 4.7(3)$ and $\log \beta_2 = 10.0(5)$, respectively, which predict that the $^{\mathbf{1}}\mathbf{A_1}^{\mathbf{DBPD}}$ species never exceeds 20% of the total

1:1 model

$$\mathbf{P} + \text{PhO} \rightleftharpoons \mathbf{A}_1 \qquad K_1 = \beta_1$$

$$\beta_1 = \frac{[\boldsymbol{A_1}]}{[\boldsymbol{P}][PhO]}$$

$$C_0 = [\boldsymbol{P}] + [\boldsymbol{A_1}]$$

$$C_{PhO} = [PhO] + [\boldsymbol{A_1}]$$

$$C_{PhO} = \frac{(C_0 - [\boldsymbol{P}])(\beta_1[\boldsymbol{P}] + 1)}{\beta_1[\boldsymbol{P}]} \qquad (1)$$

Cooperative 2:1 model

$$\mathbf{P} + 2\ \text{PhO} \rightleftharpoons \mathbf{A}_2 \qquad K_1 K_2 = \beta_2$$

$$\beta_2 = \frac{[\boldsymbol{A_2}]}{[\boldsymbol{P}][PhO]^2}$$

$$C_0 = [\boldsymbol{P}] + [\boldsymbol{A_2}]$$

$$C_{PhO} = [PhO] + 2[A_2]$$

$$C_{PhO} = 2(C_0 - [\boldsymbol{P}]) + \sqrt{\frac{C_0 - [\boldsymbol{P}]}{\beta_2[\boldsymbol{P}]}} \qquad (2)$$

Stepwise model

$$\mathbf{P} + \text{PhO} \rightleftharpoons \mathbf{A}_1 \qquad K_1 = \beta_1$$

$$\mathbf{A}_1 + \text{PhO} \rightleftharpoons \mathbf{A}_2 \qquad K_2 = \beta_2/\beta_1$$

$$\beta_1 = \frac{[\boldsymbol{A_1}]}{[P][PhO]},\ \beta_2 = \frac{[\boldsymbol{A_2}]}{[\boldsymbol{P}][PhO]^2}$$

$$C_0 = [\boldsymbol{P}] + [\boldsymbol{A_1}] + [\boldsymbol{A_2}]$$

$$C_{PhO} = [PhO] + [\boldsymbol{A_1}] + 2[\boldsymbol{A_2}]$$

$$C_{PhO} = 2(C_0 - [\boldsymbol{P}]) + \frac{(\beta_1[\boldsymbol{P}] - 1)\left(\beta_1 - \sqrt{\frac{[\boldsymbol{P}]\beta_1^2 + 4\beta_2(C_0 - [\boldsymbol{P}])}{[\boldsymbol{P}]}}\right)}{2\beta_2} \qquad (3)$$

Scheme 3 Proposed models for the formation of $^1\mathrm{A}^{\mathrm{DBPD}}$ species.

dimeric copper species in the titrations with 0–3 equivalents of **1** (Fig. 5, right). Throughout the remainder of this manuscript, all **A** species will be assumed to be 2 : 1 rather than 1 : 1 complexes.

X-ray absorption spectroscopy

X-ray absorption spectroscopy (XAS) provides further evidence of the conversion of $^{\mathrm{S}}\mathbf{P}^{\mathbf{DBPD}}$ to an $^1\mathbf{A}_2{}^{\mathbf{DBPD}}$ species. The energy of the XAS Cu K-edge varies with oxidation state with the more oxidized Cu species absorbing at higher energy.[36] Additionally, the Cu pre-edge has been shown to be a valuable tool for elucidating electronic structure, as the energy of 1s → 3d pre-edge features correlates to the Cu oxidation states.[36]

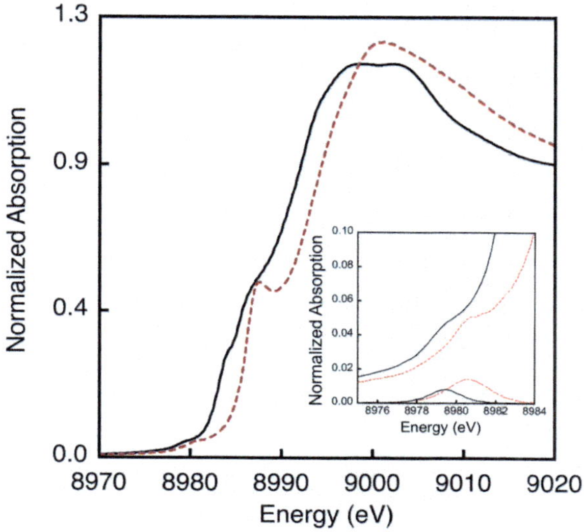

Fig. 6 XAS Cu K-edges and pre-edges (inset) for $^{S}P^{DBPD}$ (solid line) and $^{1}A_2^{DBPD}$ (dashed line, 4 mM Cu, 3 equivalents of phenolate **1**; $CF_3SO_3^-$, MeTHF). Gray = background fit.

Cu K-edge XAS is consistent with $^{1}\mathbf{A_2}^{\mathbf{DBPD}}$ as a bis(μ-oxido)-bis-copper(III) core through the characteristic Cu(III) pre-edge feature at 8980.6 eV in combination with the 2.79 Å Cu–Cu distance from the EXAFS model. The pre-edge features of $^{S}\mathbf{P}^{\mathbf{DBPD}}$ and $^{1}\mathbf{A_2}^{\mathbf{DBPD}}$ were each fitted with a single feature at 8979.4 eV and 8980.6 eV, respectively (Fig. 6). These energies fall in the range typical of Cu(II) and Cu(III) complexes of 8979 ± 0.5 eV and 8981 ± 0.5 eV, respectively. The Cu K-edge is shifted also to higher energy by *ca.* 1.5 eV relative to $^{S}\mathbf{P}^{\mathbf{DBPD}}$, in line with more oxidized Cu centers. The distinct change in Cu K-edge features suggests a predominant $^{S}\mathbf{P}^{\mathbf{DBPD}}$ to $^{1}\mathbf{A_2}^{\mathbf{DBPD}}$ conversion.

The EXAFS data for $^{S}\mathbf{P}^{\mathbf{DBPD}}$ (Fig. 7, left) were fitted well with six scattering contributions (Table 2). The first shell can be split into two waves representing the

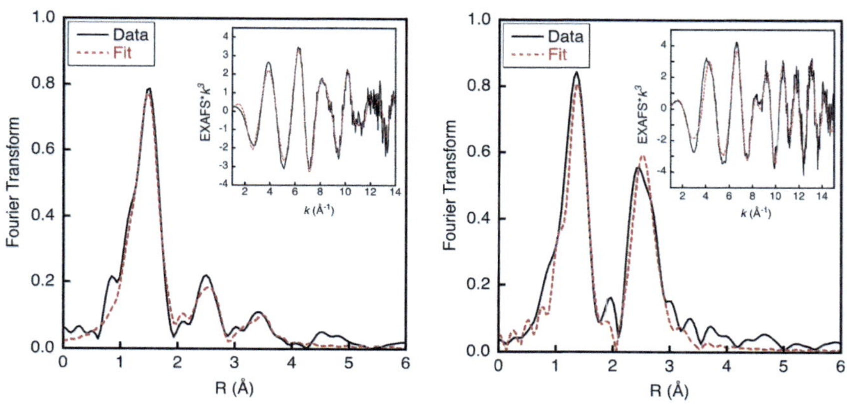

Fig. 7 EXAFS data (inset), Fourier transform data (solid line) and fits (dashed line) for $^{S}P^{DBPD}$ (left) and $^{1}A_2^{DBPD}$ (right) (3 equivalents of phenolate **1**, 4 mM Cu, $CF_3SO_3^-$, MeTHF).

Table 2 EXAFS fits of $^{S}P^{DBPD}$ and $^{1}A_2^{DBPD}$ [b]

Species	Wave	Distance (Å)	σ^2 (Å^2)	ΔE_0 (eV)	F [a]
$^{S}P^{DBPD}$	2× Cu–O	1.91	0.00280	−8.3	0.13
	2× Cu–N	2.03	0.00319		
	4× Cu–C	2.90	0.00823		
	8× Cu–C (MS)	3.03	0.00149		
	1× Cu–Cu	3.56	0.00549		
	4× Cu–Cu (MS)	3.69	0.00085		
$^{1}A_2^{DBPD}$	3× Cu–O	1.84	0.00389	−7.7	0.33
	1× Cu–N	1.99	0.00054[b]		
	1× Cu–Cu	2.79	0.00290		
	3× Cu–C	3.05	0.00201		

[a] Normalized error: $F = [\Sigma k^6(\chi_{expt} - \chi_{calc})^2/\Sigma k^6 \chi_{expt}{}^2]^{1/2}$. [b] Ref. 39.

2O and 2N atoms at 1.91 and 2.03 Å, respectively.[37] The 1.91 Å Cu–O contribution and the 3.56 Å Cu–Cu distance are consistent with other characterized ^{S}**P** species.[7] Additionally, four outer shells representing Cu–C and Cu–Cu single scattering and Cu–C and Cu–O–Cu multiple scattering complete the fit of $^{S}P^{DBPD}$.[38]

The EXAFS data of the $^{1}A_2^{DBPD}$ species (Fig. 7, right) were well fitted with four contributions: two from first shell N and O scatterers, one from a second shell Cu scatterer, and a final contribution from second shell carbon scatterers (Table 2). The first contribution is presumably attributable to the average of three Cu–O bonds at 1.84 Å: two O interactions from a $Cu_2(\mu_2\text{-}O)_2$ core and one O interaction from a ligated phenolate at each Cu center. The remaining first shell contribution from nitrogen is fitted at 1.99 Å. The remaining shells of Cu–Cu and Cu–C single scattering waves are at 2.79 and 3.05 Å, respectively. The 1.84 Å Cu–O and 2.79 Å Cu–Cu single scattering waves are consistent with a bis(μ-oxido)-bis-copper(III) core.[7]

Thermal decay of A^{DBPD}

Thermal decay of $^{1}A_2^{DBPD}$ assessed optically follows a first-order process at 153–183 K and is not influenced significantly by the presence of excess phenolate; the decay rate with 20 equivalents **1** to form $^{1}A_2^{DBPD}$ is only 5% slower than with 3 equivalents at 153 K. The activation parameters with three equivalents of phenolate are $\Delta H^{\ddagger} = 8.2(6)$ kcal mol^{-1} and $\Delta S^{\ddagger} = -25(4)$ cal K^{-1} mol^{-1} (Fig. 8, left). The small $\Delta H^{\ddagger}$ and large negative $\Delta S^{\ddagger}$ values are consistent with the activation parameters of phenolate hydroxylation of previously reported copper model complexes.[20,21,23,40,41] The decay of $^{1}A_2^{DBPD}$ (5.6×10^{-3} s^{-1}, 193 K) as compared to $^{1}A^{DBED}$ (2.3×10^{-3} s^{-1}, 193 K) is only two-fold faster, consistent with $^{1}A_2^{DBPD}$ and $^{1}A^{DBED}$ as equivalently reactive hydroxylating agents with a low free energy barrier of *ca.* 13 kcal mol^{-1} at 193 K.

The thermal stability of an A^{DBPD} species is influenced by the electron-donating ability of the ligated phenolate, with the more electron-deficient phenolate complexes being more stable.[23,34] A Hammett analysis of their decay rates formed with phenolates **1–4** at 153 K, yields a ρ value of −2.3 for a $\sigma_p{}^+$ parameter (Fig. 8, right). These values are consistent with the negative ρ values of −1.8 to −2.4 reported for phenolate hydroxylation by oxyTyr and model

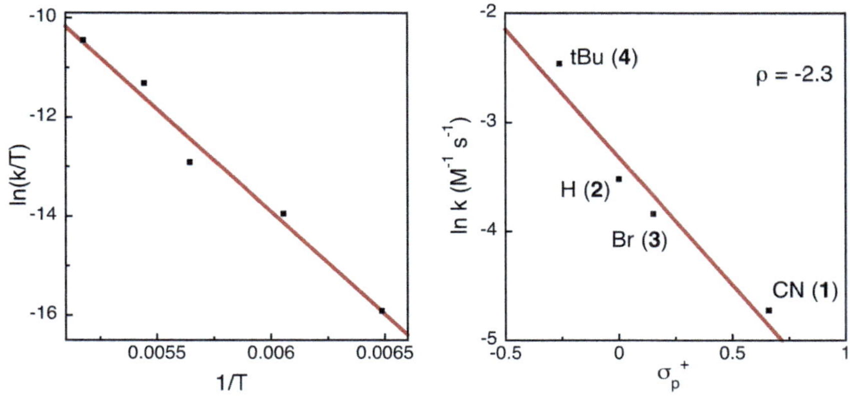

Fig. 8 Thermal decomposition of A^{DBPD} species (3 equivalents of phenolate, 1 mM Cu, $CF_3SO_3^-$, MeTHF): (left) Eyring plot for $^1A_2^{DBPD}$, 153–193 K; (right) Hammett plot with $^1A^{DBPD}$ – $^4A^{DBPD}$ species, 153 K.

complexes,[20,21,23,40,42,43] which supports an electrophilic aromatic hydroxylation mechanism.

With *ortho*-halogenated phenolates **11–14**, the $\mathbf{A^{DBPD}}$ decay is slowed dramatically compared to the *ortho*-hydrogenated phenolate counterparts. For example, the decay of $\mathbf{^{11}A_2^{DBPD}}$ ($t_{1/2} = 2000$ s, *ortho* C–H) at 183 K is nearly 10^3 times slower than $\mathbf{^4A_2^{DBPD}}$ ($t_{1/2} = 5$ s, *ortho* C–Br) under identical reaction conditions. The most electron-deficient phenolate in this study, phenolate **14**, affords the most thermally stable $\mathbf{^{14}A_2^{DBPD}}$ species with a $t_{1/2} = 2000$ s at 203 K.

Reactivity of $\mathbf{A^{DBPD}}$

$\mathbf{A^{DBPD}}$ species demonstrate efficient stoichiometric *ortho*-hydroxylation of phenolates near 200 K, with the corresponding catechols as the dominant product. A quinone product is observed only in the case of $\mathbf{^4A_2^{DBPD}}$ and $\mathbf{^{11}A_2^{DBPD}}$, with an equivalent amount of catechol; a similar quinone/catechol product mixture was found for $\mathbf{^4A^{DBED}}$, presumably due to the disproportionation of an uncharacterized semiquinone species during the acidic workup.[22,34] The hydroxylation yields correlate to phenolate electron-density. For example, hydroxylation yields of *ca.* 40–50% for electron-rich phenolates **1–7** are reduced to *ca.* 15% for the electron-deficient phenolates **8–10**. These yields are relative to the $\mathbf{^SP^{DBPD}}$ species formed, assuming a complete $\mathbf{^SP^{DBPD}}$ to $\mathbf{^mA_2^{DBPD}}$ conversion and $\mathbf{^mA_2^{DBPD}}$ as a two-electron oxidant. In addition, the hydroxylation is regioselective for the 1,2-dihydroxy product in all cases, suggesting that ligation of the phenolate in $\mathbf{A^{DBPD}}$ orientates the *ortho* position for electrophilic attack by the oxide ligands.

$\mathbf{A^{DBPD}}$ also performs dehalogenation-hydroxylation of *ortho*-dihalogenated phenolates such as **11–14**, with the formation of the corresponding catechol products. *Ortho*-defluorination of a phenol has been reported previously, however debromination and dechlorination by Cu(III)-containing **O** species is novel.[24,28] The dehalogenation–hydroxylation reaction is less facile than dehydrogenation–hydroxylation as evidenced by the intramolecular competition experiments with phenolates **9** and **10**. These phenolates are oxidized exclusively to halogenated catechols rather than a non-halogenated catechol. This stands in contrast to the reactivity of an *ortho*-fluoro substituent, which is preferentially oxidized relative to a C–H substituent.[24] Dehalogenations of poly-halogenated phenolates **12–14** are even slower and the yields of corresponding catechols are reduced, similar to the trend in the hydroxylation yields of phenolates **6–10**. No coupled phenol or X_2 (X = Cl or Br) products are observed.

DFT analysis

DFT is a powerful tool to obtain insight into the physical properties of $^{\mathbf{S}}\mathbf{P}^{\mathbf{DBPD}}$ and $\mathbf{A}^{\mathbf{DBPD}}$ and the subsequent phenolate hydroxylation mechanism. Correlations to experimental benchmarks aid in validating the choice of the functional and basis set.[22,27,28,44] Three $^{\mathbf{S}}\mathbf{P}^{\mathbf{DBPD}}$ conformers (Scheme 4, upper) were considered with the Cu-**DBPD** six-member rings in a boat, chair, and twist conformation. The boat conformation is lower in energy than the twist conformation by *ca.* 14 kcal mol^{-1} (B3LYP/BS1//B3LYP/6-31G(d), gas phase), while the chair conformation is not a minimum and optimizes to the boat conformation.

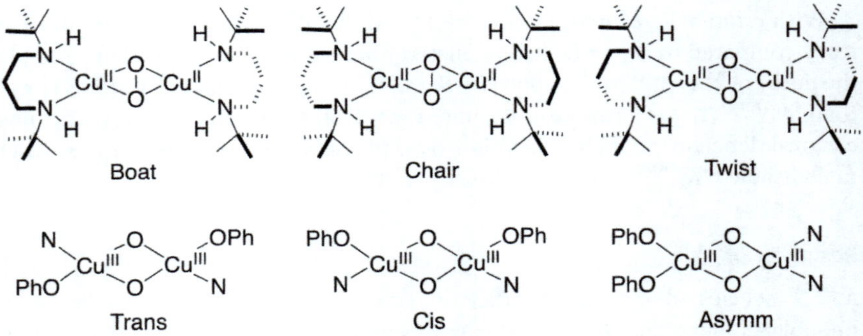

Scheme 4 Possible conformations of $^{S}P^{DBPD}$ (upper) and proposed phenolate coordination modes in $^{1}A_2{}^{DBPD}$ (lower).

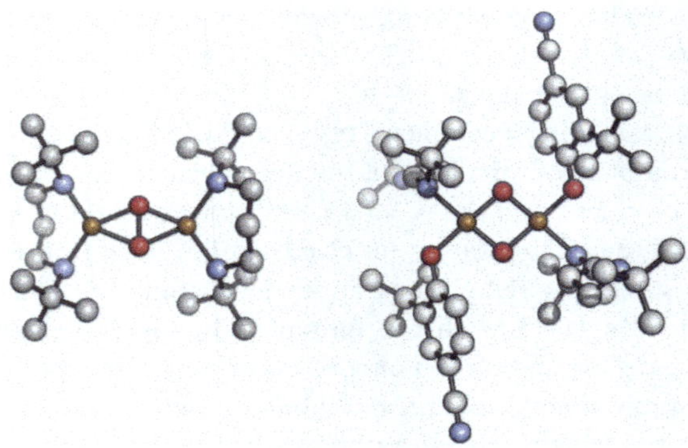

Fig. 9 DFT optimized geometries of $^{S}P^{DBPD}$ (left, boat conformation) and a proposed model of $^{1}A_{2\text{-}trans}{}^{DBPD}$ (right) (B3LYP/6-31G(d), gas phase).

Table 3 Selected experimental and calculated metrical parameters of $^{S}P^{DBPD}$ and $^{1}A^{DBPD}$

Species	Method	Cu···Cu (Å)	O···O (Å)	Cu–O (Å)	Cu–N (Å)
$^{S}\mathbf{P}^{\mathbf{DBPD}}$	EXAFS	3.56	—	1.91	2.03
	DFT	3.53	1.50	1.92	1.99
$^{1}\mathbf{A}_{2\text{-}trans}{}^{\mathbf{DBPD}}$	EXAFS	2.79	—	1.84	1.99
	DFT	2.72	2.31	1.81	1.98
$^{1}\mathbf{A}_{1}{}^{\mathbf{DBPD}}$	DFT	2.74	2.32	1.80	1.98

Three limiting phenolate configurations exist for $^{1}\mathbf{A}_{2}{}^{\mathbf{DBPD}}$ (Scheme 4, lower): *trans* phenolates on different copper centers (*trans*), *cis* phenolates on different copper centers (*cis*), and two phenolates on a single copper center without a **DBPD** ligand (*asymm*). With the *trans* $^{1}\mathbf{A}_{2\text{-}trans}{}^{\mathbf{DBPD}}$ lowest in energy, this conformation is assumed for all computational results as the bulky phenolates are best positioned away from each other. The optimized geometries of boat $^{S}\mathbf{P}^{\mathbf{DBPD}}$ and a *trans*-phenolate $^{1}\mathbf{A}_{2\text{-}trans}{}^{\mathbf{DBPD}}$ (Fig. 9, Table 3) reproduce the general experimental distances from their EXAFS analyses.

The free energy changes associated with $^{S}\mathbf{P}^{\mathbf{DBPD}} \rightleftharpoons \mathbf{O}^{\mathbf{DBPD}}$ equilibrium were computed for the dicationic $^{S}\mathbf{P}^{\mathbf{DBPD}}$ and $\mathbf{O}^{\mathbf{DBPD}}$ isomers, the monocationic $^{1S}\mathbf{P}_{1}{}^{\mathbf{DBPD}}$ and $^{1}\mathbf{A}_{1}{}^{\mathbf{DBPD}}$ isomers with one ligated phenolate, and the neutral $^{1S}\mathbf{P}_{2\text{-}trans}{}^{\mathbf{DBPD}}$ and $^{1}\mathbf{A}_{2\text{-}trans}{}^{\mathbf{DBPD}}$ isomers with two ligated *trans*-phenolates, respectively (Scheme 5). The free energy of each was corrected with a THF PCM solvation model and thermal contributions at 153 K. Without any phenolate ligation, the

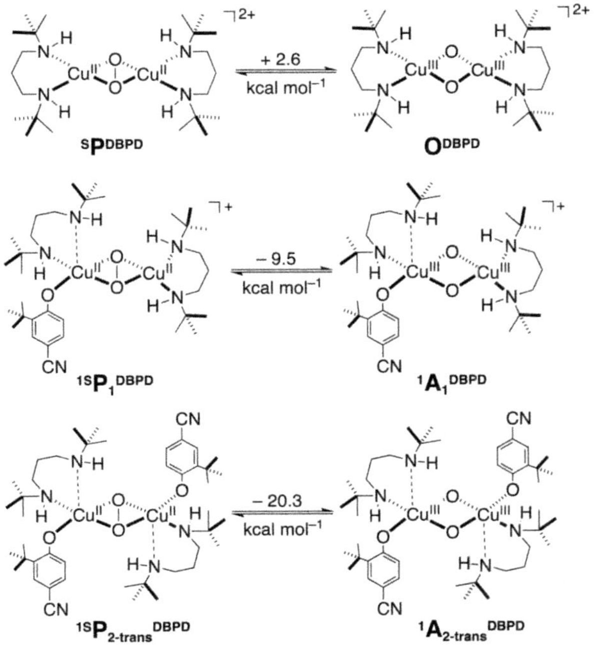

Scheme 5 Calculated free energy differences for **DBPD** copper species (PCM-BLYP/BS1//B3LYP/6-31G(d), 1 atm, 153 K).

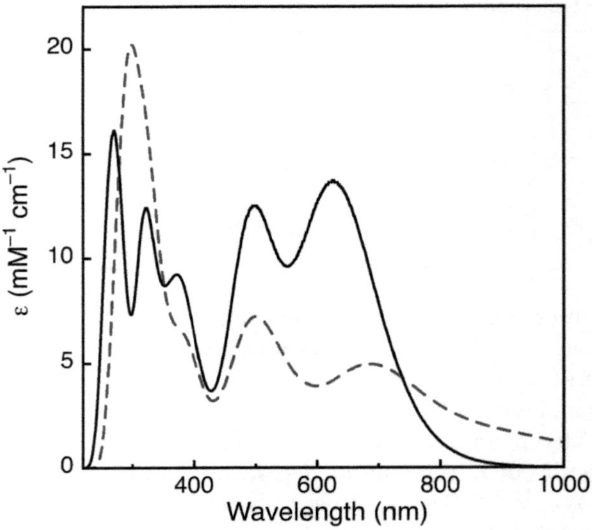

Fig. 10 Predicted UV-Vis spectra of $^1A_1{}^{DBPD}$ species (dashed line) and $^1A_{2\text{-}trans}{}^{DBPD}$ species (solid line) from TD-DFT (TD-B3LYP/BS1//B3LYP/6-31G(d), gas phase, FWHM = 3600 cm^{-1}).

$^{S}\mathbf{P}^{DBPD}$ isomer is 2.6 kcal mol^{-1} more thermodynamically stable than the $\mathbf{O}^{DBPD}$ isomer, consistent with the UV-Vis observation of an optically measurable amount of only $^{S}\mathbf{P}^{DBPD}$. Phenolate ligation in an equatorial position in the Cu_2O_2 plane, the energy-preferred geometry in all cases, biases the equilibrium to a bis(μ-oxido)-bis-copper(III) core such that $^1\mathbf{A}_1{}^{DBPD}$ is favored by *ca.* 10 kcal mol^{-1} (Scheme 5). $^1\mathbf{A}_{2\text{-}trans}{}^{DBPD}$ with two phenolates is *ca.* 20 kcal mol^{-1} more stabilized than the $^{1S}\mathbf{P}_{2\text{-}trans}{}^{DBPD}$ isomer.

TD-DFT calculated UV-Vis spectra for $^1\mathbf{A}_1{}^{DBPD}$ and $^1\mathbf{A}_{2\text{-}trans}{}^{DBPD}$ species (Fig. 10) reproduce the two experimentally observed visible bands and are qualitatively similar in the energies of the dominant transitions. The greatest differences occur in the relative oscillator strengths. The similarity of the spectra originates from the nearly equivalent donor and acceptor orbitals in both species (Fig. S1–S4†). The similarity of the optical spectra of $^1\mathbf{A}_1{}^{DBPD}$ and $^1\mathbf{A}_{2\text{-}trans}{}^{DBPD}$ poses a challenge for distinguishing these two species experimentally.

A natural transition orbital (NTO) analysis of the complicated state-to-state transitions predicted by TD-DFT calculations helps clarify the origin of these optical features.[45] The two features of the UV band predicted at 358 and 384 nm of $^1\mathbf{A}_{2\text{-}trans}{}^{DBPD}$ are each primarily composed of two transitions: O_2 σ^* → Cu d_{xy} LMCT and the phenolate π_{ip} (in-plane) → Cu d_{xy} LMCT. The two visible bands primarily consist of the phenolate π_{op} (out-of-plane) → Cu d_{xy} LMCT; the donor molecular orbitals are two different phenolate orbitals: HOMO (630 nm) and HOMO−2 (496 nm), respectively. As the predicted UV and visible bands commonly involve phenolate → Cu(III) LMCT character, significant perturbations on the spectra are anticipated with phenolate variation as observed experimentally.

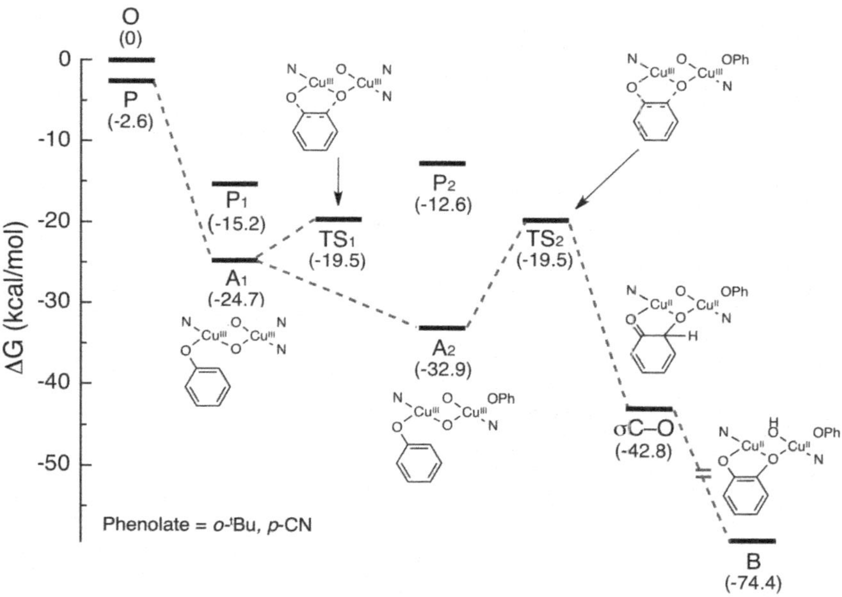

Fig. 11 DFT calculated relative free energies (kcal mol^{-1}) of $^{S}P^{DBPD}$ ⇌ O^{DBPD} and hydroxylation of phenolate 1 by $^{1}A_{1}^{DBPD}$ and $^{1}A_{2\text{-}trans}^{DBPD}$ species (PCM-BLYP/BS1//B3LYP/6-31G(d), 1 atm, 153 K). σ_{C-O}: sigma complex. B: catecholate–bis-copper(II) complex.

The free energy profile for the hydroxylation of phenolate **1** by $^{\mathbf{S}}\mathbf{P}^{\mathbf{DBPD}}$ is computed by DFT (Fig. 11). The $^{\mathbf{S}}\mathbf{P}^{\mathbf{DBPD}}$ species with one and two equatorially positioned phenolates, $^{\mathbf{1S}}\mathbf{P_1}^{\mathbf{DBPD}}$ and $^{\mathbf{1S}}\mathbf{P}_{\mathbf{2\text{-}}\boldsymbol{trans}}^{\mathbf{DBPD}}$, respectively, are significantly higher in energy than the isomeric $^{\mathbf{1}}\mathbf{A_1}^{\mathbf{DBPD}}$ and $^{\mathbf{1}}\mathbf{A}_{\mathbf{2\text{-}}\boldsymbol{trans}}^{\mathbf{DBPD}}$ species with equatorial-bonded phenolates. Transition states $\mathbf{TS_1}$ and $\mathbf{TS_2}$ reflect the *ortho*-hydroxylation pathway for $^{\mathbf{1}}\mathbf{A_1}^{\mathbf{DBPD}}$ and $^{\mathbf{1}}\mathbf{A}_{\mathbf{2\text{-}}\boldsymbol{trans}}^{\mathbf{DBPD}}$ species, respectively. Although $\mathbf{TS_2}$ is calculated to be *ca.* 5.6 kcal mol^{-1} lower than $\mathbf{TS_1}$ in enthalpy at 153 K, it is nearly isoenergetic with $\mathbf{TS_1}$ in free energy, suggesting either $^{\mathbf{1}}\mathbf{A_1}^{\mathbf{DBPD}}$ or $^{\mathbf{1}}\mathbf{A}_{\mathbf{2\text{-}}\boldsymbol{trans}}^{\mathbf{DBPD}}$ species are potential active oxidants responsible for phenolate hydroxylation.

Discussion

Phenolates bonded to a bis(μ-oxido)-bis-copper(III) core

Investigating synthetic complexes formed through the association of a phenolate to a μ-η^2:η^2-peroxo-bis-copper(II) species ($^{\mathbf{S}}\mathbf{P}$) or its isomeric bis(μ-oxido)-bis-copper(III) (**O**) is challenging due to their extreme thermal instability to either phenolate hydroxylation or other oxidative degradation pathways. In the present study, $^{\mathbf{S}}\mathbf{P}^{\mathbf{DBPD}}$ (**DBPD** = *N*,*N*′-di-*tert*-butyl-1,3-propanediamine), formed by oxygenation of [(**DBPD**)Cu(I)]$^{1+}$, isomerizes to a bis(μ-oxido)-bis-copper(III) complex upon ligation of added exogenous phenolates. These species, generically designated as $\mathbf{A}^{\mathbf{DBPD}}$, are formed with a variety of phenolates, from electron-rich to electron-poor with varying steric demands, providing the most extensive investigation of such species to date. Extreme solution temperatures of 153 K (−120 °C) – 193 K (−80 °C) suffice to slow $\mathbf{A}^{\mathbf{DBPD}}$ decay so that detailed spectroscopic and reactivity analyses are possible. The systematic variations of the steric and electronic properties of the phenolates provide an intricate probe into these $\mathbf{A}^{\mathbf{DBPD}}$ species.

The equilibrium position of the μ-η^2:η^2-peroxo-bis-copper(II) to bis(μ-oxido)-bis-copper(III) isomerization ($^{\mathbf{S}}\mathbf{P} \rightleftharpoons \mathbf{O}$) in Cu/O_2 chemistry is determined by the ligand.[7,8] In general, more donating ligands stabilize Cu(III) over Cu(II). Indeed, chelating peralkylamines,[46] diketiminates,[47] and guanidines,[48] all strongly basic nitrogen ligands, form measurable amounts only of Cu(III)-containing **O** species. **DBPD** is a chelating, secondary diamine ligand that is anticipated to be less donating and should be less capable of stabilizing the Cu(III) of an **O** core. Indeed, the initial oxygenation of $[(\mathbf{DBPD})Cu(I)]^{1+}$ at 193 K yields a side-on peroxide species, $^{\mathbf{S}}\mathbf{P}^{\mathbf{DBPD}}$, in the presence of weakly coordinating anions (SbF_6^- or $CF_3SO_3^-$); no evidence of its isomeric bis(μ-oxido)-bis-copper(III) (**O**) complex is spectroscopically observed. However, the energetic difference between these isomers is presumably small as evidenced by the closely related **DBED** ligand (**DBED** = *N*,*N*′-di-*tert*-butyl-1,2-ethanediamine), which creates an optically measurable $^{\mathbf{S}}\mathbf{P}^{\mathbf{DBED}} \rightleftharpoons \mathbf{O}^{\mathbf{DBED}}$ equilibrium biased to $^{\mathbf{S}}\mathbf{P}^{\mathbf{DBED}}$ with SbF_6^- counteranions.[34]

The facile conversion of $^{\mathbf{S}}\mathbf{P}^{\mathbf{DBPD}}$ to $\mathbf{A}^{\mathbf{DBPD}}$, and for that matter $^{\mathbf{S}}\mathbf{P}^{\mathbf{DBED}}$ to $\mathbf{A}^{\mathbf{DBED}}$, by addition of phenolate salts is rationalized readily on both thermodynamic and kinetic grounds; stronger phenolate bonds to Cu(III) result, and ligand exchange rates on copper centers are in general exceedingly fast. The dicationic $^{\mathbf{S}}\mathbf{P}^{\mathbf{DBPD}}$ complex, purposefully formed in the presence of weakly coordinating anions, reacts avidly with anionic ligands capable of forming strong σ-bonds to metals. Associative phenolate binding to $^{\mathbf{S}}\mathbf{P}^{\mathbf{DBPD}}$ through an accessible Cu(II) axial site created by simple bidentate chelation is presumably the first step towards Cu_2O_2 core isomerization. Indeed, axial association of weakly-coordinating counter-anions is evidenced in solution by slight changes in the visible color of $^{\mathbf{S}}\mathbf{P}^{\mathbf{DBPD}}$. Similar axial association is inferred for $^{\mathbf{S}}\mathbf{P}^{\mathbf{DBED}}$.[34] Structural evidence also exists for such an axial association of a carboxylate ligand in the X-ray crystal structure of $^{\mathbf{S}}\mathbf{P}^{\mathbf{spartein}}$.[49,50] If a chelating bidentate ligand is sufficiently flexible, rearrangement of the stronger-binding phenolate into an energetically preferred equatorial site of the copper is possible. Such positioning presumably triggers $^{\mathbf{S}}\mathbf{P}$ to **O** isomerization with the attendant oxidation of each copper center. Equatorial coordination of two phenolates rather than one should aid in further stabilization of the Cu(III) species, a trend supported by the presented DFT calculations (Fig. 11). The equatorial positioning of the ligated phenolate in **A** is critical in stabilizing the Cu(III) center in **A**, as d^8 metal centers have strong structural preference energies for planar coordination geometries.[51] Ligands that stabilize a $^{\mathbf{S}}\mathbf{P}$ species that are sterically congested and/or less flexible than **DBPD** and **DBED**, presumably slow or even inhibit strong phenolate binding if the rearrangement to an equatorial position is destabilized or precluded.

Composition of $\mathbf{A}^{\mathbf{DBPD}}$

$\mathbf{A}^{\mathbf{DBPD}}$ species can be formed with phenolates (**1–14**), which vary in their steric demands and electronic properties. In all cases, ≥ 2 equivalents of the phenolate per Cu_2O_2-core are required to maximize the formation of the $\mathbf{A}^{\mathbf{DBPD}}$ species at 153 K, a trend initially observed in the formation of $\mathbf{A}^{\mathbf{DBED}}$.[21] The necessity of more than 1 equivalent can be interpreted as the formation of $\mathbf{A_1}^{\mathbf{DBPD}}$, a 1 : 1 phenolate : Cu_2O_2-core with excess phenolate pushing a binding equilibrium forward; or the formation of $\mathbf{A_2}^{\mathbf{DBPD}}$, a 2 : 1 phenolate : Cu_2O_2-core. *The body of evidence from this investigation and other related investigations is that when 2–3 equivalents are required for full formation of an intermediate, a 2 : 1 complex is formed.* This is presumably the case for the **DBED** ligand.

The optical titration data of the 365 nm ligand-to-metal charge transfer (LMCT) band of $^{\mathbf{S}}\mathbf{P}^{\mathbf{DBPD}}$ with 2-*t*-butyl-4-cyano-phenolate (**1**) provides significant insight into the stoichiometry of the intermediate species formed. The titration is fit better with a cooperative 2 : 1 formation of $^{1}\mathbf{A_2}^{\mathbf{DBPD}}$ (Fig. 5), compared to a 1 : 1 formation of $^{1}\mathbf{A_1}^{\mathbf{DBPD}}$. While a stepwise model proceeding from a 1 : 1 to 2 : 1 complex fits the titration data even better than the cooperative model, it also has an additional degree of freedom. A full wavelength spectral analysis using singular value decomposition (SVD) provides comparable equilibrium constants for the 1 : 1 and cooperative 2 : 1 models as the single wavelength analysis. However, the SVD analysis fails to fit the stepwise 2 : 1 model with three independent absorbing species; one of the projected spectra consistently possesses physically unreasonable negative absorption features. The inability of SVD to decompose three independent absorbing species presumably results from the combination of the similarity of the absorption spectra of $^{1}\mathbf{A_1}^{\mathbf{DBPD}}$ and $^{1}\mathbf{A_2}^{\mathbf{DBPD}}$ species (as rationalized by NTO analysis, *v.s.*), and the significant background noise inevitable in a titration of such thermally/oxidatively sensitive species.

Using unsubstituted phenol (**6**) as the reference, the energies and the intensities of the 3 principal UV-Vis LMCT bands, characteristic of **A** species, dramatically change. Most obvious is that electron-deficient phenolates generally move the LMCT bands to higher energies. The sensitivity of these transitions to phenolate variation are fully consistent with a strong bonding interaction to a Cu(III) center, presumably through equatorial positions. Increasing steric demands of the *ortho*-substituent(s) lead to a greater angle between the two planes defined by the phenolate and Cu_2O_2 core; a larger angle correlates with greater absorption intensities in the most red-shifted LMCT bands, consistent with the band assignments as a phenolate π_{op} (out-of-plane) → Cu d_{xy} LMCT.[21,48]

Phenolate hydroxylation

The reactivity of $\mathbf{A}^{\mathbf{DBPD}}$ is dominated by phenolate hydroxylation, which proceeds efficiently even at low temperatures (153 K) for phenolates **1–6**. Since the phenolates ligate directly to the Cu(III) center(s), phenolate hydroxylation is presumed to react *via* an intramolecular process. The reaction is regioselective with only the *ortho* position to the phenolate oxygen hydroxylated. From DFT optimized structures of $^{1}\mathbf{A_2}^{\mathbf{DBPD}}$ species with equatorially ligated phenolate(s), the *ortho* C–H groups are positioned in close proximity to an oxygen atom of the Cu(III)$_2$O$_2$ core, which facilitates an efficient attack of the aromatic system. A Hammett analysis with 4 structurally similar but electronically distinct phenolates clearly supports an electrophilic attack at the carbon center as more electron-deficient phenolates are hydroxylated more slowly (Fig. 8, right).

While the spectroscopically characterized intermediate and that used in the reactivity studies is best interpreted as a 2 : 1 phenolate : Cu_2O_2 core, the hydroxylation reaction can potentially proceed through either a direct pathway through $^1A_2^{DBPD}$ or an indirect pathway through $^1A_1^{DBPD}$ according to the DFT calculations at 153 K. TS_1 and TS_2 have similar free energies (Fig. 11). In no case is the isomerization to the side-on peroxide a viable or competitive path to those through a Cu(III)-containing intermediate. The indirect pathway requires an initial dissociation of one phenolate from $^1A_2^{DBPD}$ to form $^1A_1^{DBPD}$, which is entropically favored but enthalpically unfavored. $^1A_1^{DBPD}$, as a cationic species, is a more powerful oxidant than the neutral $^1A_2^{DBPD}$ species, resulting in the hydroxylation barrier from $^1A_1^{DBPD}$ at *ca.* 5 kcal mol^{-1}, much smaller than the *ca.* 13 kcal mol^{-1} barrier from $^1A_2^{DBPD}$. Experimentally, the Eyring activation parameters suggest that the free energy barrier at 153 K should be 12 kcal mol^{-1}, in remarkable agreement with either pathway starting from $^1A_2^{DBPD}$. As the enthalpy of $^1A_2^{DBPD}$ is *ca.* 5.6 kcal mol^{-1} lower in energy than $^1A_1^{DBPD}$, one would predict that at temperatures significantly above 153 K, the indirect, dissociative pathway with a singly bonded phenolate would dominate.

Phenolate dehalogenation

The Cu(III) containing A^{DBPD} species displays novel debromination and dechlorination of *ortho*-halogenated phenolates with the formation of catecholate products. Intramolecular competition experiments and product analysis clearly indicate that C–H activation is more facile than C–Cl or C–Br activation. Company *et al.* have demonstrated that *ortho* C–F groups are more activated than *ortho* C–H or C–Cl groups for hydroxylation by phenolate-ligated **O** species.[24,28] In model studies, Karlin *et al.* observed intramolecular dehalogenation through the intermediacy of a ^{S}**P** species with the XYL-Cl and XYL-F ligands; the halogen in the ligand is substituted with a hydroxyl group to form a phenol, not a catechol, product.[52,53] Aerobic dehalogenation of phenolic compounds is known to occur in certain bacteria strains using copper and iron enzymes,[54–56] such as phenol monooxygenases and phenol oxidases. Tyr has also demonstrated dehalogenation of *ortho*-halogenated phenols, though the reaction is thought to proceed through a radical-based mechanism; the halophenols are initially hydroxylated to halogenated catechols, which are subsequently oxidatively coupled into dehalogenated bis-phenols.[57,58]

In this study, dehalogenation–hydroxylation reactions are reported for a variety of halogenated phenolates, ranging from mono- to penta-halogenated phenolates at 193 K. A radical mechanism is not probable as coupled phenol products are not observed. Although both the electrophilic and nucleophilic aromatic substitution mechanisms are possible, an electrophilic aromatic substitution as the rate limiting step is more consistent with the slowing of the dehalogenation rates with more electron-deficient phenolates. The debromination-hydroxylation of phenolate **11** yields the same 1 : 1 catechol : quinone mixture as the dehydrogenation–hydroxylation product of phenolate **4**, albeit the latter rate reaction is much faster. Presumably both reactions share an identical Cu-catecholate intermediate after the dehydrogenation or dehalogenation reaction, which subsequently forms a semiquinone species that disproportionates to the catechol : quinone mixture upon the acidic workup. Although the dehalogenation mechanism by which A^{DBPD} species proceeds is not yet clear, it differs from the radical dehalogenation mechanism of Tyr enzymes.

Conclusions

$^{S}P^{DBPD}$ is the second example[29] of a simple bidentate ligand that forms a spectroscopically faithful ^{S}P species with an ability to rapidly bind anionic substrates. Indeed, $^{S}P^{DBPD}$ reacts with a wide range of phenolates leading to a Cu_2O_2 isomerization to a bis-Cu(III)-bis-oxide species. The **DBPD** ligand presumably facilitates this conversion through its bidentate nature and flexible propane ligand backbone; phenolates initially bind to an axial site of a chelated Cu(II) center in $^{S}P^{DBPD}$, which subsequently reorganizes to place the phenolates into equatorial positions forming an A^{DBPD} species. Experimental and computational studies support that the $^{S}P^{DBPD}$ isomer is preferred thermodynamically in the presence of weakly coordinating anions, whereas A^{DBPD} species are stabilized thermodynamically by bis-phenolate coordination, as evidenced by the nearly cooperative formation of $^{1}A_2^{DBPD}$. Phenolate hydroxylation through A_2^{DBPD} is energetically preferred over compositionally related $^{S}P^{DBPD}$ species, supporting the role of the $Cu(III)_2(\mu_2\text{-}oxide)_2$ core as the active oxidant for *ortho*-phenolate hydroxylation. As the free energy barriers for the indirect and direct pathways for phenolate hydroxylation starting from A_2^{DBPD} are effectively equivalent at 153 K, the composition of the active oxidant is ambiguous. However at higher temperatures, the more entropically favored indirect pathway through a mono-phenolate species should dominate. All synthetic studies to date implicate the $Cu(III)_2$-bis-oxide species as the active oxidant of bonded phenolates to create catechols. Yet, whether this synthetic data is applicable to oxyTyr will require additional investigations.

Conflicts of interest

There are no conflicts to declare.

Acknowledgements

The following grants are gratefully acknowledged: NIH GM-50730 and NIH GM-120187 (T. D. P. S.) and CSU-Chico (E. C. W.). Portions of this research were carried out at the Stanford Synchrotron Radiation Lightsource, a national user facility operated by Stanford University on behalf of the U.S. Department of Energy, Office of Basic Energy Sciences. The SSRL Structural Molecular Biology Program is supported by the Department of Energy, Office of Biological and Environmental Research, and by the National Institutes of Health, National Center for Research Resources, Biomedical Technology Program.

References

1 E. I. Solomon, U. M. Sundaram and T. E. Machonkin, *Chem. Rev.*, 1996, **96**, 2563–2606.

2 H. Decker, R. Dillinger and F. Tuczek, *Angew. Chem., Int. Ed.*, 2000, **39**, 1591–1595.

3 Y. Matoba, T. Kumagai, A. Yamamoto, H. Yoshitsu and M. Sugiyama, *J. Biol. Chem.*, 2006, **281**, 8981–8990.

4 A. Spada, S. Palavicini, E. Monzani, L. Bubacco and L. Casella, *Dalton Trans.*, 2009, 6468–6471.

5 I. Garcia-Bosch, R. E. Cowley, D. E. Diaz, R. L. Peterson, E. I. Solomon and K. D. Karlin, *J. Am. Chem. Soc.*, 2017, **139**, 3186–3195.
6 N. Kitajima, K. Fujisawa, Y. Moro-Oka and K. Toriumi, *J. Am. Chem. Soc.*, 1989, **111**, 8975–8976.
7 L. M. Mirica, X. Ottenwaelder and T. D. P. Stack, *Chem. Rev.*, 2004, **104**, 1013–1045.
8 E. A. Lewis and W. B. Tolman, *Chem. Rev.*, 2004, **104**, 1047–1076.
9 S. Mahapatra, J. Halfen, E. Wilkinson, G. Pan, C. J. Cramer, L. Que, Jr. and W. B. Tolman, *J. Am. Chem. Soc.*, 1995, **117**, 8865–8866.
10 V. Mahadevan, J. L. DuBois, B. Hedman, K. O. Hodgson and T. D. P. Stack, *J. Am. Chem. Soc.*, 1999, **121**, 5583–5584.
11 S. Schindler, *Eur. J. Inorg. Chem.*, 2000, 2311–2326.
12 S. Itoh and Y. Tachi, *Dalton Trans.*, 2006, 4531–4538.
13 C. E. Elwell, N. L. Gagnon, B. D. Neisen, D. Dhar, A. D. Spaeth, G. M. Yee and W. B. Tolman, *Chem. Rev.*, 2017, **117**, 2059–2107.
14 J. A. Halfen, S. Mahapatra, E. C. Wilkinson, S. Kaderli, V. G. Young, L. Que Jr, A. D. Zuberbühler and W. B. Tolman, *Science*, 1996, **271**, 1397–1400.
15 V. Mahadevan, M. J. Henson, E. I. Solomon and T. D. P. Stack, *J. Am. Chem. Soc.*, 2000, **122**, 10249–10250.
16 A. Kunishita, T. Osako, Y. Tachi, J. Teraoka and S. Itoh, *Bull. Chem. Soc. Jpn.*, 2006, **79**, 1729–1741.
17 L. Santagostini, M. Gullotti, E. Monzani, L. Casella, R. Dillinger and F. Tuczek, *Chem.–Eur. J.*, 2000, **6**, 519–522.
18 S. Itoh, H. Kumei, S. Nagatomo, T. Kitagawa and S. Fukuzumi, *J. Am. Chem. Soc.*, 2001, **123**, 2165–2175.
19 G. Battaini, M. De Carolis, E. Monzani, F. Tuczek and L. Casella, *Chem. Commun.*, 2003, 726–727.
20 S. Palavicini, A. Granata, E. Monzani and L. Casella, *J. Am. Chem. Soc.*, 2005, **127**, 18031–18036.
21 L. M. Mirica, M. Vance, D. J. Rudd, B. Hedman, K. O. Hodgson, E. I. Solomon and T. D. P. Stack, *Science*, 2005, **308**, 1890–1892.
22 B. T. Op't Holt, M. A. Vance, L. M. Mirica, D. E. Heppner, T. D. P. Stack and E. I. Solomon, *J. Am. Chem. Soc.*, 2009, **131**, 6421–6438.
23 A. Company, S. Palavicini, I. Garcia-Bosch, R. Mas-Balleste, L. Que, E. V. Rybak-Akimova, L. Casella, X. Ribas and M. Costas, *Chem.–Eur. J.*, 2008, **14**, 3535–3538.
24 J. Serrano-Plana, I. Garcia-Bosch, R. Miyake, M. Costas and A. Company, *Angew. Chem., Int. Ed.*, 2014, **53**, 9608–9612.
25 A *trans*-μ-η^1:η^1-peroxodicopper(II) species coordinated with a phenolate was proposed. However, the resonance Raman spectroscopy has not been reported due to the extreme instability. See: I. Garcia-Bosch, A. Company, J. Frisch, M. Torrent-Sucarrat, M. Cardellach, I. Gamba, M. Güell, L. Casella, L. Que, X. Ribas, J. Luis and M. Costas, *Angew. Chem., Int. Ed.*, 2010, **49**, 2406–2409.
26 L. Chiang, W. Keown, C. Citek, E. C. Wasinger and T. D. P. Stack, *Angew. Chem., Int. Ed.*, 2016, **55**, 10453–10457.
27 M. Guell, J. M. Luis, P. E. M. Siegbahn and M. Sola, *JBIC, J. Biol. Inorg. Chem.*, 2009, **14**, 273–285.

28 P. Besalu-Sala, C. Magallon, M. Costas, A. Company and J. M. Luis, *Inorg. Chem.*, 2020, **59**, 17018–17027.
29 L. M. Mirica, M. Vance, D. Jackson Rudd, B. Hedman, K. O. Hodgson, E. I. Solomon and T. D. P. Stack, *J. Am. Chem. Soc.*, 2002, **124**, 9332–9333.
30 P. K. Ross and E. I. Solomon, *J. Am. Chem. Soc.*, 1990, **112**, 5871–5872.
31 P. K. Ross and E. I. Solomon, *J. Am. Chem. Soc.*, 1991, **113**, 3246–3259.
32 E. I. Solomon, P. Chen, M. Metz, S. K. Lee and A. E. Palmer, *Angew. Chem., Int. Ed.*, 2001, **40**, 4570–4590.
33 E. I. Solomon, D. E. Heppner, E. M. Johnston, J. W. Ginsbach, J. Cirera, M. Qayyum, M. T. Kieber-Emmons, C. H. Kjaergaard, R. G. Hadt and L. Tian, *Chem. Rev.*, 2014, **114**, 3659–3853.
34 L. M. Mirica, D. J. Rudd, M. A. Vance, E. I. Solomon, K. O. Hodgson, B. Hedman and T. D. P. Stack, *J. Am. Chem. Soc.*, 2006, **128**, 2654–2665.
35 M. J. Henson, P. Mukherjee, D. E. Root, T. D. P. Stack and E. I. Solomon, *J. Am. Chem. Soc.*, 1999, **121**, 10332–10345.
36 J. L. DuBois, P. Mukherjee, T. D. P. Stack, B. Hedman, E. I. Solomon and K. O. Hodgson, *J. Am. Chem. Soc.*, 2000, **122**, 5775–5787.
37 The separation of the two first-shell distances is only 0.12 Å and the theoretical separation for EXAFS waves over a *k*-range of 1–14 $Å^{-1}$ is also 0.12 Å. However, the fit value is marginally better, if two contributions are included (0.13) compared to a single averaged contribution (0.17), the disorder values are decreased reasonably with two contributions, and the separation in fit distances is not less than the theoretical resolution. Therefore, we include the split first shell fit in our analysis.
38 The Debye–Waller value for the final contribution (Cu–Cu multiple scattering) is low. The contribution, however, was modeled solely with a Cu–O–Cu three-body path, neglecting any four-body interactions at a similar distance. It seems plausible that the fit would reflect a low disorder for this wave.
39 The DW factor for the Cu–N interaction is unusually low. The validity of this fit was investigated by systematically fixing this DW factor at increasing values while allowing all remaining parameters of the fit vary. Deviations from the original fit expectedly grow as the Cu–N DW is increased. However, such deviations are small: no more than 0.01 Å in distances, a small shift in ΔE_0 (−9.0 eV), and an only slightly larger error value (0.36). We thus observe that the potential well representing the original fit is largely invariant in the dimension of the DW factor, and conclude the original fit is the best fit, especially with regard to bond distances.
40 K. D. Karlin, M. S. Nasir, B. I. Cohen, R. W. Cruse, S. Kaderli and A. D. Zuberbühler, *J. Am. Chem. Soc.*, 1994, **116**, 1324–1336.
41 S. Mahapatra, S. Kaderli, A. Llobet, Y. M. Neuhold, T. Palanche, J. A. Halfen, V. G. Young, T. A. Kaden, L. Que, A. D. Zuberbuhler and W. B. Tolman, *Inorg. Chem.*, 1997, **36**, 6343–6356.
42 S. Yamazaki and S. Itoh, *J. Am. Chem. Soc.*, 2003, **125**, 13034–13035.
43 S. Itoh and S. Fukuzumi, *Bull. Chem. Soc. Jpn.*, 2002, **75**, 2081–2095.
44 Y. F. Liu, J. G. Yu, P. E. M. Siegbahn and M. R. A. Blomberg, *Chem.–Eur. J.*, 2013, **19**, 1942–1954.
45 R. L. Martin, *J. Chem. Phys.*, 2003, **118**, 4775–4777.
46 T. D. P. Stack, *Dalton Trans.*, 2003, **10**, 1881–1889.

47 D. J. E. Spencer, N. W. Aboelella, A. M. Reynolds, P. L. Holland and W. B. Tolman, *J. Am. Chem. Soc.*, 2002, **124**, 2108–2109.
48 S. Herres-Pawlis, P. Verma, R. Haase, P. Kang, C. T. Lyons, E. C. Wasinger, U. Florke, G. Henkel and T. D. P. Stack, *J. Am. Chem. Soc.*, 2009, **131**, 1154–1169.
49 X. Ottenwaelder, D. J. Rudd, M. C. Corbett, K. O. Hodgson, B. Hedman and T. D. P. Stack, *J. Am. Chem. Soc.*, 2006, **128**, 9268–9269.
50 Y. Funahashi, T. Nishikawa, Y. Wasada-Tsutsui, Y. Kajita, S. Yamaguchi, H. Arii, T. Ozawa, K. Jitsukawa, T. Tosha, S. Hirota, T. Kitagawa and H. Masuda, *J. Am. Chem. Soc.*, 2008, **130**, 16444–16446.
51 M. Costas, X. Ribas, A. Poater, J. M. Lopez Valbuena, R. Xifra, A. Company, M. Duran, M. Sola, A. Llobet, M. Corbella, M. A. Usonn, J. Mahia, X. Solans, X. Shan and J. Benet-Buchholz, *Inorg. Chem.*, 2006, **45**, 3569–3581.
52 K. D. Karlin, R. W. Cruse, M. S. Haka, Y. Gultneh and B. I. Cohen, *Inorg. Chim. Acta*, 1986, **125**, L43–L44.
53 M. S. Nasir, B. I. Cohen and K. D. Karlin, *Inorg. Chim. Acta*, 1990, **176**, 185–187.
54 S. Fetzner, *Appl. Microbiol. Biotechnol.*, 1998, **50**, 633–657.
55 C. den Besten, P. J. van Bladeren, E. Duizer, J. Vervoort and I. M. C. M. Rietjens, *Chem. Res. Toxicol.*, 1993, **6**, 674–680.
56 S. Fetzner and F. Lingens, *Microbiol. Rev.*, 1994, **58**, 641–685.
57 J. Dec and J.-M. Bollag, *Environ. Sci. Technol.*, 1995, **29**, 657–663.
58 J. W. Park, J. Dec, J. E. Kim and J. M. Bollag, *Arch. Environ. Contam. Toxicol.*, 2000, **38**, 405–410.

Faraday Discussions

ROYAL SOCIETY OF CHEMISTRY

PAPER

Explorations of the nonheme high-valent iron-oxo landscape: crystal structure of a synthetic complex with an $[Fe_2^{IV}(\mu\text{-}O)_2]$ diamond core relevant to the chemistry of sMMOH†

Gregory T. Rohde, Genqiang Xue and Lawrence Que, Jr *

Received 3rd October 2021, Accepted 29th October 2021
DOI: 10.1039/d1fd00066g

Methanotrophic bacteria utilize methane monooxygenase (MMO) to carry out the first step in metabolizing methane. The soluble enzymes employ a hydroxylase component (sMMOH) with a nonheme diiron active site that activates O_2 and generates a powerful oxidant capable of converting methane to methanol. It is proposed that the diiron(II) center in the reduced enzyme reacts with O_2 to generate a diferric-peroxo intermediate called **P** that then undergoes O–O cleavage to convert into a diiron(IV) derivative called **Q**, which carries out methane hydroxylation. Most (but not all) of the spectroscopic data of **Q** accumulated by various groups to date favor the presence of an $Fe_2^{IV}(\mu\text{-}O)_2$ unit with a diamond core. The Que lab has had a long-term interest in making synthetic analogs of iron enzyme intermediates. To this end, the first crystal structure of a complex with a $Fe^{III}Fe^{IV}(\mu\text{-}O)_2$ diamond core was reported in 1999, which exhibited an Fe···Fe distance of 2.683(1) Å. Now more than 20 years later, a complex with an $Fe_2^{IV}(\mu\text{-}O)_2$ diamond core has been synthesized in sufficient purity to allow diffraction-quality crystals to be grown. Its crystal structure has been solved, revealing an Fe···Fe distance of 2.711(4) Å for comparison with structural data for related complexes with lower iron oxidation states.

Introduction

Nature often uses iron centers to activate dioxygen and generate high-valent oxidants that cleave the strong C–H bonds of substrates and carry out substrate functionalization into a range of products.[1] Of particular interest for this article is the soluble methane monooxygenase (sMMO), which catalyzes the hydroxylation

Department of Chemistry, Center for Metals in Biocatalysis, University of Minnesota, Minneapolis, Minnesota 55455, USA. E-mail: larryque@umn.edu

† Electronic supplementary information (ESI) available. CCDC 2111623–2111625. For ESI and crystallographic data in CIF or other electronic format see DOI: 10.1039/d1fd00066g

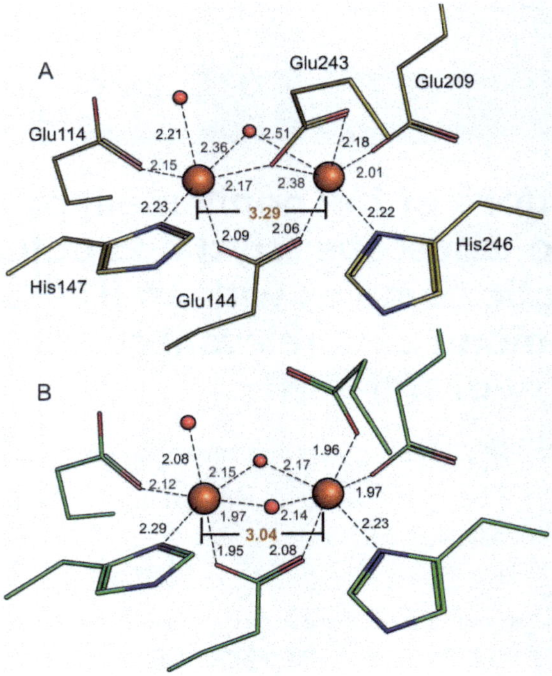

Fig. 1 Diiron active sites of diferrous (A) and diferric (B) forms of sMMOH from *Methylosinus trichosporium* OB3b based on the high-resolution XFEL studies of the enzyme bound to its regulatory component.[2c]

of methane to methanol in methanotrophic bacteria.[2] Methane has a C–H bond of 104 kcal mol^{-1}, representing the strongest bond found among aliphatic hydrocarbons. The hydroxylase component of the sMMO enzyme, sMMOH, belongs to a family of diiron enzymes in which two His and four Asp/Glu residues are bound to the diiron active site.[1,2] Fig. 1 shows the recently reported high-resolution XFEL structure of the sMMOH from *Methylosinus trichosporium* OB3b that is bound to its regulatory component.[2c] The diferrous enzyme uses two of the four active-site carboxylates as bridging ligands to the diiron(II) unit, one as a monodentate bridge and the other as a 1,3-bridge (Fig. 1A), the diferric form has the two iron(III) centers bridged by two hydroxo groups together with the 1,3-bound carboxylate found in the diferrous form (Fig. 1B).

The high-valent diiron oxidant for methane hydroxylation generated at the sMMOH active site is formed in two key steps (Scheme 1).[3] In step 1, O_2 binding to the diferrous enzyme generates an O_2-adduct called **P** or H_{peroxo}, which is best described as a (μ-1,2-peroxo)diferric intermediate commonly observed in model complexes.[4,5] In step 2, the O–O single bond of **P** is cleaved with the help of a proton to generate the diiron(IV) oxidant **Q** that is capable of cleaving the 104 kcal mol^{-1} C–H bond of methane. Both intermediates have been observed and characterized by Mössbauer spectroscopy. Intermediates **P** obtained from both *M. capsulatus* and *M. trichosporium* enzymes exhibit isomer shifts (δ) of 0.66–0.67 mm s^{-1} and quadrupole splittings (ΔE_Q) of 1.51 mm s^{-1}, values strongly indicative of high-spin ferric centers. On the other hand, the intermediate **Q**

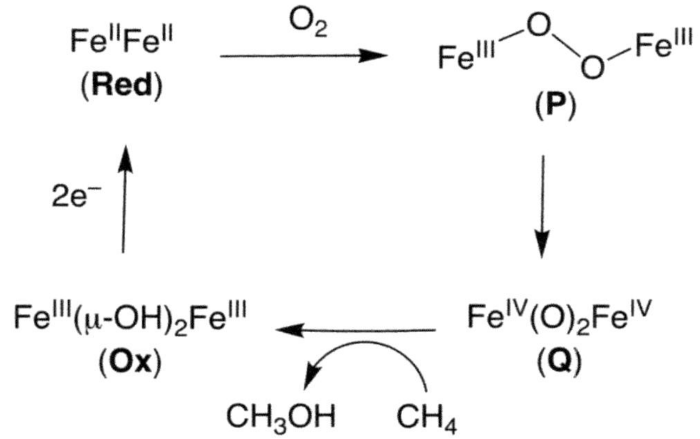

Scheme 1 Steps proposed in the activation of O_2 at the diiron active site of sMMOH and the subsequent hydroxylation of methane. Both the formation of intermediate **P** from diferrous sMMOH and the reaction of methane with intermediate **Q** leading to the release of products actually involve several steps, which have been characterized in kinetic and spectroscopic studies of the enzyme.[3]

derived from *M. trichosporium* sMMOH displays a single quadrupole doublet with an isomer shift (δ) of 0.17 mm s^{-1} and a quadrupole splitting (ΔE_Q) of 0.53 mm s^{-1}; while that from *M. capsulatus* gives rise to two distinct quadrupole doublets of equal intensity with δ values of 0.21 and 0.14 mm s^{-1}, and ΔE_Q values of 0.68 and 0.55 mm s^{-1}, respectively. These observed isomer shifts clearly reflect the formation of Fe(IV) centers in intermediate **Q**. Lastly, the decay of **P** leading to the formation of sMMOH-**Q** has been shown to be pH dependent, with O–O bond cleavage found to be facilitated by one or more protons.[1b]

Initial efforts to gain insight into the nature of the high-valent diiron center in sMMOH using X-ray absorption spectroscopy showed evidence for a 2.46 Å Fe···Fe distance in intermediate **Q**,[6a] suggesting the possibility of an $Fe_2^{IV}(\mu\text{-}O)_2$ diamond core. However, subsequent efforts in characterizing synthetic diiron complexes with $Fe_2(\mu\text{-}O)_2$ diamond cores revealed Fe···Fe distances that fall within a range of 2.6–2.8 Å,[7b,e–h] which are longer than the 2.46 Å distance deduced for sMMOH-**Q**.[6a] More recent results from DeBeer *et al.* using high-energy resolution fluorescence detected extended X-ray absorption fine structure methodology, found no evidence for such a short Fe···Fe distance, but instead supported a longer Fe···Fe distance of 3.4 Å, which could come about by isomerization of the diamond core into an open core with a terminal Fe=O unit (Scheme 2).[6b,c] On the other hand, Lipscomb and Proshlyakov *et al.*, applied resonance Raman techniques to identify a peak at 690 cm^{-1} ($\Delta\nu^{18}O = -36$ cm^{-1})[6d] that falls in the range of values associated with a $M_2(\mu\text{-}O)_2$ core vibration in synthetic complexes.[7a] This apparent disagreement between EXAFS- and resonance Raman-based interpretations has not yet been resolved. However, recent studies of sMMOH-**Q** recently reported by Solomon and coworkers, employing nuclear resonance vibrational spectroscopy, strongly favor a closed-core description for this intriguing intermediate.[6e]

Synthetic complexes that mimic the O_2-derived intermediates formed in the sMMOH active site have also contributed to our understanding of the enzyme

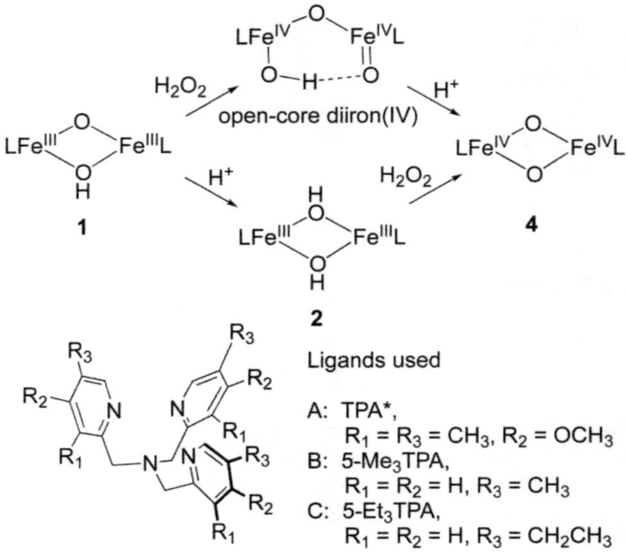

Scheme 2 Two pathways for the conversion of $[Fe_2^{III}(\mu\text{-}O)(\mu\text{-}OH)(L)_2]^{3+}$ (**1A**) to $[Fe_2^{IV}(\mu\text{-}O)_2(L)_2]^{4+}$ (**4A**) depending on the order of addition of H_2O_2 and H^+. Adding H_2O_2 to **1A** first generates $[(L)Fe^{IV}(\mu\text{-}O)(\mu\text{-}OH\cdots O)\text{–}Fe^{IV}(L)]^{3+}$, an open-core diiron(IV) species, and then converts into **4A** upon protonation in 40% yield. On the other hand, reversing the order by introducing H^+ first to form **2A** and then adding H_2O_2 increases the yield of **4** to 85%.

active site. Numerous examples of diferric-peroxo complexes have been characterized over the last three decades, with crystal structures available for five of these as well as a wealth of structural and spectroscopic data for comparison.[4,5] Notably, there are three cases in which the peroxo species has been shown to undergo 1-electron reductive cleavage of the O–O bond to generate an $Fe^{III}Fe^{IV}$ species stable enough to be characterized spectroscopically.[4d–f] However, the corresponding conversion of a synthetic diferric-peroxo species into the isoelectronic bis(μ-oxo)diiron(IV) species has been more difficult to accomplish (Scheme 1). Indeed, only recently has such a transformation been reported where $[Fe_2^{IV}(\mu\text{-}O)_2(Me_3NTB)_2]^{4+}$ (Me_3NTB = tris((1-methyl-1*H*-benzo[*d*]imidazol-2-yl)-methyl) amine) was formed in ~35% yield upon addition of the highly Lewis acidic $Sc(OTf)_3$ to $[Fe_2^{III}(\mu\text{-}O)(\mu\text{-}1,2\text{-}O_2)(Me_3NTB)_2]^{2+}$.[7j] In this paper, we describe another example of the synthesis of such a putative analog for sMMOH-**Q**, which was obtained in 85% yield from the reaction of $[Fe_2^{III}(\mu\text{-}OH)_2(TPA^*)_2]^{4+}$ (TPA* = tris(3,5-dimethyl-4-methoxypyridyl-2-methyl)amine) with stoichiometric H_2O_2. The high yield of the $[Fe_2^{IV}(\mu\text{-}O)_2(TPA^*)_2]^{4+}$ complex has allowed diffraction-quality crystals to be obtained, and the first crystal structure of a synthetic complex with an $Fe_2^{IV}(\mu\text{-}O)_2$ core to be reported herein.

Results and discussion

One-step formation of a complex with a bis(μ-oxo)diiron(IV) core

Previously, we reported that the $[Fe_2^{IV}(\mu\text{-}O)_2(TPA^*)_2]^{4+}$ complex (**4A**) could be prepared by two different routes: (a) by the nearly quantitative bulk electrolytic oxidation of the green $[Fe^{III}Fe^{IV}(\mu\text{-}O)_2(TPA^*)_2]^{3+}$ complex (**3A**) to the red **4A** complex;[7e] and (b) by the reaction of H_2O_2 with the $[Fe_2^{III}(\mu\text{-}O)(\mu\text{-}OH)(TPA^*)_2]^{3+}$ precursor (**1A**) to generate a diiron(IV) complex with an open $[(O)Fe^{IV}(\mu\text{-}O)Fe^{IV}(OH)]^{3+}$ core (**OC** $Fe^{IV}Fe^{IV}$) followed by protonation with 1 equiv. $HClO_4$ to afford **4A** in about 40% yield (Scheme 2).[7e] Despite many attempts, neither synthetic procedure afforded crystals of **4A**.

We then explored the possibility of reversing the order of addition for H_2O_2 and $HClO_4$ (Scheme 2) and indeed obtained a much higher yield of **4A**, as indicated by the higher intensity of the characteristic 875 nm band of **4A**. The addition of 1 equiv. strong acid to **1A** generated its conjugate acid **2A** ($[Fe_2^{III}(\mu\text{-}OH)_2(L)_2]^{4+}$), which could be isolated and characterized (*vide infra*). Upon treatment with 1 equiv. 70% H_2O_2 at −40 °C, **2A** was converted into **4A** in 85% yield (Fig. 2 top). The characteristic features of **4A** at 485 and 875 nm achieved maximal

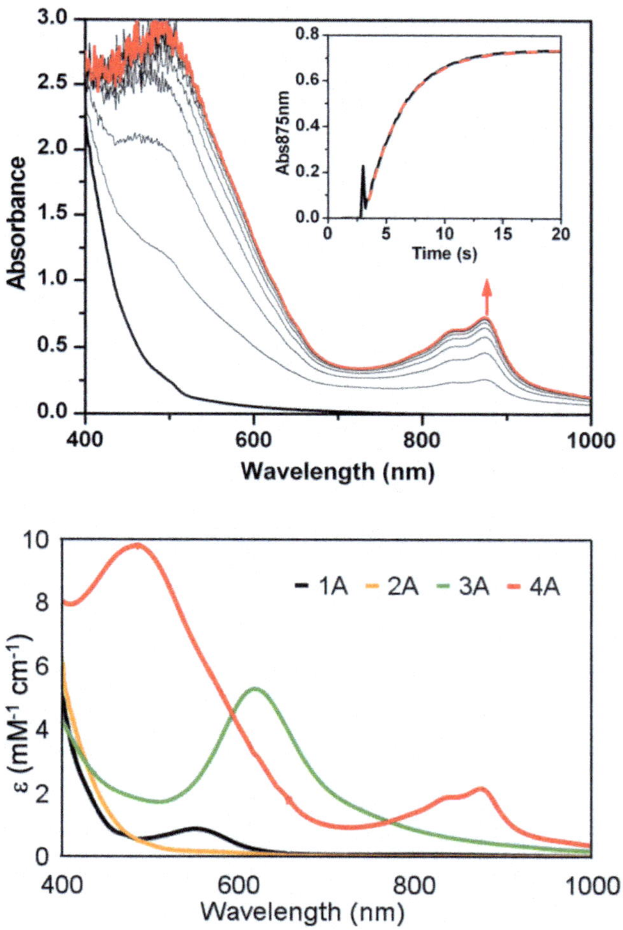

Fig. 2 (Top) Formation of **4A** from the reaction of 0.3 mM **2A** in MeCN with 1 equiv. 70% H_2O_2 at −40 °C in 85% yield as determined by UV-Vis spectroscopy. (Bottom) Comparison of electronic spectra of **1A**, **2A**, **3A**, and **4A**, showing the significantly more intense visible/near IR features of the higher-valent complexes.

intensity in less than 30 seconds. Thus, this method represents the best strategy for obtaining **4A** in high yield and purity.

Structural characterization of 1A and 2A

Of the TPA* complexes discussed in this paper, only the crystal structure of **1A** had been previously reported.[7f] However, due to crystallographic symmetry averaging, an effective inversion center was found in the crystal, so the Fe–μ-O and

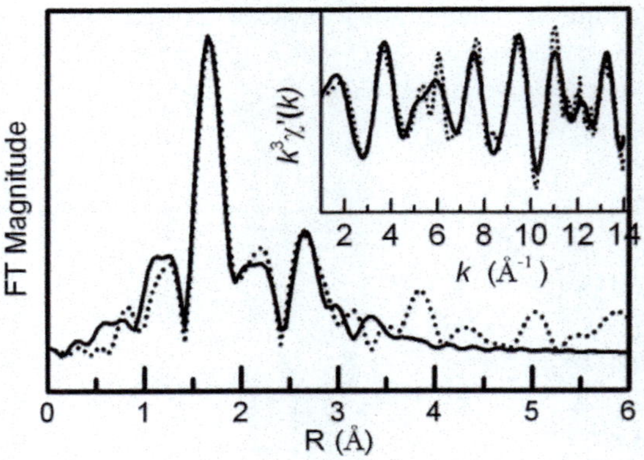

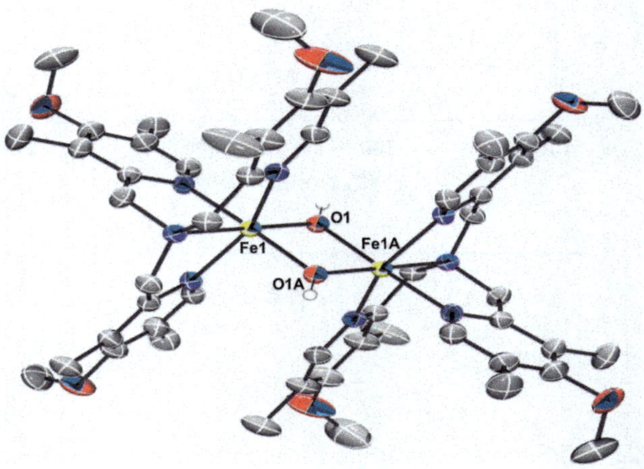

Fig. 3 (Top panel) Fourier-transformed Fe K-edge EXAFS data for $[Fe^{III}_2(\mu\text{-O})(\mu\text{-OH})(TPA^*)_2]^{3+}$ (**1A**) obtained on a 1 : 13 **1A** : boron nitride solid mixture at 20 K. Inset: $k^3\chi'(k)$ data. Fourier transform range 1–14 Å^{-1}; experimental data (dotted line) and the best fit (solid line) consisting of 1 O at 1.80 Å, 1 O at 1.97 Å, 4 N at 2.14 Å, 1 Fe at 2.80 Å, and 6 C at 3.00 Å. (Bottom panel) Thermal ellipsoid plot of **2A** ($[Fe^{III}_2(\mu\text{-OH})_2(TPA^*)_2]^{4+}$) showing 50% probability ellipsoids. Non-oxygen bound hydrogen atoms, perchlorate counterions and acetonitrile and water solvent molecules were removed for clarity. Colors of atoms: carbon, gray; oxygen, red; nitrogen, blue; iron, orange; hydrogen, green. Selected bond distances (Å) and angles (°): Fe1–O1, 1.926(2); Fe1–O1A, 1.995(2); Fe1···Fe1A, 3.0224(7); Fe–N$_{ave}$, 2.114; Fe1–O1–Fe1A, 100.86(11). See Table 2 for a comparison with other diiron complexes with $Fe_2(\mu\text{-O})_2$ diamond cores.

Table 1 EXAFS analysis of complexes having $Fe_2(\mu\text{-}O(H))_2$ diamond cores that are supported by the TPA* ligand[a]

	Fe–N			Fe–O			Fe···C			Fe···Fe		
Complex	N	R	σ^2	N	R	σ^2	N	R	σ^2	N	R	σ^2
1A	4	2.14	3.3	1	1.80	1.5	6	3.00	1.3	1	2.80	3.8
				1	1.97	4.7						
2A (solid)	4	2.10	2.3	2	1.91	5.4	4	2.95	−2.9	1	3.04	−1.2
2A (soln)	5	2.12	4.3	1	1.79	0.0	8	3.03	2.6	1	3.42	0.1
										4^m	3.57	0.8
3A	4	1.96	8.9	2	1.75	7.3	6	2.92	1.7	1	2.57	2.4
							4	2.77	0.8			

[a] **1A** = $[Fe_2^{III}(\mu\text{-}O)(\mu\text{-}OH)(TPA^*)_2]^{3+}$, **2A** = $[Fe_2^{III}(\mu\text{-}OH)_2(TPA^*)_2]^{4+}$, **3A** = $[Fe^{III}Fe^{IV}(\mu\text{-}O)_2(TPA^*)_2]^{3+}$. N = number of scatterers; R = scatterer distance listed in Å; σ^2 = mean-squared deviation in units of 10^{-3} Å^2. The superscript (m) used in one of the fits for **2A** indicates the inclusion of a multiple scattering pathway for the nearly linear Fe–O–Fe unit. See ESI for complete EXAFS analysis and methods.

Fe–μ-OH bonds were disordered over the two positions. EXAFS methods were then applied to determine Fe–μ-O and Fe–μ-OH bond distances with good accuracy, provided that there was a sufficient difference in their bond lengths relative to the resolution of the measurement. EXAFS data were then collected on a crystalline sample of **1A** with a unit cell that matched the published parameters[7f] and was diluted with boron nitride; the data thus obtained at 20 K had a resolution of 0.12 Å. The EXAFS data displayed in Fig. 3 (top panel) together with the best fit listed in Tables 1 and S1† show excellent agreement between the XAS and XRD data, with respective $Fe\text{–}N_{ave}$ distances of 2.14 Å and 2.15(2) Å and respective Fe···Fe distances of 2.80 and 2.7920(9) Å. Importantly, the two Fe–μ-O distances at 1.80 Å and 1.97 Å could be resolved and respectively assigned to the Fe–μ-O and Fe–μ-OH distances based on previously characterized oxo- and hydroxo-bridged diferric complexes.[7c–f] The unsymmetric $Fe_2^{III}(\mu\text{-}O)(\mu\text{-}OH)$ core of **1A** is thus accurately modeled by the EXAFS analysis.

Complex **2A** was synthesized from the reaction of stoichiometric amounts of $Fe(ClO_4)_3$, TPA*, and NaOH in methanol/water at room temperature. Suitable crystals of **2A** for X-ray diffraction were grown by the slow diffusion of diethyl ether into an acetonitrile solution of microcrystalline **2A**. The diamond core structure of **2A** (Fig. 3 (bottom panel) and Table 2) is similar to those of **1A** and other (μ-hydroxo)(μ-oxo)diiron(III) complexes supported by similar tripodal ligands.[7d–f] The diamond core of **2A** is in fact better described as a rhomb where the hydroxo bridge is bound unsymmetrically, with Fe–μ-OH bond distances of 1.926(2) and 1.995(2) Å (Table 2). The latter falls within the range of previously published bridging Fe–μ-OH bond distances[7d–f] and very close to the Fe–μ-OH bond distance determined for the Fe–μ-OH bond in **1A** by EXAFS (Table 1). The other Fe–μ-OH bond distance is on the short end of the range, but nevertheless is still significantly longer than the Fe–μ-O bond of **1A** determined by EXAFS (Table 1). The asymmetry of the Fe–OH–Fe unit reflects the nature of the different ligands *trans* to the hydroxo bridge in the dinuclear unit, namely a tertiary amine on one Fe and a pyridine on the other, which form Fe–N bonds of different

Table 2 Select bond distances and angles for complexes with $Fe_2(\mu\text{-}O)_2$ diamond cores based on data from XRD studies with the EXAFS-derived values listed in italics within square brackets

Complex	**0C**	**1A**	**2A**[a,b]	**3B**	**3A**	**4A**[a]	**4D**
Diiron core (L=)	$Fe_2^{III}(\mu\text{-}O)_2$ 6-Me_3TPA	$Fe_2^{III}(\mu\text{-}O)(\mu\text{-}OH)$ TPA*	$Fe_2^{III}(\mu\text{-}OH)_2$ TPA*	$Fe_2^{3.5}(\mu\text{-}O)_2$ 5-Et_3TPA	$Fe_2^{3.5}(\mu\text{-}O)_2$ TPA*	$Fe_2^{IV}(\mu\text{-}O)_2$ TPA*	$Fe_2^{IV}(\mu\text{-}O)_2$ Me_3NTB
Fe1–O1	1.842(5)	1.883(3)[c]	1.926(2)	1.805(3)	**1.781(4)**	**1.790(6)**	
		[1.80]	*[1.91]*	*[1.82]*	***[1.75]***	***[1.77]***	*[1.78]*
Fe1–O1A	1.917(4)	1.934(2)[c]	1.995(2)	1.860(3)	1.827(5)	1.809(8)	
		[1.97]					
Fe1–N_{amine}	2.192(5)	2.183(3)	2.150(3)	2.049(3)	2.012(6)	1.992(7)	
Ave Fe–N_{py}	2.258(12)[d]	2.144(9)[d]	2.103(14)[d]	2.018(13)[d]	1.989(5)[d]	1.967(8)[d]	
Ave Fe–N	2.242(34)[d]	2.154(21)[d]	2.114(26)[d]	2.026(19)[d]	1.995(13)[d]	1.973(14)[d]	
		[2.14]	[2.10]	*[2.00]*	*[1.96]*	*[1.97]*	*[1.97]*
Fe1···Fe1A	2.714(1)	2.7920(9)	3.0224(10)	2.683(1)	**2.596(1)**	**2.711(4)**	
		[2.80]	[3.04]	*[2.61]*	***[2.57]***	***[2.73]***	*[2.70]*
Fe1–O1–Fe1A	92.4(2)	94.0(1)	100.86(11)	94.1(1)	92.0(2)	97.8(4)	
O1–Fe1–O1A	87.6(2)	85.99(11)	79.14(11)	85.9(1)	88.0(2)	82.2(3)	
λ_{max} (ε_M)	470 (560)	550 (~900)		616 (6200)	620 (~6000)	485 (9800)	**600 (9000)**
	760 (80)					875 (2200)	
Raman $\nu(Fe_2O_2)$	692 (−32)	770 (−42)			661[e] (−33)	674 (−30)	**653 (−30)**
(^{18}O shift)				666 (−35)			**528 (−17)**
XRD ref.	*7b*	*7f*	This work	*7d*	This work	This work	—
Spectroscopic data	*7e*	This work, *7f*	This work	*7c*, *d* and *8b*, *c*	*7g*	*7g*	*7j*
Counter anion	ClO_4^-	ClO_4^-	ClO_4^-	ClO_4^-	ClO_4^-	ClO_4^-	Triflate

[a] Average of two molecules in the asymmetric unit. [b] An isomer of **2A** is observed in MeCN and shown to have an open $(H_2O)Fe^{III}$–O–$Fe^{III}(OH_2)$ core by EXAFS with these distances: Fe–μ-O, 1.79 Å; Fe–N_{ave}, 2.12 Å; Fe1···Fe1A, 3.42 Å (see Table 1). Its Raman spectrum shows two ^{18}O-sensitive vibrations at 835 (−40) cm^{-1} and 463 (−8) cm^{-1}; see Scheme 3. [c] From ref. *7f*; the oxo and hydroxo ligands cannot be resolved in XRD due to disorder over an inversion center. [d] Standard deviation reported from the average of Fe–N distances specified from the crystal structures. [e] Observed as a Fermi doublet.

lengths (Table 2). The average Fe–N_{amine} and Fe–N_{pyr} bond lengths of **2A** are shorter by about 0.03 Å and 0.05 Å, respectively, than those of **1A**, reflecting the lower basicity of the two hydroxo ligands of **2A** compared to the oxo and hydroxo ligands for **1A** (Table 2), making the iron centers of **2A** more Lewis acidic than those of **1A**.[7f] Interestingly, the Fe···Fe distance of **2A** is 3.0224(10) Å, the longest distance found in this series of diiron diamond core complexes (Table 2).

The solution structure of **2A** in MeCN has been investigated by EXAFS to gain insight into the nature of the diferric complex in solution, as some synthetic nonheme diferric complexes can exist either as a diamond core or an open core (Scheme 3). X-ray crystallography of **1A** by Do *et al.* showed it to have a closed core with dimensions comparable to related nonheme diiron complexes.[7d,e,g] However, the diiron core of **1A** has been observed to open in the presence of H_2O in MeCN solution,[7f] but the active form responsible for activating H_2O_2 is still likely to be the closed core form. Thus, understanding the core isomerization equilibrium is important for understanding the mechanism of H_2O_2 activation by these complexes. Therefore the solution geometry of **2A** has been investigated with several techniques in order to elucidate the structure of the diferric starting material.

EXAFS analysis of **2A** in the solid-state and as a frozen solution sample shows that the different sample states result in dramatic changes in diiron core dimensions. The Fourier transform (R' space) of the Fe K-edge EXAFS data of microcrystalline **2A** diluted with BN (1 : 20) exhibits several prominent peaks (Fig. 4 top; Tables S2 and S3†). Of note is the feature at $R + \Delta \sim 2.6$ Å with an intensity comparable to those of the first shell at $R + \Delta < 2$ Å; this feature is assigned to the other Fe in an $Fe_2^{III}(OH)_2$ diamond core with an Fe···Fe distance of 3.04 Å.[7c,e] In addition, there are two O scatterers at 1.91 Å that arise from the bridging hydroxo ligands, consistent with its crystal structure (*vide infra*).

On the other hand, the EXAFS spectrum for a 5 mM frozen solution of **2A** in anhydrous acetonitrile (Fig. 4 bottom) differs dramatically from that obtained for the solid-state sample, showing a change in the diiron core structure upon dissolution. The prominent feature at $R + \Delta \sim 2.6$ Å assigned to the Fe scatterer in the microcrystalline sample is absent and replaced by a less intense feature around 3.3 Å associated with a heavy atom scatterer that is at a longer distance and in a more flexible environment than in the crystalline sample (Table 2). The best fit for the data requires 1 O scatterer at 1.79 Å and 1 Fe scatterer at 3.42 Å, which together suggest the presence of a nearly linear Fe–O–Fe unit with an Fe–O–Fe angle of 146°. Furthermore, an acceptable fit requires the inclusion of a 3.569 Å

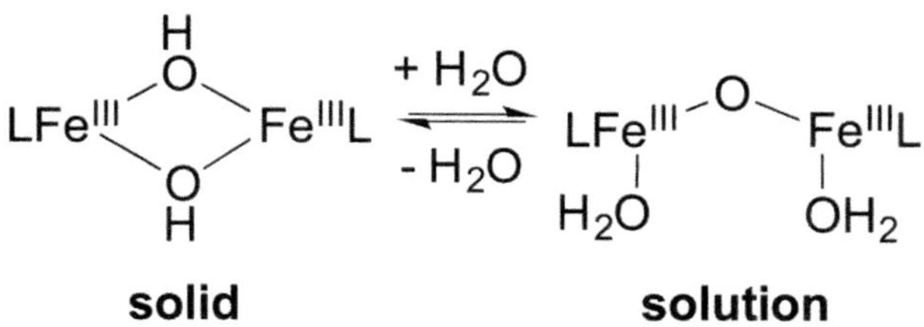

Scheme 3 Distinct structures of **2A** in the solid state and in MeCN solution, resulting in an increase in the Fe···Fe distance from 3.04 Å in the solid state to 3.42 Å in MeCN solution. Acetonitrile may replace the bound water molecules shown above.

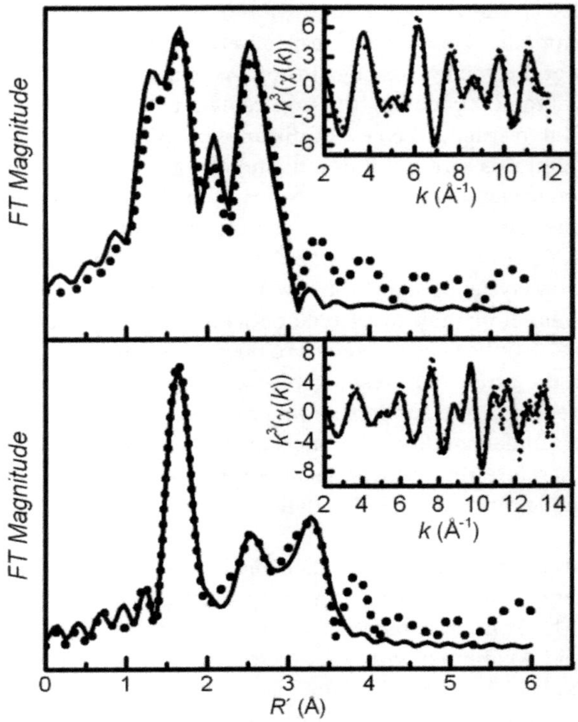

Fig. 4 Top: EXAFS data (dots) for microcrystalline **2A** diluted 1 : 20 in BN with the best fit (solid line) consisting of four shells at 1.91 Å (2 O scatterers), 2.10 Å (4 N scatterers), 2.95 Å (6 C scatterers) and 3.04 Å (1 Fe scatterer). Bottom: EXAFS data for a 5 mM solution of **2A** in MeCN with the best fit consisting of five shells at 1.79 Å (1 O scatterer), 2.12 Å (5 N/O scatterers), 3.03 Å (6 C scatterers), 3.42 Å (1 Fe scatterer) and one 3.57 Å multiple scattering pathway modeled with a Fe–O–Fe unit. In the insets are Fourier transforms of the unfiltered EXAFS experimental data for **2A** in *k*-space (dotted line) and the best fits (solid line).

multiple scattering pathway *via* the nearly linear Fe–O–Fe unit commonly found in the crystal structures of oxo-bridged diferric complexes with similar TPA-based ligands.[7d,e,g] Thus the geometry of **2A** in solution is best described as a nearly linear open core in solution as opposed to the diamond core form strongly favored in the solid state; therefore it is designated as **2A′**. On the other hand, we note that the diamond core found in the crystal structure of **1A** is conserved in solutions with low H_2O concentrations, but converts to an open core structure at higher H_2O concentrations,[7d] consistent with the lower Lewis acidity of its iron(III) centers.

Resonance Raman studies on a frozen solution of **2A′** provide additional support for the structural conclusions derived from EXAFS analysis. The ν_{sym}(Fe–O–Fe) and ν_{asym}(Fe–O–Fe) stretching frequencies of (μ-oxo)diiron(III) complexes have been found to correlate well with the Fe–O–Fe model.[8a,b] Indeed, **2A′** exhibits a feature at 835 cm^{-1} in frozen MeCN solution that can be assigned to the ν_{asym}(Fe–O–Fe) mode and shifts to 795 cm^{-1} upon ^{18}O labeling of the oxo bridge (Fig. 5). The observed downshift of ~40 cm^{-1} is in agreement with the expected shift calculated for a Fe–O diatomic oscillator according to Hooke's law. A second

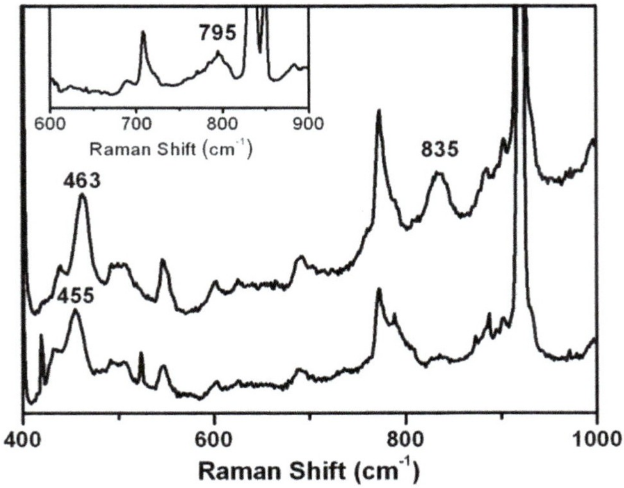

Fig. 5 Resonance Raman spectra of **2A** in MeCN collected at 77 K with 413.1 nm excitation (100 mW power). Top: ^{16}O-labeled **2A**; bottom: ^{18}O-labeled **2A**. Inset: ^{18}O-labeled **2A** spectrum obtained in CD_3CN to unmask the downshifted ^{18}O vibration.

feature is observed at 463 cm^{-1}, which downshifts by 8 cm^{-1} with ^{18}O-substitution and is consistent with its assignment to the ν_{sym}(Fe–O–Fe) mode (Fig. 6). Taken together, both vibrations indicate an open core structure based on previous studies with an Fe–O–Fe angle between 140–150°, which closely matches the angle calculated from the EXAFS scatterer distances.[7b,8] The EXAFS and resonance Raman experiments thus strongly support the open core formulation for **2A′** (Scheme 3). An H-bonded water molecule found in the unit cell of **2A** is the

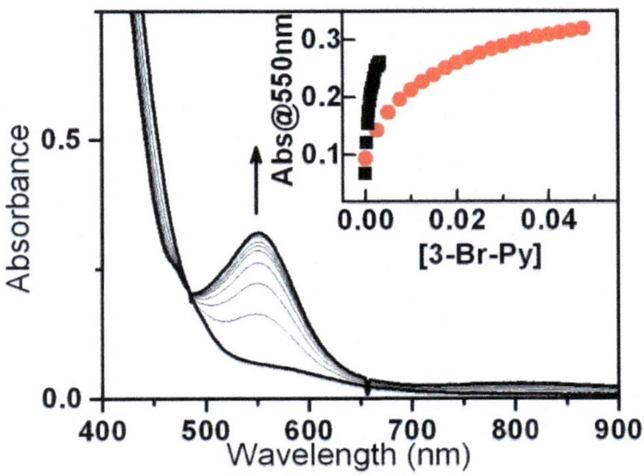

Fig. 6 Titration of a 0.3 mM solution of **2A** in MeCN with 3-bromopyridine to form its conjugate base **1A** (λ_{max} 550 nm, $\varepsilon_M \sim 900$). Inset: formation of **1A** (red circles) and **1B** (black squares) with the increase of the 550 nm band upon titration of **2A** and **2B** with 3-bromopyridine.

likely source of the second aqua ligand. While EXAFS cannot conclusively rule out MeCN binding instead of the aqua ligand, water is found bound to both Fe centers in other published structures of related complexes recrystallized from MeCN.[7g]

Because both **1A** and **2A** are able to activate H_2O_2 to form high-valent intermediates, the pK_a's of the precursors of the complexes with high-valent diiron cores have been investigated to determine the protonation states under different conditions, titrations of **2A** and **2B** in MeCN solution with small aliquots of a 3-bromopyridine reveal changes that can be monitored spectrophotometrically. Spectral data show that both **2A** and **2B** cleanly convert to **1A** and **1B**, respectively, with an increase in the concentration of 3-bromopyridine (Fig. 6, showing the transformation from **2A** to **1A**, which is representative for both). A plot of A(550 nm) *vs.* concentration of 3-bromopyridine displayed in the inset of Fig. 6 shows the pK_a of **2A** to be at least 1 pK_a unit higher than that of 3-bromopyridine (pK_a = 2.84 in H_2O). On the other hand, the pK_a of **2B** is comparable to that of 3-bromopyridine and thus about 1 pK_a unit lower than that of **2A**, consistent with having less electron-donating pyridine donors. The difference in pK_a between the diferric complexes supported by ligands **A** and **B** appears to be important. While diiron(III) complexes of other TPA-based ligands can be converted into $Fe^{III}Fe^{IV}(\mu\text{-}O)_2$ derivatives by treatment with H_2O_2,[7c,d] thus far only **2A** reacts with stoichiometric H_2O_2 to afford a high yield of **4A**. This outcome is presumably due to the presence of three electron donating substituents on each pyridine, which increase the basicity of ligand **A** to provide enough electron density to stabilize the diiron(IV) centers.

Structural characterization of 4A and 3A

Complex **4A** can be obtained by one-step synthesis from the reaction of **2A** with one equivalent of 70% H_2O_2 at −40 °C and exhibits a red color, resulting from the presence of electronic spectral features at λ_{max} 485 nm (ε_M 9800 M^{-1} cm^{-1}) and 875 nm (ε_M 2200 M^{-1} cm^{-1}). Its high yield of 85% has allowed diffraction-quality crystals of **4A** to be obtained. Crystallization of **4A** was achieved in a 3-step process by reacting 21 mg **2A** with one equivalent each of H_2O_2 and 3-bromopyridine in 2 mL butyronitrile at −40 °C for 20 minutes to generate a dark red solution, followed by cooling to −80 °C, and then layering with 1 mL diethyl ether. The dark red solution of **4A** persisted for weeks at −80 °C and afforded dark red crystals in the shape of blocks, which were stable for months in the mother liquor at −80 °C. At temperatures above −80 °C and in the presence of the mother liquor, the dark red blocks lost crystallinity and quickly decayed to a green species that is likely the 1-e^- reduced species, **3A**. To obtain a structure of **4A**, a slide containing the dark red crystals was precooled with liquid N_2, and a suitable crystal was mounted quickly on an X-ray diffractometer. The time the crystal was not on the liquid nitrogen cooled slide or in the cold stream of the cryostat was minimized to slow solvent loss and preserve crystallinity. The crystal diffracted to a resolution of 1.1 Å and was solved using direct methods (Fig. 7).

Along with one-half of two molecules of interest (both located on inversion centers), 4 perchlorate counter ions and 5.5 butyronitrile molecules were found in the asymmetric unit. The loss of long-range order above −80 °C was likely due to solvent loss rather than the decay of **4A**. The solvent molecules, along with the

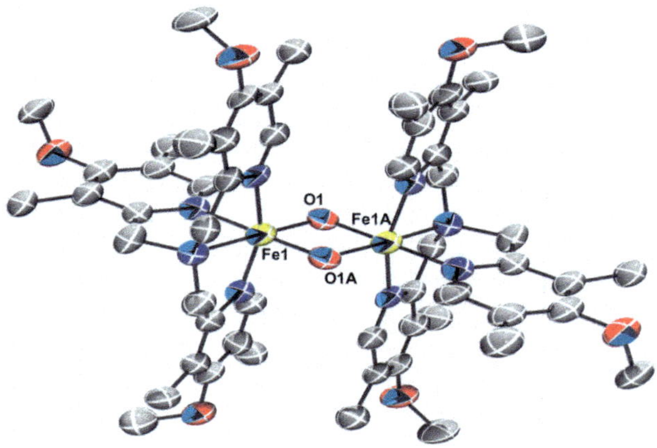

Fig. 7 Structure of the tetracation of **4A** ($[(TPA^*)_2Fe_2^{IV}(\mu\text{-}O)_2]^{4+}$) with thermal ellipsoids drawn to the 50% probability level. Perchlorate counterions, butyronitrile solvent molecules and H-atoms are removed for clarity. Atom colors: carbon, gray; oxygen, red; nitrogen, blue; iron, yellow. An inversion center is present at the center of the Fe_2O_2 core of **4A**, making one half of the molecule unique. For a list of selected bond distances (Å) and angles (°) for **4A** and related structures, see Table 2.

counter ions, create pockets that shield the tetra-cationic iron complexes from neighboring cations and give rise to a crystalline solid. After the data collection, the single crystal was transferred from the diffractometer to a microscope where it was observed to decay to a green solid, most likely the one-electron reduced **3A**. A separate batch of crystals, washed with diethyl ether at −80 °C and then dissolved in acetonitrile at −40 °C, afforded a solution that exhibited the characteristic electronic absorption and ESI-MS spectra of **4A**. The bond distances obtained from the diffraction analysis reported here are in good agreement with previously published EXAFS data;[7g] however the higher resolution of the crystal structure allowed differentiation between bonds not resolvable in the EXAFS experiment. Relevant bond distances comparing diiron complexes in different protonation and oxidation states are listed in Table 2 and discussed below.

For comparison, the crystal structure of **3A** has also been solved (Fig. S1†) from a diffraction-quality crystal obtained from butyronitrile/diethyl ether at −80 °C. The green crystals of **3A** (λ_{max} 620 nm, $\varepsilon \sim 6000$ M^{-1} cm^{-1}) are also susceptible to solvent loss, although to a lesser degree than **4A**, with one molecule of butyronitrile incorporated into the asymmetric unit. Nevertheless, mounting the crystal of **3A** onto the diffractometer required a cold mount, as described above for **4A** to maintain the crystallinity. As can be seen in Fig. 7 and S1,† **4A** and **3A** have nearly identical centrosymmetric structures. The inversion center observed for **3A** does not arise from crystallographic disorder of the iron(IV) and iron(III) centers but derives from the previously established valence-delocalized nature of the $[Fe^{III}Fe^{IV}(\mu\text{-}O)_2]^{3+}$ core, which arises from a novel double exchange mechanism.[7c] However, the differences in diiron oxidation state are manifested by changes in the iron–ligand bond distances from **3A** to **4A**. The Fe–N_{ave} distance contracts by nearly 0.03 Å from 1.995(13) Å in **3A** to 1.969(8) Å in **4A**, but the Fe–O distances in the two complexes are within the uncertainty of the measurements

(Table 2). Interestingly, the Fe···Fe distance increases substantially from 2.596(1) Å in **3A** to 2.711(4) Å in **4A**. The contraction of the first coordination sphere and expansion of the Fe···Fe distance give rise to a more distorted diamond core in **4A**, with the Fe–O–Fe angle deviating farther from the ideal 90° angle to 97.8° in **4A** *versus* 92.1° for **3A**. These crystallographic results solidify the initial structural characterization of **4A** by EXAFS, which revealed an Fe–O distance of 1.77 Å and an Fe···Fe distance of 2.73 Å.[7g] Its surprisingly longer Fe···Fe distance relative to the ~2.6 Å distance found for the one-electron reduced **3A** and **3B** may derive from two factors: (a) the unequal Fe–O bond distances and the larger Fe–O–Fe angle found in the diamond core of **4A**; and (b) the greater repulsive force between two tetracationic point charges, relative to the smaller interaction between the 3.5+ charges on the Fe atoms in the valence-delocalized **3A**.

Not surprisingly, the Fe–O and Fe–N bond distances of the mixed-valent **3A** complex (Table 2) are found to be shorter by over 0.1 Å than the corresponding diferric species **1A** and **2A** due to the 0.5 increase in the oxidation state and a decrease in the spin states of the Fe atoms. Interestingly, the Fe–$(\mu\text{-O})_2$–Fe diamond core of **3A** is also found on an inversion center, with two distinct Fe–μ-O bond lengths, supporting the valence delocalized $Fe^{3.5+}$ oxidation state previously reported for **3B**.[7c,d] The Fe–O lengths are 0.024 and 0.033 Å shorter and the average Fe–N bonds are shorter by over 0.030 Å, for **3A** compared to corresponding values for **3B**. The contraction of the Fe–N bonds is rationalized by the more Lewis basic nature of ligand **A**. The shorter Fe–(μ-O) bonds of **3A** may be a result of the O–Fe–O angle being closer to the ideal 90°. Most notably, the Fe···Fe distances for **3A** and **3B** are 2.596(1) Å and 2.683(1) Å, respectively. The shorter Fe···Fe distance is mainly due to the contraction of the Fe–O bonds and the more acute Fe–O–Fe angle, which decreased from 94.1(1)° to 92.0(2)° from **3B** to **3A**, respectively.

An 8 mM frozen solution sample of **3A** analyzed by EXAFS corroborated the structural parameters found in its crystal structure. See Table 2 for a comparison of the crystallographic and EXAFS analysis results; more details can be found in Fig. S2 and Table S4 in the ESI.† The key EXAFS results are the two Fe–O scatterers found at 1.75 Å and assigned to the two oxo bridges as well as a prominent Fe···Fe scatterer at 2.57 Å that strongly supports the persistence of the $[Fe^{III}Fe^{IV}(\mu\text{-O})_2]^{3+}$ diamond core in solution. Both of these features are in reasonable agreement with the crystallographically determined distances reported herein (Table 2).

Thus the availability of solid-state structures for the **1A-to-4A** series provides an unprecedented opportunity to examine the metrical details of the Fe_2O_2 diamond core as the diiron oxidation state increases from Fe(III)Fe(III) through Fe(III)Fe(IV) to Fe(IV)Fe(IV). The obvious differences in iron oxidation states are also accompanied by spin state changes upon oxidation, starting from the $S = 5/2$ Fe(III) centers in **1A** and **2A**, progressing *via* a valence-delocalized $S = 1/2$ Fe(III)/$S = 1$ Fe(IV) center in **3A**, and finally to the $S = 1$ Fe(IV) centers in **4A**. The diamond core structure clearly has flexibility to adjust bond lengths and angles to accommodate the differences in charge, spin and protonation states. Notably, the Fe^{III}–μ-O_{oxo} EXAFS distance of 1.80 Å for **1A** is indistinguishable within the error of EXAFS analysis (0.02 Å) from the corresponding Fe–μ-O distances determined crystallographically for the higher-valent **3A** and **4A** complexes. Thus the Fe–μ-O_{oxo} bond distance essentially remains constant for the three structures with $Fe_2(\mu\text{-O})_2$ cores in different iron oxidation states, leaving the Fe–N bonds to compensate for the changes in core charge.

Interestingly, the Fe···Fe distances exhibit a relatively broad range of values, from 2.60 (**3A**) to 3.02 Å (**2A**). The contraction of the Fe···Fe distances upon oxidation parallels the decrease in the bond distances in the first coordination sphere except with **4A.** The tetracationic complexes **4A** and **2A** have the largest Fe–O–Fe angles at 98° and 101°, respectively, while the tricationic cores of **1A** and **3A** have angles much closer to 90°. The Lewis basicity of the ligands also has a dramatic effect on the Fe···Fe distance as illustrated by a difference of almost 0.09 Å found between the Fe···Fe distances of **3A** and **3B.** As there is about a 1 pK_a unit difference in the precursor diferric compounds, the shorter Fe···Fe distance of **3A** may result from the greater electron density donated by the more Lewis basic ligand **A** than ligand **B** to mitigate the +3.5 charge on each iron center, which decreases the repulsion between the two +3.5 point charges and allows for the angles of the $Fe_2(\mu\text{-}O)_2$ core to be closer to the ideal 90°. Finally, protonating the μ-oxo ligand in **1A** to a μ-hydroxo ligand in **2A** increases the Fe···Fe distance by over 0.2 Å.

The relative accessibility of synthetic $[Fe_2(\mu\text{-}O)_2]$ diamond core complexes is demonstrated by our ability to form such high-valent species with various starting materials in different protonation and oxidation states and apply chemical or electrochemical methods $[Fe^{III}Fe^{IV}(\mu\text{-}O)_2(5\text{-}Et_3TPA)_2]^{3+}$, the first structurally characterized example of such a diamond core complex,[7d] was reported in 1995 and obtained from the reaction of H_2O_2 with its diferric precursor $[Fe_2^{III}(\mu\text{-}O)(OH)(5\text{-}Et_3TPA)_2]^{3+}$.[7c] Subsequently, the higher valent $[Fe_2^{IV}(\mu\text{-}O)_2(TPA^*)_2]^{4+}$ complex could be obtained by 1-e oxidation of its $Fe^{III}Fe^{IV}$ analog.[7g] More recently, we found that the analogous higher valent $[Fe_2^{IV}(\mu\text{-}O)_2(Me_3NTB)_2]^{4+}$ complex could be formed using a more biomimetic reaction sequence, starting by exposing $[Fe_2^{II}(\mu\text{-}OH)_2(Me_3NTB)_2]^{2+}$ to O_2 forming the O_2 adduct and then followed by treatment with $Sc(OTf)_3$ to facilitate cleavage of the O–O bond and generate the bis(μ-oxo)diiron(IV) complex[7j] (Me_3NTB is a tripodal ligand just like TPA* where the pyridines of the latter are replaced by *N*-methylbenzimidazoles). In this report we demonstrate that the corresponding diferryl complex of TPA* can be formed by the reaction of the diferric precursor **2A** with stoichiometric H_2O_2 to generate **4A** in 85% yield. These reaction conditions have led to the crystallization of $[Fe_2^{IV}(\mu\text{-}O)_2(TPA^*)_2]^{4+}$, affording the first crystal structure of a complex with an $Fe_2^{IV}(\mu\text{-}O)_2$ core.

The relative stability demonstrated for the $[Fe_2^{IV}(\mu\text{-}O)_2]$ core in these two examples may then be used to argue against the involvement of such structural motifs in biology for the hydroxylation of recalcitrant substrates such as methane. It should be noted however that the iron(IV) centers found in the two synthetic complexes characterized with $[Fe_2^{IV}(\mu\text{-}O)_2]$ cores consist of $S = 1$ $Fe^{IV}{=}O$ components, which have been shown for most cases to be less reactive than their $S = 2$ counterparts in biology. Whether the iron(IV) spin state plays such an important role in cleaving strong C–H bonds remains to be clarified, and this question is certainly worth exploring.

At present, there is no consensus on the diiron(IV) cluster structure of sMMOH-**Q**. A diamond core structure has been proposed by Proshlyakov and Lipscomb[6b] based on their observation of a band at 695 cm^{-1} by resonance Raman spectroscopy. On the other hand, more recent EXAFS analyses of DeBeer[6c,6d] disfavor the diamond-core structure and strongly support an open-core structure with an observed Fe···Fe distance of 3.4 Å. Subsequent QM/MM calculations by

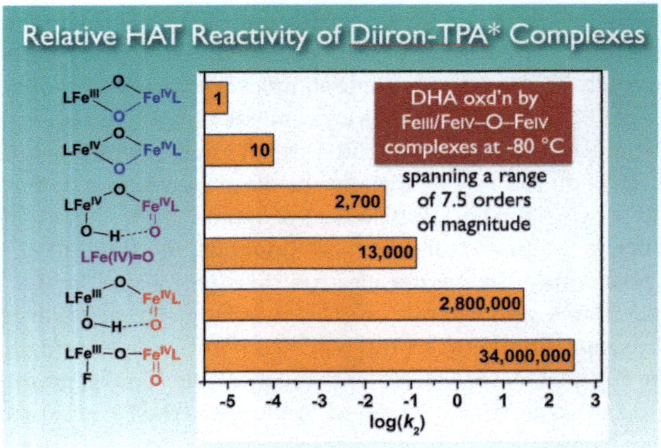

Fig. 8 Relative rates for DHA oxidation at −80 °C by a series of high-valent (μ-oxo) diiron(TPA*) complexes.[9a–c] The blue- and purple-colored Fe^{IV} atoms represent $S = 1$ centers, while the red-colored Fe^{IV} atoms correspond to $S = 2$ centers. Adapted from *Bull. Jpn. Soc. Coord. Chem.*, 2013, **62**, 30 with permission from the Bulletin of the Japan Society of Coordination Chemistry.

DeBeer and Neese[6e] are able to reconcile the observed Raman frequency with the longer Fe⋯Fe distance found by their more recent EXAFS analysis. On the other hand, Solomon and coworkers[6f] have applied nuclear resonance vibrational spectroscopy (NRVS) to probe the geometry and electronic structure of sMMOH-Q and found a subset of complexes with closed cores that can reproduce the NRVS data. Clearly further efforts are needed to clarify the issues raised above.

From the perspective of the synthetic $[Fe_2(\mu\text{-}O)_2]$ complexes we have studied, we find a structural flexibility that can be correlated with their observed reactivity.[7g,9a,b,c] Complexes **3A** and **4A**, the synthetic high-valent diamond-core complexes characterized thus far, represent the least reactive of the series of

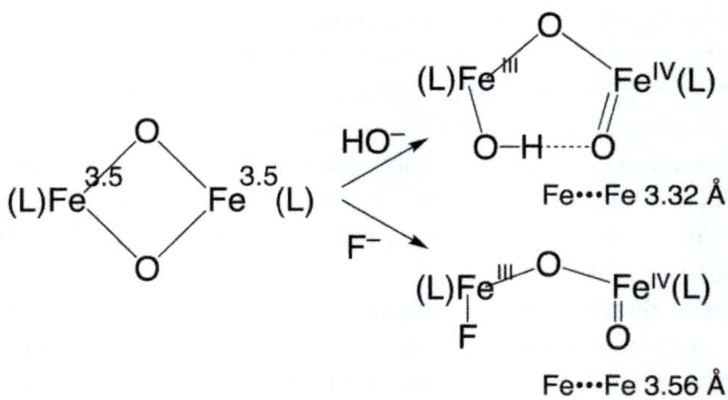

Scheme 4 Conversion of the valence-delocalized diamond core of **3A** into a valence-localized open X–Fe^{III}–O–Fe^{IV}=O core upon binding of OH^- or F^- to the Fe^{III} site to generate a more highly reactive terminal $S = 2$ Fe^{IV}=O unit.[9b,c]

complexes shown in Fig. 8 with respect to their ability to oxidize 9,10-dihydroanthracene (DHA) at −80 °C,[7g,h] with **4A** being an order of magnitude faster than **3A** because of its higher oxidation state. The mononuclear $[Fe^{IV}(O)(TPA^*)(NCMe)]^{2+}$ complex is yet another three orders of magnitude faster, demonstrating the even higher reactivity of the $S = 1$ terminal $Fe^{IV}{=}O$ unit. However, when the valence-delocalized $[Fe_2^{3.5}(\mu\text{-}O)_2]$ complex **3A** is treated with an equivalent of hydroxide, a valence-localized open-core adduct is formed (Scheme 4) that has an HAT rate 6.5 orders of magnitude faster than **4A**.[9a,b] The HAT rate becomes another 10-fold faster with the binding of fluoride (Scheme 4).[9c] The significant increase in HAT reactivity observed for **3A** + OH and **3A** + F stems from the unmasking of a terminal $Fe^{IV}{=}O$ unit upon HO^- or F^- binding and conversion into a high-spin $Fe^{IV}{=}O$ unit, both factors contributing to a much higher reactivity than found for a bridged oxo unit. Similar changes are observed for the hydroxide adduct, but hydrogen bonding of the O–H proton to the terminal $Fe^{IV}{=}O$ unit decreases its HAT reactivity by approximately an order of magnitude. These proposed structural changes were corroborated by EXAFS analysis, which showed differences in the Fe···Fe distances for the two adducts, namely 3.56 Å for a nearly linear Fe–O–Fe unit in the fluoride adduct and 3.32 Å for the hydroxide adduct with an Fe–O–Fe angle of 130°.[9c]

The $M_2(\mu\text{-}O)_2$ diamond-core reactivity story has been recently extended to include later first-row transition metal centers. Hildebrandt, Driess, and Ray have described a series of $[M^{III}(\mu\text{-}O)_2(M^{III})']^{2+}$ complexes of β-diketiminate ligands,[9e]

Table 3 Related oxo-bridged dimetal complexes of interest

Complex	M···M′ distance[a] (Å)	k_2 for oxdidation in $M^{-1}\ s^{-1}$ [b]	Ref.
$[Fe^{III}Fe^{IV}(\mu\text{-}O)_2(5\text{-}Et_3\ TPA)_2]^{3+}$	2.683(1)	5×10^{-4}, −30 °C, MeCN	7g
$[Fe^{III}Fe^{IV}(\mu\text{-}O)_2(TPA^*)_2]^{3+}$	2.596(2) (this work)	1×10^{-5}, −80 °C [1.6×10^{-4}, −40 °C] in 3 : 1 CH_2Cl_2/MeCN	9a
		4×10^{-5}, −30 °C, MeCN	7h
$[Fe_2^{IV}(\mu\text{-}O)_2(TPA^*)_2]^{4+}$	2.711(4) (this work)	1×10^{-4}, −80 °C [1.6×10^{-3}, −40 °C] in 3 : 1 CH_2Cl_2/MeCN	9a
$[Fe^{IV}(O)(TPA^*)(NCMe)]^{2+}$	—	28, −80 °C [~450, −40 °C] in 3 : 1 CH_2Cl_2/MeCN	9a
$[Fe_2^{IV}(\mu\text{-}O)_2(Me_3NTB)_2]^{4+}$	*2.70*	*7×10^{-4}, −40 °C, MeCN*	7j
$[Fe^{IV}(O)(Me_3NTB)(NCMe)]^{2+}$	—	*9.4×10^{2}, −40 °C, MeCN*	7j
$[Co^{III}Co^{III}(L)_2]^{2+}$, L = β-ketimidate	2.6715(5)	*6×10^{-3}, −65 °C, toluene*	9e
$[Co^{III}Ni^{III}(L)_2]^{2+}$, L = β-ketimidate	*2.72* (Co edge) *2.78* (Ni edge)	*3.4×10^{-2}, −65 °C, toluene*	9e
$Ni^{III}Ni^{III}(L)_2]^{2+}$, L = β-ketimidate		*2×10^{-2}, −65 °C, toluene*	9e
$[Cu^{III}Ni^{III}(L)_2]^{2+}$, L = β-ketimidate	*2.81* (at Cu and Ni edges)	*5×10^{-4}, −65 °C, toluene*	9e
$[Co^{III}Co^{IV}(\mu\text{-}O)_2(TPA)_2]^{3+}$	*2.78*	*1.5, −60 °C, MeCN*	9f

[a] M···M′ distances obtained by crystallography are shown in plain text, while those derived from EXAFS analysis are listed in italics. [b] DHA oxidation rates are shown in plain text, while those for CHD oxidation are shown in italics. Values at −40 °C shown in square brackets are extrapolated from −80 °C by doubling the value every 10 °C.

with NiFe, NiCo, NiNi and CoCo centers that carry out both nucleophilic and electrophilic oxidations; within the context of this paper, we just focus on their reported electrophilic reactivity with CHD. As listed in Table 3, these complexes exhibit a range of rates spanning two orders of magnitude, from 10^{-4} to 10^{-2} M^{-1} s^{-1} at −60 °C, many of which are more reactive than the corresponding high-valent diiron diamond core complexes described in this paper. Even more significant is the recent work of Wang and coworkers who have synthesized and characterized the $[Co^{III}Co^{IV}(\mu\text{-}O)_2(TPA)_2]^{3+}$ complex.[9f] Their EXAFS analysis of this complex has revealed a Co⋯Co distance of 2.78 Å that is somewhat longer than the distances found crystallographically for its $Fe^{III}Fe^{IV}$ counterparts, 2.683(1) Å for the 5-Et_3TPA complex[7d] and 2.596(2) Å for the TPA* complex (reported in this paper). Most interestingly, the $Co^{III}Co^{IV}$ complex oxidizes DHA with $k_2 = 1.5$ M^{-1} s^{-1} at −60 °C, which is about 4 orders of magnitude more reactive than the related $[Fe^{III}Fe^{IV}(\mu\text{-}O)_2(TPA^*)_2]^{3+}$ complex after correcting for the temperature difference. This observation makes the $[Co^{III}Co^{IV}(\mu\text{-}O)_2(TPA)_2]^{3+}$ complex the most reactive in this series of diamond core complexes described to date and bodes well for more interesting oxidative chemistry to come.

Conflicts of interest

There are no conflicts of interest to declare.

Acknowledgements

We are grateful to the U. S. National Institutes of Health for support of our work *via* grants R01 GM-38767 and R35 GM-131721 to L. Q., Jr. XAS data were collected on beamline X3B at the National Synchrotron Radiation Light Source, which is supported by the U.S. Department of Energy under Contract No. DE-AC02-98CH10886. Use of beamline X3B was made possible by the Center for Synchrotron Biosciences grant, P30-EB-00998, from the National Institute of Biomedical Imaging and Bioengineering. Dedicated to Prof. John D. Lipscomb for his friendship and his brilliant career in studying methane monooxygenase and other nonheme iron oxygenases.

References

1 (*a*) A. J. Jasniewski and L. Que, Jr, *Chem. Rev.*, 2018, **118**, 2554; (*b*) C. E. Tinberg and S. J. Lippard, *Acc. Chem. Res.*, 2011, **44**, 280; (*c*) R. Banerjee, J. C. Jones and J. D. Lipscomb, *Annu. Rev. Biochem.*, 2019, **88**, 409.

2 sMMOH crystal structures: (*a*) A. C. Rosenzweig, C. A. Frederick, S. J. Lippard and P. Nordlund, *Nature*, 1993, **366**, 537; (*b*) A. C. Rosenzweig and S. J. Lippard, *Acc. Chem. Res.*, 1994, **27**, 229; (*c*) V. Srinivas, R. Banerjee, H. Lebrette, J. C. Jones, O. Aurelius, I.-S. Kim, C. C. Pham, S. Gul, K. D. Sutherlin, A. Bhowmick, J. John, E. Bozkurt, T. Fransson, P. Aller, A. Butryn, I. Bogacz, P. Simon, S. Keable, A. Britz, K. Tono, K. S. Kim, S.-Y. Park, S. J. Lee, J. Park, R. Alonso-Mori, F. D. Fuller, A. Batyuk, A. S. Brewster, U. Bergmann, N. K. Sauter, A. M. Orville, V. K. Yachandra, J. Yano, J. D. Lipscomb, J. Kern and M. Högbom, *J. Am. Chem. Soc.*, 2020, **142**, 14249.

3 Early studies of sMMOH intermediates: (*a*) B. G. Fox, M. P. Hendrich, K. K. Surerus, K. K. Andersson, W. A. Froland, J. D. Lipscomb and E. Münck, *J. Am. Chem. Soc.*, 1993, **115**, 3688; (*b*) S. K. Lee, B. G. Fox, W. A. Froland, J. D. Lipscomb and E. Münck, *J. Am. Chem. Soc.*, 1993, **115**, 6450; (*c*) S. K. Lee, J. C. Nesheim and J. D. Lipscomb, *J. Biol. Chem.*, 1993, **268**, 21569; (*d*) K. E. Liu, D. Wang, B. H. Huynh, D. E. Edmondson, A. Salifoglou and S. J. Lippard, *J. Am. Chem. Soc.*, 1994, **116**, 7465; (*e*) K. E. Liu, A. M. Valentine, D. Wang, B. H. Huynh, D. E. Edmondson, A. Salifoglou and S. J. Lippard, *J. Am. Chem. Soc.*, 1995, **117**, 10174.

4 Synthetic diferric-peroxo complexes not crystallographically characterized: (*a*) Y. Dong, Y. S. Ménage, B. A. Brennan, T. E. Elgren, H. G. Jang, L. L. Pearce and L. Que, Jr, *J. Am. Chem. Soc.*, 1993, **115**, 1851; (*b*) N. Kitajima, N. Tamura, H. Amagai, H. Fukui, Y. Moro-oka, Y. Mizutani, T. Kitagawa, R. Mathur, K. Heerwegh, C. A. Reed, C. R. Randall, L. Que, Jr and K. Tatsumi, *J. Am. Chem. Soc.*, 1994, **116**, 9071; (*c*) Y. Hayashi, T. Kayatani, H. Sugimoto, M. Suzuki, K. Inomata, A. Uehara, Y. Mizutani, T. Kitagawa and Y. Maeda, *J. Am. Chem. Soc.*, 1995, **117**, 11220; (*d*) Y. Dong, Y. Zang, L. Shu, E. C. Wilkinson, L. Que, Jr, K. Kauffmann and E. Münck, *J. Am. Chem. Soc.*, 1997, **119**, 12683; (*e*) J. R. Hagadorn, L. Que, Jr and W. B. Tolman, *J. Am. Chem. Soc.*, 1998, **120**, 13531; (*f*) D. Lee, J. DuBois, D. Petasis, M. P. Hendrich, C. Krebs, B. H. Huynh and S. J. Lippard, *J. Am. Chem. Soc.*, 1999, **121**, 9893; (*g*) M. Kodera, M. Itoh, K. Kano, T. Funabiki and M. Reglier, *Angew. Chem., Int. Ed.*, 2005, **44**, 7104; (*h*) M. Kodera and K. Kano, *Bull. Chem. Soc. Jpn.*, 2007, **80**, 662; (*i*) A. T. Fiedler, X. Shan, M. P. Mehn, J. Kaizer, S. Torelli, J. R. Frisch, M. Kodera and L. Que, Jr, *J. Phys. Chem. A*, 2008, **112**, 13037; (*j*) M. A. Cranswick, K. K. Meier, X. Shan, A. Stubna, J. Kaizer, M. P. Mehn, E. Münck and L. Que, Jr, *Inorg. Chem.*, 2012, **51**, 10417; (*k*) M. Kodera, Y. Kawahara, Y. Hitomi, T. Nomura, T. Ogura and Y. Kobayashi, *J. Am. Chem. Soc.*, 2012, **134**, 13236; (*l*) M. Kodera, T. Tsuji, T. Yasunaga, Y. Kawahara, T. Hirano, Y. Hitomi, T. Nomura, T. Ogura, Y. Kobayashi, P. K. Sajith, Y. Shiota and K. Yoshizawa, *Chem. Sci.*, 2014, **5**, 2282; (*m*) J. S. Pap, M. A. Cranswick, É. Balogh-Hergovich, G. Baráth, M. Giorgi, G. T. Rohde, J. Kaizer, G. Speier and L. Que, Jr, *Eur. J. Inorg. Chem.*, 2013, 3858.

5 Crystallographically characterized diferric-peroxo model complexes: (*a*) Y. Dong, S. Yan, V. G. Young, Jr and L. Que, Jr, *Angew. Chem., Int. Ed.*, 1996, **35**, 618; (*b*) K. Kim and S. J. Lippard, *J. Am. Chem. Soc.*, 1996, **118**, 4914; (*c*) T. Ookubo, H. Sugimoto, T. Nagayama, H. Masuda, T. Sato, K. Tanaka, Y. Maeda, H. Okawa, Y. Hayashi, A. Uehara and M. Suzuki, *J. Am. Chem. Soc.*, 1996, **118**, 701; (*d*) X. Zhang, H. Furutachi, S. Fujinami, S. Nagatomo, Y. Maeda, Y. Watanabe, T. Kitagawa and M. Suzuki, *J. Am. Chem. Soc.*, 2005, **127**, 826.

6 Characterization of sMMOH-**Q**: (*a*) L. Shu, J. C. Nesheim, K. Kauffmann, E. Münck, J. D. Lipscomb and L. Que, Jr, *Science*, 1997, **275**, 515; (*b*) R. Banerjee, Y. Proshlyakov, J. D. Lipscomb and D. A. Proshlyakov, *Nature*, 2015, **518**, 431; (*c*) R. G. Castillo, R. Banerjee, C. J. Allpress, G. T. Rohde, E. Bill, L. Que, Jr, J. D. Lipscomb and S. DeBeer, *J. Am. Chem. Soc.*, 2017, **139**, 18024; (*d*) G. Cutsail III, R. Banerjee, A. Zhou, L. Que, Jr, J. D. Lipscomb and S. DeBeer, *J. Am. Chem. Soc.*, 2018, **140**, 16807; (*e*) C. E. Schulz, R. G. Castillo, D. A. Pantazis, S. DeBeer and F. Neese, *J. Am. Chem. Soc.*, 2021, **143**, 6560; (*f*)

A. B. Jacobs, R. Banerjee, D. E. Deweese, A. Braun, J. T. Babicz, Jr, L. B. Gee, K. D. Sutherlin, L. H. Böttger, Y. Yoda, M. Saito, S. Kitao, Y. Kobayashi, M. Seto, K. Tamasaku, J. D. Lipscomb, K. Park and E. I. Solomon, *J. Am. Chem. Soc.*, 2021, **143**, 16007.

7 Properties of synthetic diiron diamond core complexes related to sMMOH-**Q**: (*a*) L. Que, Jr and W. B. Tolman, *Angew. Chem., Int. Ed.*, 2002, **41**, 1114; (*b*) Y. Zang, Y. Dong, L. Que, K. Kauffmann, Jr and E. Münck, *J. Am. Chem. Soc.*, 1995, **117**, 1169; (*c*) Y. Dong, H. Fujii, M. P. Hendrich, R. A. Leising, G. Pan, C. R. Randall, E. C. Wilkinson, Y. Zang, L. Que, Jr, B. G. Fox, K. Kauffmann and E. Münck, *J. Am. Chem. Soc.*, 1995, **117**, 2778; (*d*) H.-F. Hsu, Y. Dong, L. Shu, V. G. Young, Jr and L. Que, Jr, *J. Am. Chem. Soc.*, 1999, **121**, 5230; (*e*) H. Zheng, Y. Zang, Y. Dong, V. G. Young, Jr and L. Que, Jr, *J. Am. Chem. Soc.*, 1999, **121**, 2226; (*f*) L. H. Do, G. Xue, L. Que, Jr and S. J. Lippard, *Inorg. Chem.*, 2012, **51**, 2393; (*g*) G. Xue, D. Wang, R. De Hont, A. T. Fiedler, X. Shan, E. Münck and L. Que, Jr, *Proc. Natl. Acad. Sci. U. S. A.*, 2007, **104**, 20713; (*h*) G. Xue, A. T. Fiedler, M. Martinho, E. Münck and L. Que, Jr, *Proc. Natl. Acad. Sci. U. S. A.*, 2008, **105**, 20615; (*i*) S. A. Stoian, G. Xue, E. L. Bominaar, L. Que, Jr and E. Münck, *J. Am. Chem. Soc.*, 2014, **136**, 1545; (*j*) S. Banerjee, A. Draksharapu, P. M. Crossland, R. Fan, Y. Guo, M. Swart and L. Que, Jr, *J. Am. Chem. Soc.*, 2020, **142**, 4285.

8 Vibrational data: (*a*) J. Sanders-Loehr, W. D. Wheeler, A. K. Shiemke, B. A. Averill and T. M. Loehr, *J. Am. Chem. Soc.*, 1989, **111**, 8084; (*b*) E. C. Wilkinson, Y. Dong, Y. Zang, H. Fujii, R. Fraczkiewicz, R. S. Czernuszewicz and L. Que, Jr, *J. Am. Chem. Soc.*, 1998, **120**, 955; (*c*) A. J. Skulan, M. A. Hanson, H.-f. Hsu, L. Que, Jr and E. I. Solomon, *J. Am. Chem. Soc.*, 2003, **125**, 7344.

9 HAT Reactivity of **Q**-related models: (*a*) G. Xue, R. De Hont, E. Münck and L. Que, *Nat. Chem.*, 2010, **2**, 400; (*b*) G. Xue, A. Pokutsa and L. Que, Jr, *J. Am. Chem. Soc.*, 2011, **133**, 16657; (*c*) G. Xue, C. Geng, S. Ye, A. T. Fiedler, F. Neese and L. Que, Jr, *Inorg. Chem.*, 2013, **52**, 3976; (*d*) S. Kundu, F. F. Pfaff, E. Miceli, I. Zaharieva, C. Herwig, S. Yao, E. R. Farquhar, U. Kuhlmann, E. Bill, P. Hildebrandt, H. Dau, M. Driess, C. Limberg and K. Ray, *Angew. Chem., Int. Ed.*, 2013, **52**, 5622; (*e*) X. Engelmann, S. Yao, E. R. Farquhar, T. Szilvási, U. Kuhlmann, P. Hildebrandt, M. Driess and K. Ray, *Angew. Chem., Int. Ed.*, 2017, **56**, 297; (*f*) Y. Li, S. Handunneththige, E. R. Farquhar, Y. Guo, M. R. Talipov, F. Li and D. Wang, *J. Am. Chem. Soc.*, 2019, **141**, 20127.

Faraday Discussions

ROYAL SOCIETY OF CHEMISTRY

DISCUSSIONS

Small molecule activation and synthetic analogues: general discussion

Leigh Aldous, Peter Comba, Serena DeBeer, Abhishek Dey, Apparao Draksharapu, Carole Duboc, Shinobu Itoh, Kenneth Karlin, Subrata Kundu, Rocío López Domene, Jean-Didier Maréchal, Shyamalava Mazumdar, Rabindranath Mukherjee, David Parker, Anca Pordea, Lawrence Que, Sankar Prasad Rath, Peter Sadler, Chivukula Sastri, Siegfried Schindler, Volker Schünemann, Sayam Sen Gupta, Edward I. Solomon and T. Daniel P. Stack

DOI: 10.1039/d2fd90012b

Leigh Aldous opened the discussion of the paper by Carole Duboc: Since the CoFe has a lower overpotential for H_2 production, the current will be much lower than those of NiFe and FeFe (even if they have the same kinetic parameters). So, was there *any* evidence for electrocatalytic ability, *e.g.* even at slow scan rates? This Co–Fe complex was in theory very promising compared to Ni–Fe and Fe–Fe because of the more positive reduction potential, but in my mind it needs to have much better parameters to have any current due to the lower overpotential. So, have you tried hour-long reduction reactions, *e.g.* by reducing with cobaltocene? Does it have any activity at all, or is no H_2 ever produced?

Carole Duboc answered: The CoFe complex can be reduced by one electron and its reduced form has been characterized. We have experiments showing that it cannot directly react with H^+ and needs to be further reduced, but this time at a very low potential. If we use a stronger acid, the reduced species is not stable. With the Et_3NH^+ acid, the complex reacts with a proton after a double reduction but the blank is very similar so we cannot really say that this is a catalyst.

Anca Pordea asked: What do you expect the reduction mechanism (H^+ to H_2) to be after the reduction of Co(II) to Co(I)? Do you expect formation of a hydride or of a bridging hydride? Is it possible that Co–H does not form, instead the ligand gets protonated by Et_3NH^+?

Carole Duboc replied: Currently, we have no calculation to support a proposed mechanism. In acetonitrile under homogeneous conditions, the CoFe is reduced first into Co(I)Fe(II) and then it needs a second reduction to react with H^+. The comparison with the NiFe and FeFe complexes suggests that a hydride could be generated at that time. But there is neither experimental evidence nor calculations to support this hypothesis.

Peter Sadler enquired: Please tell me what is known about the detailed mechanism for H_2 production. Is the reaction sequential? How does the efficiency of the reaction depend on the solvent. What is the origin of the hydride (H^-) and the proton (H^+)? Is the solvent the source of the hydride and/or proton? If not, where do they come from?

Carole Duboc answered: For the two NiFe and FeFe catalysts, theoretical calculations have been performed, and experimentally we have characterized a hydride species with the NiFe complex. When the MFe complexes are reduced enough to be able to react with H^+, they react together and can form hydride species. In acetonitrile, the proton source is ET_3NH^+ under homogeneous conditions. When the reaction is carried out in acidic aqueous solutions, the complexes are first immobilized at the electrode. Under such experimental conditions, we do not know if hydride species are still generated during the catalysis.

The mechanism is sequential and the sequence of one electron reduction and protonation varies as a function of the complex and of the experimental conditions (strength of the acid).

Abhishek Dey commented: The Co(0)Fe(II) is quite active on the electrode at pH 4.

Carole Duboc replied: Interestingly, when the CoFe is immobilized by simple physiadsorption on graphite, at very low pH, it displays catalytic activity for H_2 production because under such conditions it becomes more stable at low pH, and the first reduced species can react with directly with H^+.

Shyamalava Mazumdar remarked: What is the rationale for the lower redox potential of the NiFe complex? Does this change in redox potential directly correlate to the catalytic activity?

Carole Duboc replied: Currently, theoretical calculations are being carried out to understand why CoFe, which displays the highest redox potential for the first reduction process, is less active. We do not have an answer yet.

Shyamalava Mazumdar said: Thank you for the nice talk. When you replace Fe by Ni in the Ni–Fe complex, is the geometry changed around the metal centre and how does this affect the properties of the catalyst?

Carole Duboc replied: Indeed, the geometry is not exactly the same. In the solid state, there are two CO in the FeFe complex, one semi bridging between the two iron ions and one terminally bound to the Fe site with the N_2S_2 ligand. In addition, in diluted acetonitrile solutions, we observed that one CO is removed and the corresponding complex displays a bridging CO (instead of a terminally bound CO at the Fe site in the NiFe complex). However, the reactivity and the mechanism is similar for both complexes (the NiFe being a little bit more performant).

Subrata Kundu asked: Although the NiFe, FeFe, and CoFe models are more relevant to hydrogenase activity, what about the synthetic accessibility to the other heterobimetallic complexes, such as a CuFe complex of the given ligand? Is it also possible to obtain NiMn, FeMn and CoMn complexes through the employment of $CpMn(CO)_3$ precursors instead of using $[CpFe(CO)_2(thf)]BF_4$?

Carole Duboc responded: We are starting to work on MMn complexes, starting with $MnCp(CO)_3$ bricks. Since we have fully characterized the $Cu(II)N_2S_2$ complex, we will also try to look at the synthesis of the CuFe complex. Thank you for the idea.

Subrata Kundu asked: The CoFe and related heterodinuclear complexes with thiolate as well as redox-active bipyridine ligands are truly fascinating. Is there any attempt to employ these complexes for the activation of NO_x species? While such a heterodinuclear approach in nitrite reduction/NO_x activation is limited to complexes such as CoMg,[1] it would be exciting to investigate similar chemistry at the heterodinuclear complexes of your paper (DOI: 10.1039/d1fd00085c).

1 C. Uyeda and J. C. Peters, *J. Am. Chem. Soc.*, 2013, **135**, 12023–12031.

Carole Duboc responded: Thank you for the idea. This is something that I will keep in mind.

Rocío López Domene commented: You mimic hydrogenases such as the Fe–Co complex. They are just a complex – they do not have protein scaffold *etc.* I have also been interested in mimicking these type of hydrogenases. Have you thought to try to coordinate your complex to some protein scaffolds? Would you think that incorporating these complexes with coordinant residues, such as cysteines, would lead to obtaining new catalytic properties? I observed a higher catalytic activity when coordinating with a protein scaffold. Also, this is relevant to the solvent. If you coordinate to proteins, they will become water soluble.

Carole Duboc replied: Thank you for your question. The first coordination sphere of these MFe complexes is already full so it will difficult to add one more coordination *via* a cysteine. To stabilize these complexes, we made the choice to immobilize them on graphite. We are not expert in artificial metalloenzymes but this is another interesting strategy.

Shinobu Itoh commented: The intense absorption band at 950 nm of the Co(I)Fe(II) complex shown in Fig. 3 of the paper (DOI: 10.1039/d1fd00085c) seems to be attributed to inter-valence charge transfer. Do you have any comment on that?

Carole Duboc responded: You are right, but we do not have any comment on that.

Sankar Prasad Rath opened a general discussion of the paper by Sayam Sen Gupta: Have you done spectroelectrochemistry around potentials of 0.4 and 0.8 V? Also, at pH 12, is it still oxo or hydroxo?

Sayam Sen Gupta answered: At pH > 12, we generally see the oxoiron(IV) intermediate. If we do spectroelectrochemistry at pH < 11, we end up with a μ-O-Fe^{IV}_2 dimer.

Abhishek Dey enquired: When you are forming this intermediate, can you see the formation of the iron species using spectroelectrochemistry? FTIR could be useful to probe the intermediates formed. Have you tried SPEC-FTIR? It will be good at detecting anything that forms at the electrode surface.

Sayam Sen Gupta responded: That is a good suggestion. We have not tried SPEC-FTIR. That would be useful in determining what is happening at the surface.

Apparao Draksharapu queried: Does the CV of (bTAML)Fe(III) suggest the existence of a Fe(V/IV)–Cl redox event? If yes, is it possible to trap the Fe(V)–Cl species?

Sayam Sen Gupta answered: Yes, we can trap the Fe^{IV}–Cl. In fact, we can synthesize it independently and we do have a crystal structure of it.

Apparao Draksharapu commented: With regards to ligand *vs.* metal oxidation – can you alter the ligands so you can track them in the cyclic voltammetry? If it is ligand-based oxidation, changing the ligand will show this. EPR can tell you if it is metal-based or ligand-based. Also, Raman will tell you if the oxidation is at the ligand or metal.

Sayam Sen Gupta responded: Yes, that is correct. Earlier studies indicate (from EPR/Mössbauer/Raman) that most oxidation is metal-based when you substitute the Ph-ring.[1] However, with substitution with more electron-rich substituents, such as –OMe, we do see a ligand-based oxidation (unpublished data).

S. Pattanayak, D. R. Chowdhury, B. Garai, K. K. Singh, A. Paul, B. B. Dhar and S. S. Gupta, *Chem.–Eur. J.*, 2017, **23**, 3414–3424.

Sankar Prasad Rath opened the discussion of the paper by Chivukula Sastri: Have you observed any ring cleavage products in the reaction mixture?

Chivukula Sastri answered: We have not seen any ring cleavage products during the oxidation of any of the halophenols that we have used.

Abhishek Dey said: For characterization of the intermediate, if the metal is Fe(IV), it would produce a phenoxyl, not a phenolate. Or do you have an Fe(III) bound phenoxyl? Does the EPR support any of these? Also, you may consider doing Mössbauer and Raman to confirm the nature of the species.

Chivukula Sastri answered: What we believe is that the Fe^{IV}=O species will do an initial H atom abstraction from the phenol to give rise to a phenoxyl radical. This phenoxyl radical we are expecting binds to the Fe center, giving rise to Fe^{III}–O–Ph type species. We have tried to do IR with ^{18}O labeled Fe^{IV}=O species but did not get any definitive information. We guess that the O atom in the Fe–O–Ph species is not from the oxo oxygen atom. If that is the case, Resonance Raman experiments may not yield any further information. As suggested, Mössbauer should give us a better understanding of the ligand field around the Fe and this should further be explored. We are also looking to get a decent ESI-MS for such a stable intermediate. So far we have been unsuccessful on that front too.

Peter Comba commented: These are very interesting observations. I found the idea of a hepta-coordinate intermediate interesting, and for high-spin iron(III) (or iron(II)) this should be unproblematic – for iron(IV), the sterics would probably prevent 7-coordination. However, the stability of any 7-coordinate intermediate must depend on the sterics of the supporting ligand. Are there any observations or ideas to support this? The question of stability of a putative intermediate of importance should also strongly influence the product selectivity.

Also, it would of course be interesting to trap this species (which seems to be rather stable even at room temperature) and get further spectroscopic information, possibly in combination with computational analyses.

Chivukula Sastri responded: The new intermediate is stable at room temperature for quite some time. However, we failed to get a good mass spectrometry to characterize it further. Also, as suggested, we are now also running DFT calculations with this system and also with the native N4Py ligand (with the native N4Py ligand being less sterically crowded). We totally agree with you that the product selectivity is being governed by the stability of the new intermediate in the ~500 nm region. As commented by you, the formation and the stability is indeed dependent on the sterics of the substrate. We did not see such formation when we employed trimethyl or tri-*tert*-butyl phenols. Even bulkier halogens like bromo cause enough sterics, but for F derivatives we are able to trap the intermediate *in situ*.

Sayam Sen Gupta enquired: Can the phenol replace the –OH bound to the Fe(III) center to form the phenolate?

Chivukula Sastri answered: We are not sure if phenol is replacing the OH or if it is forming a 7 coordinate intermediate.

Shinobu Itoh said: The KIE of 26 is unprecedentedly huge for the hydrogen atom abstraction reaction from phenol derivatives. I wonder if you have any explanations for that. The temperature dependence of the KIE might give you some information about the tunneling effect.

Chivukula Sastri replied: The KIE of 26 is really abnormal for phenols. But it should be noted that most of the KIEs were reported for methyl/*t*-butyl substituted phenols. In the enzymatic reactions for dehalogenation reactions, as reported by Prof. Dawson and his group,[1–3] this number range is not very uncommon. Furthermore, as suggested by you, the variable temperature KIE will surely give us insights, but the challenge is that at higher temperatures we are seeing the formation of the new adduct (phenolate bound) which overrides the Fe(IV)=O decay profiles. Thus, to address this, we are now looking at the native N4Py system, which has a more open ligand architecture to support the formation of the adducts and, thus, will help us to evaluate the rate of formation of the adduct *vs.* rates of decay of the Fe(IV)=O species.

1 R. Davydov, R. L. Osborne, M. Shanmugam, J. Du, J. H. Dawson and B. M. Hoffman, *J. Am. Chem. Soc.*, 2010, **132**, 14995–15004.
2. J. Du, X. Huang, S. Sun, C. Wang, L. Lebioda and J. H. Dawson, *Biochemistry*, 2011, **50**, 8172–8180.
3. J. Du, M. Sono and J. H. Dawson, *Biochemistry*, 2010, **49**, 6064–6069.

Shinobu Itoh remarked: Usually, values are between 1 and 2, but you get 26. That is amazing. So, is this due to the kinetic tunnelling effect?

Chivukula Sastri answered: We can only attribute this to the tunneling effect to explain such a high KIE.

Shinobu Itoh commented: Why do you see such a huge difference in this value?

Chivukula Sastri answered: The KIE of 26 is really abnormal for phenols. But it should be noted that most of the KIE were reported for methyl/*t*-butyl substituted phenols.

Apparao Draksharapu remarked: Could your product be a catechol-bound Fe(III) species? Does the UV spectrum indicate the existence of either Fe(III) phenolate or Fe(III) catecholate?

Chivukula Sastri responded: There is no evidence of catechol binding to Fe in our case. What we see is a simple signature spectrum of Fe–O–Ph species in the ~500 nm region.

Sankar Prasad Rath enquired: Have you done EPR at low temperature with time?

Chivukula Sastri answered: We have not recorded time dependent EPR.

Sayam Sen Gupta commented: The intramolecular attack of –OH to form the *p*-quinone is not possible. If water is responsible, how would you prove it?

Chivukula Sastri replied: We believe there could be two alternatives. One which we proposed is that there is a 7 coordinate Fe center where Fe has OH as well as the phenolate bound to it. Due to the orientation of the phenol and the OH, we believe as of now that this OH attacks the *ortho* position of the phenol to form *ortho* quinone. As we reported in the paper (DOI: 10.1039/d1fd00064k), we see a mixture of both *ortho* and *para* quinone. *Para* quinone is formed from normal oxidation, as shown in scheme 1 in the paper (DOI: 10.1039/d1fd00064k). We have done ^{18}O labeled experiments but in the mass spectrum we could not get any clear indication for the same.

Sayam Sen Gupta enquired: What is the yield of the quinones formed after the reaction? How much C–O or C–C bond forming reactions occur?

Chivukula Sastri said: We have not evaluated the absolute yields as we were more interested in the ratio of *ortho vs. para* quinones. Usually, they are quantitative in nature. Regarding the C=O *vs.* C–C bond formation reactions, I am not sure if I understood the question correctly. We did not see any C–C bond products. The only products that we saw were quinones.

Abhishek Dey stated: To probe the reactivity, one of ways could be to get TEMPO-H. The TEMPO radical does not bind; really, it is poorly coordinating. This could be followed with EPR. Straightforward Fe(III) hydroxide and EPR would be very different from one you are seeing, so you could follow it.

Chivukula Sastri responded: This is a nice suggestion. We will look into it and get back with more details once we record the EPR with TEMPO-H.

Rabindranath Mukherjee observed: Your starting reaction was between Fe(IV)-oxo and halophenols, generating phenoxyl radical and Fe(III)-hydroxo species. Did you use 2,4,6-tri-*tert*-butylphenol to detect the phenoxyl radical?

Chivukula Sastri responded: The reactions between iron oxo and methyl/*t*-butyl phenol do not lead to the formation of the new intermediate in the ~500 nm region. Because of the bulk factor of the methyl and the *t*-butyl group, we expect that the phenoxy radical is unable to bind to the Fe center. So, what we see is actually the normal decay of oxo species and we do see the formation of the 2,4,6-tri-*tert*-butyl radical, as one would expect to see.

Rabindranath Mukherjee opened the discussion of the paper by Siegfried Schindler: Curiosity-driven exploratory research using ozone as an oxidant is fantastic. Considering availability and handling issues, how do you foresee the general use of ozone in metal–dioxygen chemistry?

Siegfried Schindler responded: It will be a good alternative possibility to synthesize "oxygen adduct" complexes that might not be obtained otherwise.

Abhishek Dey asked: With metal thioether complexes, with ozone, which is a strong oxidant system, are you not worried about sulfoxidation? That's something we see a lot with Ni complexes.

Siegfried Schindler responded: We are aware of it, however so far we have not observed it yet. Furthermore, we are not worried so much because our intention was first to learn more about the mechanism of the ozone reaction and working with temperatures as low as −140 °C we can either suppress these reactions or slow them down to a large extent.

Jean-Didier Maréchal opened the discussion of the paper by Dan Stack: How feasible would it be for natural tyrosinases to bind two phenols at the copper sites in a similar way to what your models show?

Dan Stack responded: We do not think that tyrosinases would bind two phenolates. Our model complexes with two phenolates bonded to the Cu_2O_2 core

(2 : 1 species) are showing what is thermodynamically most stable at −125 °C. Our DFT results suggest that the free energy barrier for the oxidation of a bonded phenolate from the 2 : 1 species can occur either directly or through a 1 : 1 species by the dissociation of one phenolate followed by hydroxylation of the ligated phenolate. The latter would be much more in line with the anticipated mechanism of tyrosinases. However, we are unable at this time to distinguish between the two pathways with these synthetic systems.

Sayam Sen Gupta queried: What happens if you use phenol instead of phenolates as a substrate? Does the phenol do HAT?

Dan Stack answered: If the 2,4-di-*tert*-butyl-phenol is added it slowly reacts with the peroxo intermediate to form the bis-phenol product, which we interpret as a HAT reactivity and radical coupling.

Siegfried Schindler commented: We sometimes try to hydroxylate methane or benzene. So far, we have not been successful. Have you tried benzene for hydroxylation?

Dan Stack answered: We have not observed any benzene or toluene hydroxylation.

Sayam Sen Gupta asked: Is there any difference in the decomposition of the Cu(III) intermediate when you use deuterated phenol instead of phenol?

Dan Stack replied: There is no significant kinetic isotope effect.

Kenneth Karlin asked: You didn't get a chance to talk about the chemistry of the dehalogenation of phenols, which was mentioned in your paper (DOI: 10.1039/d1fd00071c). Can you summarize the aspects of this dehalogenation? Does it occur in a stoichiometric manner and, if so, what is that stoichiometry?

Dan Stack answered: The stoichiometry of the intermediate formed with the 2,4,6-trihalogenated phenolates is the same as with other phenolates – slightly more than two equivalents of the phenolates are required, which we interpret as two phenolates to one Cu_2O_2 core. The 2,4,6-trihalogenated phenolates hydroxylate at a significantly slower rates than phenolates with an *ortho* C–H group and the yields of the dehalogenated catecholates is significantly less. The intramolecular competition between a C–H and C–Br shows exclusive C–H hydroxylation, in line with the reactivity described above.

Lawrence Que asked: What happens to the phenol? What does it dehalogenate to?

Dan Stack responded: The major product identified with the 3 different 2,4,6-trihalogenated phenolates are 4,6-dihalogenated catechols (*e.g.* 3,5-dichlorobenzene-1,2-diol) upon acidic workup.

Lawrence Que said: So, it is a tyrosinase mimic (in the conversion of a phenol-like species to a catechol)?

Dan Stack replied: These model complexes activate O_2 and regiospecifically insert an oxygen atom into a phenolate substrate to create a catechol product – this reaction is similar to that performed by tyrosinases. We doubt that the two bonded phenolate species is a species in the reaction pathway operative in tyrosinases, however, the DFT calculations indicate that for the model complexes, we are unable to discern whether the one or two bonded phenolate species is the active oxidant; the free energy barriers starting from the two bonded phenolate species are indistinguishable. If tyrosinase operates through a bonded phenolate, then the one bonded phenolate complex would be mimicking the reaction pathway of tyrosinases in our opinion.

Rabindranath Mukherjee commented: From experimental and theoretical studies, it has been established that the two isomers of side-on peroxo-dicopper(II) and bis-oxido-dicopper(III) differ only slightly in energy. Although the existence of bis-oxido-dicopper(III) has been firmly established in model studies, why has its existence so far not been detected in tyrosinase?

Dan Stack replied: If the bis-oxido-dicopper(III) species is involved in the tyrosinase mechanism, it could be that its rate of decay is faster than its formation. Under such conditions, its detection would be difficult.

Abhishek Dey opened a general discussion of the paper by Lawrence Que: The iron–iron distance for the [3,4] complex is shorter than that of the [4,4] complex – is this the same with the Fe–O–Fe angle? Is there any chance that the spin state is different between the two complexes or you have some sort of electron delocalisation?

Lawrence Que responded: The [4,4] complex is an antiferromagnetically coupled pair of $S = 1$ Fe(IV) centers, while the [3,4] complex is a valence-delocalized $S = 3/2$ complex [$S = 1/2$ Fe(III)/$S = 1$ Fe(IV)], just like the previously described 5-R_3TPA complexes.[1,2]

1 Y. Dong, H. Fujii, M. P. Hendrich, R. A. Liesing, G. Pan, C. R. Randall, E. C. Wilkinson, Y. Zhang, L. Que Jr., B. G. Fox, K. Kauffmann and E. Münck, *J. Am. Chem. Soc.*, 1995, **117**, 2778–2792.
2 H.-F. Hsu, Y. Dong, V. G. Young and L. Que, *J. Am. Chem. Soc.*, 1999, **121**, 5230–5237.

Serena DeBeer commented: It is nice to see these structures! The changes you see upon going from the Fe(IV)Fe(IV) to the Fe(III)Fe(IV) seems somewhat counter-intuitive – in particular the shorter Fe–Fe bond in the Fe(III)Fe(IV). Is spin state also playing a role? Could you comment on the local spin states of the irons, the AF/F coupling and what the structural implications are?

Lawrence Que answered: Both you and Abhishek Dey have raised this intriguing point in this conversation. These differences may arise from a difference in their Fe–Fe interactions. The [3,4] species is a valence-delocalized complex consisting of an $S = 1/2$ Fe(III) center and an $S = 1$ Fe(IV) center that has an overall spin of $S = 3/2$. The [4,4] complex is likely to be an antiferromagnetically coupled species. Clearly, these are aspects that require further scrutiny.

Serena DeBeer remarked: In your paper (DOI: 10.1039/d1fd00066g), you indicate that the discrepancy between the EXAFS and Raman of sMMO-Q has not been resolved, and I just wanted to point out that together with Frank Neese's group, we have shown using QM/MM calculations[1] that both open and closed core models for Q are consistent with the resonance Raman. However, the EXAFS data clearly favour an open core. Ed Solomon[2] also recently showed using cluster DFT models that the Raman is not necessarily indicative of either a close of open core, though they favour a close core based on the NRVS. In any case, I think it is clear that there is an ongoing controversy regarding the nature of Q and it is important to take all available spectroscopic data into account.

1 C. E. Schulz, R. G. Castillo, D. A. Pantazis, S. DeBeer and F. Neese, *J. Am. Chem. Soc.*, 2021, **143**, 6560–6577.
2 A. B. Jacobs, R. Banerjee, D. E. Deweese, A. Braun, J. T. Babicz Jr., L. B. Gee, K. D.Sutherlin, L. H. B ottger, Y. Yoda, M. Saito, S. Kitao, Y. Kobayashi, M. Seto, K. Tamasaku, J.D. Lipscomb, K. Park and E. I. Solomon, *J. Am. Chem. Soc.*, 2021, **143**, 16007–16029.

Lawrence Que replied: I agree with your comments and have added a paragraph in our article (DOI: 10.1039/d1fd00066g) addressing the points you have raised.

Edward I. Solomon asked: This is in reference to Serena DeBeer's comment on Larry Que's comment on the 690 cm^{-1} vibration in the resonance Raman spectrum of Int. Q in sMMO. Yes, it can be associated with an "open core", an HO–Fe1(IV)–oxo–Fe2(IV)=O species, where the stretch of the oxo-bridge is more localized on Fe1. However, one needs to also consider the shift observed with the $^{16/18}O_2$ mixed isotope, which is 1/2 that of the pure isotopic shift. This means that both of the labeled oxygen atoms must be involved in this vibration, the second being the Fe1(IV)–OH. However, in NRVS (nuclear resonance vibrational spectroscopy) studies,[1] there is a vibration at 605 cm^{-1} that shifts to $\sim$580 cm^{-1} upon cryoreduction and shows an $^{18}O_2$ isotope down shift. In an open core structure, this would be the Fe1(IV)–OH stretch and it is at too low an energy to significantly mix with the 690 cm^{-1} mu-oxo vibration. The observed intermediate magnitude of the isotope shift of the 690 cm^{-1} vibration really argues against the open core oxo/hydroxo/mu-oxo bridged 2Fe(IV) structure.

1 A. B. Jacobs, R. Banerjee, D. E. Deweese, A. Braun, J. T. Babicz Jr., L. B. Gee, K. D. Sutherlin, L. H. Böttger, Y. Yoda, M. Saito, S. Kitao, Y. Kobayashi, M. Seto, K. Tamasaku, J. D. Lipscomb, K. Park and E. I. Solomon, *J. Am. Chem. Soc.*, 2021, **143**, 16007–16029.

Lawrence Que answered: Clearly, more work needs to be done to resolve these conflicting interpretations.

Serena DeBeer said: I would like to note that our proposed open core model can be reconciled with the X-ray absorption data, Mössbauer data, EXAFS data and also the resonance Raman data (including the $^{16}O_2/^{18}O_2$ isotope shifts).[1] We note that the interpretation of the NRVS as a closed core is at odds with the Fe–Fe distance from the EXAFS. Furthermore, it is important to acknowledge that vibrational frequencies can be significantly affected by the protein environment. This is properly taken into account in the QM/MM calculations but not in the cluster models that lack essential hydrogen bonding and long range electrostatic interactions. Hence, QM/MM calculations of the NRVS spectra will be an important next step in reconciling the available data. Furthermore, we acknowledge that the mixed $^{16}O/^{18}O$ isotope labels for the Raman calculations need to be explored within our QM/MM models. In this context, it is important to note that the full calculation of the *resonance* Raman spectrum of a protein is outside the scope of what density functional theory (or any other method that can be applied to systems of this size and complexity) can do. Hence, perfect agreement with the vibrational experimental data may not be expected, and emphasises the need for taking all available experimental data into account – with a clear acknowledgement of how accurately theory is able to predict a given experimental observable.

Perhaps it is useful to clarify the perceived ambiguity in the EXAFS distance. It is true that there is some uncertainty in our distance due to presence of a biferrous component, which I will discuss in detail below. Importantly, what the EXAFS fits of MMO-Q show is a fitted Fe–Fe distance of $\sim$3.35 Å. This is a longer Fe–Fe distance than that observed in either MMO-ox or MMO-red and provides experimental evidence that MMO-Q contains an Fe–Fe distance that is longer than that of either MMO-ox or MMO-red. There is no evidence for a shorter $\sim$2.6–2.8 Å Fe–Fe distance that would be present in a closed core model. The previously reported short Fe–Fe distance in MMO-Q[2] has now clearly been demonstrated to be a result of background iron contamination.[3] Please let me briefly elaborate on our EXAFS fitting protocol. In order to establish that we have a robust protocol, HERFD-EXAFS was measured on oxidized (MMO-ox), reduced (MMO-red) and rapid freeze quench trapped Q (MMO-Q) sMMO samples. All data were measured and fitted from $k = 2\text{–}11$ Å^{-1}, giving a resolution of 0.17 Å for similar scatterers. The fitted Fe–Fe distances for MMO-ox and MMO-red were 3.05 Å and 3.25 Å, respectively, in excellent agreement with both the crystallography and previous EXAFS analyses – highlighting the strength of EXAFS in establishing metal–metal distances. In the case of MMO-Q, the samples are intrinsically a mixture of species, as is the case for all spectroscopic studies on Q. All samples were loaded into Mössbauer/XAS cells and the Mössbauer spectra were used to assess the composition of each sample. Typical MMO-Q samples were comprised of $\sim$50% Q, 40% red and 10% ox. This varies somewhat from preparation to preparation, but allows us to assess the ability to robustly and quantitatively deconvolute the XAS (which has been demonstrated in ref. 3 and 4). Due to the resolution of the EXAFS data, Fe–Fe vectors within 0.17 Å or less of each other will not be separable. The fits of MMO-Q, as noted above, fit to an Fe–Fe distance $\sim$3.35 Å. Given the 3.25 Å Fe–Fe vector that is also present from MMO-red, this clearly indicates that the fitted Fe–Fe distance is a mixture of the MMO-red and MMO-Q Fe–Fe vectors and that the MMO-Q vector must be longer than that of MMO-red. We thus estimate a $\sim$3.4 Å distance, but acknowledge that there is some uncertainty in the exact number. Importantly, there is no evidence for any shorter Fe–Fe vectors that would provide support for the closed core model. Here, it is important to note that the $1/R^2$ dependence of the Fourier transformed EXAFS signal means a short Fe–Fe vector at $\sim$2.7 Å would have a signal 1.6 times more intense than an analogous Fe–Fe vector at $\sim$3.4 Å. This point highlights that if a short Fe–Fe vector is present in MMO-Q, it should be detectable even at 50% conversion.

Regarding the comment that both open and closed core models are consistent with the XAS pre-edges, I must disagree with this statement. While Solomon and coworkers state that there are a range of both open and core models that could be consistent with the pre-edge intensity of Q (ref. 5), they utilized a value of 28 units of experimental pre-edge intensity (based on ref. 2). However, the 1997 data have been shown (based on joint studies of Que, Lipscomb and DeBeer, see ref. 3) to have suffered from background iron contamination. This issue impacts not only the EXAFS but also the pre-edge shape and intensity. As a result, the value used in Solomon and co-workers in their 2021 *JACS* correlation does not accurately reflect the pre-edge area of Q. The value from the 2017 HERFD paper is 42 units, correcting for the differences between standard and HERFD XAS; a value of 35 units of intensity would have been more appropriate for the intended correlation. Utilizing this value indicates that only their open core models are consistent with the XAS pre-edge intensities of MMO-Q. In summary, I feel that the XAS, EXAFS, Mössbauer and Raman consistently support an open core model. There are, however, details regarding the mixed isotope labeling, as well as the interpretation of the NRVS which await further verification. We hope to soon bring additional calculations and experimental data to this lively discussion!

1 C. E. Schulz, R. G. Castillo, D. A. Pantazis, S. DeBeer and F. Neese, *J. Am. Chem. Soc.*, 2021, **143**, 6560–6577.
2 L. Shu, J. C. Nesheim, K. Kauffmann, E. Münck, J. D. Lipscomb and L. Que Jr., *Science*, 1997, **275**, 515–518.
3 G. E. Cutsail III, R. Banerjee, A. Zhou, L. Que Jr., J. D. Lipscomb and S. Debeer, *J. Am. Chem. Soc.*, 2018, **140**, 16807–16820.
4 R. G. Castillo, R. Banerjee, C. J. Allpress, G. T. Rohde, E. Bill, L. Que Jr., J. D. Lipscomb and S. DeBeer, *J. Am. Chem. Soc.*, 2017, **139**, 18024–18033.
5 A. B. Jacobs, R. Banerjee, D. E. Deweese, A. Braun, J. T. Babicz Jr., L. B. Gee, K. D.Sutherlin, L. H. B ottger, Y. Yoda, M. Saito, S. Kitao, Y. Kobayashi, M. Seto, K. Tamasaku, J. D. Lipscomb, K. Park and E. I. Solomon, *J. Am. Chem. Soc.*, 2021, **143**, 16007–16029.

Edward I. Solomon responded: Both the closed and opened core structures are consistent with the Mössbauer and XAS pre-edge intensity of Q. The Fe–Fe distance is ambiguous as there is ~50% bi-ferrous present is the Q samples. The $^{16}O^{18}O$ mixed isotope shift being 1/2 that of the $^{16}O_2$ to $^{18}O_2$ isotope shift of the 690 cm^{-1} resonance Raman vibration is not consistent with an open core, oxo=Fe(IV)–O–Fe(IV)–OH structure and, since this vibration would be localized on the core, it would be reasonably described by a cluster calculation that includes H-bonds. Finally, resonance Raman intensity relates to excited state distortions which derive from the change in bonding in going to the excited state, giving the resonance enhancement. For this open core, this would involve charge transfer transitions from the terminal hydroxo and bridging oxo to the same Fe and lead to the enhancement of vibrations involving these ligand–Fe bonds.

Volker Schünemann enquired: What is your evidence that the spin states of the solid state Fe(IV)–Fe(IV) compound are $S = 1$ for the irons as it has been observed by Mössbauer spectroscopy in solution? Is that based on the crystal structure and the iron–ligand distances? Did you do some Mössbauer work on the isolated complex? I am asking because I am eager to know whether the Mössbauer parameters are the same in the isolated complex and in solution.

Lawrence Que answered: We have not carried out Mössbauer studies on the crystals themselves, but have no reason to doubt the conclusions from the frozen solution results reported in ref. 7g of the paper (DOI: 10.1039/d1fd00066g, given below as ref. 1).

1 G. Xue, D. Wang, R. De Hont, A. T. Fiedler, X. Shan, E. Münck and L. Que, Jr, *Proc. Natl. Acad. Sci. U. S. A.*, 2007, **104**, 20713.

Sankar Prasad Rath asked: Can you comment on the high stability of low spin iron(V)–oxo? What is the origin of such stability?

Lawrence Que responded: I am guessing that you are asking about the stability of low-spin iron(IV)–oxo, on which I have done much work, and not about iron(V)–oxo species, on which I have done much less work. The impression that nonheme low-spin iron(IV)–oxo complexes are relatively stable derives from experimental observations reported by the various groups that work in this field. On the other hand, the much fewer examples of high-spin iron(IV)–oxo complexes that have been described document their much higher reactivity.

Sorry, but I cannot provide you with much deeper insights.

Siegfried Schindler remarked: Have you ever tried methane to see if hydroxylation occurs with your complex?

Lawrence Que replied: No, the strongest C–H bond we have ever attempted is that from cyclohexane. Methane is not as easy to set up to do for us.

Kenneth Karlin commented: In your presentation, you earlier on mentioned that the reaction of an Fe(II) complex of the Me_3-NTB ligand produced a mu-oxo, mu-peroxo diiron(III) complex. Do you know how that occurs, *i.e.*, what are the steps in the course of reaction? I think this is or could be interesting and revealing.

Lawrence Que answered: As we have discussed in ref. 1, the Fe(II)(Me_3NTB) precursor was treated with the base DBU in the presence of some water and then exposed to O_2. This procedure resulted in the formation of the mu-oxo, mu-1,2-peroxo diiron(III) complex *via* a putative $[Fe(II)_2(OH)_2(Me_3NTB)_2]^{2+}$ complex.

1 S. Banerjee, A. Draksharapu, P. M. Crossland, R. Fan, Y. Guo, M. Swart and L. Que Jr.*J. Am. Chem. Soc.*, 2020, **142**, 4285.

Dan Stack queried: Have you or your collaborators ever tried to estimate the redox potentials of the high-valent monomeric and dimeric iron complexes by DFT?

Lawrence Que responded: This is a good question and an effort worth attempting.

Abhishek Dey asked: I wonder if there has ever been a comparison of high valent species of different metals and their reactivities? With $Cu(III)_2O_2$ *vs.* $Fe(IV)_2O_2$ being discussed in back to back talks, one has to wonder about the relative reactivities and the nature of the oxidant. It would be a useful comparison for people starting out in the field. Do you have any thoughts?

Lawrence Que commented: Is there the complication of spin state in iron chemistry though?

Abhishek Dey replied: Yes. Which is what makes it interesting. However, the community seems to be converging on the right spin state.

Dan Stack asked: Are you concerned about the relative reactivity of the Cu_2O_2 and Fe_2O_2 cores?

Abhishek Dey replied: Yes. I wonder if a comparison between these oxidants has ever be made. Their electronic structures ought to be different at a fundamental level, but these are oxidants nonetheless. If one were to decide which bioinorganic system to take inspiration from when designing catalysts for different reactions (*e.g.* aromatic hydroxylation, epoxidation *etc.*), is there any literature available comparing these different oxidants?

Dan Stack remarked: You have commented on the significant reactivity of mononuclear high-valent high-spin iron complexes. Do you have any estimate of the redox potentials of the high-valent Fe_2O_2 dimers to monomers?

Abhishek Dey answered: No, I don't. I would imagine them to be in the same ball park as that of Cu_2O_2 as both can be derived from O_2.

Conflicts of interest

There are no conflicts to declare.

Faraday Discussions

PAPER

O_2 reduction by iron porphyrins with electron withdrawing groups: to scale or not to scale†

Sk Amanullah, Paramita Saha and Abhishek Dey *

Received 22nd October 2021, Accepted 2nd November 2021
DOI: 10.1039/d1fd00076d

Iron porphyrins are synthesized by systematically introducing electron withdrawing groups (EWGs) on pyrroles to evaluate the relationship between rate (k) and overpotential (η). The results indicate that while EWGs lead to a rise in the thermodynamic $Fe^{III/II}$ reduction potential (E^0), the potential of the O_2 reduction reaction (ORR) does not scale with E^0. More importantly, the iron porphyrins with higher E^0 show an order of magnitude higher rate of ORR than unsubstituted iron tetraphenyl porphyrin. This contests the scaling relationship often offered to predict rates of ORR by iron porphyrins based on their E^0. Mechanistic investigations reveal that the rate-determining step (rds) of ORR change between these iron porphyrins with EWG's, as the pK_a and E^0 of several key intermediate species likely change on altering the macrocycle. These results suggest that linear dependence of log(rate) on E^0 or η may only be valid for complexes where the rds of ORR remains the same.

Introduction

The oxygen reduction reaction (ORR), combining O_2, electrons, and protons to produce water, is a clean and renewable technology used in a fuel cell to convert chemical energy to electrical energy.[1,2] The development of earth-abundant metal-based catalysts which can catalyze the reaction with a high rate and lower overpotential is warranted.[3] Nature uses cytochrome c oxidase (CcO), a heme-copper enzyme, to catalyze the transformation of O_2 to water in the terminal step of the electron transport chain during respiration.[4,5] CcO can catalyze the transformation at ~0.35 V *vs.* NHE,[6] close to the thermodynamic potential of O_2 to water, *i.e.*, 0.82 V at pH 7;[6] while the synthetic systems can operate at much higher overpotential. Therefore, the factors responsible for the efficiency of CcO are essential to understand. One of the significant and conspicuous factors lies in the O_2 binding site of

School of Chemical Science, Indian Association for the Cultivation of Science, 2A Raja SC Mullick Road, Kolkata, 700032, India. E-mail: abbeyde@gmail.com

† Electronic supplementary information (ESI) available. CCDC 2116511. For ESI and crystallographic data in CIF or other electronic format see DOI: 10.1039/d1fd00076d

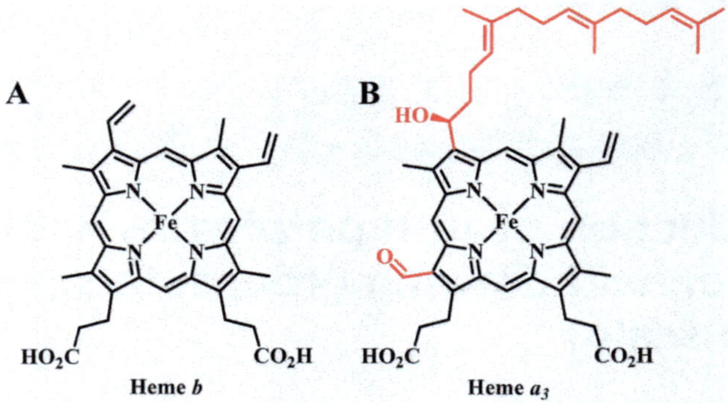

Fig. 1 Structure of (A) heme b and (B) heme a_3. The electron-withdrawing groups in heme a_3 have been highlighted with red bonds.

the CcO, which has a heme a_3 cofactor. Unlike the most common heme (*i.e.*, heme b), heme a_3 has two electron-withdrawing groups (EWGs), *viz.*, a hydroxyfarnesyl group at C_3 and a formyl group at the C_{18} position of the heme (Fig. 1).[7,8] The large increase in the $Fe^{III/II}$ reduction potential may be attributed to the EWGs, as postulated from the electronic structure of the model iron porphyrins with EWGs.[9,10]

The facile and selective reduction of O_2 to water is always a daunting task. For iron-porphyrins with a tetraphenylporphyrin ligand framework, under homogeneous conditions, the protonation of the $Fe^{III}_{superoxy}$ intermediate is proposed to be the rate-determining step (Scheme 1A).[11] The pK_a of the protonated superoxide intermediate, $[Fe^{III}(porphyrin)(\dot{O_2}H)]^+$ is responsible for the facile reduction of O_2 to water. Alternatively, under heterogeneous conditions, *in situ* SERRS-RDE has indicated that the O–O bond cleavage of Fe^{III}–OOH/Fe^{II}–OOH species is the rate-determining step for most iron porphyrins.[12] With a neutral axial ligand (*e.g.*,

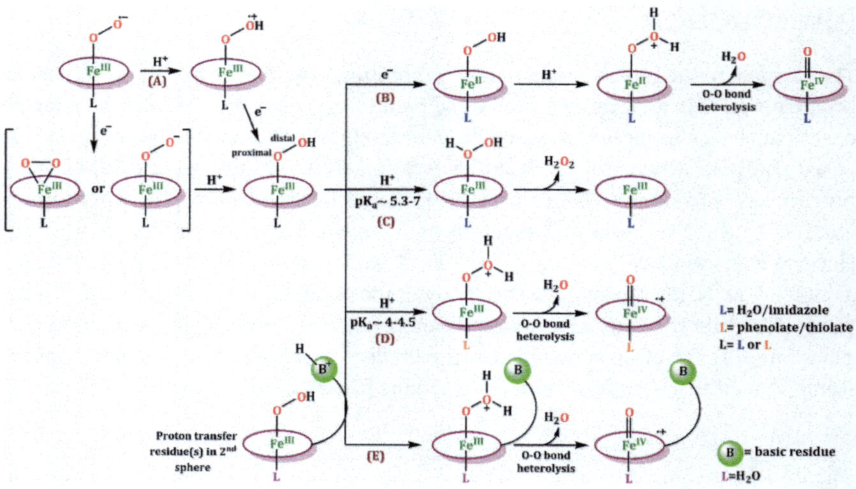

Scheme 1 Schematic representation of possible ORR pathways by iron porphyrin complexes with different axial ligation. The oval ring represents a porphyrin ligand and L represents a proximal axial ligand bound to iron.

water or imidazole), the O–O bond cleaves from the Fe^{II}–OOH state (Scheme 1B),[13,14] while with an anionic axial ligand (*e.g.*, thiolate or phenolate), the O–O bond cleaves from the Fe^{III}–OOH state (Scheme 1D).[15] Due to the "push" effect of the anionic ligands, the pK_a of the Fe^{III}–OOH species becomes sufficiently high to get protonated and induce heterolytic O–O bond cleavage. While another electron transfer is required for neutral axial ligands to increase the pK_a of the Fe^{III}–OOH to induce the protonation and heterolytic O–O bond cleavage. In the presence of a strong acid source, Fe^{III}–OOH species can also be protonated. However, installing a local proton source (or a basic moiety that can modulate the proton transfer) close to the active site, can exert a "pull" effect inducing protonation to the Fe^{III}–OOH state (Scheme 1E).[16] The selectivity for the $2e^-/2H^+$ reduction to H_2O_2 entails from the protonation to the proximal oxygen of the Fe^{III}–OOH species (Scheme 1C); while the protonation to the distal oxygen leads to the $4e^-/4H^+$ reduction to H_2O.

Mayer and co-workers have developed a scaling relationship for the soluble molecular electrocatalytic ORR, considering the rates (turnover frequencies, TOF) and efficiencies (overpotentials, η).[17] The TOFs correlate with the overpotential, suggesting a fast TOF is a consequence of high overpotential, which can be modulated by tuning the $E_{1/2}$ of the catalysts using different ancillary ligands or changing the catalyst's protonation state with different solvents and solution acidity. However, this scaling relationship had failed in iron porphyrins where distal proton transfer residues were installed.[16] Similarly, it has recently been shown from the same group that a catalyst with polycationic porphyrin framework (iron αβαβ-tetra(*o*-*N*,*N'*,*N''*-trimethylanilinium)porphyrin), can enhance anionic ligand binding and can change the $E_{1/2}$.[17] Also, the pK_a of the acid source can tune the $E_{1/2}$ and η_{eff}. However, the relationship holds true across a wide range of porphyrins with different *meso*-substituents on the ring.

Recently, we have shown that a unique iron-porphyrinoid with EWGs and saturated pyrrole centres, can model cytochrome c containing nitrite reductases (Cd_1NiR).[18] The lower backbonding to the π* orbitals of NO leads to the formation of a weak Fe–NO bond and the lower $BDFE_{N-H}$ of the $\{FeHNO\}^8$ renders the PCET from $\{FeNO\}^7$ to $\{FeHNO\}^8$ disfavoured, thereby leading to the release of NO from the $\{FeNO\}^7$ state, which is unexpected because the $\{FeNO\}^7$ species usually possesses a very strong Fe–NO bond with a $K_d \sim 10^{-9}$.[17] However, a similar porphyrin with only the EWGs can tune the $\{FeNO\}^{7/8}$ reduction potential and can undergo the PCET. Therefore, the modification in the porphyrin-core can cause a drastic change in the electronic structure of the porphyrins, rendering them suitable for inaccessible reactivities. In pursuit of developing molecular electrocatalysts for ORR, in this manuscript, iron-porphyrins with different EWGs are used to investigate ORR reactivities. While the $E_{1/2}$ varies expectedly with the inclusion of EWGs, it does not correlate with the overpotential. The results suggest that the overpotential for ORR can only be tuned up to a limited extent, beyond which the $4e^-/4H^+$ multi-step ORR follows a different mechanistic pathway. Therefore, the scaling relationship cannot be universal.

Results and discussion

Synthesis

FeTPP,[17] FeDEsP,[19] and FeTEsP [18] were synthesized following the reported protocols. Free-base tetraphenylporphyrin ligand (H_2TPP) was synthesized using

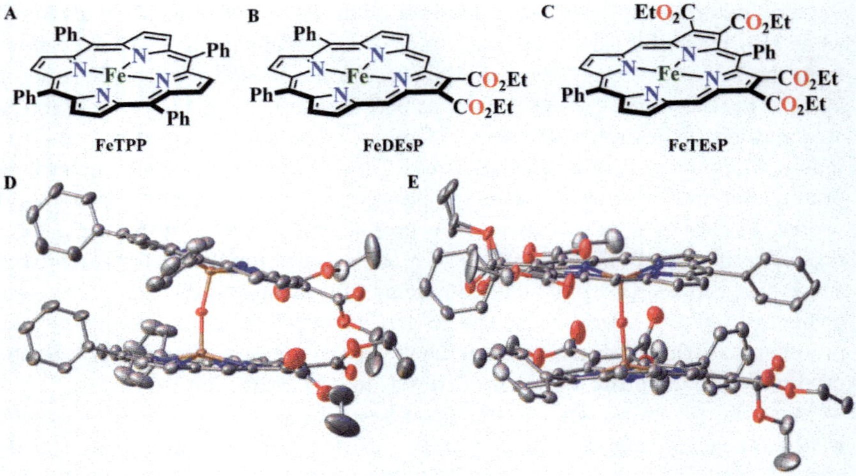

Fig. 2 Molecular structure of the representative molecules for this work: (A) FeTPP; (B) FeDEsP; (C) FeTEsP. ORTEP diagrams of the corresponding μ-oxo dimers of the complexes: (D) FeDEsP and (E) FeTEsP. C: grey, N: blue, O: red, Fe: brown, hydrogens are omitted for clarity. Ellipsoids are drawn with a 50% probability level.

the Adler–Longo method through the propionic acid condensation of pyrrole and benzaldehyde. The diester porphyrin ligand (H_2DEsP) was synthesized from the condensation of 2,5-diformyl-diester pyrrole and phenyltripyrromethane, in the presence of trifluoroacetic acid under a high-dilution technique. The tetraester porphyrin ligand (H_2TEsP) was synthesized from the propionic acid condensation of 2,10-diformyl-tetraester phenyldipyrromethane and 5-phenyldipyrromethane. The ligands were metallated using $FeBr_2$ and a non-coordinating base, 2,4,6-collidine, under an inert atmosphere.

Structure

FeDEsP. Greenish purple-colored needle-like crystals of FeDEsP were grown from the slow evaporation of a THF solution. The SCXRD analysis revealed that it was a μ-oxo dimer (Fig. 2D and Table S1†) and crystallized in a monoclinic system with centrosymmetric $P2_1/c$ space group. The complex showed a square pyramidal geometry with an average Fe–O and Fe–$N_{pyrrole}$ distance of 1.76 Å and 2.07 Å, respectively.

FeTEsP. Purple-colored plate-like crystals were grown through slow diffusion of hexane to the chloroform solution of FeTEsP. It was crystallized in a centrosymmetric $P2_1/c$ space group as a μ-oxo dimer (Fig. 2E and Table S1†). The complex showed a square pyramidal geometry with an average Fe–O and Fe–$N_{pyrrole}$ distance of 1.75 Å and 2.07 Å, respectively.

Cyclic voltammetry

Under homogeneous conditions in organic solvent. The complexes were transformed into the corresponding triflate salts (please see the Experimental details section in the ESI† for the synthetic details). In the cyclic voltammograms (CVs) of the complexes, FeTPP-OTf, FeDEsP-OTf, and FeTEsP-OTf, show a quasi-reversible $Fe^{III/II}$ reduction couple at −0.63 V, −0.52 V, and −0.47 V *vs.* $Fc^{+/0}$ (Fig. 3A). Therefore, the $Fe^{III/II}$ reduction potential of FeDEsP-OTf is ~110 mV more positive than that of FeTPP-OTf. Replacing two β-pyrrolic hydrogens from the porphyrin ring with two electron-withdrawing ester groups shift the reduction potential to ~110 mV more positive. This is in agreement with our previous findings that introducing electron-withdrawing groups (EWGs) to the β-pyrrolic positions of the porphyrin renders the Fe^{III}-centre more electropositive.[10,19,20] Also, lowering the energy of the π* orbitals allows greater π-backbonding from the filled $d_{xz/yz}$ orbitals of the Fe^{II}-porphyrin. Upon the addition of two more ester groups in FeTEsP-OTf, the $Fe^{III/II}$ reduction potential is shifted by another ~50 mV. However, the CV shows additional peaks likely due to the dissociation of the triflate anion (Fig. 3A, red line marked with *). Thus, cyclic voltammetry was performed with the bis-imidazole complexes. The complexes show quasi-reversible $Fe^{III/II}$ processes at −0.59 V, −0.44 V, and −0.33 V *vs.* $Fc^{+/0}$ (Fig. 3B). The $Fe^{III/II}$ reduction potentials of FeDEsP-Im_2 and FeTEsP-Im_2 are ~150 mV and ~260 mV, respectively, more positive than that of FeTPP-Im_2. The peak currents (i_p) vary linearly with the square root of the scan rate ($\nu^{1/2}$), suggesting diffusion-

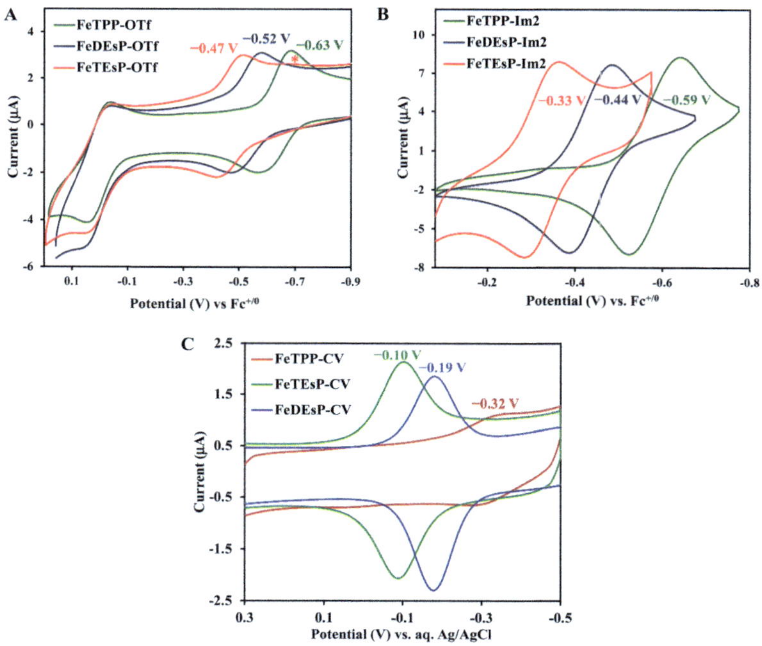

Fig. 3 Cyclic voltammograms of the complexes under N_2 saturation; recorded using aqueous Ag/AgCl in 4 M KCl as reference and platinum counter electrode in (A) a solution of the triflate-salts in DMF; (B) a solution of the bis-imidazole complexes in DMF, with 100 mM tetrabutylammonium hexafluorophosphate ($TBAPF_6$) as the supporting electrolyte, and glassy carbon working electrode. Ferrocene was added as an internal reference; (C) pH 7 phosphate buffer with 100 mM KPF_6 as supporting electrolyte. The complexes were physiadsorbed onto an edge-plane graphite (EPG) working electrode.

controlled processes (Fig. S1†). The diffusion coefficients (D) for the bis-imidazole complexes of FeTPP, FeDEsP, and FeTEsP are 6.16×10^{-6}, 5.58×10^{-6}, and 1.22×10^{-7}, respectively (see the ESI†).

Under heterogeneous conditions in aqueous solvent. Cyclic voltammograms of the triflate complexes, FeTPP-OTf, FeDEsP-OTf, and FeTEsP-OTf, physiadsorbed on an EPG electrode in deoxygenated pH 7 phosphate buffer, show a reversible $Fe^{III/II}$ reduction couple at −0.32 V −0.19 V, and −0.10 V *vs.* aq. Ag/AgCl in 4 M KCl (Fig. 3C). Therefore, FeDEsP possesses ~130 mV higher $Fe^{III/II}$ reduction potential than that of FeTPP. And, FeTEsP shows ~90 mV higher $Fe^{III/II}$ reduction potential than that of FeDEsP. Therefore, the trend observed in the organic solvent is mirrored in the aqueous solvent. By integrating the redox waves, the surface coverage of FeTPP, FeDEsP, and FeTEsP were found to be $(3.43 \pm 0.22) \times 10^{-11}$, $(1.43 \pm 0.28) \times 10^{-11}$, and $(2.35 \pm 0.35) \times 10^{-11}$ mol cm^{-2}, respectively. The peak-currents (i_p) vary linearly with the scan rate (ν), suggesting the catalysts are attached to the surface (Fig. S2†).[17] The plot of $E°$ *versus* pH for the complexes (Fig. S3†) shows a slope of ~60 mV pH^{-1}, suggesting that both these processes involve $1e^-/1H^+$ proton-coupled electron transfer (PCET). All the complexes studied here have the triflate anion ($^-$OTf) as the counterion bound to

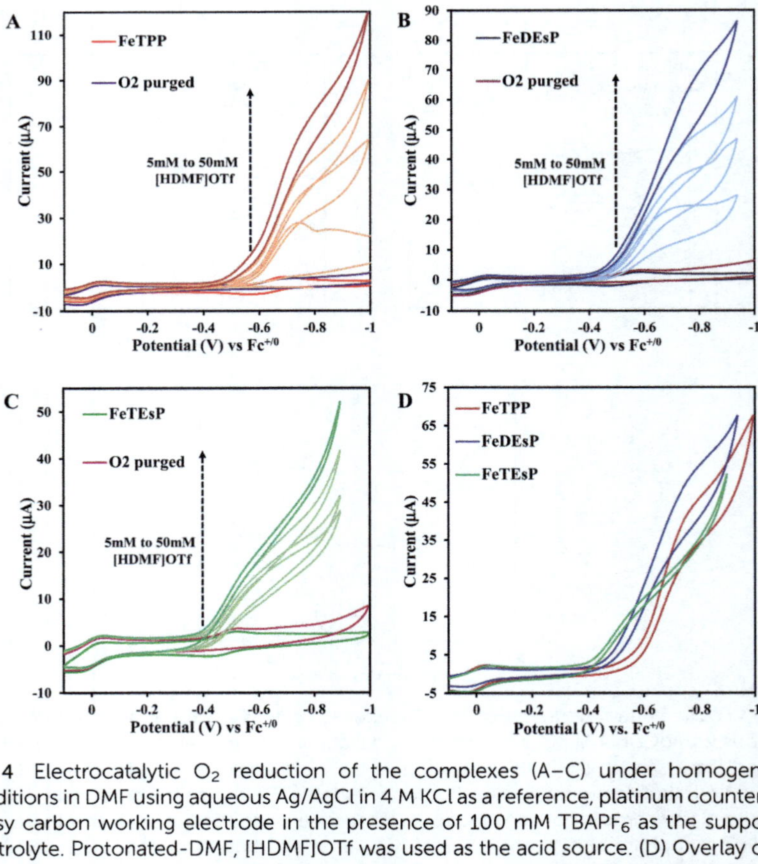

Fig. 4 Electrocatalytic O_2 reduction of the complexes (A–C) under homogeneous conditions in DMF using aqueous Ag/AgCl in 4 M KCl as a reference, platinum counter, and glassy carbon working electrode in the presence of 100 mM $TBAPF_6$ as the supporting electrolyte. Protonated-DMF, [HDMF]OTf was used as the acid source. (D) Overlay of the CV of the complexes under aerobic conditions in the presence of 20 mM [HDMF]OTf.

Table 1 Rate of the ORR at different acid concentrations (under homogeneous conditions) and pH (under heterogeneous conditions)

From foot-of-the-wave analysis: more likely to be the protonation of $Fe^{III}_{superoxy}$					
FeTPP		FeDEsP		FeTEsP	
$[H^+]$ (mM)	k (s^{-1})	$[H^+]$ (mM)	k (s^{-1})	$[H^+]$ (mM)	k (s^{-1})
5	90.26	5	266.24	5	18.32
10	171.91	10	691.76	10	24.41
20	302.27	30	1654.08	20	36.97
30	426.92	50	2545.47	30	45.54
40	541.84	80	3922.07	40	51.51
50	662.94	100	4312.02	50	54.58
100	1134.62	120	6131.38		
From K–L analysis					
FeTPP		FeDEsP		FeTEsP	
pH	k_{cat} (M^{-1} s^{-1})	pH	k_{cat} (M^{-1} s^{-1})	pH	k_{cat} (M^{-1} s^{-1})
7	6.53×10^6	7	1.43×10^7	7	1.98×10^6
—	—	9	9.37×10^6	9	2.70×10^6
—	—	11	1.77×10^7	11	2.33×10^6

the resting ferric complex. However, in the aqueous medium, the counter anion has been exchanged with the hydroxide anion (^-OH). Therefore, the reduction of the Fe^{III}–OH is associated with the concomitant protonation of the hydroxide ligand to water, which is evident from the slope of the $E°$ *versus* pH plot (Fig. S3†).

Electrocatalytic oxygen reduction

Under homogeneous conditions in organic solvent. The cyclic voltammograms of the triflate anion bound complexes under O_2 saturated DMF have a small change in current compared to that of under deaerated DMF. However, when an acid (protonated DMF, [HDMF]OTf) was added, the CV showed a large catalytic current corresponding to electrocatalytic O_2 reduction (Fig. 4A–C). The rate observed for FeTPP is in agreement with previous assessments.[17] As compared to FeTPP, the onset potential as well as the E_{ORR} (obtained from the first-derivative of the cathodic wave) shift to a more positive potential in the case of FeDEsP. FeTEsP shows a similar positive shift in the onset potential and E_{ORR} from FeDEsP (Fig. 4D). Therefore, the trend in the $Fe^{III/II}$ reduction potential has been mirrored in the homogeneous ORR.

The rate of the O_2 reduction reaction obtained from the foot-of-the-wave (FOWA) analysis[17] (Fig. S4† and Table 1, please also see the ESI†) suggests that under the same acid concentration (50 mM), the rate is highest for FeDEsP (2545 s^{-1}), followed by FeTPP (663 s^{-1}), and lowest for FeTEsP (54 s^{-1}). This result deviates from the scaling relationship reported previously by Mayer and co-workers, suggesting "a higher over-potential in ORR should be accompanied

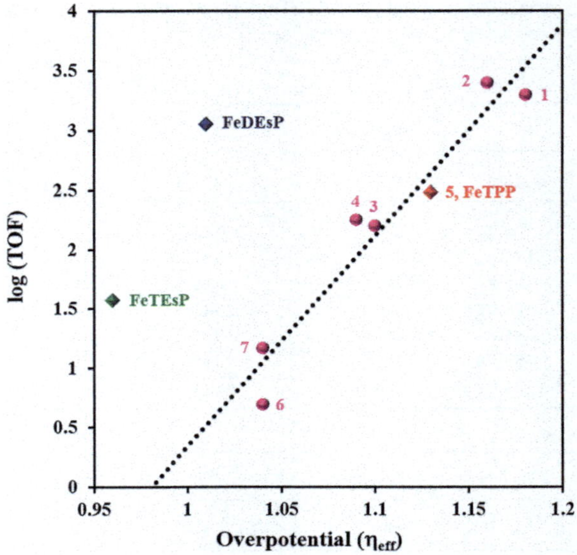

Fig. 5 log(TOF) as a function of the catalyst-specific effective overpotential (η_{eff}) with 20 mM [HDMF]OTf under 1 atm O_2 in DMF in the presence of 100 mM $TBAPF_6$ as the supporting electrolyte. FeTPP, FeDEsP, and FeTEsP are represented as a red diamond, blue diamond, and green diamond, respectively. The pink circles indicate the compounds reported by Mayer and co-workers under the same conditions, where the porphyrins differ by the *meso*-substitutions:[17] (1) Ph-2-CO_2H; (2) Ph-2-CO_2Me; (3) Ph-4-OMe; (4) Ph-4-SO_3H; (5) Ph(FeTPP); (6) Ph-4-Br; (7) Ph-4-CO_2H.

with higher rate" as FeDEsP has an order of magnitude higher rate than FeTPP despite having >100 mV lower overpotential. In the log(TOF) *versus* effective overpotential (η_{eff}) plot (Fig. 5), it is evident that under the same conditions, FeDEsP and FeTEsP deviate from the linear relationship, while other iron-complexes with tetraphenylporphyrin ligand framework, maintain a linear correlation.[17]

Under heterogeneous conditions in aqueous solvent. The catalysts physiadsorbed on the electrode immersed in an air-saturated pH 7 buffer. While sweeping the potential of the $Fe^{III/II}$ reduction potential, the CV was replaced by a large catalytic wave corresponding to the reduction of O_2 (Fig. 6A–C). However, in the case of FeTPP, the potential corresponding to the maximum catalytic current is at the $Fe^{III/II}$ reduction potential (Fig. 6A). On the contrary, for FeDEsP, the current reaches the maximum before the $Fe^{III/II}$ reduction potential (Fig. 6B), and hence it can be considered to be involved in "total catalysis", as suggested by Savéant and co-workers.[21,22] However, FeTEsP reaches the maximum current at ~100 mV lower potential than that of its $Fe^{III/II}$ reduction potential (Fig. 6C). Linear sweep voltammetry (LSV) of these catalysts at different rotation rates shows a substrate diffusion-limited catalytic O_2 reduction current below 100 mV *vs.* Ag/AgCl (satd. KCl), indicative of mass transfer limited O_2 reduction. These current increases with increase in the rotation rates following the Koutecky–Levich (K–L) equation, $I^{-1} = i_K(E)^{-1} + i_L^{-1}$, where $i_K(E)$ is the potential dependent kinetic current and i_L is the Levich current. The Levich current, i_L expressed as $0.62nFA$

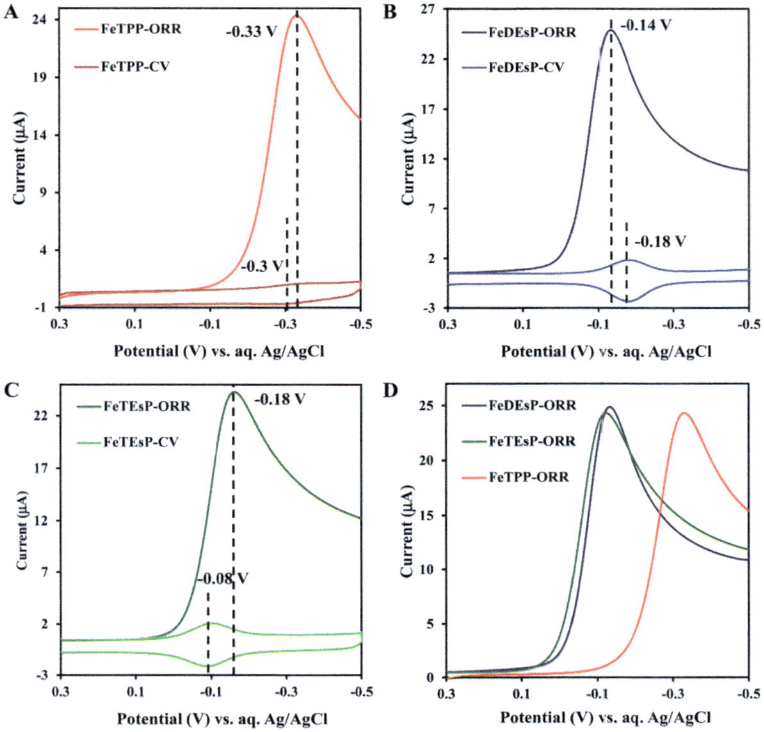

Fig. 6 Electrocatalytic O_2 reduction of the complexes (A–C) under heterogeneous conditions in aerobic pH 7 phosphate buffer using aqueous Ag/AgCl in 4 M KCl as a reference, and platinum counter in the presence of 100 mM KPF_6 as the supporting electrolyte. The catalysts were physiadsorbed on the EPG electrode. The $Fe^{III/II}$ reduction potentials and the potentials of the peak current during catalysis are shown as dotted lines. (D) Overlay of the linear-sweep voltammetry (LSV) data under aerobic conditions.

$[O_2](D_{O_2})^{2/3}\omega^{1/2}\nu^{-1/6}$, where n is the number of electrons transferred to the substrate, A is the macroscopic area of the disk (0.096 cm^2), $[O_2]$ is the concentration of O_2 in an air-saturated buffer (0.26 mM) at 25 °C, D_{O_2} is the diffusion coefficient of O_2 (1.8×10^{-5} cm^2 s^{-1}) at 25 °C, ω is the angular velocity of the disk, and ν is the kinematic viscosity of the solution (0.009 cm^2 s^{-1}) at 25 °C.[23,24] From the slope of the I^{-1} *versus* the inverse square root of the angular velocity of the disk ($\omega^{-1/2}$), the number of electrons transferred to the substrate can be obtained. The slopes obtained from the experimental data for the complexes are almost identical to the theoretical slope predicted for a $4e^-$ O_2 reduction process (Fig. S5†), suggesting the complexes undergo complete reduction of O_2 to water. The second-order rate constants of ORR catalysis (k_{cat}) can also be obtained from the intercept of this K–L plot using the equation: $i_K = nFAk_{cat}\Gamma_{cat}[O_2]$, where Γ_{cat} is the surface coverage of the catalysts on the electrode surface. The k_{cat} values are obtained to be $(6.53 \pm 0.5) \times 10^6$, $(1.43 \pm 0.4) \times 10^7$, and $(1.98 \pm 0.2) \times 10^6$ M^{-1} s^{-1} for complexes FeTPP, FeDEsP, and FeTEsP, respectively (Table 1, ESI†). Hence, the trend FeDEsP ≫ FeTPP > FeTEsP in the rate of ORR under homogeneous conditions is mirrored in the heterogeneous condition, *i.e.*, FeDEsP can catalyze

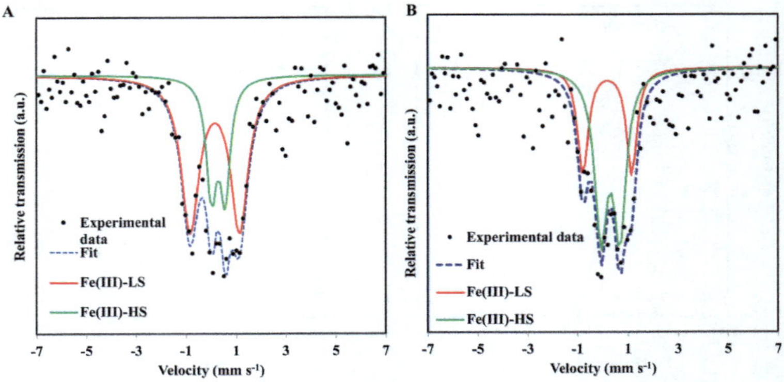

Fig. 7 Mössbauer spectra of (A) FeDEsP-superoxy and (B) FeTEsP-superoxy complexes (10 mM), collected in THF solution of the complex at 100 K.

the reaction at a lower overpotential but at a higher rate. Also, FeTEsP, despite the fact that it possesses ~100 mV higher $Fe^{III/II}$ reduction potential than FeDEsP, can reduce O_2 to water at the same potential FeDEsP catalyzes the reaction (Fig. 6D).

Mechanistic investigations

O_2 reduction is a multi-step reaction that requires both oxygen and H^+. Generally, the reaction proceeds *via* O_2 binding to ferrous iron porphyrins, which is a fast step, resulting in the formation of an iron dioxygen adduct. This adduct is a diamagnetic species where a low-spin Fe^{III} porphyrin is anti-ferromagnetically coupled to a $O_2^{-\bullet}$ species; generally described as Fe^{III}–$O_2^{-\bullet}$.[25–28] To investigate this, FeDEsP and FeTEsP were reduced to their ferrous forms. Their reaction with O_2 is followed at −40 °C (liquid nitrogen/acetonitrile bath) to trap and characterize the species formed. Upon bubbling the O_2 gas into the solution of Fe^{II}DEsP at −40 °C for 3 minutes, the Mössbauer spectra possess a major (~69.3%) species with $\delta_{iso} = 0.12 \pm 0.02$ mm s^{-1} and $\Delta E_Q = 1.93 \pm 0.04$ mm s^{-1} (Fig. 7A). This corresponds to a low-spin Fe(III) species, which is likely the Fe^{III}–$O_2^{-\bullet}$ species. This species is prone to solvolysis to form a high spin Fe^{III}porphyrin which is the other species observed having a $\delta_{iso} = 0.28 \pm 0.02$ mm s^{-1} and $\Delta E_Q = 0.68 \pm 0.03$ mm s^{-1}. Similarly, the spectra for the superoxy intermediate of FeTEsP possess $\delta_{iso} = 0.16 \pm 0.03$ mm s^{-1} and $\Delta E_Q = 1.93 \pm 0.06$ mm s^{-1} (Fig. 7B) along with the final Fe^{III}porphyrin product with $\delta_{iso} = 0.28 \pm 0.03$ mm s^{-1} and $\Delta E_Q = 0.75 \pm 0.05$ mm s^{-1}. Therefore, both the complexes form a low-spin $Fe^{III}_{superoxy}$ intermediate. However, the isomer shift of the Fe^{III}–$O_2^{-\bullet}$ species of FeDEsP and FeTEsP is much lower than that of myoglobin and other porphyrins with tetraphenyl backbone, consistent with the higher Lewis acidity of the metal centres.[29–33]

The difference between FeTPP, FeDEsP, and FeTEsP observed in organic solutions deserve special attention as the systematic introduction of the EWG results in a lowering in overpotential as well as an increase in the rate of ORR, contrary to the expectations from the scaling relationship. Previous investigations have indicated that the protonation of the Fe^{III}–$O_2^{-\bullet}$ species is likely to be the rate-determining step (rds) in these cases.[17,34] The rate of ORR in organic medium

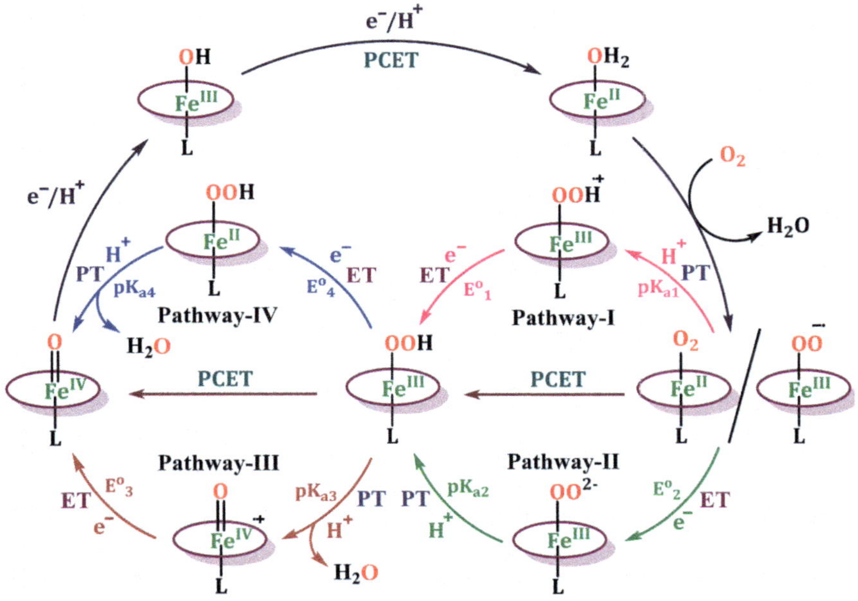

Scheme 2 Plausible routes of the $4e^-/4H^+$ reduction of O_2 to water. Where, E°_1, E°_2, E°_3, and E°_4 represent $E^\circ_{\text{Fe(III) protonated-superoxy/hydroperoxy}}$, $E^\circ_{\text{Fe(III) superoxy/peroxy}}$, $E^\circ_{\text{compound-I/compound-II}}$, $E^\circ_{\text{Fe(III) hydroperoxy/Fe(II) hydroperoxy}}$ processes, respectively; pK_{a1}, pK_{a2}, pK_{a3}, ad pK_{a4} represent the pK_a of $Fe^{III}_{\text{superoxy}}$, Fe^{III}_{peroxy}, $Fe^{III}_{\text{hydroperoxy}}$, and $Fe^{II}_{\text{hydroperoxy}}$ species, respectively. The oval ring represents a porphyrin ligand and L represents a proximal axial ligand bound to iron.

was determined under different $[H^+]$ and the results show that for FeDEsP and FeTPP, the rate of ORR increases linearly with $[H^+]$ (Table 1). A ten fold increase in rate is achieved with a ten fold increase in $[H^+]$, indicating a first-order dependence of the rate with $[H^+]$, suggesting that the rds involves a H^+. FeTEsP on the other hand shows much lower dependence of the rate of ORR on $[H^+]$ indicating that the nature of TS is different in FeTEsP relative to FeDEsP and FeTPP. The rate of ORR is also determined in aqueous media at different pH's (Table 1). The data show that a change of 4 units of pH, which corresponds to a change in 4 orders of magnitude in proton concentration, does not change the rate substantially. Thus, the TS of FeDEsP is different in organic solvent, where its 1st order with $[H^+]$, relative to aqueous medium where the rate is almost $[H^+]$ independent.

Thus, the minimal dependence of ORR rate on $[H^+]$ by FeTEsP in organic medium is not only interesting but also revealing in that the EWG's must be offsetting the redox as well as acid–base properties of the $Fe^{III}–O_2^{-\bullet}$ species of these porphyrin ligands differently. And the fact that none of these shows any $[H^+]$ dependence in near neutral aqueous conditions is an interesting departure from the mechanism offered for ORR by iron porphyrins in organic solvents and is likely due to the pK_a's of the intermediate species involved and is discussed in later sections. The $Fe^{III}–O_2^{-\bullet}$ species, once formed during ORR in organic solvents, is then further reduced to Fe^{III}–OOH species. This process requires an electron and a proton and can follow either a proton transfer (PT) followed by an electron transfer (ET) pathway (PTET) or an ETPT or a proton-coupled electron

Table 2 Theoretical prediction for the $Fe^{III}_{superoxy/peroxy}$ reduction potential ($E°$ in V) and pK_a, in water

Complexes	$\Delta E°$ (mV)	ΔpK_a
FeTPP	—	—
FeDEsP	130	−1.79
FeTEsP	190	−2.47

transfer (PCET). Iron tetraphenyl porphyrin derivatives have been proposed to follow a PTET pathway in the presence of a strong acid.[11] Under homogeneous conditions, after the initial ET, Fe^{II}TPP undergoes O_2 binding to form the Fe^{III}–$O_2^{-\bullet}$ species. The protonation of the $Fe^{III}_{superoxy}$ species is proposed to be the rate-determining step due to the very high activation barrier (pathway-I in Scheme 2). The barrier is proposed to stem from the requirement of a pre-association complex involving the acid, solvent, and the $Fe^{III}_{superoxy}$ species – an entropically disfavored trimolecular step.[34] The differences in $[H^+]$ dependence observed in FeTEsP relative to FeDEsP must originate from the electronic structure and reactivity of their Fe^{III}–$O_2^{-\bullet}$ species. The Mössbauer data on this species did not reveal substantial differences in the electronic structure. Electronic structure calculations can help analyze the effect of the EWG on the bound axial ligands.

Geometry-optimized density functional theory (DFT) calculations were performed on these complexes to understand the thermodynamics of the $1e^-/1H^+$ Fe^{III}–$O_2^{-\bullet}$ to Fe^{III}–OOH process such as the reduction potential ($E°$) and pK_a (Table 2). The $E°$ represents the one-electron reduction potential of $Fe^{III}_{superoxy}$ and Fe^{III}_{peroxy} process, which is obtained from Gibb's free energy ($\Delta G°$) change in the electron transfer equilibrium between $Fe^{III}_{superoxy}$ and Fe^{III}_{peroxy}. The pK_a is obtained from the change in $\Delta G°$ of the protonation equilibrium between Fe^{III}_{peroxy} to

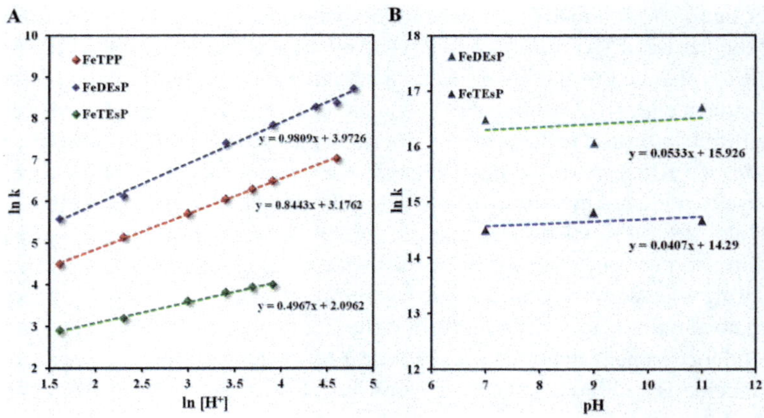

Fig. 8 (A) The plot of the logarithm of rate of ORR under homogeneous conditions *versus* the logarithm of the concentration of acid ([HDMF]OTf) under 1 atm O_2 in DMF in the presence of 100 mM $TBAPF_6$ as the supporting electrolyte. (B) The plot of logarithm of 2nd order rate constant of ORR under heterogeneous conditions *versus* pH of the solution under 1 atm O_2 in phosphate.

$Fe^{III}_{hydroperoxy}$ species. Clearly, successive addition of the EWGs, the pK_a of the $Fe^{III}_{superoxy}$ and the Fe^{III}_{peroxy} species, decreases. Note that the pK_a of Fe^{III}–OOH species have been shown to be between 5.3–7 in aqueous media,[14,35] whereas the pK_a of $O_2^{-\bullet}$ is 4.8 (ref. 36) which is expected to be lowered substantially on binding to Fe^{III}, such that a Fe^{III}–$O_2^{-\bullet}$ species should have substantially lower pK_a than a Fe^{III}–OOH species.[37–39] Therefore, if the acid used here in an organic solvent can protonate the Fe^{III}–$O_2^{-\bullet}$ species, it can easily protonate the Fe^{III}–OOH species which is why the rds for ORR is the former step, as the later step must be much faster. Given the fact that the rate of ORR by FeDEsP is an order of magnitude faster than FeTPP, while the pK_a of the Fe^{III}–$O_2^{-\bullet}$ species is lower in the former, FeDEsP cannot have the same rate-determining step, *i.e.*, the protonation of the $Fe^{III}_{superoxy}$ species. On the contrary, the $Fe^{III}_{superoxy/peroxy}$ reduction potential has been increased by ~130 mV. Thus, the electron transfer to the $Fe^{III}_{superoxy}$ species should be more facile in FeDEsP relative to FeTPP to form the Fe^{III}_{peroxy} species.

From the linear free-energy relationship, rate constant (k) and the concentration of acid can be related by the equation: $\ln k = \alpha \ln[H^+] + c$; where, c is a constant, and α estimates the participation of H^+ in the rate-limiting transition state, and it varies from 0–1.[40] It is evident (Fig. 8A) that α is close to 1 in the case of FeDEsP and FeTPP and only 0.5 for FeTEsP, suggesting the rds for FeDEsP and FeTPP must be a PT step. For FeTPP, the protonation of the $Fe^{III}_{superoxy}$ species is the rds, for FeDEsP, which also shows linear dependence on $[H^+]$, it has to be a different PT step which is the rds. The protonation of the $Fe^{III}_{hydroperoxy}$ species, which is not the rds in ORR by FeTPP, should be at least an order of magnitude faster and is likely to be the rds in ORR by FeDEsP. Due to the presence of two EWGs, the pK_a of the $Fe^{III}_{superoxy}$ species is calculated to be lower by almost 2 units relative to FeTPP. Considering the fact that the protonation of the $Fe^{III}_{superoxy}$ by [HDMF]OTf was already determined to be the rds for ORR in organic solvents and computed to have ~10 kcal mol^{-1} barrier,[34] PT from the same acid may no longer be feasible for $Fe^{III}_{superoxy}$ species of FeDEsP. Therefore, the mechanism of reduction of Fe^{III}–$O_2^{-\bullet}$ to Fe^{III}–OOH changes from rate-determining PTET in FeTPP to a facile ETPT in FeDEsP (pathway-II in Scheme 2). For FeDEsP, the rds is the protonation of the $Fe^{III}_{hydroperoxy}$ species (pathway-III in Scheme 2), which likely undergoes heterolytic O–O bond cleavage to form compound-I. The protonation of the proximal oxygen of the Fe^{III}–OOH species leads to the release of H_2O_2 (*i.e.*, partial reduction of O_2 to H_2O_2), while the protonation on the distal oxygen results in O–O bond cleavage (*i.e.*, complete reduction of O_2 to H_2O).[41–43] Previous experiments using FeTPP based iron porphyrins show substantial selectivity towards complete reduction of O_2 to H_2O.[34] Thus, FeDEsP breaks away from the scaling relation and exerts an order of magnitude higher rate than that of FeTPP because the modification of the porphyrin ligand has led to change in the rds.

Now, in the case of FeTEsP, the pK_a of the $Fe^{III}_{superoxy}$ species is even lower than that of FeDEsP. But the $Fe^{III}_{superoxy/peroxy}$ reduction potential is even higher. Hence there is likely a facile ET followed by a facile PT on the Fe^{III}_{peroxy} species to generate the $Fe^{III}_{hydroperoxy}$ species. Due to the presence of four EWGs, the Lewis acidity of the iron-centre in FeTEsP is higher than both FeTPP as well as FeDEsP. Hence, the pK_a of the species should be lower than that of FeDEsP. Note that protonation to the $Fe^{III}_{hydroperoxy}$ species cannot be the rds in FeTEsP, because α is substantially less that 1.0 suggesting lower proton-dependence in the rate-limiting transition state (Fig. 8A). It is likely that either a PCET or an ETPT to the Fe^{III}–OOH species is likely to be the rds of ORR for FeTEsP (pathway IV, Scheme 2).

Under heterogeneous conditions, the dependence of ORR rate on proton concentration (pH) is complicated by the fact that the selectivity changes from $4e^-/4H^+$ at neutral and basic pH to $2e^-/2H^+$ in acidic pH. The pK_a of FeTPP based species with neutral trans-axial ligands is determined to be 5.3–7.[14,35] Similarly, the pK_a of the $Fe^{III}_{superoxy}$ species would be very low in an aqueous solvent and pK_a of free $O_2^{-•}$ is ~4.8.[36] Furthermore, due to the presence of the EWGs in FeDEsP and FeTEsP, the pK_a of both $Fe^{III}_{superoxy}$ and $Fe^{III}_{hydroperoxy}$ will be lowered even further such that protonation of these species will not be possible around neutral pH. Thus, in the range where the ORR is selective for $4e^-/4H^+$, the 2nd order ORR rate (from K–L analysis) does not show any noticeable change (Fig. 8B) on changing the [H^+] by four orders of magnitude (*i.e.*, four units of pH). Thus, the rds are almost independent of [H^+] in an aqueous medium for both FeDEsP and FeTEsP. Our previous SERRS-RDE experiments revealed that in FeTPP based iron-porphyrins bearing water or any neutral axial ligand, the heterolysis of the O–O bond is the rds, and it proceeds through an ETPT process (pathway IV in Scheme 2) through the protonation of the $Fe^{II}_{hydroperoxy}$ species.[13,14] It is likely to be the same case here for FeDEsP and FeTEsP. Accordingly, at pH 7, there is a linear correlation between the logarithm of rate and the overpotential of ORR, suggesting an ET step is indeed the rate-determining step (Fig. S6†).

For the scaling relationship between overpotential and log(rate) proposed to work, the mechanism investigated for the series remains the same, *i.e.*, for the series, the rds remains the same.[17] When the rds is the protonation of the $Fe^{III}_{superoxy}$ intermediate, the pK_a of the species is governed by the Lewis acidity of iron, which in turn also dictates the $E_{1/2}$, resulting in a linear correlation between $E_{1/2}$ and log k. However, as the nature of rds changes, the scaling relationship will not be followed. Not only is the modification of the ligand framework, as demonstrated here, changing the metal and, as recently demonstrated, introducing a functional group, can exert a through-space effect on the metal which breaks the scaling relationship as all of these lead to change of the rds.[44,45] Based on such experimental deviations, Costentin and Saveant have recently suggested that the scaling relationship proposed by Mayer and co-workers is unlikely to be very general.[44]

Conclusion

The scaling relationship for the electrocatalytic ORR developed by Mayer and co-workers holds for a series of complexes. However, where the electronic structure of the system has been significantly affected, either by the modification in the porphyrin-core structure or through the active participation of the second coordination sphere, the rds of the ORR, a multi-step reaction, changes. In those situations, the $E°$, pK_a and intra-molecular PT rates may be tuned abruptly, resulting in a change in the rds of the reaction and a departure from a scaling relation established assuming a specific step to be the rds. Attempting to summarize a multi-step reaction on the basis of a single scaling descriptor is difficult and sometimes erroneous as the rds changes between different systems. Mechanistic investigations, either *in situ* or in solutions using different experimental conditions *via operando* trapping of the reactive intermediates, is vital for the proper understanding of the factors that control the thermodynamics, kinetics, and the selectivity of ORR.

Conflicts of interest

There are no conflicts to declare.

Acknowledgements

This research was funded by the Department of Science and Technology Grant EMR-008063 and STR/2019/000081. S. Amanullah and P. Saha would like to acknowledge a fellowship from IACS-institute fellowship program and UGC, India, respectively.

References

1 M. L. Pegis, C. F. Wise, D. J. Martin and J. M. Mayer, *Chem. Rev.*, 2018, **118**, 2340–2391.
2 E. I. Solomon and S. S. Stahl, *Chem. Rev.*, 2018, **118**, 2299–2301.
3 A. A. Gewirth, J. A. Varnell and A. M. DiAscro, *Chem. Rev.*, 2018, **118**, 2313–2339.
4 S. Ferguson-Miller and G. T. Babcock, *Chem. Rev.*, 1996, **96**, 2889–2908.
5 K. Muramoto, K. Ohta, K. Shinzawa-Itoh, K. Kanda, M. Taniguchi, H. Nabekura, E. Yamashita, T. Tsukihara and S. Yoshikawa, *Proc. Natl. Acad. Sci.*, 2010, **107**, 7740–7745.
6 P. L. Dutton, D. F. Wilson and C.-P. Lee, *Biochemistry*, 1970, **9**, 5077–5082.
7 W. S. Caughey, G. A. Smythe, D. H. O'Keeffe, J. E. Maskasky and M. I. Smith, *J. Biol. Chem.*, 1975, **250**, 7602–7622.
8 T. Tomitake, A. Hiroshi, Y. Eiki, T. Takashi, Y. Hiroshi, S.-I. Kyoko, N. Ryosuke, Y. Rieko and Y. Shinya, *Science*, 1995, **269**, 1069–1074.
9 S. Amanullah, A. Singha and A. Dey, *Coord. Chem. Rev.*, 2019, **386**, 183–208.
10 S. Amanullah, P. Saha, R. Saha and A. Dey, *Inorg. Chem.*, 2019, **58**, 152–164.
11 D. J. Martin, C. F. Wise, M. L. Pegis and J. M. Mayer, *Acc. Chem. Res.*, 2020, **53**, 1056–1065.
12 S. Chatterjee, K. Sengupta, B. Mondal, S. Dey and A. Dey, *Acc. Chem. Res.*, 2017, **50**, 1744–1753.
13 K. Mittra, S. Chatterjee, S. Samanta and A. Dey, *Inorg. Chem.*, 2013, **52**, 14317–14325.
14 S. Chatterjee, K. Sengupta, S. Samanta, P. K. Das and A. Dey, *Inorg. Chem.*, 2015, **54**, 2383–2392.
15 S. Chatterjee, K. Sengupta, S. Samanta, P. K. Das and A. Dey, *Inorg. Chem.*, 2013, **52**, 9897–9907.
16 S. Bhunia, A. Rana, P. Roy, D. J. Martin, M. L. Pegis, B. Roy and A. Dey, *J. Am. Chem. Soc.*, 2018, **140**, 9444–9457.
17 M. L. Pegis, B. A. McKeown, N. Kumar, K. Lang, D. J. Wasylenko, X. P. Zhang, S. Raugei and J. M. Mayer, *ACS Cent. Sci.*, 2016, **2**, 850–856.
18 S. Amanullah and A. Dey, *Chem. Sci.*, 2020, **11**, 5909–5921.
19 S. Amanullah, P. K. Das, S. Samanta and A. Dey, *Chem. Commun.*, 2015, **51**, 10010–10013.
20 S. Amanullah and A. Dey, *JBIC, J. Biol. Inorg. Chem.*, 2019, **24**, 437–442.

21 J. M. Savéant and K. B. Su, *J. Electroanal. Chem. Interfacial Electrochem.*, 1984, **171**, 341–349.

22 C. Costentin and J.-M. Savéant, *Elements of Molecular and Biomolecular Electrochemistry*, 2019, pp. 81–181.

23 J.-M. Savéant, *Elements of Molecular and Biomolecular Electrochemistry: an Electrochemical Approach to Electron Transfer Chemistry*, Wiley-Interscience, Hoboken, NJ, 2006.

24 L. R. Bard and A. J. Faulkner, *Electrochemical Methods, Fundamentals and Applications*, John Wiley and Sons, New York, 1980.

25 A. Singha, P. K. Das and A. Dey, *Inorg. Chem.*, 2019, **58**, 10704–10715.

26 L. Pauling and C. D. Coryell, *Proc. Natl. Acad. Sci.*, 1936, **22**, 210–216.

27 L. Pauling, *Nature*, 1964, **203**, 182–183.

28 J. J. Weiss, *Nature*, 1964, **202**, 83–84.

29 K. Duerr, J. Olah, R. Davydov, M. Kleimann, J. Li, N. Lang, R. Puchta, E. Hübner, T. Drewello, J. N. Harvey, N. Jux and I. Ivanović-Burmazović, *Dalton Trans.*, 2010, **39**, 2049–2056.

30 R. Garcia-Serres, R. M. Davydov, T. Matsui, M. Ikeda-Saito, B. M. Hoffman and B. H. Huynh, *J. Am. Chem. Soc.*, 2007, **129**, 1402–1412.

31 P. G. Debrunner, in *Physical Bioinorganic Chemistry Series: Iron Porphyrins Part III*, ed. H. B. Lever and A. B. P. Gray, VCH Publishers, New York, 1989, vol. 4, pp. 137–234.

32 J. Li, B. C. Noll, A. G. Oliver, C. E. Schulz and W. R. Scheidt, *J. Am. Chem. Soc.*, 2013, **135**, 15627–15641.

33 N. Godbout, L. K. Sanders, R. Salzmann, R. H. Havlin, M. Wojdelski and E. Oldfield, *J. Am. Chem. Soc.*, 1999, **121**, 3829–3844.

34 M. L. Pegis, D. J. Martin, C. F. Wise, A. C. Brezny, S. I. Johnson, L. E. Johnson, N. Kumar, S. Raugei and J. M. Mayer, *J. Am. Chem. Soc.*, 2019, **141**, 8315–8326.

35 S. Mukherjee, S. Bandyopadhyay, S. Chatterjee and A. Dey, *Chem. Commun.*, 2014, **50**, 12304–12307.

36 E. L. G. Samuel, D. C. Marcano, V. Berka, B. R. Bitner, G. Wu, A. Potter, R. H. Fabian, R. G. Pautler, T. A. Kent, A.-L. Tsai and J. M. Tour, *Proc. Natl. Acad. Sci.*, 2015, **112**, 2343–2348.

37 E. A. Mohamed, Z. N. Zahran and Y. Naruta, *Chem. Commun.*, 2015, **51**, 16900–16903.

38 T. Ohta, P. Nagaraju, J.-G. Liu, T. Ogura and Y. Naruta, *JBIC, J. Biol. Inorg. Chem.*, 2016, **21**, 745–755.

39 H. Kim, P. J. Rogler, S. K. Sharma, A. W. Schaefer, E. I. Solomon and K. D. Karlin, *J. Am. Chem. Soc.*, 2020, **142**, 3104–3116.

40 J. E. Leffler, *Science*, 1953, **117**, 340–341.

41 B. Meunier, S. P. de Visser and S. Shaik, *Chem. Rev.*, 2004, **104**, 3947–3980.

42 W. Zhang, W. Lai and R. Cao, *Chem. Rev.*, 2017, **117**, 3717–3797.

43 S. Dey, B. Mondal, S. Chatterjee, A. Rana, S. Amanullah and A. Dey, *Nat. Rev. Chem.*, 2017, **1**, 98.

44 C. Costentin and J.-M. Savéant, *J. Am. Chem. Soc.*, 2018, **140**, 16669–16675.

45 D. J. Martin, B. Q. Marcado and J. M. Mayer, *Sci. Adv.*, 2021, **6**, eaaz3318.

Faraday Discussions

ROYAL SOCIETY OF CHEMISTRY

PAPER

Comparative analysis of lanthanide excited state quenching by electronic energy and electron transfer processes†

David Parker, [*ab] Jack D. Fradgley, [a] Martina Delbianco, [a] Matthieu Starck, [a] James W. Walton [a] and Jurriaan M. Zwier [c]

Received 27th August 2021, Accepted 5th November 2021
DOI: 10.1039/d1fd00059d

The relative sensitivities of structurally related Eu(III) complexes to quenching by electron and energy transfer processes have been compared. In two sets of 9-coordinate complexes based on 1,4,7-triazacyclononane, the Eu emission lifetime decreased as the number of conjugated sensitising groups and the number of unbound ligand N atoms increased, consistent with photoinduced electron transfer to the excited Eu(III) ion that is suppressed by N-protonation. Quenching of the Eu $^5D_0$ excited state may also occur by electronic energy transfer, and the quenching of a variety of 9-coordinate complexes by a cyanine dye with optimal spectral overlap occurs by an efficient FRET process, defined by a Förster radius (R_0) value of 68 Å and characterised by second rate constants in the order of 10^9 M^{-1} s^{-1}; these values were insensitive to changes in the ligand structure and to the overall complex hydrophilicity. Quenching of the Eu and Tb excited states by energy transfer to Mn(II) and Cu(II) aqua ions occurred over much shorter distances, with rate constants of around 10^6 M^{-1} s^{-1}, owing to the much lower spectral overlap integral. The calculated R_0 values were estimated to be between 2.5 to 4 Å in the former case, suggesting the presence of a Dexter energy transfer mechanism that requires much closer contact, consistent with the enhanced sensitivity of the rate of quenching to the degree of steric shielding of the lanthanide ion provided by the ligand.

Introduction

The sensitised emission of lanthanide luminescence has been studied in detail because of the wide range of applications that have arisen in developing luminescence bioassays, in devising targeted optical imaging agents and in the creation of selective analytical probes and sensors for a wide range of analytes.[1–4] The

[a]Department of Chemistry, Durham University, South Road, Durham, DH1 3LE, UK. E-mail: david.parker@dur.ac.uk

[b]Department of Chemistry, Hong Kong Baptist University, Kowloon Tong, Hong Kong

[c]Cisbio Bioassays, 30200 Codolet, BP 84175, France

† Electronic supplementary information (ESI) available: Photophysical data and selected spectra. See DOI: 10.1039/d1fd00059d

salient properties of the sensitising chromophore have been considered in depth, and key design criteria have emerged.[5–7] In particular, careful consideration has been given to the photophysical properties of the sensitiser, notably its molar absorptivity (high ε) at the excitation wavelength, the energies of its internal charge transfer (ICT), singlet and triplet states and the size of the energy gap that determines the facility of inter-system crossing. A small S_1–T_1 energy gap favours a fast rate of intersystem crossing, k_{ISC}, which typically needs to be $>10^9\ s^{-1}$ to compete with k_f, which is of the order of $3 \times 10^8\ s^{-1}$. These features determine the choice of the excitation wavelength (λ_{exc}), the efficiency of triplet excited state formation and the energetic feasibility of the overall process that allows intramolecular energy transfer to proceed, thereby populating the lanthanide excited state.

A mechanistic scheme (Scheme 1) can be put forward that characterises the competing pathways in sensitised lanthanide emission.[7] The salient excited states have a transient existence and are subject to different radiative or non-radiative decay processes, each defined by a rate constant. A downhill energy cascade requires that the sensitising excited state donor lies close in energy, but more than about 10 k_BT (2050 cm^{-1} at ambient temperature) higher than the accepting lanthanide excited state. In this way, any back energy transfer process is not prone to thermal activation. The excited states possess notably different lifetimes: the relaxed S_1 and ICT states live for a few nanoseconds, the triplet state may often exist for a few microseconds, but the lanthanide excited state is the longest lived, with lifetimes ranging from the microsecond (*e.g.* Dy, Yb, Nd, Er) to the millisecond (Eu, Tb) domain. The long lifetime and large pseudo-Stokes' shifts that epitomise sensitised lanthanide luminescence have allowed the use of time-gated methods of data acquisition for spectroscopy or

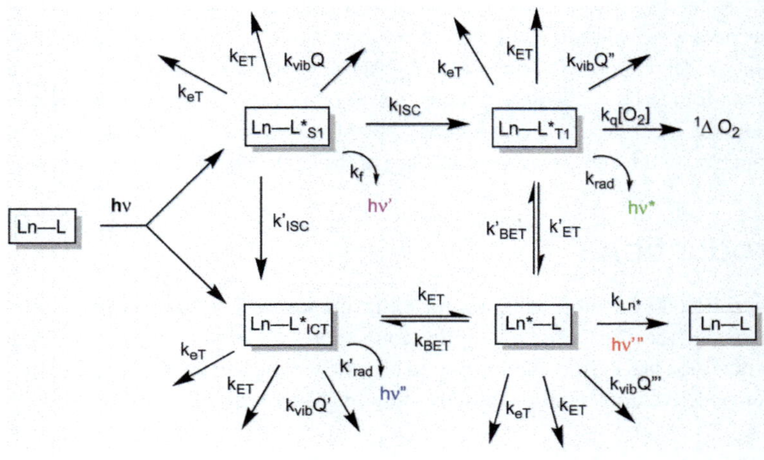

Scheme 1 Mechanistic pathways in sensitised lanthanide luminescence, where k_{eT} and k_{ET}/k_{BET} refer to electron and forward/back energy transfer processes, k_{ISC} refers to the rate constant of inter-system crossing, k_{vib} relates to the vibrational energy transfer to neighbouring energy matched oscillators and k_q is the second order rate constant characterising the quenching of a ligand triplet state by molecular oxygen, forming singlet oxygen.

microscopy, averting problems associated with Rayleigh light scattering, self-absorption or auto-fluorescence.

The non-radiative deactivation of the lanthanide excited state may occur by three main pathways involving the transfer of charge, electronic energy or vibrational energy. The key aspects of each of these quenching processes have been reported and considered in detail previously.[7–10] In this article, examples of charge and electronic energy transfer are compared and considered in the context of enhancing our understanding of the mechanism of the underlying photo-physical processes, thereby informing the design of bright lanthanide complexes that are inherently resistant to excited state deactivation. The examples described here relating to charge transfer are of particular interest to the creation of pH-responsive probes that may permit the tracking of receptor internalisation into cells.[11] And, with regard to energy transfer, the efficiency of Förster resonance energy transfer (FRET)[12] from an excited lanthanide ion to a strongly absorbing cyanine dye forms the basis of many commercial time-resolved luminescence bioassays[13] and is contrasted here with the inefficiency of energy transfer to weakly absorbing transition metal aqua ions.

Results and discussion

Charge transfer quenching of the Eu excited state

Intramolecular electron transfer processes from the HOMO of a proximate electron rich aryl group or a nitrogen lone pair orbital to the Eu $^5D_0$ excited state provide examples where the Eu emission lifetime can be reduced. Consider the following series of complexes, [EuL$^{1-6}$] (Fig. 1).[14–16] The first three Eu complexes, [EuL$^{1-3}$], contain increasing numbers of electron rich chromophores with a simple methyl group in the upper aryl ring and are compared with the behaviour of [EuL$^{4-6}$], where a diethylamino substituent is present in the upper phenyl ring that is not in direct conjugation with the alkyne bond.[16]

A typical Eu emission spectrum (Fig. 2) highlights the large pseudo-Stokes' shift and the characteristic spectral fingerprint of europium(III) luminescence for this series of complexes. The emission from the europium(III) $^5D_0$ excited state decays mono-exponentially. The lifetime for each complex was measured in water and values are compared in Table 1. With [EuL$^{1-3}$], there is a small decrease in the lifetime of the emission as the number of chromophores increases, suggesting that charge transfer from the electron rich sensitising moiety quenches the Eu excited state to some degree.[8,17] Of course, the overall quenching process will also involve vibrational energy transfer to some degree, *e.g.* to higher overtones of the alkyne stretch, aryl ring vibrations or C–H stretches. In the former case, the triple bonds are located about 7 Å from the metal centre, and given the r^{-6} dependence of vibrational energy transfer, such a process is likely to be very inefficient. In the former case, this has been established through many examples involving deuterated ligands.[17] For each europium(III) complex, the metal-based emission lifetime was found to be independent of complex concentration over the range 1 to 50 μM, and did not change on degassing the solution to remove oxygen, in both acidic and basic media.

The Weller equation[18] (eqn (1)) considers the free energy of activation for an electron transfer process and can be applied to assess the feasibility of the quenching of the europium excited state by intramolecular electron transfer from

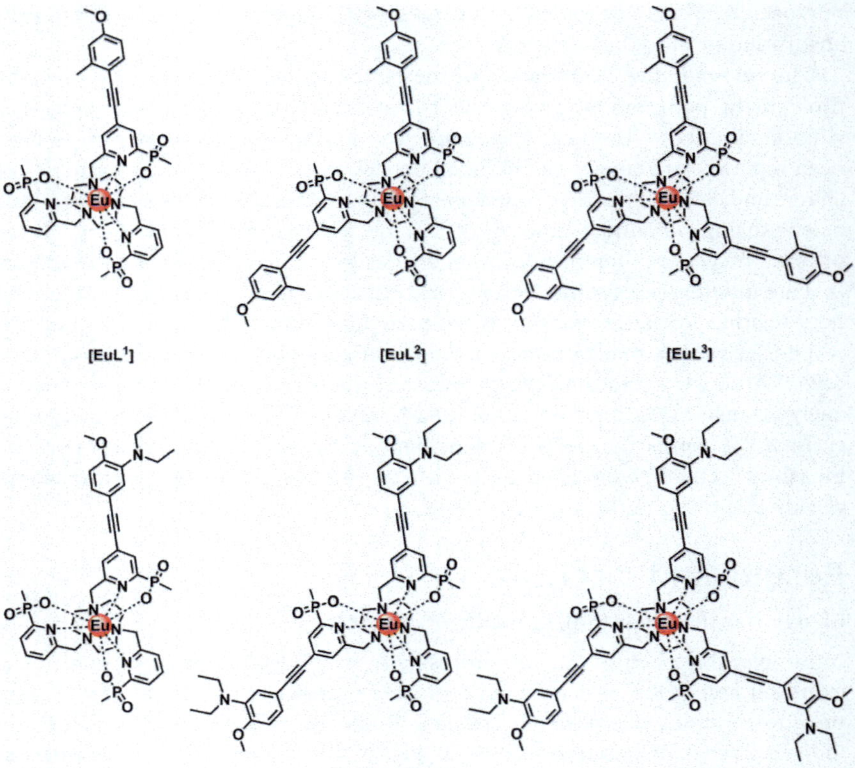

Fig. 1 Europium(III) complexes with different numbers of aryl–alkynyl sensitising groups that are pH independent [EuL^{1-3}] and pH sensitive [EuL^{4-6}] in their emission behaviour.[14–16]

the highest filled orbital, *e.g.*, the lone pair orbital on N present in the conjugate base form of [EuL^{4-6}]. The likelihood of the electron transfer event is predicted for photo-induced electron transfer processes.

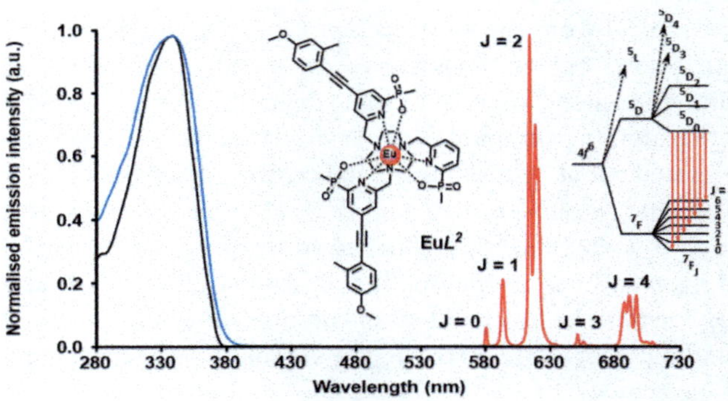

Fig. 2 Absorption (blue), excitation (black) and emission (λ_{exc} 355 nm; red) spectra of [EuL^2], showing the five major transitions to the ^7F_n ground state from the europium $^5D_0$ excited state, (295 K, 0.1 M NaCl).

Table 1 Europium excited state lifetimes for $[EuL^{1-6}]$[14–16]

Complex	$\tau_{Eu}{}^{a}$/ms	Comment
$[EuL^1]$	1.39	Small lifetime reduction with increasing number of electron rich chromophores
$[EuL^2]$	1.27	
$[EuL^3]$	1.21	
$[EuL^4]$	0.53^b	Electron transfer quenching increases with the number of N-substituted chromophores
$[EuL^5]$	0.34^b	
$[EuL^6]$	0.25^b	
$[EuHL^4]^+$	1.16^c	N-protonation suppresses electron transfer quenching
$[EuH_2L^5]^{2+}$	1.00^c	
$[EuH_3L^6]^{3+}$	0.84^c	

^{a} 295 K, 0.1 M NaCl, H_2O. ^{b} pH 8, 0.1 M NaCl, H_2O. ^{c} pH 4, 0.1 M NaCl, H_2O; pK_a values are 6.75, 6.30 and 6.21, respectively for $[EuH_nL^{4-6}]^+$.

$$\Delta G_{eT} = nF([E_{ox} - E_{red}] - E_S - e^2/\varepsilon_r)\ \text{J mol}^{-1} \quad (1)$$

In the original description, E_{ox} is the oxidation potential of a donor (*e.g.* a nitrogen lone pair or the HOMO of an electron rich aromatic group), E_{red} is the reduction potential of the acceptor (*e.g.* an aryl group or a Ln^{3+} ion), E_S is the singlet excited state energy in eV and e^2/ε_r is an attractive energy term related to the formation of a radical ion pair; this term is often less than +0.2 eV. This equation estimates the free energy of activation for the electron transfer step involving a particular excited state highlighting the facility of photo-induced electron transfer to those lanthanide ions with smaller E_{red} potential terms, where the Ln^{3+} oxidation state is relatively stabilised, *e.g.* Eu^{3+}, Yb^{3+} and to a lesser extent, Sm^{3+}.

The irreversible one electron oxidation potential of a typical *N,N*-dialkylaniline, such as *N,N*-dimethylaniline, is +0.76 V in acetonitrile (*vs.* standard calomel electrode),[19] which rises to about 1.0 V for a trialkylamine such as *N,N*-dimethylbenzylamine, where the nitrogen lone pair is not in conjugation with the aromatic ring.[20] The reduction potential of the Eu^{3+} ion when coordinated by an anionic polydentate ligand that stabilises the +3 oxidation state lies in the range −0.8 to −1.1 V, based on literature values[8,21] in MeCN, and can be estimated to be −1.0 V here. The energy of the Eu^{3+} $^5D_1$ and $^5D_0$ excited states lie at 19 100 and 17 220 cm^{-1}; the latter energy value is equivalent to +2.13 eV or 206 kJ mol^{-1}.

$$\Delta G_{eT} = F[(0.76 + 1.0) - 2.13\text{–}0.2] = -0.57,\ F = -55\ \text{kJ mol}^{-1} \quad (2)$$

Given these values, the free energy for the electron transfer process from the π-conjugated N lone pair orbital to the Eu^{3+} ion is energetically favourable by 55 kJ mol^{-1} (eqn (2)). Such an analysis assumes that the excited state reduction potential of the europium ion is the same as it is in the ground state.

The Weller analysis shows that the Eu excited state is prone to quenching by an intramolecular electron transfer process associated with the nitrogen lone pair of the conjugated diethylamino group. Next, the behaviour of the Eu(III) complexes of ligands L^4–L^6 is considered; each of these complexes exhibits a pronounced pH dependence of emission intensity and lifetime, following excitation of the ligand chromophore and intramolecular energy transfer to the Eu ion; their behaviour has been described in detail previously.[16] The large lifetime and emission intensity variations in these complexes has led to the creation of targeted luminescent pH probes for cells being developed to monitor receptor internalisation and endosomal acidification in real time.[11]

The rate of proton transfer to and from nitrogen in aqueous solution (typically 10^{10} s^{-1}) occurs much more quickly than both the rate of decay of the excited europium ion (Table 1, 10^3 s^{-1}) and the intermediate ligand ICT or triplet excited states. Furthermore, the rate of electron transfer from the lone pair orbital to the excited Eu ion is faster than the rate of energy transfer populating the Eu* $^5D_0$ state, and much faster than the rate of decay of the Eu excited state itself. Thus, during the long lifetime of the Eu* $^5D_0$ excited state in solution, over the pH range of 4 to 8, deprotonation of the amine group occurs readily, allowing fast electron transfer to occur from the unprotonated chromophore to the Eu ion, quenching the Eu excited state and shortening the 'time-averaged' observed lifetime. Such an effect is more likely to occur with increasing numbers of chromophores bearing the amine group, consistent with the observation that the tris-amine complex, [EuL^6], has the shortest measured emission lifetime at pH 8 (0.25 ms, Table 1), and the complex with only one amine-containing chromophore, [EuL^4] has the longest, both when protonated at pH 4 (1.16 ms) and as its conjugate base at pH 8 (0.53 ms).

In a key control experiment, comparing behaviour with sensitised emission after excitation of the chromophore at the isosbestic wavelength of 332 nm, the same pH dependence of the Eu emission lifetime was observed for [EuL^{4-6}], following direct excitation of the Eu ion at 397 nm. Thus, it is the Eu $^5D_0$ excited state that is being quenched by charge transfer in each of these cases, and not an intermediate ligand excited state.

Efficient energy transfer involving a cyanine dye acceptor

The rate of Förster resonance energy transfer (FRET), k_T, from a donor (D) to an acceptor (A) is commonly described by eqn (3),[12,22]

$$k_T = \frac{1}{\tau_0}\left(\frac{R_0}{r}\right)^6 \quad (3)$$

where τ_0 is the lifetime of the donor in absence of the acceptor and R_0 is the Förster distance, *i.e.* the distance at which FRET is 50% efficient. The distance R_0 has a particular value for each D–A pair, given by eqn (4),

$$R_0 = (Jk^2\Phi_0\eta^{-4})^{1/6} \times (9.79 \times 10^2) \quad (4)$$

in which J is the spectral overlap integral (in cm^3 M^{-1}), k^2 is an orientation factor (2/3), ϕ_0 is the quantum yield of the donor in the absence of transfer, and η is the refractive index of the medium. The spectral overlap integral, J, can be calculated from the donor emission spectrum and the acceptor absorption spectrum:

$$J(\lambda) = \frac{\int F(\lambda)\varepsilon(\lambda)\lambda^4 d\lambda}{\int F(\lambda)d\lambda} \quad (5)$$

where $F(\lambda)$ is the fraction of donor emission at wavelength λ and $\varepsilon(\lambda)$ is the extinction coefficient of the acceptor at wavelength λ, in units of M^{-1} cm^{-1}.

Three neutral highly luminescent Eu(III) complexes, [EuL^{7-9}] (Fig. 3) were chosen as donors, as they form well defined charge neutral complexes and their photophysical properties have been investigated in detail; they are amongst the brightest Eu(III) coordination complexes reported that absorb strongly beyond 340 nm in protic media.[2,14,15,24] The acceptor chosen for the energy transfer studies was the commercially available cyanine dye, **1** (Fig. 3), which also possesses no overall charge at neutral pH. It was selected because it and its analogues have been used in many commercial Eu-based FRET assays, and it has a large extinction coefficient at 647 nm (270 000 M^{-1} cm^{-1}). The emission spectral form of the three Eu(III) complexes presents a near-optimal profile for efficient energy transfer to this cyanine dye. Indeed, the intense hypersensitive emission band associated with the $\Delta J = 2$ transition ($^5D_0$ to $^7F_2$ at 610–620 nm) provides a very good spectral overlap with the dye acceptor absorption band (Fig. 4). Furthermore, the low intensity of Eu emission at 670 nm permits separate monitoring of the two emissive species without significant spectral overlap.

Studies of energy transfer in methanol

In the rapid diffusion limit, energy transfer from a donor to an acceptor is assumed to obey pseudo-first order kinetics (eqn (6)–(9)).[25,26] Hence, a plot of τ_0/τ *vs.* [Q] gives a straight line with a slope equal to k_2/k_0, allowing the second order

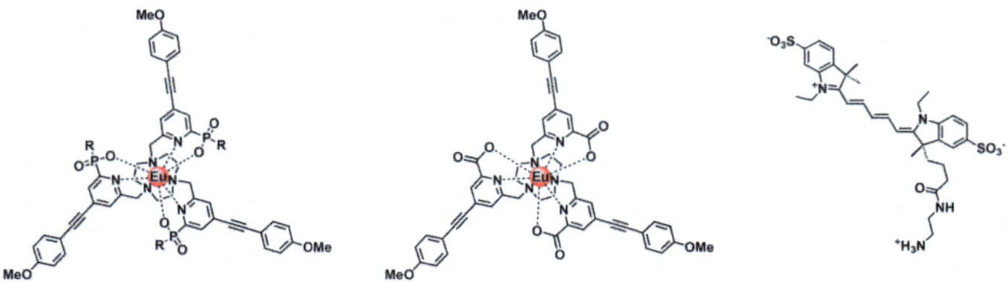

Fig. 3 Structures of [EuL^{7-9}] with different coordinated anionic groups and of the cyanine dye, **1**.

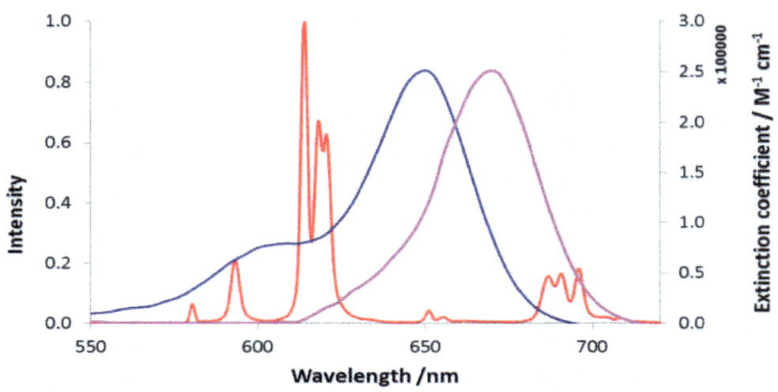

Fig. 4 Absorbance (blue) and emission (purple) spectra of the cyanine dye, **1**, compared to the [EuL^7] emission spectrum (red), (MeOH, 295 K).

rate constant for energy transfer, k_2, and the Stern–Volmer quenching constant, typically expressed as $K_{SV}{}^{-1}$, to be estimated.

$$1/\tau_0 = k_0 \tag{6}$$

$$1/\tau = k_{obs} \tag{7}$$

$$k_{obs} = k_0 + k_2[Q] \tag{8}$$

$$\tau_0/\tau = k_{obs}/k_0 = 1 + k_2/k_0[Q] = 1 + K_{SV}[Q] \tag{9}$$

The quenching of the three Eu(III) complexes by the cyanine dye, **1**, was examined by observing changes in the Eu(III) $^5D_0$ excited state lifetime as a function of the added acceptor concentration, over the range of 0.3 to 5 μM using 5 μM solutions of the Eu(III) complex. The values of k_2 were found to be 0.64, 0.57 and 1.40×10^9 M^{-1} s^{-1} for [EuL^{7-9}] respectively (Fig. S1, Tables S1† and 2). The order of the rate constants revealed a slightly enhanced rate of energy transfer with the carboxylate derivative, compared to those of the two phosphinate complexes, which gave very similar results within experimental error. The spectral overlap integrals (J) and, hence, the Förster radii (R_0) for each Eu(III) complex with this acceptor were also estimated in MeOH using eqn (4) and (5), respectively. The shorter lifetime of the less sterically shielded tris-carboxylate complex, [EuL^9], was accompanied by a slightly larger second order rate constant, k_2. The similar emission quantum yields and spectral overlap integrals, J, meant that the Förster radius, R_0, for each complex was the same, within experimental error, *i.e.* 68 (±1) Å.

Study of energy transfer in water

Due to the extremely low water solubility of [EuL^7] and the carboxylate analogue [EuL^9] it was only possible to carry out the quenching studies in water (50 mM HEPES, 50 mM NaCl, pH = 7.4) for the P–Me derivative, [EuL^8]. It was decided to examine different MeOH/water compositions to check if there was any significant solvent effect.

The results (Fig. S2† and Table 3) revealed a 2.8 fold increase in the value of k_2 in water compared to that in MeOH, tentatively related to the higher dielectric constant of water. No differences in k_2 were observed between 100% MeOH and 50 : 50 MeOH/H_2O. The absence of variation in the mixed solvent can be ascribed

Table 2 Förster radii calculated for the three Eu(III) complexes, (MeOH, 295 K)[a,b]

	[EuL^7]	[EuL^8]	[EuL^9]
J/M^{-1} cm^3	1.016×10^{-12}	1.184×10^{-12}	1.109×10^{-12}
ϕ	0.52	0.43	0.48
τ_0/ms	1.26	1.18	0.95
R_0/nm	6.81	6.77	6.86
k_2/M^{-1} $s^{-1} \times 10^9$	0.64	0.57	1.40

[a] The spectral overlap integral was calculated for the region 550 < λ < 720 nm. The refractive index for MeOH is 1.328, compared to 1.333 for H_2O. Errors on τ_0, and ϕ_{em} are ±10%. [b] The values of the Stern–Volmer quenching constants under these conditions, $K_{SV}{}^{-1}$ are: 1.24 μM, 1.49 μM and 0.75 μM for [EuL^{7-9}], respectively.

Table 3 Quenching studies for [EuL^8] in mixed protic solvents. Values ($\pm$5%) were recorded at 295 K in: 100% 50 mM HEPES buffer, 50 mM NaCl, pH = 7.4; 50% 50 mM HEPES buffer, 50 mM NaCl and 50% MeOH; 100% MeOH; λ_{ex} = 332 nm

Solvent	Gradient	τ_0/ms	k_2/M^{-1} s^{-1} $\times 10^9$ ($\pm$0.05)
MeOH	0.62	1.18	0.57
MeOH : H_2O (50 : 50)	0.67	1.11	0.56
H_2O	1.64	1.03	1.59

to the specific solvation of the complex by methanol. Soper has shown that in binary mixtures of water and alcohols, local clusters of alcohol molecules occur (or *vice versa*), so that the more hydrophobic Eu complex is likely to undergo specific solvation by a methanol cluster.[27]

A variety of structurally similar hydrophilic Eu(III) complexes was studied (ESI, Fig. S3†) in water, examining energy transfer to the cyanine dye. No significant dependence of the rate of energy transfer on the nature and number of the coordinating anionic donor groups was evident. Neither was any correlation found between complex hydrophilicity (comparing log *P* values) and the second order rate constant, k_2, characterising the rate of energy transfer quenching.

Inefficient energy transfer involving Mn^{2+} and Cu^{2+} aqua ions

The quenching of lanthanide emission by metal ions is an important aspect in bioassays because ions like Mn^{2+} are used in millimolar concentrations as cofactors in certain kinases, and therefore understanding their quenching behaviour can help in the development of optimal probes.[28] Sporadic reports have described the quenching of the emission of various Eu and Tb complexes in water by transition metal ion salts.[29–34] In these studies, the lanthanide complexes that were examined were often of ill-defined speciation and the influence of competitive lanthanide ion dissociation complicated some of the systems studied, such that both static and dynamic quenching processes were sometimes observed. Typically, quenching by copper(II) ions was found to be faster than that by Cr^{3+},

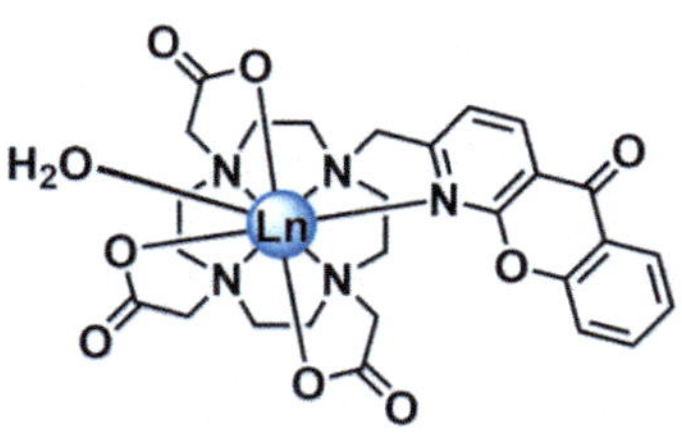

[LnL^{10}(H_2O)]

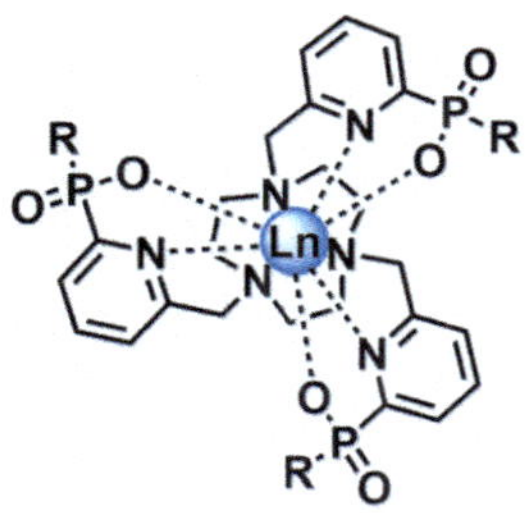

[LnL^{11}] R = Ph
[LnL^{12}] R = Me

Mn^{2+}, Co^{2+} and Ni^{2+} aqua ions, with dynamic quenching postulated to occur by an energy transfer process, with second order rate constants of about 10^6 to 10^7 M^{-1} s^{-1} for copper(II) quenching. Much less work has been reported on quenching by Mn^{2+} salts, apart from comments about the enhanced sensitivity to the nature of the ligand in lanthanide coordination complexes.[34]

In empirical screening assays, the emission intensity and lifetime of Eu and Tb complexes has often been measured in the presence of a million fold excess of $MnCl_2$, serving as a simple means of assessing the ability of the complex to resist intermolecular quenching of the lanthanide excited state. Such preliminary screens had shown that the charge neutral lanthanide complexes of L^{10}, possessing an azaxanthone sensitiser, were quenched rather efficiently by Mn^{2+},[34] whereas the charge neutral complexes of L^{11} and L^{12} with phosphinate groups were much more resistant to quenching. The latter two complexes are 9-coordinate and adopt a tricapped trigonal prismatic geometry[35] in which the lanthanide ion is much more shielded (Ph > Me) from the environment compared to [LnL^{10}], where a more open square anti-prismatic coordination geometry occurs, with a water molecule capping the open axial position.[36]

A solution of $MnCl_2$ was added to a solution of [$Eu\cdot L^{10}$] (H_2O, 100 μM complex, 0.1 M HEPES, 0.1 M NaCl, pH 7.4, 295 K) to give final Mn^{2+} concentrations of 0.1 and 10 mM. Upon the addition of Mn^{2+}, both the emission and excited state lifetime of [EuL^{10}] decreased, and there was no change in the Eu emission spectral form. The ratios I_0/I and τ_0/τ, where I and τ refer to the intensity and lifetime of Eu emission, respectively, were very similar, consistent with predominant dynamic quenching of the Eu excited state. To understand better which excited state is quenched, the emission spectrum of [EuL^{10}] was recorded at each Mn^{2+} concentration, following excitation of the sensitising group ($\lambda_{ex} = 328$ nm) or after direct excitation of the Eu^{3+} ion ($\lambda_{ex} = 397$ nm). Despite the low signal to noise ratio following direct excitation, it was apparent that the spectral form was not changing and the extent of quenching was the same within experimental error, *i.e.* independent of the mode of excitation. Such behaviour is consistent with dynamic quenching of the lanthanide excited state, rather than any sensitiser excited state.

Luminescence titrations were also carried out with [EuL^{10}] and [TbL^{10}], measuring their emission spectra and lifetimes in the presence of increasing concentrations of Mn^{2+}. The intensity of emission decreases with increasing [Mn^{2+}]. To quantify the rate of quenching, Stern–Volmer plots were examined (Fig. 4). The variation of either I_0/I or τ_0/τ with [Mn^{2+}] gave very similar slopes, consistent with predominant dynamic quenching. From the gradients of the slopes, Stern–Volmer quenching constants, ${K_{SV}}^{-1}$, of 1.22 and 0.16 mM were calculated for [EuL^{10}] and [TbL^{10}], respectively. The plots showed no deviation from linearity, consistent with a single bimolecular quenching process (Fig. 5).

Stern–Volmer quenching constants were calculated for the three pairs of Tb and Eu(III) complexes to allow a comparative analysis (Table 4). The tri(phenylphosphinate) C_3 symmetric complexes, [EuL^{11}] and [TbL^{11}],[36] and their methylphosphinate analogues, [LnL^{12}], form the core structures in the EuroTracker™ series of emissive lanthanide probes.[2,14,15,23,24] In each case, the Tb^{3+} species was found to be more susceptible to quenching by Mn^{2+}. Such behaviour is not consistent with a quenching mechanism involving electron transfer to the Ln* state, as the reduction of Tb^{3+} is highly energetically unfavourable. According to

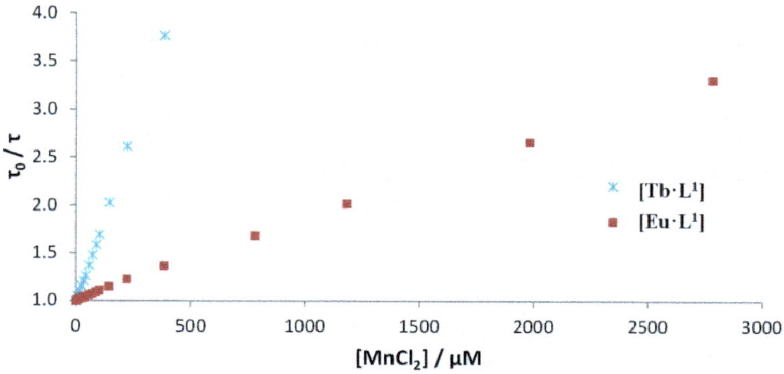

Fig. 5 Stern–Volmer plots for the quenching of [EuL[10]] and [TbL[10]] by $MnCl_2$ (H_2O, 20 μM complex, 0.1 M HEPES, 0.1 M NaCl, pH 7.4, 295 K).

Table 4 Stern–Volmer constants and second order rate constants for the Mn^{2+} and Cu^{2+} quenching of selected Eu^{3+} and Tb^{3+} complexes (H_2O, 20 μM complex, 0.1 M HEPES, 0.1 M NaCl, pH 7.4, 295 K)

	$K_{SV}{}^{-1}$/mM	
Complex	Mn^{2+} (aq.)	Cu^{2+} (aq.)
[EuL[10]]	1.22	0.08
[TbL[10]]	0.16	0.18
[EuL[11]]	12.3	3.78
[TbL[11]]	4.00	8.50
[EuL[12]]	7.69	2.30
[TbL[12]]	1.81	3.75
	k_2/M^{-1} s^{-1} × 10^6	
Complex	Mn^{2+} (aq.)	Cu^{2+} (aq.)
[EuL[10]]	1.44	21.9
[TbL[10]]	3.43	3.05
[EuL[11]]	0.06	0.19
[TbL[11]]	0.08	0.04
[EuL[12]]	0.08	0.28
[TbL[12]]	0.21	0.10

the Weller equation, even the photo-induced electron transfer from Mn^{2+} to Eu^{3+} is energetically unfavourable, with a ΔG_{ET} = +22 kJ mol^{-1} (assuming that $E_{ox}(Mn^{2+}) = +1.51$ V, $E_{red}(Eu^{3+}) = -1.0$ V, $E_{Ln*} = +2.13$ eV and $e^2/\varepsilon_r = 0.15$). The only plausible alternative mechanism involves energy transfer, which is highly dependent upon the distance between the donor lanthanide complex and the d block aqua complex acceptor.

Parallel quenching studies were carried out with aqueous $CuCl_2$ solutions to investigate any differences in the ability of the two first row transition metals to deactivate the Ln* excited state. An electron transfer quenching process was also not thermodynamically feasible in this case. Under the same conditions as for the Mn^{2+} quenching experiments, Cu^{2+} was found to reduce both the emission intensity and the excited state lifetime of each complex to similar extents. The Stern–Volmer quenching constants and second order rate constants were determined and compared with the values measured for Mn^{2+} quenching (Table 4). In the presence of Cu^{2+}, the Eu complexes were quenched more efficiently than by Mn^{2+}. There was around a threefold increase in the rate constant for [EuL^{11}] and a 15 fold increase for the more sterically accessible complex, [EuL^{10}]. The observed behaviour contrasts with quenching by Mn^{2+}, where slightly higher rate second order constants were found for the Tb^{3+} analogues. The smallest rate constant values were found for the Eu/Tb complexes of the most shielded ligand, L^{11}. The highest rate constant (2.2×10^7 M^{-1} s^{-1}) was found in the quenching of [EuL^{10}] by the aqua Cu^{2+} ion; this value is of the same magnitude as the highest rate constants reported earlier, *e.g.* with the rather kinetically labile anionic tris-chelate Eu(III) complex of oxy-diacetate.[33]

For energy transfer to occur between Ln* and Mn^{2+}, some spectral overlap is required between the donor emission and acceptor absorption spectra. The absorption spectrum of $MnCl_2$ has an absorption maximum at 520 nm that arises from the spin-forbidden d–d transition of the high-spin d^5 Mn^{2+} aqua ion, resulting in a very low molar extinction coefficient ($\varepsilon = 0.018$ M^{-1} cm^{-1}). The emission spectra of [EuL^{11}] and [TbL^{11}] are also shown (Fig. 6), and reveal that spectral overlap exists between both Ln^{3+} emission spectra and the Mn^{2+} absorption spectrum. It is also apparent that the spectral overlap is greater for the Tb^{3+} complex, explaining the observed higher values of k_2 and lower values of ${K_{SV}}^{-1}$ for Tb^{3+} complexes of a common ligand. The very low molar extinction coefficient of Mn^{2+} and the very small spectral overlap integral seem to disfavour the hypothesis that Förster energy transfer quenching is operative. However, the excited state lifetimes of these Ln^{3+} complexes are relatively long (in the millisecond range), and over this period many diffusive encounters can take place between Mn^{2+} ions and the lanthanide excited state, increasing the probability of the energy transfer process.[37]

The absorption spectrum of $CuCl_2$ was also measured in water and revealed the characteristic broad d–d transition of the aqua ion, centred at 810 nm. In contrast to the absorption of the Mn^{2+} aqua ion, the corresponding Cu^{2+} transition is spin allowed, resulting in a higher molar extinction coefficient (11 M^{-1} cm^{-1}), compared to that of the Mn^{2+} aqua ion.[38] The Cu^{2+} absorption band tails into the spectral region of Eu^{3+} emission and to a lesser extent Tb^{3+} emission (Fig. 6), consistent with the higher rate of quenching of the Eu complexes by copper aqua ions and supporting the hypothesis of an energy transfer quenching mechanism. In summary, and compared to Mn^{2+}, the greater molar extinction coefficient of the Cu^{2+} aqua ion and the better spectral overlap with Eu^{3+} emission explain the higher rate of quenching by Cu^{2+} for each Eu(III) complex studied here.

The spectral overlap integrals (J) and, hence, the Förster radii (R_0) for each donor lanthanide complex with Mn^{2+} and Cu^{2+} acceptors were estimated in water, using eqn (4) and (5), respectively (Table 5). The estimated values of R_0 are much smaller for the Mn^{2+} aqua ion acceptor, consistent with the feeble spectral overlap integral associated with its low molar extinction coefficient, which is 600 times smaller than that for Cu^{2+} and 15 million times smaller than that for the near IR

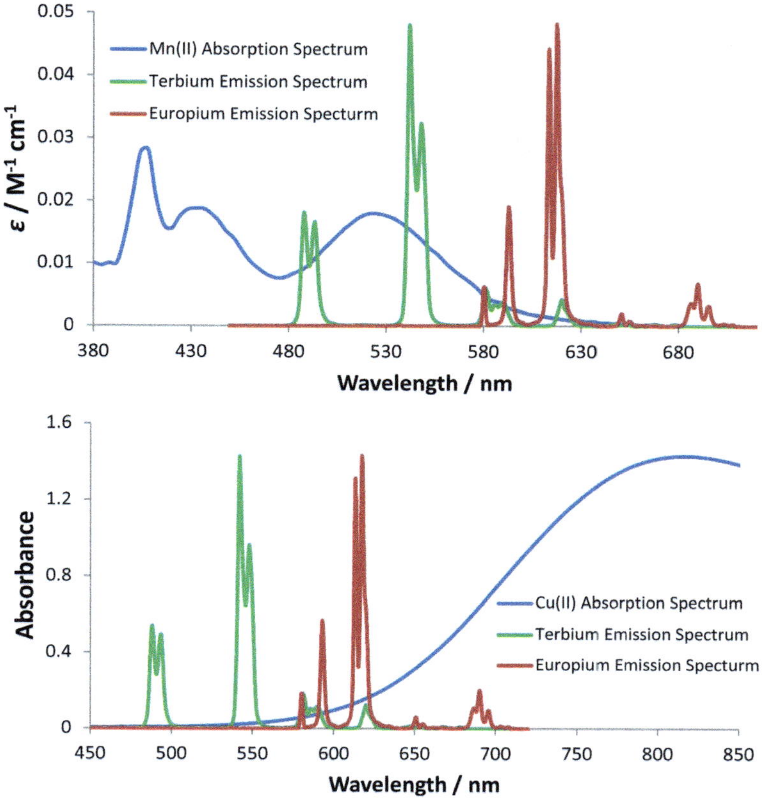

Fig. 6 (upper) Molar extinction coefficient for $MnCl_2$ (blue) as a function of wavelength (H_2O, pH 7.4, 295 K); (lower) absorption spectrum of $CuCl_2$ (blue) as a function of wavelength (H_2O, pH 7.4, 295 K). In each case, emission spectra for [TbL^{11}] (green) and [EuL^{11}] (red) are shown, permitting the analysis of the spectral overlap integrals.

Table 5 Calculated values of the Förster radii (R_0) using eqn (2) and (3) for energy transfer from the lanthanide complex donor to the transition metal aqua ion acceptor

	R_0/Å (calc.)	
Complex	Mn^{2+} (aq.)	Cu^{2+} (aq.)
[EuL^{10}]	2.4	9.9
[TbL^{10}]	4.0	7.6
[EuL^{11}]	3.5	11.6
[TbL^{11}]	4.0	8.3
[EuL^{12}]	3.4	12.1
[TbL^{12}]	4.4	9.2

dye, **1**, discussed above, for which the R_0 values were around 68 Å. Indeed, the short distances found in this case (R_0 between 2.8 to 4.4 Å for Mn^{2+} quenching) fall within the range where the mechanism of energy transfer is collisional, operates over a short encounter distance and may involve the Dexter energy exchange mechanism[39] and not only Förster resonance energy transfer into the copper $^2T_{2g}$ orbital, which is more likely to operate with the aqua copper(II) ion.[40]

Summary and conclusions

The lanthanide excited state may be subject to quenching by electron (charge) transfer and by vibrational and electronic energy transfer processes, leading to a reduction in its lifetime. For Eu(III) complexes, a ligand to metal charge transfer state may lie close in energy to the Eu $^5D_0$ excited state, leading to potential energy surface crossing. Such a situation is more prevalent in cationic complexes with ligands that are charge neutral and do not stabilise the Eu(III) oxidation state as efficiently as those containing hard anionic ligands, such as O or N donors.[8] More commonly, fast electron transfer occurs from a ligand HOMO, such as a π or lone pair orbital, exemplified here with a set of six Eu(III) complexes where the lifetime was shortest for the complexes with the most high lying π orbitals or the most N atoms. In the latter case, N-protonation suppressed photoinduced electron transfer and the pH dependence of the Eu emission lifetime was identical when populating the $^5D_0$ state by direct excitation or *via* the more efficient sensitised emission pathway that involves ligand excited states. Such observations confirm that it is the Eu excited state that is being quenched directly in these cases. Consideration was given to the utility of semi-empirical calculations of the energy of the key ligand, charge transfer and europium excited states involved in such a quenching process. However, the limited accuracy of such calculations, as exemplified in recent examples using DFT and modified Judd–Ofelt theory,[41,42] meant that such work was not undertaken for these complexes.

Electronic energy transfer was examined from a Eu donor to a cyanine dye acceptor with which very good spectral overlap occurs. The calculated Förster radii (R_0) values did not change significantly with the structure of the Eu(III) complexes, each of which gave rise to very similar Eu emission spectral finger-print. Thus, with R_0 values averaging around 68 Å, the nature of the ligand anionic groups and the aryl sensitiser substituents did not affect this long range FRET process. Nor was any correlation found between complex hydrophilicity and the value of the second order rate constant, k_2, that characterises the rate of energy transfer quenching. In contrast, energy transfer to the Cu(II) and especially the Mn(II) aqua ion was much more inefficient, with rate constants 1000 times smaller, owing to the much smaller spectral overlap integral. In these and earlier reported cases, the steric shielding created by the ligand plays a key role in defining the sensitivity to quenching, especially with Mn(II) aqua ions for which more open coordination complex structures show the greatest sensitivity to Mn(II) quenching.[34] Such behaviour is consistent with a predominant short range collisional quenching mechanism, and suggests that a Dexter energy transfer mechanism operates, at least in part. It requires wave-function overlap, which occurs most efficiently when the distance between the excited Eu ion and the quenching acceptor species is minimised.

Conflicts of interest

There are no conflicts to declare.

Acknowledgements

We thank EPSRC for grant (EP/L01212X/1) and studentship support.

References

1 S. E. Bodman and S. J. Butler, *Chem. Sci.*, 2021, **12**, 2716–2734.

2 E. Mathieu, A. Sipos, E. Demeyere, D. Phipps, D. Sakaveli and K. E. Borbas, *Chem. Commun.*, 2018, **54**, 10021–10035.

3 A. B. Aletti, D. M. Gillen and T. Gunnlaugsson, *Coord. Chem. Rev.*, 2018, **354**, 98–120.

4 M. C. Heffern, L. M. Matosziuk and T. J. Meade, *Chem. Rev.*, 2014, **114**, 4496–4539.

5 S. J. Butler, M. Delbianco, L. Lamarque, B. K. McMahon, E. R. Neil, R. Pal, D. Parker, J. W. Walton and J. M. Zwier, *Dalton Trans.*, 2015, **44**, 4791–4803.

6 E. G. Moore, A. P. S. Samuel and K. N. Raymond, *Acc. Chem. Res.*, 2009, **42**, 542–552.

7 D. Parker, J. D. Fradgley and K.-L. Wong, *Chem. Soc. Rev.*, 2021, **50**, 8193–8213.

8 D. Kovacs, E. Mathieu, S. R. Kiraev, E. Demeyere, A. Sipos and K. E. Borbas, *J. Am. Chem. Soc.*, 2020, **142**, 13190–13200.

9 A. Beeby, I. M. Clarkson, R. S. Dickins, S. Faulkner, D. Parker, L. Royle, A. S. de Sousa, J. A. G. Williams and M. Woods, *J. Chem. Soc., Perkin Trans. 2*, 1999, 493–503.

10 M. W. Mara, D. S. Tatum, A.-M. March, G. Doumy, E. G. Moore and K. N. Raymond, *J. Am. Chem. Soc.*, 2019, **141**, 11071–11081.

11 J. D. Fradgley, M. Starck, M. Laget, E. Bourrier, E. Dupuis, L. Lamarque, E. Trinquet, J. M. Zwier and D. Parker, *Chem. Commun.*, 2021, **57**, 5814–5817.

12 R. Algar, N. Hildebrandt, S. S. Vogel and I. L. Medintz, *Nat. Methods*, 2019, **16**, 815–829.

13 J. M. Zwier, H. Bazin, L. Lamarque and G. Mathis, *Inorg. Chem.*, 2014, **53**, 1854–1866.

14 M. Soulié, F. Latzko, E. Bourrier, V. Placide, S. J. Butler, R. Pal, J. W. Walton, P. L. Baldeck, B. LeGuennic, C. Andraud, J. M. Zwier, L. Lamarque, D. Parker and O. Maury, *Chem.–Eur. J.*, 2014, **20**, 8636–8646.

15 S. J. Butler, L. Lamarque, R. Pal and D. Parker, *Chem. Sci.*, 2014, **5**, 1750–1756.

16 M. Starck, J. D. Fradgley, R. Pal, J. M. Zwier, L. Lamarque and D. Parker, *Chem.–Eur. J.*, 2021, **27**, 766–777.

17 M. T. Berry, P. S. May and H. Xu, *J. Phys. Chem.*, 1996, **100**, 9216–9222.

18 A. Weller, *Pure Appl. Chem.*, 1968, **16**, 115–124.

19 H. Yang, D. O. Wipf and A. J. Bard, *J. Electroanal. Chem.*, 1992, **331**, 913–924.

20 M. Sheykan, S. Khani, M. Abbasnia, S. Shaabanzadeh and M. Joafshan, *Green Chem.*, 2017, **19**, 5940–5948.

21 D. Parker, *Coord. Chem. Rev.*, 2000, **205**, 109–130.

22 H. Sahoo, *J. Photochem. Photobiol., C*, 2011, **12**, 20–30.

23 J. W. Walton, A. Bourdolle, S. J. Butler, M. Soulie, M. Delbianco, B. K. McMahon, R. Pal, H. Puschmann, J. M. Zwier, L. Lamarque, O. Maury, C. Andraud and D. Parker, *Chem. Commun.*, 2013, **49**, 1600–1602.

24 M. Starck, R. Pal and D. Parker, *Chem.–Eur. J.*, 2016, **22**, 570–580.

25 C. F. Meares and T. G. Wensel, *Acc. Chem. Res.*, 1984, **17**, 202–209.

26 D. D. Thomas, W. F. Carlsen and L. Stryer, *Proc. Natl. Acad. Sci. U. S. A.*, 1978, **75**, 5746–5750.
27 S. Dixit, J. Crain, W. C. K. Poon, J. L. Finney and A. K. Soper, *Nature*, 2002, **416**, 829–832.
28 M. J. Knape, M. Ballez, N. C. Burghardt, B. Zimmermann, D. Bertinetti, A. P. Kornev and F. W. Herberg, *Metallomics*, 2017, **9**, 1576–1584.
29 W. DeW Horrocks Jr, B. Holmquist and B. L. Vallee, *Proc. Natl. Acad. Sci. U. S. A.*, 1975, **72**, 4764–4768.
30 D. T. Cronce and W. DeW Horrocks, *Biochemistry*, 1992, **31**, 7963–7969.
31 D. H. Metcalf, J. P. Bolender, M. S. Driver and F. S. Richardson, *J. Phys. Chem.*, 1993, **97**, 553–564.
32 M. A. Kessler, *Anal. Chim. Acta*, 1998, **364**, 125–129.
33 B. C. Barja, A. Remorino, M. J. Roberti and P. F. Aramendia, *J. Argent. Chem. Soc.*, 2005, **93**, 81–96.
34 D. Parker, J. W. Walton, L. Lamarque and J. M. Zwier, *Eur. J. Inorg. Chem.*, 2010, 3961–3966.
35 J. W. Walton, R. Carr, N. H. Evans, A. M. Funk, A. M. Kenwright, D. Parker, D. S. Yufit, M. Botta, S. De Pinto and K.-L. Wong, *Inorg. Chem.*, 2012, **51**, 8042–8051.
36 P. A. Atkinson, K. S. Findlay, F. Kielar, R. Pal, R. A. Poole, H. Puschmann, S. L. Richardson, P. A. Stenson, A. L. Thompson, J. Yu and D. Parker, *Org. Biomol. Chem.*, 2006, **4**, 1707–1722.
37 L. Stryer, D. D. Thomas and C. F. Meares, *Annu. Rev. Biophys. Bioeng.*, 1982, **11**, 203–222.
38 S. R. Qiu, B. C. Wood, B. R. Ehrmann, S. G. Demos, P. E. Miller, K. I. Schaffers, T. I. Suratwala and R. K. Brow, *Phys. Chem. Chem. Phys.*, 2015, **17**, 18913–18923.
39 P. A. Brayshaw, J.-C. G. Bünzli, P. Froidenaux, J. M. Harrowfield, Y. Kim and A. N. Sobolev, *Inorg. Chem.*, 1995, **34**, 2068–2076.
40 K. Szyszka, S. Targonska, A. Lewinska, A. Watras and R. J. Wiglusz, *Nanomaterials*, 2021, **11**, 464–482.
41 A. G. Cosby, J. J. Woods, P. Nawrocki, T. J. Sorensen, J. J. Wilson and E. Boros, *Chem. Sci.*, 2021, **12**, 9442–9451.
42 G. Hovhannesyn, V. Boudon and M. Lepers, *J. Lumin.*, 2022, **241**, 118456.

Faraday Discussions

PAPER

Can you break the oxo-wall? A multiconfigurational perspective

Asmita Sen and Gopalan Rajaraman *

Received 16th October 2021, Accepted 8th November 2021
DOI: 10.1039/d1fd00072a

The concept of the "oxo-wall" was conceived about 60 years ago by Harry B. Gray, and has been found to be related to the non-existence of high-valent M–oxo species in the +IV oxidation state in a tetragonal geometry beyond group 8 in the periodic table. Several efforts have been made in the past decades to test and find examples that violate this theory. Several claims of violation in the past were attributed to the difference in the geometries/coordination number and, therefore, these are not examples of true violation. In recent years, substantial efforts have been undertaken to synthesise a true $Co^{IV}{=}O$ species with various ligand architectures. $Co^{IV}{=}O$ and $Co^{III}{-}O^{\bullet}$ are electromers and, while they are interchangeably used in the literature; the former violates the oxo-wall while the latter does not. The possibility that these two species could exist in various proportions similar to resonating structures has not been considered in detail in this area. Furthermore, there have been no attempts to quantify such mixing. In this direction, we have employed density functional theory (DFT) and *ab initio* CASSCF/NEVPT2 methods to quantify the co-existence of $Co^{IV}{=}O$ and $Co^{III}{-}O^{\bullet}$ isomeric species. By thoroughly studying six different metal-oxo species, we affirm that the nature of such electromeric mixing is minimal/negligible for $Fe^{IV}{=}O$ and $Mn^{IV}{=}O$ species – both are pre-oxo-wall examples. By studying four different ligand architectures with Co–oxo species, our results unveil that the mixing of $Co^{IV}{=}O \leftrightarrow Co^{III}{-}O^{\bullet}$ is substantial in all geometries, with dominant $Co^{IV}{=}O$ species favourable for the $S = 3/2$ intermediate spin state. The percentage of the $Co^{III}{-}O^{\bullet}$ species is enhanced substantially for the $S = 1/2$ low-spin state. We have attempted to develop a tool to estimate the percentage of the $Co^{III}{-}O^{\bullet}$ species using various structural parameters. Among those tested, a linear relationship between % of $Co^{III}{-}O^{\bullet}$ and a bond length based ratio is found ($R_d = \frac{d_{(Co-O)}}{d_{(Co-N_{ax})}}$, where $d_{(Co-O)}$ and $d_{(Co-Nax)}$ are the axial Co–O and $Co{-}N_{ax}$ bond lengths in Å, respectively). It is found that the higher the R_d, the greater the $Co^{III}{-}O^{\bullet}$ character will be and the geometrically portable correlation developed offers a way to qualitatively compute the % of $Co^{III}{-}O^{\bullet}$ character for unknown geometries.

Department of Chemistry, Indian Institute of Technology Bombay, Mumbai, 400076, India. E-mail: rajaraman@chem.iitb.ac.in

Introduction

The most important biological oxidative processes, water oxidation during photosynthesis in photosystem II and oxygen reduction in respiration, have been found to be catalysed by multiply bonded high-valent terminal Mn/Fe–oxo species.[1,2] Apart from these naturally occurring enzymatic reactions, high-valent terminals, as well as bridging Fe/Mn-dependent synthetic models mimicking these metalloenzymes, have been proposed to be key intermediates in the oxidative transformations of various organic molecules. In the past few decades, these active intermediates have been synthesised, spectroscopically characterised, and their mechanism in activating C–H/O–H/N–H bonds has been explored.[3–9] For instance, various biomimicking high-valent $Fe^{IV}{=}O$ dependent models of haem and non-haem metalloenzymes were found to play an active role in O_2 activation. Similarly, in photosystem II, a high-valent Mn–oxo species was found to catalyse the most important O–O bond formation step, although this species is elusive. However, a number of monomeric or dimeric high-valent biomimic Mn–oxo species have been synthesised, characterised, and proven to be superior to Fe catalysts. Furthermore, early transition metals, such as Cr, V and Mo, have been investigated to understand the electronic effect on the catalytic transformations of several metalloenzymes and design a robust catalyst.[10] Although the chemistry of the first-row early transition metals with terminal oxo groups has been well-established, for the late transition metals, such as Co, Ni and Cu elements, high oxidation state (+4) terminal oxo species are rare and are non-existent for a tetragonal geometry. An invisible wall, namely the "oxo-wall", has been proposed to separate the Fe–Ru–Os and Co–Rh–Ir triads in the periodic table.[11–13]

The "oxo-wall" was conceived about 60 years ago in Copenhagen by Harry B. Gray, who proposed the existence of a V^{IV}–oxo triple bond ($V{\equiv}O$) for a d^1 vanadyl ion (VO^{2+}), employing a modified molecular orbital diagram including oxo-metal π-bonding.[11,12,14] Later, this triple bond ($M{\equiv}O$) formulation was identified in d^0, d^1, d^2 (low-spin) monomeric M–oxo tetragonal complexes including CrO^{3+}, MoO^{3+} *etc.*[15,16] The modified molecular orbital diagram in these tetragonal terminal M–oxo species is composed of a lowest-lying M–O σ-bonding orbital, followed by the degenerate M–O π-bonding orbitals, a nonbonding metal d_{xy}, orbital, two degenerate M–O π*-antibonding orbitals ($d_{xz/yz}$), a σ*-antibonding $d_{x^2-y^2}$ orbital and a highest lying M–O σ*-antibonding d_{z^2} orbital [σ_z (M–O) < π_x (M–O) = π_y (M–O) < nb (d_{xy}) < $\pi_x^*(d_{xz}) = \pi_y^*(d_{xz}) < \sigma^*(d_{x^2-y^2}) < \sigma^*$ (d_{z^2}), Scheme 1]. Apart from the M–O σ and π-bonding orbitals, for d^0, d^1, d^2 (low-spin) systems, the first two d-electrons occupy the nonbonding d_{xy} orbital, yielding a π bond-order of 2. This value decreases to 3/2, 1 and 1/2 if we move from d^3 to d^5 electronic configurations, as the additional d-electrons in these configurations enter π*-antibonding d_{xz} and d_{yz} orbitals. Thus, enhancing the electron occupancy in π*-orbitals reduces the π bond order in tetragonal d^5-$M^{IV}{=}O$ complexes (Scheme 1). The d-orbital splitting pattern of C_{4v} symmetric $L_5M(O)$ complexes with more than 5 electrons in the metal d-orbitals leads to a M–O π-bond order of 0.5, making the oxo group extremely basic, and it undergoes either protonation or dimerisation or is attacked by electrophiles.[16] To obtain high M–O π-bond multiplicity for elements beyond Fe, the oxidation state needs to be very large ($\geq$+5), making the

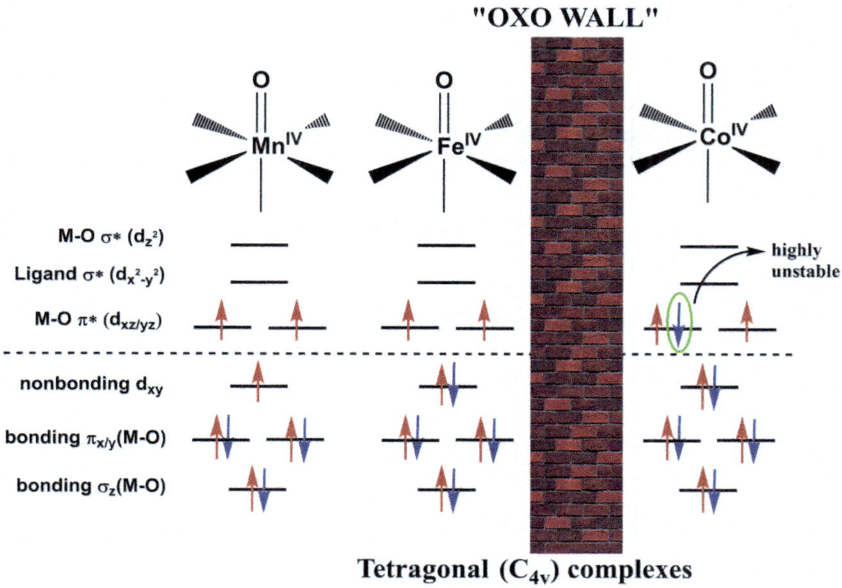

Scheme 1 Pictorial representation of the "oxo-wall" in tetragonal (C_{4v}) high-valent Fe/Mn/Co–oxo complexes.

complex unstable and susceptible to the elimination of O_2 or H_2O_2. The summation of all these effects is to limit the metal oxidation state to as high as +4 and at the same time limit to a lower number of d-electrons, generating a wall between groups 8 and 9 in the periodic table. On the left-hand side (V–Fe), the d^4 electronic configuration leads to an oxidation state of +4 or lower, whereas on the right side of the periodic table, the formal oxidation state should be +5 or higher to form M–O multiple bonds.

There are many reports where the reactivity of $Fe^{IV}{=}O$ species has been attributed to its valence isomer $Fe^{III}–O^•$ (oxyl-Fe^{III} species), which gets stabilised upon the approach of electrophiles.[17] For $Mn^{IV}{=}O$ species, on the other hand, a complex set of spin-state reactivities has emerged where the role of $Mn^{III}–O^•$ (oxyl–Mn^{III} species) in dictating the reactivity is ruled out.[18] It is clear from these contrasting examples that understanding the nature of metal multiple bond oxo species is important to decipher their reactivity. Although adequate attention has been paid to the structural perspective of high-valent tetragonal (C_{4v}) Fe/Mn–O complexes with multiple bonds, there are only limited studies on the species' stability and electronic structure on the other side of the wall. In particular, several recent reports have focused their attention on the formation of $Co^{IV}–O$ or its valence isomer $Co^{III}–O^•$ species and very little is been explored with respect to their structure and bonding, which has high relevance to many catalytic processes, including water oxidation.[19] The oxo-wall theory suggests that the formation of multiple bonds is unstable between Co and the oxo moiety when the Co-centre is in a +4 oxidation state (d^5-system). There are efforts to seek violation of oxo-wall theory in this respect, and there are claims and counterclaims.[19] As these species are extremely transient in nature, establishing their structure and bonding is challenging, especially given that $Co^{IV}{=}O$ and its valence isomer $Co^{III}{-}O^{\bullet}$ can not be easily distinguished unless sophisticated experiments, such as EPR/ENDOR experiments, are carried out with labelled oxygen atoms.[20] Furthermore, there have also been attempts to stabilise formal $Co^{IV}{=}O$ species in other coordination environments, such as a trigonal planer, square pyramidal *etc.* The Nam and Ray groups have photochemically synthesised a square pyramidal $[(13\text{-TMC})Co^{IV}(O)]^{2+}$ complex that has been characterised as a formal $Co^{IV}{=}O$ species. Furthermore, it is hard to rule out the possibility of $Co^{IV}{=}O$ and its valence isomer $Co^{III}{-}O^{\bullet}$ species co-existing in a mixed form with various proportions. Such a possibility challenges various experimental, spectral and analytic interpretations, which often assume the existence of only one species in solution.

To understand the delicate balance between the two valence isomers $Co^{IV}{=}O$ and $Co^{III}{-}O^{\bullet}$ in relation to the oxo-wall theory proposed by Gray, we have performed density functional theory (DFT) and *ab initio* CASSCF/NEVPT2 calculations on a series of tetragonal $L_5M(O)$ high-valent Co–oxo complexes, namely, $[(15\text{-TMC})Co(O)(CH_3CN)]^{2+}$ (**1**), $[(13\text{-TMC})Co(O)(CH_3CN)]^{2+}$ (**2**), and $[(N4Py)Co(O)]^{2+}$ (**3**) complexes (where 15-TMC is 1,5,9,12-tetramethy-1,5,9,12-tetraazacyclotridodecane, 13-TMC is 1,4,7,10-tetramethy-1,4,7,10-tetraazacyclotridecane and N4Py is *N*,*N*-bis(2-pyridylmethyl)-*N*-bis(2-pyridyl)methylamine)]. Furthermore, we have extended our studies to $[(13\text{-TMC})Co(O)]^{2+}$ (**4**), $[(N4Py)Fe(O)]^{2+}$ (**5**) and $[(N4Py)Mn(O)]^{2+}$ (**6**) to thoroughly analyse and understand the existence of the electromers $M^{IV}{=}O$ *vs.* $M^{III}{-}O^{\bullet}$ in these complexes. Our study aims to answer the following questions: (i) before and after the wall, is there a $M^{IV}{=}O \leftrightarrow M^{III}{-}O^{\bullet}$ equilibrium or co-existence of these species in proportions? (ii) Does the individual spin state play a role in breaking the wall? (iii) Does the geometry/ligand architecture control the character of the $Co^{IV}{=}O/Co^{III}{-}O^{\bullet}$ species? (iv) Is a black-box tool to quantify the $Co^{III}{-}O^{\bullet}$ character for an unknown structure possible?

Computational details

Density functional theory (DFT) calculations

The geometries of all complexes in the present study were optimised by employing density functional theory (DFT) methods using the Gaussian 16 suite of programs.[21] An unrestricted dispersion-corrected B3LYP hybrid density functional[22,23] (B3LYP-D3), along with a double-ξ quality LanL2DZ[24] with a Los Alamos effective core potential basis set for the metal and 6-31G* formalism for the rest of the atoms (H, C, N and O),[25] was used for the geometric optimisation, as well as frequency calculations in the gas phase. The gas-phase energetics were refined further by single-point calculations using an all-electron def2-TZVP basis set in the same B3LYP-D3 level of theory. The solvation energies were integrated using the solvation model based on the density (SMD) solvation model[26] in acetone solvent using the same methodology. The final energies were the B3LYP-D3 computed Gibbs free energy corrected solvation energies. The energy correction terms were obtained from the vibrational frequency calculations in the gas phase.

Ab initio calculations

All *ab initio* calculations were carried out using the ORCA 4.2.1 program.[27] For these calculations, the scaler-relativistic effect was incorporated using the

Douglas–Kroll–Hess (DKH) Hamiltonian.[28] The starting guess orbitals were found by performing ROKS calculations on the DFT optimised structures, as well as several points obtained from the scanning along the M–O as well as the M–N_{ax} bonds using the BP86 functional, DKH-def2-TZVP basis set for the metal, DKH-def2-TZVP(-f) for the directly coordinated N and O atoms with the metal and the DKH-def2-SVP basis set for the rest of the atoms (C and H). For further CASSCF/NEVPT2 calculations, a similar basis set as ROKS was used. For all calculations, an active space of CAS(n,8) was employed using electrons in five metal d-orbitals and three oxo p-orbitals, including two oxo p_π (π_x and π_y) orbitals and one oxo p_σ (σ_z) orbital. For the d^4 and d^5 metal centres, 5 quintet, 45 triplet and 50 singlet roots and one sextet, 24 quartets and 75 doublet roots were considered, respectively. By these calculations, the electronic wavefunctions were expanded in all possible configuration state functions by the distribution of a number of electrons in the total number of active spaces by employing the complete active space self-consistent-field (CASSCF) method. As we have employed state-average CASSCF calculations to compute the per cent of various electromers, the per cent of individual contribution is likely to alter if we expand the reference space or vary the number of roots employed. A limited benchmark performed with expanded reference space reveal that the variations are within ± 3 per cent.

Results and discussion

Ab initio quantification of $Co^{IV}{=}O$ and $Co^{III}{-}O^{\bullet}$ electromerism

In the following section, we have attempted to comprehend how the so-called oxo-wall operates in various high-valent octahedral (O_h) or pseudo octahedral M–oxo complexes beyond group 8 in the periodic table. To quantify the co-existence of isomeric $Co^{IV}{=}O$ and $Co^{III}{-}O^{\bullet}$ forms, we choose the complexes possessing an octahedral [$L_5M(O)$] type of geometry with the Co centre in a high-oxidation state (+III or +IV). These contain either (i) a tetradentate ligand with two axial positions occupied by the oxo ligand and one relatively loosely bound N-donor ligand, such as $[(15\text{-TMC})Co(O)(CH_3CN)]^{2+}$ (**1**), $[(13\text{-TMC})Co(O)(CH_3CN)]^{2+}$ (**2**) complexes, or (ii) a pentacoordinate ligand where the axial position opposite to the oxo ligand is strongly bound to the metal, *e.g.* the $[(N4Py)Co(O)]^{2+}$ (**3**) complex. All three complexes are found to have a pseudo-C_s symmetry (Chemcraft calculated R = 0.661, 0.392 and 0.000, respectively).[29] The tetragonal C_{4v} symmetry is lifted in complexes **1** and **2** by modification of the cyclam ligand (1,4,8,11-tetraazacyclotetradecane). In the 15-TMC ligand architecture, an extra methylene group is added, and in 13-TMC (1,4,7,10-tetramethyl-1,4,7,10-tetraazacyclotridecane) a methylene group is removed. The symmetry in complex **3** is lifted due to the presence of a tertiary methyl group connected to the apical N-atom. Complex **3** has an ideal C_s symmetry, while those of the other two deviate significantly (Fig. 1).

Complex **1** is found to have an $S = 3/2$ intermediate state as the ground state, with the high-spin $S = 5/2$ and low-spin $S = 1/2$ states lying at 64.0 kJ mol^{-1} and 13.7 kJ mol^{-1}, respectively. In contrast, for complexes **2** and **3**, an $S = 1/2$ low-spin state is the ground state. The energy gaps between the ground state and other states *viz.* $S = 5/2$ and 3/2 spin states for **2** are very small (7.8 kJ mol^{-1} and 22.9 kJ mol^{-1}, respectively) while these spin-states lie comparatively higher in

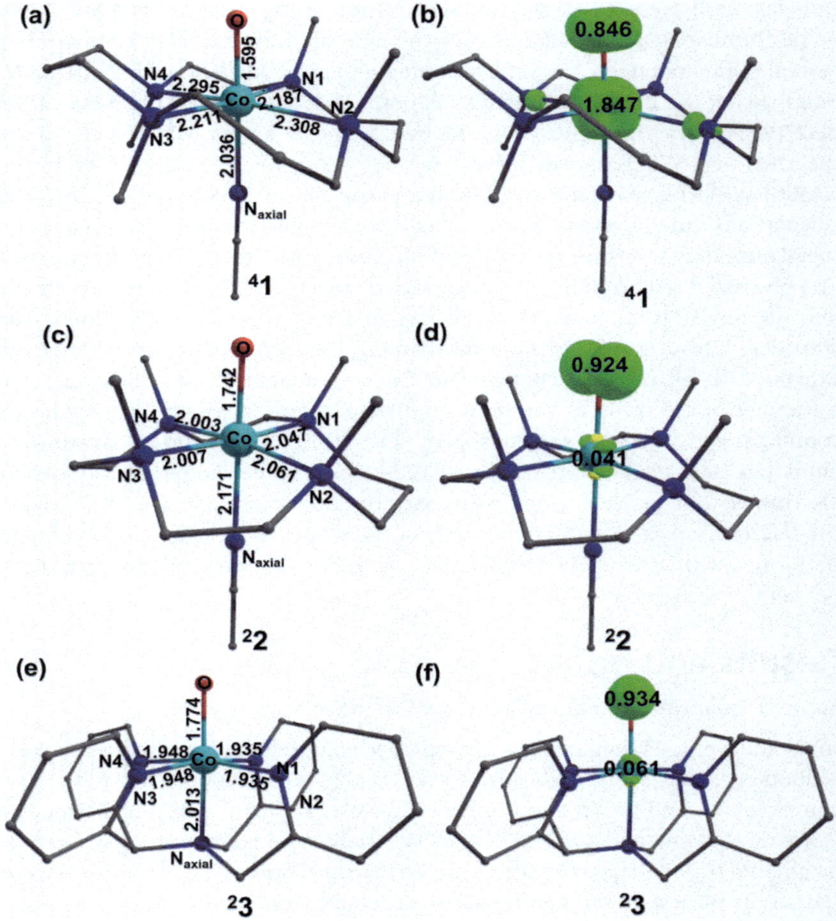

Fig. 1 (a, c and e) B3LYP-D3 optimised structures and (b, d and f) spin density plots of the ground state of complexes **1–3**.

energy in complex **3** (99.9 kJ mol^{-1} and 32.1 kJ mol^{-1}, respectively). In the ground state optimised geometry, the respective Co–O and Co–N_{ax} bond distances are found to be 1.595 Å and 2.036 Å; 1.742 Å and 2.171 Å; and 1.770 Å and 2.013 Å respectively in **1–3**. The other four Co–N_{eq} (Co–N1, Co–N2, Co–N3 and Co–N4) bonds in **1–3** are 2.187 Å, 2.308 Å, 2.211 Å, 2.295 Å; 2.047 Å, 2.061 Å, 2.007 Å, 2.003 Å and 1.935 Å, 1.935 Å, 1.948 Å, 1.948 Å, respectively. The bonding parameters point towards the highest symmetric geometry in complex **3** and a lowest one in **1**.[30] Although experimental geometrical parameters are unavailable for complexes **1–3**, we compared the geometries with those of structurally similar species reported. In particular, the geometry of complex **3** resembles another Co^{IV} species reported, and the Co–N_{avg} distances reported for this species agree with our calculated results.[31] Furthermore, the DFT computed Co–O bond stretching frequency values are calculated to be 886 cm^{-1}, 621 cm^{-1} and 615 cm^{-1} for **1–3**, respectively. The Co–O bond stretching frequency for a true Co^{IV}=O has been reported to be 840 cm^{-1}.[32] Benchmarking the above results with this value suggests that complex **1** is likely to be closer to an ideal $Co^{IV}{=}O$ species, while the values for species **2** and **3** deviate significantly. This suggests a strong deviation for **2** and **3** from an ideal $Co^{IV}{=}O$ form.

In complexes **1–3**, the DFT computed spin densities on the (Co : O) pairs are (1.847 : 0.846), (0.041 : 0.924) and (0.061 : 0.934), respectively. The spin-up α-electron density of ~2 in **1** and ~0 in **2** and **3** on the metal indicates a +III oxidation state on Co rather than a +IV oxidation state, which always gives a residual spin value close to 5.0/3.0/1.0 depending on the nature of the spin states (high-spin, intermediate spin or low spin, respectively). Secondly, the DFT methodology employed tends to overestimate the delocalisation of spin densities, resulting in larger than expected spin densities on the M=O oxyl O atom. This is one of the well-established limitations of DFT methods.[33]

Aside from the spin densities, the charges on the metal centre are also often underestimated by DFT methods, with the metallic centre exhibiting a significant loss of charge with a concomitant gain on the ligands.[34,35] Additionally, for $Mn^{III/IV/V}{=}O$ species, Borovik and co-workers have eliminated the possibility of $Mn{-}O^{\bullet}$ character with the help of EPR spectroscopy, albeit DFT calculations seemingly indicate a strong radical character at the oxyl centre.[36] All of these arguments suggest that using DFT calculated spin densities to assess the true nature of $Co^{IV}{=}O$ *vs.* $Co^{III}{-}O^{\bullet}$ species is not reliable. Furthermore, the DFT methods are a single determinant approach, and the chances of having a multiconfigurational character for the ground state within or among these two species cannot be accounted for by these methods. Furthermore, when the metal–oxo bond is elongated far from the equilibrium geometry, this is likely to stabilise the high-spin state as the ground state, with a significant radical character generated at the oxyl oxygen. Earlier method assessments of various functionals, including B3LYP against the RCCSD(T) method, have revealed that the structure and energetics match better between the two methods for the ground state geometries. However, for the excited state (in this case, $S = 2$ $Fe^{IV}{=}O$ species), a larger deviation between B3LYP and *ab initio* RCCSD(T) was found.[37]

As the radical character at the oxyl oxygen atom is crucial to prove/disprove the oxo-wall theory, a robust method that does not suffer from the above limitations is required. Although several accurate wavefunction (WF) based methods are available, many sophisticated methods are computationally expensive, limiting their usage on realistic molecules. For these reasons, we turn to CASSCF/NEVPT2 methods.[38,39] The CASSCF calculations are multiconfigurational self-consistent field methods, where the number of determinants used for the expansion of the CI vector is defined by the division of the orbitals in subspaces. The orbitals and electrons are chosen, so that those that contribute most are used to explain the system. The total number of electrons is distributed equally among the active space, resulting in a non-integer number of electrons. The CASSCF/NEVPT2 method is also plagued by computational cost, reference space problems and issues with the dynamic correlation. Moreover, the optimised geometry is taken from the DFT method and, therefore, is not a true minimum in these WF based methods. However, apart from these pros and cons, the *ab initio* CASSCF/NEVPT2 method is believed to give a better description of the wavefunction of a system if an appropriate reference space is chosen. Hence, in this study, we have performed *ab initio* CASSCF/NEVPT2 calculations to assess and understand the existence of these two isomeric forms.

Our CASSCF/NEVPT2 results show that the DFT calculated ground states and energy ordering of the high-lying states in **1–3** agree with each other, offering confidence in the results. Herein, we have employed a CAS(11,8) active space, including five metal d orbitals and three oxo 2p orbitals (p_x, p_y and p_z), to understand the nature of the dominant electromeric form. The NEVPT2 calculated $S = 5/2$ and $S = 1/2$ states are lying at 117.8 kJ mol^{-1} and 93.0 kJ mol, 126.0 kJ mol^{-1} and 55.2 kJ mol, and 167.0 kJ mol^{-1} and 99.4 kJ mol^{-1} for complexes **1–3**, respectively.

At the equilibrium Co–O bond lengths of 1.595 Å for species **1**, the state-average CAS(11,8) SCF calculation yield follows the electronic configuration $\sigma_{(p_z+d_z{}^2)^2}\,\pi_{(p_y+d_{yz})^2}\,\pi_{(p_x+d_{xz})^2}\,\delta_{(d_{xy})^2}\,\pi^*_{(p_y+d_{yz})^1}\,\pi^*_{(p_x+d_{xz})^1}\,\delta^*_{(d_{x^2-y^2})^1}\,\sigma^*_{(p_z+d_{z^2})^0}$. Here, the p_x, p_y and p_z orbitals correspond to the oxyl oxygen atom. This electronic configuration corresponds to an ideal Co^{IV}=O species, and this is found to contribute 77.5% to the ground state wavefunction. Additionally, the $\sigma_{(p_z+d_z{}^2)^1}\,\pi_{(p_y+d_{yz})^2}\,\pi_{(p_x+d_{xz})^2}\,\delta_{(d_{xy})^2}\,\pi^*_{(p_y+d_{yz})^1}\,\pi^*_{(p_x+d_{xz})^1}\,\delta^*_{(d_{x^2-y^2})^1}\,\sigma^*_{(p_z+d_{z^2})^1}$ configuration, which corresponds to the Co^{III}–O$^•$ species, contributes 3.1% to the ground state wavefunction. Similarly, for complexes **2** and **3**, the CAS(11,8) SCF calculations at equilibrium Co–O distances of 1.742 Å and 1.774 Å, respectively, result in $\sigma_{(p_z+d_z{}^2)^2}\,\pi_{(py+d_{yz})^2}\,\pi_{(p_x+d_{xz})^2}\,\pi^*_{(p_y+d_{yz})^1}\,\delta_{(d_{xy})^2}\,\pi^*_{(p_x+d_{xz})^2}\,\delta^*_{(d_{x^2-y^2})^0}\,\sigma^*_{(p_z+d_{z^2})^0}$ and $\sigma_{(p_z+d_z{}^2)^2}\,\pi_{(p_y+d_{yz})^2}\,\pi_{(p_x+d_{xz})^2}\,\pi^*_{(p_x+d_{xz})^1}\,\pi^*_{(p_y+d_{yz})^2}\,\delta_{(d_{xy})^2}\,\delta^*_{(d_{x^2-y^2})^0}\,\sigma^*_{(p_z+d_{z^2})^0}$ configurations, respectively, which correspond to true Co^{IV}=O species, contributing 70.3% and 64.2% to the total wavefunction, respectively. Furthermore, the following Co^{III}–O$^•$ configuration $\sigma_{(p_z+d_{z^2})^2}\,\pi_{(p_y+d_{yz})^2}\,\pi_{(p_x+d_{xz})^1}\,\delta_{(d_{xy})^2}\,\pi^*_{(p_y+d_{yz})^1}\,\pi^*_{(p_x+d_{xz})^2}\,\delta^*_{(d_{x^2-y^2})^0}\,\sigma^*_{(p_z+d_{z^2})^0}$ and $\sigma_{(p_z+d_{z^2})^2}\,\pi_{(p_y+d_{yz})^2}\,\pi_{(p_x+d_{xz})^1}\,\pi^*_{(p_y+d_{yz})^2}\,\delta_{(d_{xy})^2}\,\pi^*_{(p_x+d_{xz})^2}\,\delta^*_{(d_{x^2-y^2})^0}\,\sigma^*_{(p_z+d_{z^2})^0}$ contribute 14.3%, and 20%, respectively, to the ground state wavefunction. Thus, these results suggest that the Co^{IV}=O contribution is highest in species **1**, while the Co^{III}–O$^•$ contribution is highest in species **3**. Furthermore, the calculations unequivocally suggest that in all three species both Co^{IV}=O and Co^{III}–O$^•$ species states are admixed at the ground state.

True Co^{IV}=O *vs.* Co^{III}–O$^•$ species: the role of spin-states

Furthermore, to understand the role of the Co–O bond length on dictating the stabilisation of the two valence isomers, Co^{IV}=O and Co^{III}–O$^•$ species, we perform a relaxed potential energy surface scan along the Co–O bond coordinate for all three (**1–3**) species using DFT methods. All scan graphs consist of three segments, the minimum energy point, which corresponds to the equilibrium structure of the complex (optimised global minima), in both left and right sides of the graph with gradual decrease and increases in the Co–O bond distance, the energy of the system is found to be increasing steadily. The compression is, however, steeper, while elongation demands lesser energy strain (Fig. 2). Using the ground state geometry obtained from the DFT calculations, we proceed further for CASSCF/NEVPT2 calculations on each scan point at an interval of 0.04 Å. Also, we have estimated vertical excitation energy of different spin states to understand the potential energy landscape. Fig. 3 depicts the CASSCF/NEVPT2 calculated energy ordering of $S = 5/2$ (blue curve), 3/2 (green curve) and 1/2 (red curve) spin-states of species **1–3** along the Co–O bond coordinate. From the equilibrium point, as the Co–O bond-length

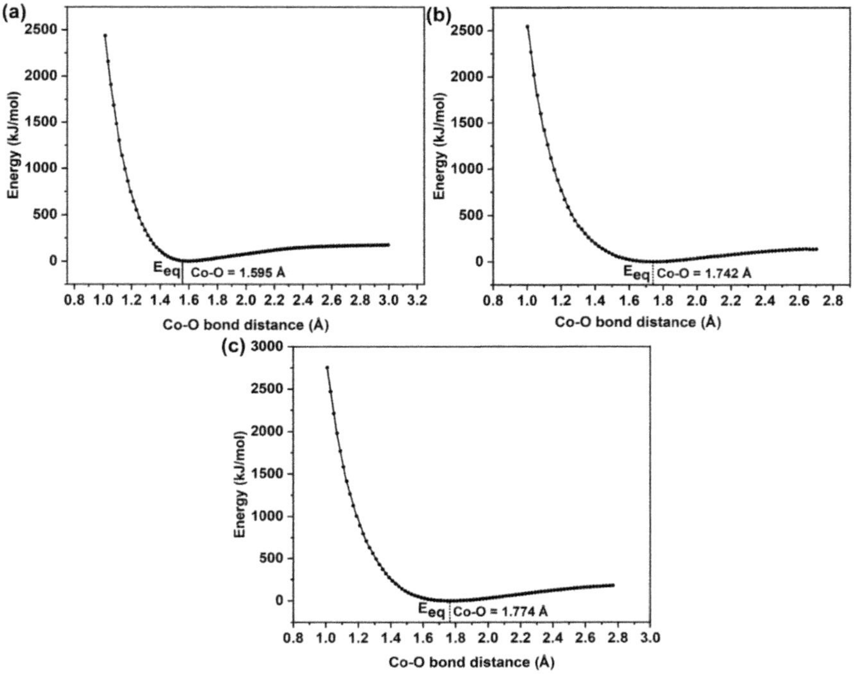

Fig. 2 DFT estimated potential energy scan surfaces of the lowest-energy states of (a–c) **1–3** as a function of the Co–O distance.

elongates/compresses, the energy increases gradually for all spin states yielding a parabolic behaviour. For **1**, below the Co–O distance of 1.59 Å, the $S = 3/2$ state remains the ground state. However, elongation of the Co–O distance not only bring the excited states closer but also stabilise them as the high-spin ground state beyond the Co–O distance of 1.87 Å, suggests spin-crossover type behaviour for a very long bond length. The trend in the energetics of species **2** and **3** with respect to the Co–O bond distance is found to be similar. The ground state of $S = 1/2$ remains unaltered throughout the Co–O bond elongation for both **2** (except Co–O=2.1 Å) and **3**. As we compress the Co–O bond, the energy gap among all three spin states decreases and at 1.58 Å (1.53 Å) distance for species **2** (**3**), $S = 3/2$ is stabilised as the ground state. Therefore, in **2** and **3**, the ground state of the $S = 1/2$ spin remains unaltered for the Co–O bond distance of $\sim$1.8 $\pm$ 0.2 Å, below/above which the $S = 3/2$ state becomes the ground state.

To understand the role of spin states and their correlation with the existence of $Co^{IV}{=}O$ *vs.* $Co^{III}–O^{\bullet}$ species, the underlying configurations computed were analysed. Fig. 4 depicts the changes in the contributions arising from the $Co^{IV}{=}O$ and $Co^{III}–O^{\bullet}$ characters at various points. For species **1**, compression of the Co–O distance from the equilibrium point enhances the $Co^{IV}{=}O$ electromeric character, which reaches as high as $\sim$90% at 1.35 Å; this is accompanied by a concomitant reduction in the $Co^{III}–O^{\bullet}$ character. Elongation of the Co–O bond from the equilibrium geometry enhances the $Co^{III}–O^{\bullet}$ character, which reaches $\sim$30% at 1.87 Å with a concomitant reduction of the $Co^{IV}{=}O$ species. The elongation of Co–O bond lengths beyond this point leads to the enhancement of the

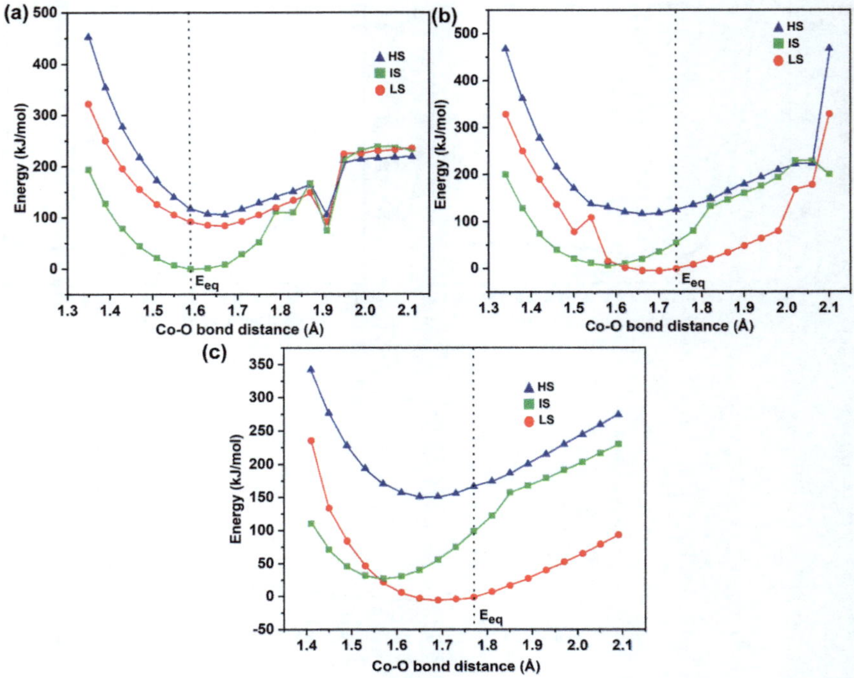

Fig. 3 CASSCF/NEVPT2 estimated potential energy surfaces scan of (a–c) **1–3** as a function of the Co–O distance.

$Co^{IV}{=}O$ and reduction of the $Co^{III}–O^{•}$ species, though the nature of the species at these long Co–O bond lengths rather reflects $Co^{IV}\cdots O^{2-}$ type species.

A similar trend is followed for species **2** and **3**. Compression of the Co–O bond results in a steady decrease in the % of $Co^{III}–O^{•}$ species and, below 1.58 Å for **2** and 1.57 Å for **3**, the$Co^{IV}{=}O$ species dominate. As the Co–oxo bond distance decreases below these points, the ground state wavefunctions of these species comprise >75% of the $Co^{IV}{=}O$ character with <4% of the oxyl-radical species. With the elongation, the $Co^{III}–O^{•}$ character increases as high as ~46% for **2** and 31% for **3** at ~2.00 Å. Beyond a Co–O distance of 2.00 Å, the contribution from the $Co^{IV}{=}O$ species increases again.

To correlate the nature of $Co^{IV}{=}O$ *vs.* $Co^{III}–O^{•}$ species with the spin states, Fig. 3 and 4 need to be compared. The following points emerge with respect to species **1**: (i) as we compress the Co–O bond, the $\Delta(E_{3/2} - E_{1/2})$ gap decreases and so does the $Co^{III}–O^{•}$ character. This is accompanied by the concomitant increase of $Co^{IV}{=}O$ character. The ground state remains $S = 3/2$ in these points with dominant $Co^{IV}{=}O$ species, indicating that the $S = 3/2$ spin state favours the $Co^{IV}{=}O$ species. (ii) As we elongate the Co–O bond, the $\Delta(E_{3/2} - E_{1/2})$ gap decreases and this is accompanied by an increase in the $Co^{III}–O^{•}$ radical character, with a concomitant decrease of $Co^{IV}{=}O$ species. This is expected as the Co–O bond's elongation cleaves the corresponding π-bond, leading to a favourable $Co^{III}–O^{•}$ electromer. At 1.87 Å, an $S = 1/2$ state becomes the ground state, and coincidentally at this point, the $Co^{III}–O^{•}$ character is the highest. This suggests that $S = 1/2$ favours $Co^{III}–O^{•}$ species. (iii)

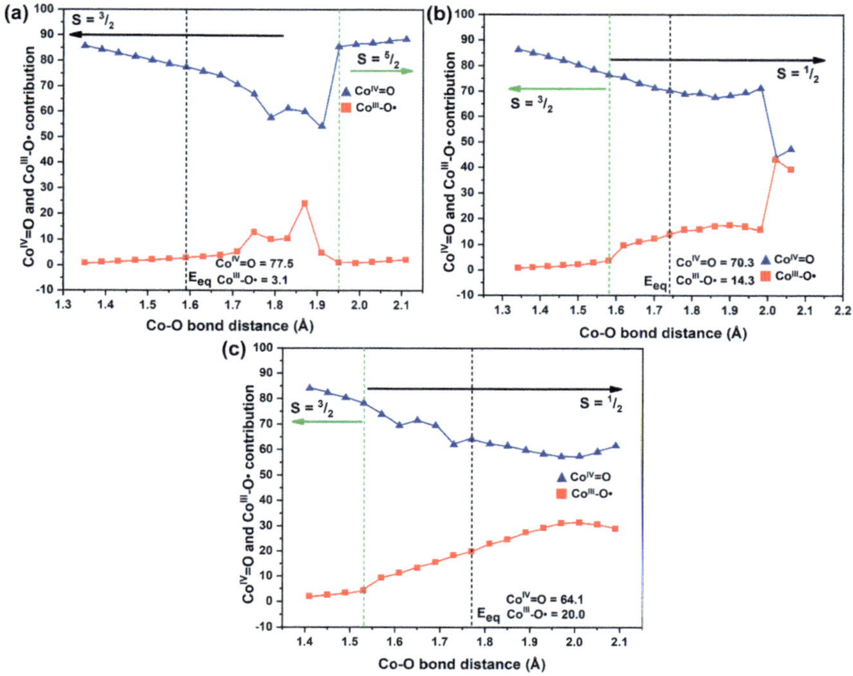

Fig. 4 Percentage of the $Co^{IV}{=}O$ and $Co^{III}{-}O^{\bullet}$ species changes in (a–c) **1–3** as a function of the Co–O distance.

Beyond a Co–O distance of 1.95 Å, an $S = 5/2$ state becomes the ground state, and this is accompanied by a substantial increase of $Co^{IV}{=}O$ species, though a closer look at the electronic structure reveals a $Co^{IV}\cdots O^{2-}$ species due to relatively longer bond lengths. While this is not confirmed, an $S = 5/2$ state with a longer Co–O bond length seemingly favours $Co^{IV}{=}O$ species. For species **2** and **3**, the following points emerge: (i) both have the ground state of $S = 1/2$, and compression of the Co–O bond decreases the $\Delta(E_{1/2} - E_{3/2})$ gap, increasing the $Co^{IV}{=}O$ character and decreasing the $Co^{III}{-}O^{\bullet}$ character. Below ~1.55 Å, an $S = 3/2$ becomes the ground state, and this leads to a sharper increase in the $Co^{IV}{=}O$ character (see the change in the slope in Fig. 4b and c beyond the dotted green line). (ii) As we elongate the Co–O bond, an $S = 1/2$ state stabilises in most of the points with a larger $\Delta(E_{1/2} - E_{3/2})$ gap, and this leads to a significant increase in the $Co^{III}{-}O^{\bullet}$ character. This is consistent with our assessment in species **1**.

To rationalise the spin-state dependency on the preference for $Co^{IV}{=}O$ *vs.* $Co^{III}{-}O^{\bullet}$ species, qualitative MO diagrams based on DFT computed MOs and energies were plotted and are shown in Fig. 5a–c. Fig. 5a shows the MO diagram for the $Co^{III}{-}O^{\bullet}$ species with Co^{III} in the low-spin state. Our calculations reveal that the degeneracies of all five metal d-orbitals and three oxo p-orbitals have lifted in the presence of a ligand field, with the bonding $\pi_{(p_x+d_{xz})}$ orbital having a spin-up α-electron which is the lowest in energy, with a 45% contribution from the oxygen p_x orbital, followed by the $\sigma_{(p_z+d_{z^2})}$-α > $\sigma_{(p_z+d_{z^2})}$-β > $\pi_{(p_y+d_{yz})}$-α > $\pi_{(p_y+d_{yz})}$-β > $\pi_{(p_y+d_{yz})}$-α > $\pi_{(p_y+d_{yz})}$-β > $\pi_{(p_x+d_{xz})}$-β orbitals lying at 0.3 eV, 0.5 eV, 0.8 eV, 0.9 eV and 1.6 eV, respectively. These bonding molecular orbitals are composed of

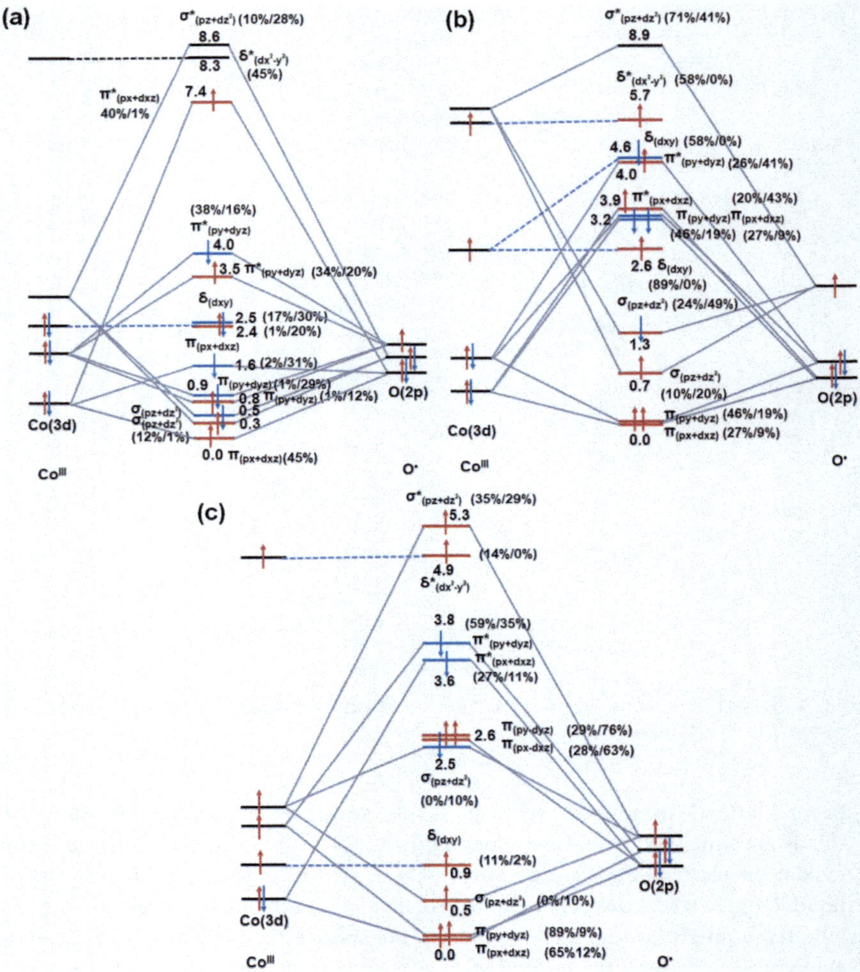

Fig. 5 Molecular orbital diagram of Co=O unit (a) low-spin ($S = {}^{1}/_{2}$), (b) intermediate spin ($S = {}^{3}/_{2}$) and (c) high spin ($S = {}^{5}/_{2}$) state in species **3**. The first value in the parentheses is the % of p-character, and the second value is the % of d-character in the corresponding MO.

contributions from the metal 3d and oxygen 2p orbitals. For instance, the bonding $\pi_{(p_x+d_{xz})}$ MO with a spin-down β-electron consists of 31% d-character and merely 2% p-character from oxygen. The $\delta_{(d_{xy})}$ orbital is non-bonding and contains a pair of electrons, lying at 2.4–2.5 eV. This non-bonding molecular orbital is followed by $\pi^*_{(p_y+d_{yz})}$-α(3.5 eV) > $\pi^*_{(p_y+d_{yz})}$-β(4.0 eV) > $\pi^*_{(p_x+d_{xz})}$-α(7.4 eV) > $\delta^*_{(d_{x^2-y^2})}$ (8.3 eV) > $\sigma^*_{(p_z+d_z)^2}$ (8.4 eV). According to the energy ordering, the electron filling is as follows: $[\pi_{(p_x+d_{xz})}\text{-}\alpha]^1 > [\sigma_{(p_z+d_z{}^2)}\text{-}\alpha]^1 > [\sigma_{(p_z+d_z{}^2)}\text{-}\beta]^1 > [\pi_{(p_y+d_{yz})}\text{-}\alpha]^1 > [\pi_{(p_y+d_{yz})}\text{-}\beta]^1 > [\pi_{(p_x+d_{xz})}\text{-}\beta]^1 > [\delta_{(d_{xy})}\text{-}\alpha]^1 > [\delta_{(d_{xy})}\text{-}\beta]^1 > [\pi^*_{(p_y+d_{yz})}\text{-}\alpha]^1 > [\pi^*_{(p_y+d_{yz})}\text{-}\beta]^1 > [\pi^*_{(p_x+d_{xz})}\text{-}\alpha]^1 > [\delta^*_{(d_{x^2-y^2})}] > [\sigma^*_{(p_z+d_{z^2})}]$. From the orbital contribution values from the metal 3d and oxygen 2p orbitals, it is noted that the bonding and antibonding orbitals are heavily mixed with the contributions from the metal and oxo centre.

The singly occupied antibonding molecular orbital, $\pi^*_{(p_x+d_{xz})}$, is found to be composed of 40% p-character with a mere 1% d-character, indicating that the unpaired electron is located on an oxygen 2p orbital, reflecting a strong radical character, as expected for the Co^{III}–O$^{\bullet}$ species. A qualitative bond order analysis, if performed, suggests that for this species, the σ bond has a bond order of $\left(\frac{2-0}{2}=1\right)$ and the π bond has $\left(\frac{4-3}{2}=0.5\right)$, suggesting an overall bond order of 1.5. For Co^{IV}=O with $S = 3/2$ (intermediate spin state), the MO diagram (Fig. 5b) reveals that the additional unpaired electron found in the $\pi^*_{(p_x+d_{xz})}$ orbital earlier is absent, and this electron in fact is present in the non-bonding $\delta^*_{(d_{x^2-y^2})}$ orbital. This enhances the π bond order, and here the bond order is $\left(\frac{2-0}{2}=1\right)$ and the π bond order is 1.0 $\left(\frac{4-2}{2}=1\right)$, suggesting an overall bond order of 2.0. For the Co^{IV}=O species with $S = 5/2$ spin state (high-spin) at the Co centre, the MO diagram (Fig. 5c) reveals that the scenario is similar to the first case, with the only change being that the $\sigma^*_{(p_z+d_{z^2})}$ orbital stabilises significantly due to very weak antibonding interactions. Due to these changes, the bond order is as follows: the σ bond order is $\left(\frac{2-1}{2}=0.5\right)$ and the π bond order is $\left(\frac{4-2}{2}=1.0\right)$, suggesting an overall bond order of 1.5.

A tool to predict the percentages of Co^{III}–O$^{\bullet}$ *vs.* Co^{IV}=O species in unknown examples

After analysing the dependency of the Co^{IV}=O/Co^{III}–O$^{\bullet}$ equilibrium on the Co–O bond length, we have performed a potential energy surface scan along the Co–N_{ax} bond in **2** and **3**. As we have stated earlier, in complex **2**, a loosely bound CH_3CN molecule is attached in the apical position opposite the Co–oxo unit. In contrast, the opposite apical position in **3** is occupied by one of the amine arms of the N4Py ligand. Fig. 6 shows the variation in energy in the different spin states as a function of the Co–N_{ax} bond length, where the spin-state splittings and the nature of the ground state are unaltered. For both **2** and **3**, as we elongate the Co–N_{ax}

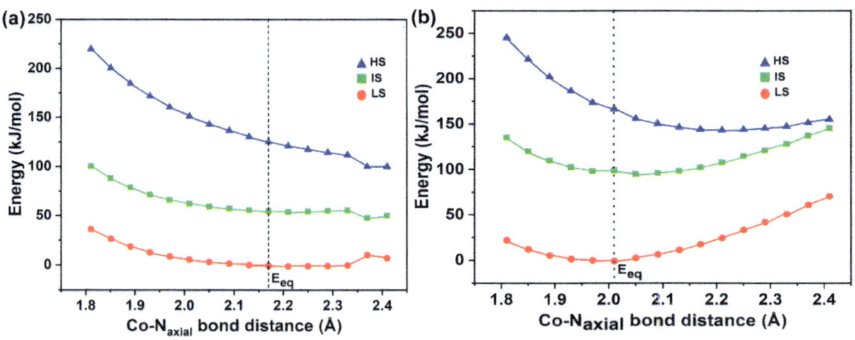

Fig. 6 Potential energy surfaces of the lowest energy states of (a and b) **1–2** as a function of the Co–N_{ax} distance.

distance, the energies of the individual spin states initially decrease and then increase slowly (Fig. 6). Furthermore, CAS(11,8) calculations on the stationary points in an interval of 0.04 Å on the Co–N_{ax} scan surface reveal that in complex **2**, with the decrease in the Co–N_{ax} bond distance, the oxyl radical character increases and reaches 20.2% at a Co–N_{ax} distance of 1.81 Å and *vice versa*. In **3**, a similar trend is followed, but the extent of the rise of the oxyl radical character with a reduction of the Co–N_{ax} bond distance is comparatively significant. Therefore, the existence of the Co^{III}–O$^{\bullet}$ isomer is directly dependent on the Co–O as well as the Co–N_{ax} bond lengths. Previous reports show that the M═O anti-bonding interactions were found to push the ligand d_{z^2} orbital present in the axial position to the M═O moiety and result in M–L_{ax} bond elongation; this push–pull effect is known to influence the reactivity of Fe^{IV}═O species.[40,41] Considering these points, we attempt to develop a tool based on the structural parameters to estimate approximately the % of Co^{III}–O$^{\bullet}$ species for any high-valent (+III/+IV) Co-oxo complexes by looking at the ratio of Co–O/Co–N_{ax}, as shown below,

$$\%\mathrm{Co^{III}{-}O^{\bullet}} \propto \frac{d_{(\mathrm{Co-O})}}{d_{(\mathrm{Co-N_{ax}})}}$$

Our calculations reveal that the higher the $\frac{d_{(\mathrm{Co-O})}}{d_{(\mathrm{Co-N_{ax}})}}$ ratio (termed as R_d), the greater the Co^{III}–O$^{\bullet}$ contribution will be. This relationship shown above yields a linear correlation, as shown in Fig. 7, for both **2** and **3**, with varying slopes. We fit the points for a linear equation (Fig. 7), which yield a very good fit (R^2 value exceeding 0.98). Using the parameters derived, one can calculate the % of Co^{III}–O$^{\bullet}$ species for any given complex. As DFT calculations often yield this structural information, the electromeric ratio for an unknown species can be calculated directly.

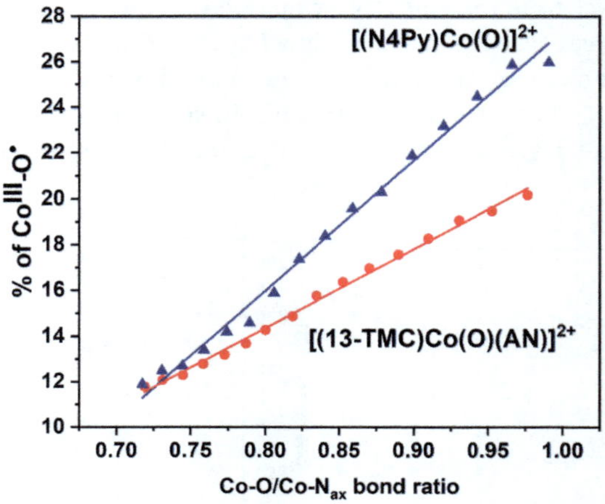

Fig. 7 % of the Co^{III}–O$^{\bullet}$/Co^{IV}═O ratio changes in (a–c) **1–3** as a function of the Co–oxo/Co–N_{axial} ratio. Red and blue points are for the $[(13\text{-TMC})Co(O)(CH_3CN)]^{2+}$ and $[(N4Py)Co(O)]^{2+}$ complexes, respectively. The linear fit is performed using the $y = mx + C$ equation, where the slopes are 34.69 ± 0.63 and 56.94 ± 1.38, respectively. The intercept values are −13.37 ± 0.53 and −29.55 ± 1.17 for the respective red and blue graphs, respectively. For both cases, the *R*-squared values are greater than 0.99.

Co^{III}–O$^{\bullet}$ *vs.* CO^{IV}=O species in five-coordinate square-pyramidal geometries?

When the axial CH_3CN ligand in $[(13\text{-TMC})Co^{IV}(O)(CH_3CN)]^{2+}$ elongates to an infinite distance, it will result in the formation of a square pyramidal $[(13\text{-TMC})Co^{IV}(O)]^{2+}$ complex.[19] This metastable monomeric non-haem complex was photocatalytically synthesised and characterised by Nam and co-workers. Various spectroscopic studies have been performed to establish the presence of the Co^{IV}=O core in this molecule. DFT calculations on this molecule show that the molecule possesses an intermediate $S = 3/2$ state as the ground state, with the low-spin and high-spin states lying 15.2 kJ mol^{-1} and 46.0 kJ mol^{-1} higher in energy. The ground state of $S = 3/2$ remains unchanged in the CASSCF/NEVPT2 calculations. A quartet ground state agrees with the experimental observation. The Co–O and Co–N_{eq} bonding parameters in the ground $S = 3/2$ state are 1.684 Å, 2.036 Å, 2.057 Å, 2.041 Å and 2.127 Å (Fig. 8). A CAS(11,8) calculation on this molecule predicts the presence of 69% Co^{IV}=O character along with a small % of Co^{III}–O$^{\bullet}$ character (7.0%) in the ground state. The ground state wavefunction is dominated by the $\sigma_{(p_z+d_{z^2})^2}\,\pi_{(p_x+d_{xz})^2}\,\pi_{(p_y+d_{yz})^2}\,\delta_{(d_{xy})^2}\,\pi^*_{(p_x+d_{xz})^1}\,\pi^*_{(p_y+d_{yz})^1}\,\delta^*_{(d_{x^2-y^2})^1}\,\sigma^*_{(p_z+d_{z^2})^0}$ configuration (55.3%) $\sigma_{(p_z+d_{z^2})^2}\,\pi_{(p_x+d_{xz})^2}\,\pi_{(p_y+d_{yz})^2}\,\delta_{(d_{xy})^2}\,\pi^*_{(p_y+d_{yz})^1}\,\pi^*_{(p_x+d_{xz})^1}\,\delta^*_{(d_{x^2-y^2})^0}\,\sigma^*_{(p_z+d_{z^2})^1}$ and $\sigma_{(p_z+d_{z^2})^1}\,\pi_{(p_x+d_{xz})^2}\,\pi_{(p_y+d_{yz})^2}\,\delta_{(d_{xy})^2}\,\pi^*_{(p_y+d_{yz})^1}\,\pi^*_{(p_x+d_{xz})^1}\,\delta^*_{(d_{x^2-y^2})^1}\,\sigma^*_{(p_z+d_{z^2})^1}$ configuration is less, 13.2% and 7.0% respectively, indicating the presence of Co^{IV}=O character in this complex.

From the earlier detailed analysis and MO diagram presented, it is clear that the dominant Co^{IV}=O character present in this complex is due to the stabilisation of the $S = 3/2$ ground state. Then, to assess the linear correlation estimated, we take the computed Co–O distance of 1.684 Å and approximate the non-existent axial ligand to 3.0 Å. This yields the % of the Co^{III}–O$^{\bullet}$ character of 6.1 and this is similar to the 7.0% obtained from the CASSCF calculations. This suggests the portability of this relationship even to a square pyramidal geometry, offering confidence in the proposed tool to calculate the Co^{III}–O$^{\bullet}$ radical character.

Valance electromerism before the oxo-wall?

To compare and contrast the M=O *vs.* M–O$^{\bullet}$ species before the oxo-wall, we have extended our calculations to Mn and Fe-analogues of $[(N4Py)Co(O)]^{2+}$ species.[3,42]

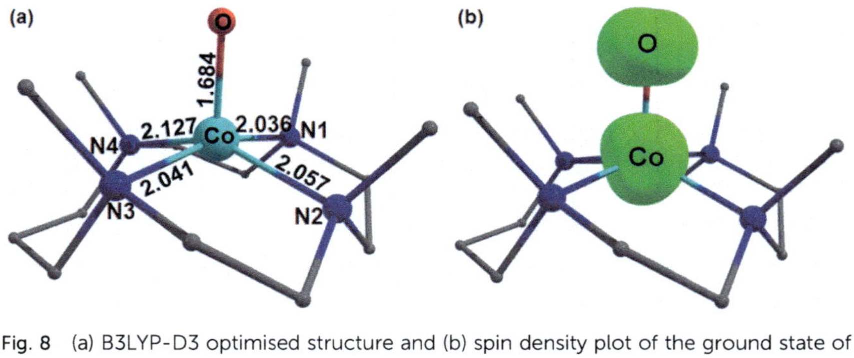

Fig. 8 (a) B3LYP-D3 optimised structure and (b) spin density plot of the ground state of complex **4**.

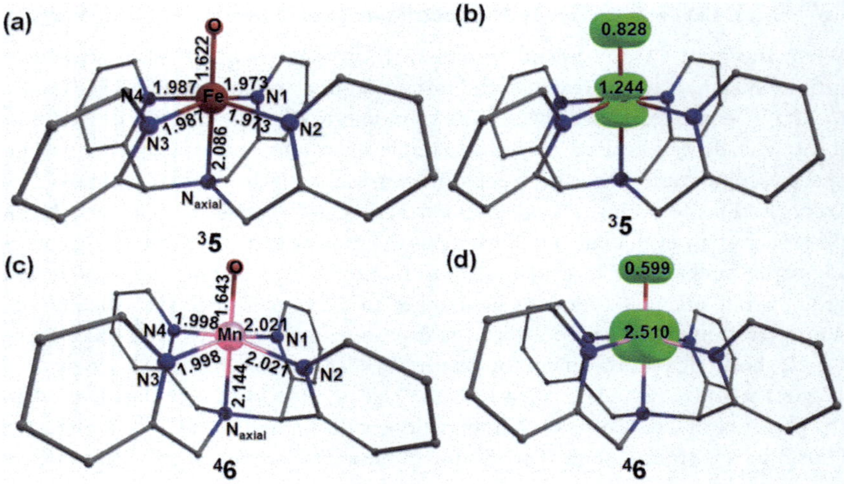

Fig. 9 (a and c) B3LYP-D3 optimised structures and (b and d) spin density plots of the ground state of complexes **5–6**, respectively.

For both the $[(N4Py)Fe(O)]^{2+}$ (**5**) and $[(N4Py)Mn(O)]^{2+}$ (**6**) species, an $S = 3/2$ and an $S = 1$ spin state are found to be the ground states, respectively,[43] and this is in agreement with the experiments and previous DFT calculations. The other spin states lie higher in energy (103.2 kJ mol^{-1} and 126.4 kJ mol^{-1} for the $S = 2$ and $S = 0$ states in $[(N4Py)Fe(O)]^{2+}$, respectively and 38.2 kJ mol^{-1} for the $S = 1/2$ spin state in $[(N4Py)Mn(O)]^{2+}$). Here also, DFT calculations show an overdelocalisation problem, where the spin density values on the metal (Fe/Mn) and on the oxyl moiety are 1.244 and 0.828, and 2.510 and 0.599, respectively (Fig. 9). The CAS(10,8)/NEVPT2 and CAS(9,8)/NEVPT2 calculations reveal that the energy orderings of the spin states are unaltered. For the $[(N4Py)Mn(O)]^{2+}$ species, a CAS calculation unveils a 74.6% contribution from the $\sigma_{(p_z+d_z{}^2)^2}\ \pi_{(p_y+d_{yz})^2}\ \pi_{(p_x+d_{xz})^2}\ \delta_{(d_{xy})^1}\ \pi^*_{(p_y+d_{yz})^1}\ \pi^*_{(p_x+d_{xz})^1}\ \delta^*_{(d_{x^2-y^2})^0}\ \sigma^*_{(p_z+d_{z^2})^0}$ configuration, corresponding to $Mn^{IV}{=}O$ species with a small (~8%) contribution from the $\sigma_{(p_z+d_z{}^2)^1}\ \pi_{(p_y+d_{yz})^2}\ \pi_{(p_x+d_{xz})^2}\ \delta_{(d_{xy})^1}\ \pi^*_{(p_y+d_{yz})^1}\ \pi^*_{(p_x+d_{xz})^1}\ \delta^*_{(d_{x^2-y^2})^0}\ \sigma^*_{(p_z+d_{z^2})^1}$ configuration that corresponds to the Mn^{III}–$O^{\bullet}$ species. Similarly, the ground state wavefunction of the $[(N4Py)Fe(O)]^{2+}$ species consist of 74% of the $\sigma_{(p_z+d_{z2})^2}\ \pi_{(p_y+d_{yz})^2}\ \pi_{(p_x+d_{xz})^2}\ \delta_{(d_{xy})^2}\ \pi^*_{(p_x+d_{xz})^1}\ \pi^*_{(p_y+d_{yz})^1}\ \delta^*_{(d_{x^2-y^2})^0}\ \sigma^*_{(p_z+d_{z^2})^0}$ configuration ($Fe^{IV}{=}O$) and ~7% of the $\sigma_{(p_z+d_{z2})^2}\ \pi_{(p_y+d_{yz})^1}\ \pi_{(p_x+d_{xz})^2}\ \delta_{(d_{xy})^2}\ \pi^*_{(p_x+d_{xz})^1}\ \pi^*_{(p_y+d_{yz})^1}\ \delta^*_{(d_{x^2-y^2})^0}\ \sigma^*_{(p_z+d_{z^2})^1}$ configuration (Fe^{III}–$O^{\bullet}$). This reveals that before group 8, the complexes are stable with the M=O formalism with a very small/insignificant contribution from their oxyl-radical counterpart.

Take home message

In the present work, we have employed the density functional theory (DFT) and *ab initio* methods to study one of the outstanding problems in biomimic chemistry that has existed for a very long time, *i.e.* the existence of double bond character beyond the group 8 molecules. The originally proposed oxo-wall theory is strictly

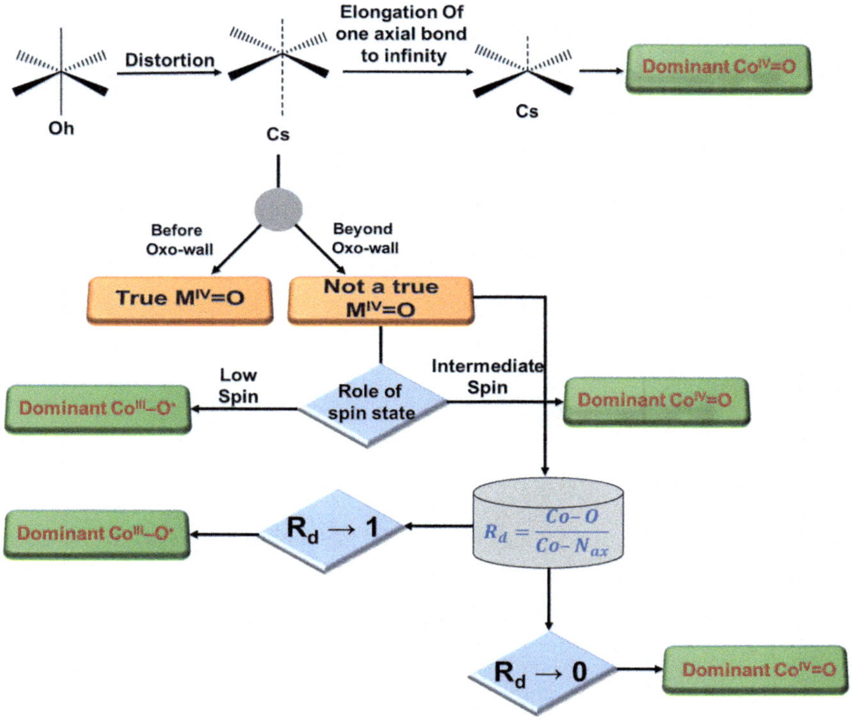

Scheme 2 Take home message of the work presented here.

applicable to tetragonal complexes, and for other geometries or coordination numbers, the oxo-wall does not strictly forbid the formation of ideal $M^{IV}{=}O$ species. Over the last few years, several synthetic groups have employed different ligand architecture to synthesise a true $Co^{IV}{=}O$ species. Although this would not violate the oxo-wall theory proposed by Ballhausen and Gray, the possibility that the $Co^{IV}{=}O$ species could exist along with its electromer Co^{III}–$O^{\bullet}$ has not been explored and has been the subject of this study. We arrived at the following take-home messages by thoroughly studying six different metal–oxo species (see Scheme 2).

(i) Prior to the wall (before group 8), the high-valent (+4) terminal M–oxo species exist in their formal $M^{IV}{=}O$ form. The ground state wavefunction is composed of ~80% of the $M^{IV}{=}O$ character with very small/negligible M^{III}–$O^{\bullet}$ character.

(ii) After the oxo-wall (beyond group 8), our calculations reveal a drastic decrease of the formal $Co^{IV}{=}O$ character, accompanied by a concomitant increase of isomeric Co^{III}–$O^{\bullet}$ species. This suggests that irrespective of the geometry, obtaining a true $Co^{IV}{=}O$ species (100% $Co^{IV}{=}O$ character and 0% Co^{III}–$O^{\bullet}$) is hard.

(iii) Secondly, we look at the dominant character of $Co^{IV}{=}O$ *vs.* Co^{III}–$O^{\bullet}$ under various conditions. Our calculations reveal that a formal $Co^{IV}{=}O$ species is dominant if an $S = 3/2$ spin state is the ground state and isomeric Co^{III}–$O^{\bullet}$ species composition is substantial if an $S = 1/2$ low-spin state is the ground state.

(iv) By analysing thoroughly various parameters, we propose that $\%\mathrm{Co^{III}}\text{–}\mathrm{O^{\bullet}} \propto \frac{d_{(\mathrm{Co-O})}}{d_{(\mathrm{Co-N_{ax}})}}$ ($d_{(\mathrm{Co-O})}$ and $d_{(\mathrm{Co-N_{ax}})}$ are the distances of the axial Co–O and Co–N_{ax} bonds in Å), and this relationship yields a linear correlation for the two different Co complexes studied, offering a chance to directly calculate the % of $\mathrm{Co^{III}}$–$\mathrm{O^{\bullet}}$ character in any unknown system. Our calculations suggest that the larger the bond ratio, the higher the $\mathrm{Co^{III}}$–$\mathrm{O^{\bullet}}$ character will be, with values closer to zero yielding an ideal $\mathrm{Co^{IV}{=}O}$ species. Surprisingly, the proposed correlation is portable even to square pyramidal geometries to assess the percentage of $\mathrm{Co^{III}}$–$\mathrm{O^{\bullet}}$ character.

Conflicts of interest

There are no conflicts to declare.

Acknowledgements

We thank DST and SERB (CRG/2018/00430; DST/CSA-03/2018-10; SB/SJF/2019-20/12) for funding.

References

1 D. G. Nocera, *Acc. Chem. Res.*, 2012, **45**, 767–776.
2 M. Wikström, K. Krab and V. Sharma, *Chem. Rev.*, 2018, **118**, 2469–2490.
3 M. Costas, M. P. Mehn, M. P. Jensen and L. Que, *Chem. Rev.*, 2004, **104**, 939–986.
4 C. Krebs, D. Galonic Fujimori, C. T. Walsh and J. M. Bollinger Jr, *Acc. Chem. Res.*, 2007, **40**, 484–492.
5 W. J. Song, M. S. Seo, S. D. George, T. Ohta, R. Song, M. J. Kang, T. Tosha, T. Kitagawa, E. I. Solomon and W. Nam, *J. Am. Chem. Soc.*, 2007, **129**, 1268–1277.
6 A. Borovik, *Chem. Soc. Rev.*, 2011, **40**, 1870–1874.
7 R. Kumar, B. Pandey, A. Sen, M. Ansari, S. Sharma and G. Rajaraman, *Coord. Chem. Rev.*, 2020, **419**, 213397.
8 A. Sen, N. Vyas, B. Pandey and G. Rajaraman, *Dalton Trans.*, 2020, **49**, 10380–10393.
9 S. Rana, J. P. Biswas, A. Sen, M. Clémancey, G. Blondin, J.-M. Latour, G. Rajaraman and D. Maiti, *Chem. Sci.*, 2018, **9**, 7843–7858.
10 A. Gunay and K. H. Theopold, *Chem. Rev.*, 2010, **110**, 1060–1081.
11 H. B. Gray and J. R. Winkler, *Acc. Chem. Res.*, 2018, **51**, 1850–1857.
12 H. B. Gray, *Chem. Int.*, 2019, **41**, 16–19.
13 V. A. Larson, B. Battistella, K. Ray, N. Lehnert and W. Nam, *Nat. Rev. Chem.*, 2020, **4**, 404–419.
14 C. J. Ballhausen and H. B. Gray, *Inorg. Chem.*, 1962, **1**, 111–122.
15 H. B. Gray and C. R. Hare, *Inorg. Chem.*, 1962, **1**, 363–368.
16 J. R. Winkler and H. B. Gray, *Molecular Electronic Structures of Transition Metal Complexes I*, 2011, pp. 17–28.
17 S. Ye and F. Neese, *Proc. Natl. Acad. Sci. U. S. A.*, 2011, **108**, 1228–1233.
18 K.-B. Cho, S. Shaik and W. Nam, *J. Phys. Chem. Lett.*, 2012, **3**, 2851–2856.

19 B. Wang, Y.-M. Lee, W.-Y. Tcho, S. Tussupbayev, S.-T. Kim, Y. Kim, M. S. Seo, K.-B. Cho, Y. Dede and B. C. Keegan, *Nat. Commun.*, 2017, **8**, 1.

20 H. Kurreck, B. Kirste and W. Lubitz, *Angew. Chem., Int. Ed.*, 1984, **23**, 173–194.

21 M. J. Frisch, G. W. Trucks, H. B. Schlegel, G. E. Scuseria, M. A. Robb, J. R. Cheeseman, G. Scalmani, V. Barone, G. A. Petersson, H. Nakatsuji, X. Li, M. Caricato, A. V. Marenich, J. Bloino, B. G. Janesko, R. Gomperts, B. Mennucci, H. P. Hratchian, J. V. Ortiz, A. F. Izmaylov, J. L. Sonnenberg, D. Williams-Young, F. Ding, F. Lipparini, F. Egidi, J. Goings, B. Peng, A. Petrone, T. Henderson, D. Ranasinghe, V. G. Zakrzewski, J. Gao, N. Rega, G. Zheng, W. Liang, M. Hada, M. Ehara, K. Toyota, R. Fukuda, J. Hasegawa, M. Ishida, T. Nakajima, Y. Honda, O. Kitao, H. Nakai, T. Vreven, K. Throssell, J. A. Montgomery Jr, J. E. Peralta, F. Ogliaro, M. J. Bearpark, J. J. Heyd, E. N. Brothers, K. N. Kudin, V. N. Staroverov, T. A. Keith, R. Kobayashi, J. Normand, K. Raghavachari, A. P. Rendell, J. C. Burant, S. S. Iyengar, J. Tomasi, M. Cossi, J. M. Millam, M. Klene, C. Adamo, R. Cammi, J. W. Ochterski, R. L. Martin, K. Morokuma, O. Farkas, J. B. Foresman and D. J. Fox, *Gaussian 16, Revision C.01*, Gaussian, Inc., Wallingford CT, 2016.

22 R. H. Hertwig and W. Koch, *Chem. Phys. Lett.*, 1997, **268**, 345–351.

23 F. A. Hamprecht, A. J. Cohen, D. J. Tozer and N. C. Handy, *J. Chem. Phys.*, 1998, **109**, 6264–6271.

24 L. E. Roy, P. J. Hay and R. L. Martin, *J. Chem. Theory Comput.*, 2008, **4**, 1029–1031.

25 V. A. Rassolov, J. A. Pople, M. A. Ratner and T. L. Windus, *J. Chem. Phys.*, 1998, **109**, 1223–1229.

26 A. V. Marenich, C. J. Cramer and D. G. Truhlar, *J. Phys. Chem. B*, 2009, **113**, 6378–6396.

27 F. Neese, *Wiley Interdiscip. Rev.: Comput. Mol. Sci.*, 2018, **8**, e1327.

28 M. Reiher, *Theor. Chem. Acc.*, 2006, **116**, 241–252.

29 G. Zhurko and D. Zhurko, 2009, URL: http://www.chemcraftprog.com.

30 D. Kim, J. Cho, Y.-M. Lee, R. Sarangi and W. Nam, *Chem. - Eur. J.*, 2013, **19**, 14112.

31 W. K. Lo, C. E. Castillo, R. Gueret, J. Fortage, M. Rebarz, M. Sliwa, F. Thomas, C. J. McAdam, G. B. Jameson and D. A. McMorran, *Inorg. Chem.*, 2016, **55**, 4564–4581.

32 M. Zhang, M. De Respinis and H. Frei, *Nat. Chem.*, 2014, **6**, 362–367.

33 M. Atanasov, P. Comba, B. Martin, V. Müller, G. Rajaraman, H. Rohwer and S. Wunderlich, *J. Comput. Chem.*, 2006, **27**, 1263–1277.

34 T. Z. Gani and H. J. Kulik, *J. Chem. Theory Comput.*, 2016, **12**, 5931–5945.

35 E. R. Johnson, A. Otero-De-La-Roza and S. G. Dale, *J. Chem. Phys.*, 2013, **139**, 184116.

36 R. Gupta, T. Taguchi, B. Lassalle-Kaiser, E. L. Bominaar, J. Yano, M. P. Hendrich and A. Borovik, *Proc. Natl. Acad. Sci. U. S. A.*, 2015, **112**, 5319–5324.

37 H. Chen, W. Lai and S. Shaik, *J. Phys. Chem. Lett.*, 2010, **1**, 1533–1540.

38 P. E. Siegbahn, J. Almlöf, A. Heiberg and B. O. Roos, *J. Chem. Phys.*, 1981, **74**, 2384–2396.

39 J. Olsen, *Int. J. Quantum Chem.*, 2011, **111**, 3267–3272.

40 Z. Ma, N. Nakatani, H. Fujii and M. Hada, *Phys. Chem. Chem. Phys.*, 2020, **22**, 12173–12179.
41 R. Wang and S. P. de Visser, *J. Inorg. Biochem.*, 2007, **101**, 1464–1472.
42 D. F. Leto, R. Ingram, V. W. Day and T. A. Jackson, *Chem. Commun.*, 2013, **49**, 5378–5380.
43 D. Kumar, H. Hirao, L. Que and S. Shaik, *J. Am. Chem. Soc.*, 2005, **127**, 8026–8027.

Faraday Discussions

ROYAL SOCIETY OF CHEMISTRY

PAPER

Binding of the substrate analog methanol in the oxygen-evolving complex of photosystem II in the D1-N87A genetic variant of cyanobacteria†

Vidmantas Kalendra,[a] Krystle M. Reiss,[b] Gourab Banerjee,[b] Ipsita Ghosh,[b] Amgalanbaatar Baldansuren,[a] Victor S. Batista,[b] Gary W. Brudvig *[b] and K. V. Lakshmi *[a]

Received 22nd November 2021, Accepted 21st December 2021
DOI: 10.1039/d1fd00094b

The solar water-splitting protein complex, photosystem II (PSII), catalyzes one of the most energetically demanding reactions in nature by using light energy to drive a catalyst capable of oxidizing water. The water oxidation reaction is catalyzed at the Mn_4Ca-oxo cluster in the oxygen-evolving complex (OEC), which cycles through five light-driven S-state intermediates (S_0–S_4). A detailed mechanism of the reaction remains elusive as it requires knowledge of the delivery and binding of substrate water in the higher S-state intermediates. In this study, we use two-dimensional (2D) hyperfine sublevel correlation spectroscopy, in conjunction with quantum mechanics/molecular mechanics (QM/MM) and density functional theory (DFT), to probe the binding of the substrate analog, methanol, in the S_2 state of the D1-N87A variant of PSII from *Synechocystis* sp. PCC 6803. The results indicate that the size and specificity of the "narrow" channel is altered in D1-N87A PSII, allowing for the binding of deprotonated ^{13}C-labeled methanol at the Mn4(IV) ion of the catalytic cluster in the S_2 state. This has important implications on the mechanistic models for water oxidation in PSII.

Introduction

The photosynthetic protein photosystem II (PSII) catalyzes one of the most energetically demanding reactions in nature by using light energy to drive a catalyst capable of oxidizing water.[1–3] The light-driven four-electron water oxidation reaction takes place at the catalytic Mn_4Ca-oxo cluster in the oxygen-

[a]*Department of Chemistry and Chemical Biology, The Baruch '60 Center for Biochemical Solar Energy Research, Rensselaer Polytechnic Institute, Troy, New York, 12180, USA. E-mail: lakshk@rpi.edu; Tel: +1-518-698-7976*

[b]*Department of Chemistry, Yale University, New Haven, Connecticut, 06520, USA. E-mail: gary.brudvig@yale.edu; Tel: +1-203-432-5202*

† Electronic supplementary information (ESI) available. See DOI: 10.1039/d1fd00094b

evolving complex (OEC) of PSII (Fig. 1).[4–7] The core of the cluster is a distorted cubane comprising three Mn (Mn1–Mn3), four O^{2-} (O1–O3 and O5), and one Ca^{2+} ion. The fourth 'dangler' Mn ion, Mn4, is connected to the cubane by two μ-O bridges, O4 and O5. The shape of the cluster resembles a distorted chair, with the cubane serving as the base and the Mn4 ion serving as the back of the chair. The distorted shape is important as its flexible nature enables the cluster to undergo structural rearrangements during the catalytic cycle. Additionally, the cluster is coordinated by six carboxylates and one histidine residue, D1-H332, which is coordinated to the Mn1 ion.[4,8–10]

It is known that the OEC cycles through five light-driven charge-storage or S-state intermediates (S_0–S_4) as it accumulates oxidizing equivalents to oxidize water, where S_1 is the dark state and S_4 is the final transient state that leads to O–O bond formation and oxygen evolution.[11] In recent years, conventional and time-resolved X-ray crystallography,[4,5,12–16] X-ray absorption fine structure (EXAFS),[7,17–20] electron paramagnetic resonance (EPR),[10,21–29] FTIR[30–33] and quantum mechanics/molecular mechanics (QM/MM)[34–38] investigations have provided information on the structure and oxidation states of the metal ions in the OEC during the S-state cycle. The recent X-ray crystal structures have enabled analyses of the location and possible function of water molecules in the OEC.[4,5,13–16] There are four water-derived terminal ligands (W1–W4) coordinated to the Mn_4Ca-oxo cluster in the dark-stable S_1 state. The binding of these ligands has been studied by several methods, including magnetic resonance,[21,39] Fourier-

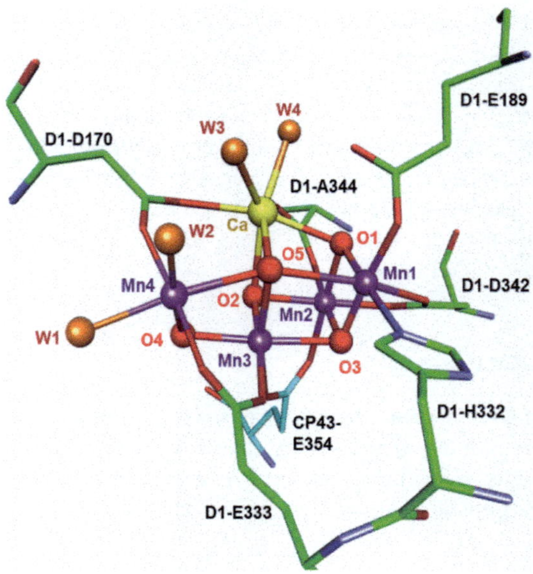

Fig. 1 The Mn_4Ca-oxo cluster of PSII as observed in the X-ray crystal structure of PSII.[4] The core of the cluster consists of a distorted cubane comprising three Mn, one Ca^{2+} and four μ-O^{2-} ions. A fourth dangler Mn is linked to the cubane through a μ-O^{2-} ion. There are four water-derived ligands (W1–W4) bound to the cluster, where W1 and W2 are coordinated to the Mn4 ion, while W3 and W4 are bound to the Ca^{2+} ion. Also shown are the amino acid ligands bound to the cluster.

transform infrared spectroscopy (FTIR) [32,40] and time-resolved membrane-inlet mass spectrometry (TR-MIMS).[41,42] Based on these studies, it has been concluded that the two substrate waters are delivered to the OEC in the S_3 to S_0 and S_2 to S_3 transitions. TR-MIMS has identified two types of exchangeable substrate waters, W_f and W_s, which are in fast and slow exchange with the bulk solvent, respectively; W_s has been proposed to be a μ-oxo bridge between the Ca and Mn ion and W_f as a terminal ligand coordinated to Mn,[41,43] although these assignments remain unclear.[44] There are additional water molecules that form an extensive hydrogen (H)-bonded network around the Mn_4Ca-oxo cluster that are positioned by amino acids that interact with the μ-oxo bridges or water ligands of the cluster, including the D1-N87, D1-S169, D1-D61, CP43-R357 and D1-H337 residues in cyanobacterial PSII.[2,4,5]

Although high-resolution X-ray crystallography, EXAFS and theoretical studies have provided models for the catalytic Mn_4Ca-oxo cluster, the mechanism of delivery and binding of substrate water, participation of the protein environment in substrate activation and the structure of the higher S_3 and S_4 intermediates remain heavily debated. This is important as the OEC is considered to be a promising blueprint for the development of artificial catalysts for water splitting that can generate clean and renewable energy from sunlight.[45–48] Currently, there is intense interest in determining the mechanism for the delivery of the substrate (H_2O) and products (H^+ and O_2) to and from the Mn_4Ca-oxo cluster in the OEC during water oxidation.[49,50] Based on computational analyses of the X-ray crystal structures, there are three putative channels that are suggested to be associated with the OEC: the "narrow", "broad" and "large" channels.[50] The narrow and broad channels are proximal to the Mn4 ion, while the large channel is close to the Ca^{2+} ion in the Mn_4Ca-oxo cluster. The barrier for the entry of water molecules in the narrow channel is suggested to be lower than that of the broad channel[50] and molecular dynamics (MD) simulations have suggested that there may be strict control of accessibility and ordered substrate binding at the active site that is critical in regulating the complex chemistry of water oxidation.[51,52]

Current proposals suggest that the S_2 to S_3 state transition of the OEC involves the binding of a water ligand at the Mn_4Ca-oxo cluster with the oxidation of a Mn ion.[37] However, there is considerable debate on the origin and binding mode of the water molecule. There are primarily two theoretical models proposed for the delivery of water in the S_2 to S_3 transition.[37] In the first model, a water molecule, WX, is delivered through the narrow channel and binds to the dangler Mn4 ion through a carousel mechanism. In the second model, WX is delivered through the large channel that is proximal to the Ca^{2+} ion and binds to the Mn1 ion through a pivot mechanism. However, recent cryo-electron microscopy studies have shown that the "large" channel is not conserved in PSII from *Synechocystis* sp. PCC 6803.[53] There is an urgent need for spectroscopic data on the function of the channels and the coordination environment of the Mn ions in order to determine the precise pathway for water delivery in the OEC. PSII is a unique case where the substrate and solvent molecules are identical and, therefore, it has been a substantial challenge to differentiate between them while probing the mechanism of substrate delivery and binding in the OEC.

There are two experimental approaches that have previously been employed to investigate the delivery of the substrate: (i) to probe the water binding sites by studying the indirect effects of the site-directed mutagenesis of amino acid residues and (ii) to employ small molecule analogs, such as ammonia, hydroxylamine and methanol, to probe access to the Mn_4Ca-oxo cluster in the OEC.[54–62] However, there have been few attempts to interrogate the high-resolution electronic and geometric structure of the binding of small molecule analogs associated with the various channels of site-directed variants of cyanobacterial PSII.[57] This is important as elucidation of the mechanism of water oxidation requires precise knowledge of how, when and where substrate waters bind at the Mn_4Ca-oxo cluster during the S-state cycle.

The X-ray crystal structures reveal that the asparagine-87 residue in the D1 polypeptide, D1-N87, is located in the "narrow" channel and is proximal to serine, arginine and aspartic acid residues (D1-S169, CP43-R357 and D1-D61, respectively) in PSII from *Thermosynechococcus vulcanus*.[4,50] The D1-N87 residue has been suggested to be involved in the delivery of water molecules, as well as proton transport in the OEC.[50,63] Moreover, D1-N87 indirectly interacts with the water molecule, WX, and is likely hydrogen (H)-bonded to the D1-S169 residue that was implicated in substrate binding in the OEC.[64,65] It is interesting that, despite the highly-conserved sequence of the OEC across species, D1-N87 in cyanobacteria is replaced by an alanine, D1-A87, in PSII from higher plants.[50,66] It has been suggested that an important consequence of the replacement of D1-N87 by alanine is that the shape and flexibility of the narrow channel is wider in higher plants, while also removing a H-bonding residue in it.[50] Some of us have previously observed that the D1-N87A variant of *Synechocystis* PCC 6803 shows comparable S-state cycling efficiency to wild-type (WT) cyanobacterial and spinach PSII.[67] However, it displays altered chloride-binding properties and pH dependence that mimic those in spinach PSII.

Based on the location and H-bonding pattern of the D1-N87 residue, it is possible that it is also involved in the delivery of substrate water at the Mn_4Ca-oxo cluster in the S_2 to S_3 transition. More specifically, D1-N87 could regulate the size and selectivity of the putative "narrow" channel. This is important as it is plausible that this is the only channel available for water delivery and its narrow path prevents access of larger molecules that could potentially react with the OEC. Additionally, D1-N87 could participate in maintaining the well-defined H-bonding network of water molecules in and around the OEC, which regulates the entry and binding of water ligands. Previous studies involving the binding of small molecule substrate analogs in the OEC have indicated that although methanol competes with the binding of water, its binding does not inhibit oxygen evolution. More recently, Britt and coworkers have demonstrated that isotope-labeled methanol can be used as a substrate analog to probe the binding of substrate molecules in the OEC. Using electron-spin-echo envelope modulation (ESEEM), hyperfine sublevel correlation (HYSCORE) and electron nuclear double resonance (ENDOR) spectroscopy of CD_3OH and $^{13}CH_3OH$, these studies probed the binding of methanol through the measurement of electron–nuclear hyperfine couplings of the ^2H and ^{13}C atom(s) with the paramagnetic Mn_4Ca-oxo cluster in the S_2 state of spinach PSII.[58,59] In particular, the use of ^{13}C-methanol offered a selective probe of the location of the methyl group that allowed for the evaluation of possible methanol binding sites in spinach PSII.

In the present study, using two-dimensional (2D) hyperfine sublevel correlation (HYSCORE) spectroscopy, we present direct evidence of the binding of ^{13}C-methanol to the catalytic Mn_4Ca-oxo cluster in the S_2 state of the OEC of D1-

N87A PSII from *Synechocystis* sp. PCC 6803. 2D HYSCORE spectroscopy[68] is ideally suited to probe the binding of small molecule analogs as the hyperfine interactions of magnetic nuclei with the unpaired electron spin in the S_2 state are detected in 2D frequency space, which dramatically improves the spectral resolution. Moreover, the nuclear frequencies are correlated in the two dimensions, which crucially simplifies the analysis of the experimental spectra. In conjunction with HYSCORE spectroscopy, we also employ QM/MM and density functional theory (DFT) methods to model the possible binding sites of methanol and identify the most likely binding site by obtaining a best match of the experimental and calculated ^{13}C hyperfine couplings in the S_2 state of D1-N87A PSII.

Results and discussion

Pulsed EPR spectroscopy of ^{13}C-methanol-treated D1-N87A PSII

The S_2 state of the OEC of PSII from higher plants and cyanobacteria can be photo-accumulated by steady-state illumination at cryogenic temperatures (180–200 K),[69] which induces the one-electron oxidation of the S_1 to the S_2 state. We and others have demonstrated that the S_2 state displays distinct low- and high-spin isomers with $g \sim 2.0$ 'multiline' and $g > 4.0$ EPR signals, respectively.[22,69–71] Shown in Fig. 1SA and B† are the EPR spectra of the S_2 state of wild-type (WT) and D1-N87A PSII from *Synechocystis* sp. PCC 6803 treated with 5% (v/v) ^{13}C-methanol. The S_2 state multiline EPR spectrum is centered at a g value of ~2.0 with an overall spectral width of 200 mT and contains 12–14 hyperfine peaks that are characteristic of the low-spin isomer of the S_2 state. The multiline signal has been shown to arise from the electron–nuclear hyperfine interactions of the effective electron spin ($S = 1/2$) with the four equivalents of ^{55}Mn nuclear spins ($I = 5/2$) of the Mn_4Ca-oxo cluster in the OEC. The spectra are in agreement with previously published multiline signals of the S_2 state.

In principle, the binding of ^{13}C-methanol in the OEC should lead to additional splittings in the spectrum from hyperfine interactions of the ^{13}C nuclear spin(s) ($I = 1/2$) with the paramagnetic Mn_4Ca-oxo cluster in the S_2 state. However, the ^{13}C hyperfine interactions are much weaker than the ^{55}Mn couplings in the OEC and are masked by the inhomogeneous line width of the multiline peaks. Therefore, in the present study, we probe the ^{13}C hyperfine couplings of ^{13}C-methanol-treated PSII by 2D HYSCORE spectroscopy, which has the ability to resolve the weak hyperfine interactions of the ^{13}C nuclei that are coupled to the unpaired electron spin in the S_2 state.

Shown in Fig. 2A and B are the experimental and simulated ^{13}C HYSCORE spectra of the S_2 state of D1-N87A PSII from *Synechocystis* sp. PCC 6803 treated with 5% (v/v) ^{13}C-methanol. As can be seen in Fig. 2A, we observe a pair of off-diagonal cross-peaks, C^I, at (5.4, 2.1) and (2.1, 5.4) MHz in the (+,+) quadrant of the HYSCORE spectrum. The cross-peaks are centered at the ^{13}C Zeeman frequency of 3.6 MHz (at a magnetic field of 335 mT) and arise from the hyperfine interactions of the ^{13}C atom (nuclear spin, $I = 1/2$) of $^{13}CH_3OH$ that is bound to the OEC in the S_2 state. The separation between the pair of cross-peaks C^I is ~3.5 MHz, which is proportional to the isotropic hyperfine coupling, A_{iso}, of the ^{13}C atom of bound methanol.

We performed numerical simulations of the experimental spectrum (Fig. 2B) to obtain a quantitative measure of the isotropic, A_{iso}, and anisotropic, T,

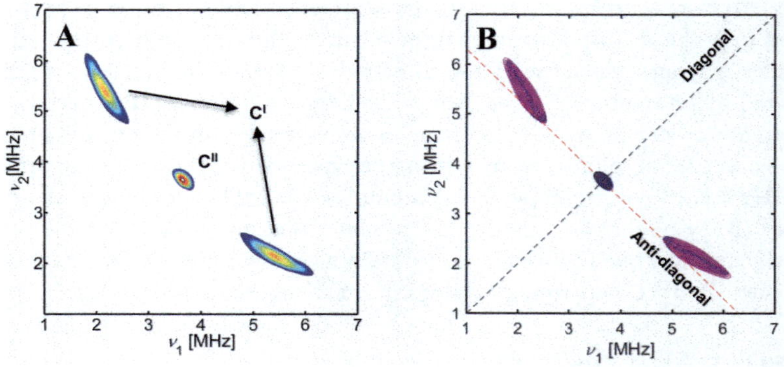

Fig. 2 Contour-plot representation of the (+,+) quadrant of the (A) experimental and (B) overlay of the numerically simulated (pink) and experimental (purple) 2D ^{13}C HYSCORE spectrum of the ^{13}C-methanol bound S_2 state of the OEC of the D1-N87A variant of PSII from *Synechocystis* sp. PCC 6803. The spectrum was acquired at a magnetic field of 335.0 mT and temperature of 5 K. The dashed black and red lines represent the diagonal (defined by $\nu_1 = \nu_2$) and anti-diagonal (defined by $|\nu_1 \pm \nu_2| = 2 \times {}^{13C}\nu$, where ${}^{13C}\nu$ is the carbon nuclear Larmor frequency), respectively.

hyperfine interactions and rhombicity parameter, δ (Table 1). As can be seen in Fig. 2A and B, there is excellent agreement between the experimental and simulated line shapes of the cross-peaks, C^I. The value of the A_{iso} coupling of 3.63 ± 0.15 MHz that was obtained from the numerical simulations (Table 1) is in agreement with the qualitative estimate predicted from the separation of the cross-peaks. The anisotropic hyperfine coupling, T, of 1.65 ± 0.11 MHz is similar in magnitude to that of the protons of a water-derived ligand coordinated to the Mn4 ion.[39] Interestingly, the cross-peaks display deviations from the anti-diagonal of the spectrum, shown as a dashed red line in Fig. 2B. It has previously been noted that such cross-peak distortions in HYSCORE spectroscopy could originate from a distribution of hyperfine parameters or strain in orientationally-disordered protein samples or frozen solutions, where the hyperfine strain can be modeled using a distribution function. In order to account for the hyperfine strain observed in C^I, we employed a weighted Gaussian-type function to model the distorted line shape by assigning the center of distribution and the full-width at half maximum (FWHM) in the simulations (standard deviation, σ, of 0.3 MHz).

Additionally, the experimental spectrum in Fig. 2A also reveals a cross-peak, C^{II}, on the diagonal which arises from very weak hyperfine couplings (A_{iso} < 0.1

Table 1 Isotropic and anisotropic hyperfine parameters obtained from numerical simulations of the 2D ^{13}C HYSCORE spectrum of the S_2 state of the OEC of the ^{13}C-methanol bound D1-N87A variant of PSII from *Synechocystis* sp. PCC 6803. The hyperfine coupling constants are in units of MHz

Nucleus	A_{iso}	T	A_x	A_y	A_z
^{13}C	3.63 ± 0.15	1.65 ± 0.11	2.4 ± 0.14	1.5 ± 0.12	7.0 ± 0.16

MHz) of the effective electron spin ($S = 1/2$) with ambient non-bonded ^{13}C nuclei that are not directly coordinated to the Mn ions of the Mn_4Ca-oxo cluster in the S_2 state.[58,59] Therefore, C^{II} was not included in the numerical simulations in Fig. 2B. It is important to note that both of the cross-peaks, C^{I} and C^{II}, were absent in the S_2 state of untreated D1-N87A PSII (Fig. 3SA and B†) and ^{13}C-methanol-treated WT PSII from *Synechocystis* sp. PCC 6803 (Fig. 4S†), while C^{I} was not observed in the S_2 state of ^{13}C-methanol-treated PSII from spinach (spectrum not shown).

In order to assess the effects of methanol binding on the primary and secondary sphere of ligands to the Mn_4Ca-oxo cluster in the S_2 state, we examined the ^{14}N hyperfine interactions of ^{13}C-methanol bound D1-N87A PSII. Shown in Fig. 2SA and B† are the experimental and simulated 2D ^{14}N HYSCORE spectra of the ^{13}C-methanol bound S_2 state of D1-N87A PSII. The ^{14}N hyperfine and quadrupolar couplings in Table 1S† indicate that the hyperfine couplings of the nitrogen atoms are identical to those observed in the S_2 state of untreated WT PSII,[9,10,72] indicating that magnetic interactions with the D1-H332 and CP43-R357 residues remain unchanged in the S_2 state of ^{13}C-methanol bound D1-N87A PSII.

QM/MM and DFT studies of methanol binding in the oxygen-evolving complex

We obtained ^{13}C A_{iso} couplings of 3.63 ± 0.15 MHz and <0.1 MHz from the 2D HYSCORE spectra of the ^{13}C-methanol bound S_2 state of D1-N87A PSII. While the A_{iso} coupling of <0.1 MHz can be attributed to non-bonded ^{13}C nuclei, the larger ^{13}C A_{iso} of 3.63 ± 0.15 MHz suggests that methanol is directly coordinated to a paramagnetic Mn ion in the S_2 state of the OEC of D1-N87A PSII.[58] As evidenced in previous studies, direct structural assignment of the specific binding sites in the OEC based solely on the measured hyperfine parameters is limited in scope. However, it is possible to interpret the EPR parameters in conjunction with quantum mechanical models to gain insight into the binding of ligands in the OEC.[73–75]

Since the binding of methanol increases the energy separation between the S_2 state spin isomers,[54,76–78] stabilizing the $S = 1/2$ ground spin state of the S_2 state, we used QM/MM methods to optimize structural models of the low-spin $g \sim 2.0$ open cubane conformer of the Mn_4Ca-oxo cluster to evaluate the possibility of a methoxy group or methanol interacting with the Mn4(IV) or Mn1(III) ion. The QM/MM studies yielded three plausible models in which methanol binds to the cluster in the protonated or deprotonated methoxy form (Fig. 3A–C). In the first model, Mn4-W1, a methoxy group displaces the water-derived ligand W1 that is coordinated to the Mn4(IV) ion in the low-spin $g \sim 2.0$ isomer of the S_2 state, binding at a distance of 2.1 Å (Fig. 3A). In this case, the methoxy group is accepting a hydrogen bond from the protonated D1-D61 residue. The methyl group is between the backbone of the D1-S169 residue and the water molecule, W9, where one of the methyl hydrogen atoms donates a short (1.5 Å) hydrogen bond to W9. In the second model, Mn4-W2, a methoxide group displaces the water-derived ligand W2, binding to Mn4(IV) at a distance of 1.8 Å, accepting a hydrogen bond from the water molecule, W5 (Fig. 3B). The methyl group is located less than 4 Å from the D1-V185 residue, where it is in contact with the rigid side chain of valine and one of the methyl hydrogen atoms donates a hydrogen bond (2.1 Å) to the water molecule, W8.

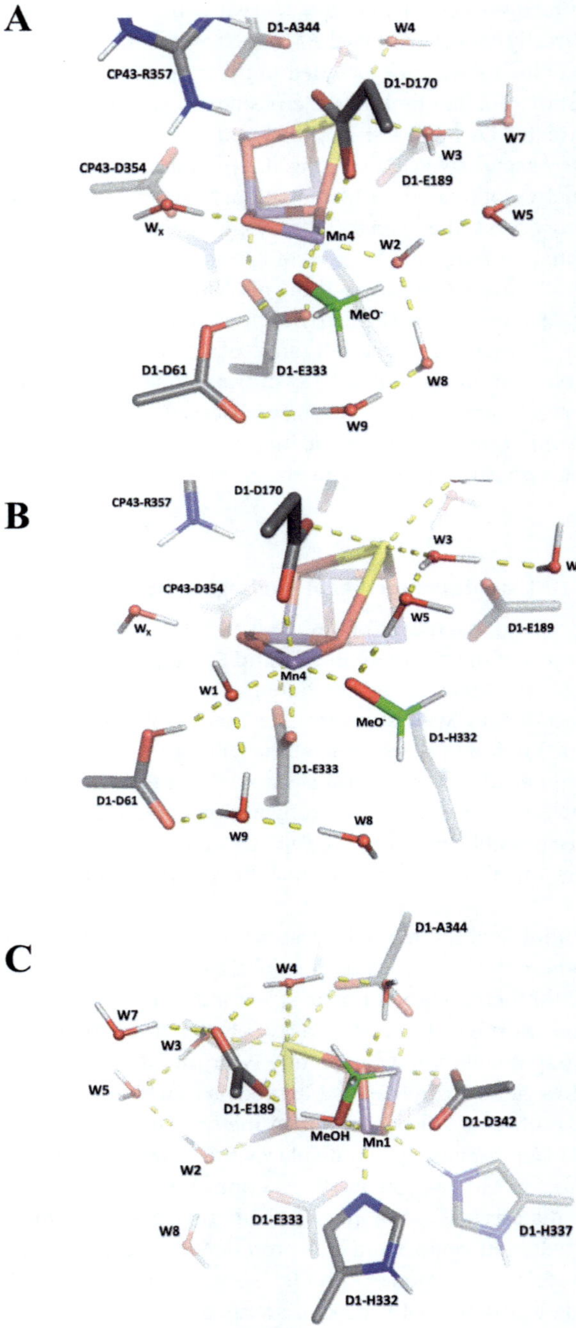

Fig. 3 QM/MM models of methanol binding at a Mn ion of the Mn_4Ca-oxo cluster in the S_2 state. The (A) Mn4-W1 model where a methoxy group replaces the water ligand W1 at the Mn4(IV) ion, (B) Mn4-W2 model where a methoxy group replaces the water ligand W2 at the Mn4(IV) ion and (C) Mn1-methanol model with methanol binding at the Mn1(III) ion. The manganese, calcium, carbon, nitrogen, oxygen and hydrogen atoms are shown in purple, yellow, grey, blue, red and white color, respectively.

In contrast to the two Mn4 models, methanol displaces D1-E189 ligand in the third model, Mn1-MeOH, binding to the Mn1(III) ion and donating a short (1.4 Å) hydrogen bond to the displaced D1-E189-Oε2 atom (Fig. 3C). The Oε2 atom in turn displaces Oε1 to bind to the calcium ion of the OEC. In this case, Oε1 is no longer coordinated to a metal ion, and instead inserts itself between the waters, W3 and W7, disrupting the H-bonded water tetramer in the OEC. The methyl group of the methanol points towards the large channel and does not appear to form any notable interactions. We observed that the geometric parameters of the Mn and Ca ions of the Mn_4Ca-oxo cluster, as well as the amino acid ligands of all three QM/MM models, are consistent with the X-ray structures of PSII.[4] It is important to note that while methanol can readily access the Mn4 ion in the Mn4-W1 and Mn4-W2 models, it has very limited access to the Mn1 ion due to steric limitations of the site. Previous studies by Britt and coworkers also considered the binding of methanol either in μ-oxido positions replacing either the $O4^{2-}$ or $O5^{2-}$ ion or the W3 water coordinated at the Ca^{2+} ion.[58] We did not consider a bridging μ-oxido as we do not observe large alterations in the ^{14}N hyperfine couplings and effective ^{55}Mn couplings. Moreover, the large ^{13}C A_{iso} value of 3.63 ± 0.15 MHz indicates that the methanol is directly coordinated to a paramagnetic Mn, which limits the possibility of its binding at the Ca^{2+} ion. However, it is possible that the A_{iso} of <0.1 MHz obtained from the experimental data represents the binding of a second methanol/methoxy group at the Ca^{2+} ion.[58,62] As we and others have previously demonstrated, the best way to validate the QM/MM structural models of the methanol-bound S_2 state is by the prediction of the ^{13}C hyperfine interactions.[63,73]

After validating our predictions of ^{13}C hyperfine couplings using broken-symmetry density functional theory (BS-DFT) methods to reproduce the hyperfine interactions of select computational models described by Retegan and Pantazis,[63] we calculated the ^{13}C hyperfine couplings for the three QM/MM models described above (Table 2) for comparison with the experimental hyperfine couplings shown in Table 1. We found that the calculated isotropic and anisotropic hyperfine couplings of Mn4-W2 are consistent with the results of the HYSCORE studies. As can be seen in Table 2, the BS-DFT calculations of the model with a methoxy group coordinated at the W2 position of the Mn4(IV) ion yielded ^{13}C isotropic and anisotropic hyperfine couplings of 3.23–3.37 MHz and 1.89–1.90 MHz, respectively, which are close to the experimental ^{13}C A_{iso} and T values of 3.63 ± 0.15 and 1.65 ± 0.11 MHz in Table 1, respectively. Moreover, the ^{13}C hyperfine couplings of the Mn4-W2 model are also in the range of the previous

Table 2 Normalized ^{13}C-hyperfine constants for methanol/methoxy binding in the OEC that were calculated using a TPSSh functional with the EPR-II or ZORA/TZVP basis set. The hyperfine coupling constants are in units of MHz

	EPR-II		ZORA/TZVP	
QM/MM model	A_{iso}	T	A_{iso}	T
Mn4-W2	3.23	1.89	3.37	1.90
Mn4-W1	0.86	2.09	0.33	2.98
Mn1-MeOH	1.53	1.78	1.89	1.75

model W2-1d[63] in a study where the authors had selected twenty representative structures that covered a range of possibilities with respect to the interaction of methanol with the OEC. These models included structures where methanol was either protonated or deprotonated when binding as a first-sphere ligand. In general, it was found that the ^{13}C A_{iso} and T values of methoxy binding were dramatically larger (5.73–4.06 MHz and 1.53–2.96 MHz for A_{iso} and T, respectively) than the corresponding values for methanol binding, which is consistent with the larger couplings observed in the present study.

It is interesting that the results obtained with ^{13}C-methanol binding in the S_2 state of D1-N87A PSII in this study are different from those of spinach PSII. While the HYSCORE spectra of the ^{13}C-methanol bound S_2 state of D1-N87A PSII yield off-diagonal cross-peaks with a ^{13}C A_{iso} coupling of 3.63 $\pm$ 0.15 MHz, previous ^{13}C-methanol binding studies in spinach PSII displayed weak ^{13}C couplings that were not resolved by X-band HYSCORE spectroscopy.[58] Further Q-band electron nuclear double resonance (ENDOR) measurements of the samples yielded A_{iso} and T values of 0.05 $\pm$ 0.02 MHz and 0.27 $\pm$ 0.05 MHz, respectively. Based on the nearly negligible ^{13}C A_{iso} coupling, it was inferred that methanol was not binding along the Jahn–Teller axis of the Mn1(III) ion and the small anisotropic coupling T suggested that the methyl group was not part of a ligand coordinated at a Mn ion in the S_2 state.[58,62] Isosurface plots using projection factors relevant for the low-spin S_2 state indicated that methanol could replace the W3 water ligand coordinated at the Ca^{2+} ion of the cluster in spinach PSII. Further detailed analyses of the ^{13}C hyperfine interaction data from a series of quantum chemical models that included the explicit binding of methanol at different sites indicated that it was not a direct ligand, but was likely situated at the end-point of a water channel associated with the $O4^{2-}$ bridge in the OEC of spinach PSII.[63] This observation was supported by FTIR difference and time-resolved IR[79] and proton matrix ENDOR[62] spectroscopies.

The ^{13}C A_{iso} value of 3.63 $\pm$ 0.15 MHz in the ^{13}C-methanol bound S_2 state of D1-N87A PSII is roughly comparable to the signals that were obtained from the binding of ^{13}C methanol to Mn(III)Mn(IV)salpn (where, salpn = *N,N′*-bis(3,5-dichlorosalicylidene)-1,3-diamino-2-hydroxypropane).[58] The cross-peaks from methanol-bound Mn(III)Mn(IV)salpn were numerically simulated by using moderate A_{iso} and T couplings of 0.65 $\pm$ 0.05 MHz and 1.25 $\pm$ 0.05 MHz, respectively. In the case of Mn(III)Mn(IV)salpn, the degree of curvature of the correlation ridge was taken to be diagnostic of a large anisotropic hyperfine interaction,[80] where $T > A_{iso}$. However, we observed an opposite trend with ^{13}C-methanol binding, where the A_{iso} coupling of 3.63 $\pm$ 0.15 MHz is larger than the anisotropic component of 1.65 $\pm$ 0.11 MHz.

Previous ^{13}C ENDOR measurements of the S_2 state in which either the alanine carboxylate carbons or all the carbon atoms were ^{13}C-labeled showed A_{iso} couplings of 1.2 MHz and 1.2–2.0 MHz for the D1 C-terminus alanine ligand and uniformly ^{13}C-labeled ligands, respectively.[81] Subsequent studies using BS-DFT calculations suggested that symmetrically bridging carboxylate groups generally provided larger anisotropic hyperfine couplings and rhombicities than terminal or Mn–Ca bridging ligands.[82] The interpretation of the ENDOR results and computational models for the BS-DFT calculations originated prior to the high-resolution X-ray crystal structure of PSII.[4] Most recently, calculations based on models derived from the 1.9 Å resolution X-ray structure[4] indicated multiple A_{iso}

couplings in the range of 1.3–4.3 MHz for the carboxylate ligands in the S_2 state of the OEC.[83]

By combining previous pulsed EPR and computational analyses of methanol[58,63] and ammonia binding,[57,61,84–86] it was suggested that there is a unique channel active in the delivery of substrate analogues to the OEC. In the literature, this channel is referred to as the "narrow" channel,[87] "E, F" channel,[88] "channel 2",[89] "O4" or "Path 3" channel.[90] Ho and Styring had initially suggested that this channel would be the most permeable to methanol.[87] Molecular dynamics simulations by Bruce and coworkers demonstrated that this channel connects the OEC directly with the lumenal surface of PSII at the cavity formed by the extrinsic PsbO and PsbU proteins, and it has the lowest free energy barrier for water permeation (peak activation energy *ca.* 9 kcal mol^{-1}) than all other channels.[89] Moreover, recent experiments demonstrated that NH_3 binds to the Mn4 ion in the S_2 state of the OEC.[57,84–86] Although the conformation of the cluster is fundamentally different when NH_3 and MeOH bind to the OEC, it appears that both analogs act as a direct ligand to the Mn4 ion. The narrow channel that is suggested as the methanol access pathway could be the same one that delivers ammonia to the Mn4 ion, supporting the notion that this is the only directly solvent-accessible manganese site of the OEC.[63] Additionally, studies of the $S_2 \rightarrow S_3$ transition and proposals for water binding in the OEC (in both the carousel and pivot mechanism),[37] have suggested that Mn4 is the only directly accessible Mn ion of the cluster and that substrate inclusion to the OEC occurs through initial binding at this site. This is in agreement with the present study, where computational analysis of the ^{13}C HYSCORE experimental data indicates that a deprotonated methoxy group binds at the W2 position at the Mn4(IV) ion in the S_2 state of D1-N87A PSII. Further studies are in progress to understand the mechanistic implications for the water oxidation reaction in the OEC.

Previous comparisons of the primary sequence of the D1 polypeptide of cyanobacteria and spinach revealed that the substitution of the D1-N87 residue by an alanine in spinach allows for the binding of methanol in the OEC. This offered a structure-based proposal for the species-dependent response to methanol. Moreover, it was found that the D1-N87A mutation made chloride binding weaker and the dissociation of chloride more facile, suggesting that the replacement of D1-N87 by alanine alters the access of small molecules and ions from the lumen to the OEC.[67] Subsequently, detailed computational analyses by Pantazis and coworkers converged upon an atomistic model that was consistent with experimental observations of methanol binding in spinach PSII.[63] However, it did not directly explain why methanol binding was less pronounced in cyanobacteria. There were no species-dependent substitutions for any of the amino acid residues in the previous computational models, and hence it is improbable that fundamentally different access modes exist between species. Hence, the model must have been equally valid for both cyanobacteria and higher plants, if methanol can reach this site and adopt these positions in all organisms. The narrow channel was suggested to be the structural origin of the species-dependent response to methanol and it is proposed to act as the substrate delivery channel. Within the context of the present work, the above ^{13}C HYSCORE, QM/MM and DFT analyses establish that the substitution of the D1-N87 residue by alanine leads to the delivery and binding of methanol in the S_2 state of cyanobacterial PSII.

Conclusions

In summary, the use of ^{13}C HYSCORE spectroscopy in conjunction with QM/MM and DFT provides evidence that ^{13}C-methanol is coordinated to the OEC in the S_2 state of D1-N87A PSII from *Synechocystis* sp. PCC 6803. Previous studies have shown that the D1-N87 residue in cyanobacterial PSII is required for the efficient S-state cycling of the OEC and the replacement of D1-N87 by an alanine residue alters the Cl^--binding properties, such that D1-N87A PSII mimics spinach PSII. The results of the present study confirm that the size and specificity of the "narrow" channel is altered in D1-N87A PSII, allowing for the binding of deprotonated ^{13}C-labeled methanol to the Mn4 ion of the catalytic cluster, which has important implications for the mechanistic models for water oxidation in PSII.

Author contributions

V. K., G. B., I. G. and K. V. L. conducted the experiments and analyzed the data, K. M. R. and A. B. performed QM/MM and DFT calculations, V. K., G. B., I. G., G. W. B. and K. V. L. designed the experiments and V. K., K. M. R., A. B., G. B., I. G., V. S. B., G. W. B. and K. V. L. wrote the paper.

Conflicts of interest

The authors declare no conflicts of interests.

Acknowledgements

This work was supported by the U.S. Department of Energy, Office of Basic Energy Sciences, Photosynthetic Systems Program under the contract DE-FG02-07ER15903 (K. V. L.), DE-FG0205ER15646 (G. W. B.) and DE-SC0001423 (V. S. B.). The authors would like to thank Professor R. J. Debus (University of California, Riverside) for the D1-N87A variant of *Synechocystis* sp. PCC 6803 and the Center for Computational Innovations (CCI) at Rensselaer Polytechnic Institute for computational resources.

References

1 D. J. Vinyard and G. W. Brudvig, Progress Toward a Molecular Mechanism of Water Oxidation in Photosystem II, *Annu. Rev. Phys. Chem.*, 2017, **68**, 101–116.

2 J.-R. Shen, The Structure of Photosystem II and the Mechanism of Water Oxidation in Photosynthesis, in *Annu. Rev. Plant Biol.*, ed. S. S. Merchant, 2015, pp. 23–48.

3 J. P. McEvoy and G. W. Brudvig, Water-splitting Chemistry of Photosystem II, *Chem. Rev.*, 2006, **106**, 4455–4483.

4 Y. Umena, K. Kawakami, J. R. Shen and N. Kamiya, Crystal Structure of Oxygen-evolving Photosystem II at a Resolution of 1.9 Angstrom, *Nature*, 2011, **473**, 55–65.

5 M. Suga, F. Akita, M. Sugahara, M. Kubo, Y. Nakajima, T. Nakane, K. Yamashita, Y. Umena, M. Nakabayashi, T. Yamane, T. Nakano, M. Suzuki, T. Masuda, S. Inoue, T. Kimura, T. Nomura, S. Yonekura, L. J. Yu,

T. Sakamoto, T. Motomura, J. H. Chen, Y. Kato, T. Noguchi, K. Tono, Y. Joti, T. Kameshima, T. Hatsui, E. Nango, R. Tanaka, H. Naitow, Y. Matsuura, A. Yamashita, M. Yamamoto, O. Nureki, M. Yabashi, T. Ishikawa, S. Iwata and J. R. Shen, Light-induced Structural Changes and the Site of O═O Bond Formation in PSII Caught by XFEL, *Nature*, 2017, **543**, 131.

6 J. Yano, J. Kern, Y. Pushkar, K. Sauer, P. Glatzel, U. Bergmann, J. Messinger, A. Zouni and V. K. Yachandra, High-resolution Structure of the Photosynthetic Mn_4Ca Catalyst from X-ray Spectroscopy, *Philos. Trans. R. Soc. London, Ser. B*, 2008, **363**, 1139–1147.

7 J. Yano and V. K. Yachandra, Where Water is Oxidized to Dioxygen: Structure of the Photosynthetic Mn_4Ca Cluster From X-ray Spectroscopy, *Inorg. Chem.*, 2008, **47**, 1711–1726.

8 R. J. Debus, Protein Ligation of the Photosynthetic Oxygen-evolving Center, *Coord. Chem. Rev.*, 2008, **252**, 244–258.

9 R. J. Debus, K. A. Campbell, W. Gregor, Z. L. Li, R. L. Burnap and R. D. Britt, Does Histidine 332 of the D1 Polypeptide Ligate the Manganese Cluster in Photosystem II? An Electron Spin Echo Envelope Modulation Study, *Biochemistry*, 2001, **40**, 3690–3699.

10 S. Milikisiyants, R. Chatterjee, A. Weyers, A. Meenaghan, C. Coates and K. V. Lakshmi, Ligand Environment of the S_2 State of Photosystem II: A Study of the Hyperfine Interactions of the Tetranuclear Manganese Cluster by 2D ^{14}N HYSCORE Spectroscopy, *J. Phys. Chem. B*, 2010, **114**, 10905–10911.

11 B. Kok, B. Forbush and M. McGloin, Cooperation of Charges in Photosynthetic O_2 Evolution. 1. A Linear 4-Step Mechanism, *Photochem. Photobiol.*, 1970, **11**, 457–475.

12 I. D. Young, M. Ibrahim, R. Chatterjee, S. Gul, F. D. Fuller, S. Koroidov, A. S. Brewster, R. Tran, R. Alonso-Mori, T. Kroll, T. Michels-Clark, H. Laksmono, R. G. Sierra, C. A. Stan, R. Hussein, M. Zhang, L. Douthit, M. Kubin, C. de Lichtenberg, L. V. Pham, H. Nilsson, M. H. Cheah, D. Shevela, C. Saracini, M. A. Bean, I. Seuffert, D. Sokaras, T. C. Weng, E. Pastor, C. Weninger, T. Fransson, L. Lassalle, P. Brauer, P. Aller, P. T. Docker, B. Andi, A. M. Orville, J. M. Glownia, S. Nelson, M. Sikorski, D. L. Zhu, M. S. Hunter, T. J. Lane, A. Aquila, J. E. Koglin, J. Robinson, M. N. Liang, S. Boutet, A. Y. Lyubimov, M. Uervirojnangkoorn, N. W. Moriarty, D. Liebschner, P. V. Afonine, D. G. Waterman, G. Evans, P. Wernet, H. Dobbek, W. I. Weis, A. T. Brunger, P. H. Zwart, P. D. Adams, A. Zouni, J. Messinger, U. Bergmann, N. K. Sauter, J. Kern, V. K. Yachandra and J. Yano, Structure of Photosystem II and Substrate Binding at Room Temperature, *Nature*, 2016, **540**, 453.

13 A. Tanaka, Y. Fukushima and N. Kamiya, Two Different Structures of the Oxygen-evolving Complex in the Same Polypeptide Frameworks of Photosystem II, *J. Am. Chem. Soc.*, 2017, **139**, 1718–1721.

14 J. Kern, R. Chatterjee, I. D. Young, F. D. Fuller, L. Lassalle, M. Ibrahim, S. Gul, T. Fransson, A. S. Brewster, R. Alonso-Mori, R. Hussein, M. Zhang, L. Douthit, C. de Lichtenberg, M. H. Cheah, D. Shevela, J. Wersig, I. Seuffert, D. Sokaras, E. Pastor, C. Weninger, T. Kroll, R. G. Sierra, P. Aller, A. Butryn, A. M. Orville, M. Liang, A. Batyuk, J. E. Koglin, S. Carbajo, S. Boutet, N. W. Moriarty, J. M. Holton, H. Dobbek, P. D. Adams, W. Bergmann, N. K. Sauter, A. Zouni,

J. Messinger, J. Yano and V. K. Yachandra, Structures of the Intermediates of Kok's Photosynthetic Water Oxidation Clock, *Nature*, 2018, **563**, 421–425.

15 M. Suga, F. Akita, K. Yamashita, Y. Nakajima, G. Ueno, H. Li, T. Yamane, K. Hirata, Y. Umena, S. Yonrkura, L.-J. Yu, H. Mrakami, T. Nomura, T. Kimura, M. Kubo, S. Baba, T. Kumasaka, K. Tono, M. Yabashi, H. Isobe, K. Yamaguchi, M. Yamamoto, H. Ago and J.-R. Shen, An Oxyl/oxo Mechanism for Oxygen-oxygen Coupling in PSII Revealed by an X-ray free-electron Laser, *Science*, 2019, **366**, 334–338.

16 M. Ibrahim, T. Fransson, R. Chatterjee, M. H. Cheah, R. Hussein, L. Lassalle, K. D. Sutherlin, I. D. Young, F. D. Fuller, S. Gul, I. Kim, P. S. Simon, C. de Lichtenberg, P. Chernev, I. Bogacz, C. C. Pham, A. M. Orville, N. Saichek, T. Norther, A. Batyuk, S. Carbajo, R. Alonso-Mori, K. Tono, S. Owada, A. Bhowmick, A. S. Bolotovsky, P. D. Adams, N. K. Sauter, U. Bergmann, A. Zouni, J. Messinger, J. Kern, V. K. Yachandra and J. Yano, Untangling the Sequence of Events During the S_2 to S_3 Transition in Photosystem II and Implications for the Water Oxidation Mechanism, *Proc. Natl. Acad. Sci. U. S. A.*, 2020, **117**, 12624–12635.

17 K. Sauer, J. Yano and V. Yachandra, X-ray Spectroscopy of the Photosynthetic Oxygen-evolving Complex, *Coord. Chem. Rev.*, 2008, **252**, 318–335.

18 J. Yano, J. Kern, K. Sauer, M. J. Latimer, Y. Pushkar, J. Biesiadka, B. Loll, W. Saenger, J. Messinger, A. Zouni and V. K. Yachandra, Where Water is Oxidized to Dioxygen: Structure of the Photosynthetic Mn_4Ca Cluster, *Science*, 2006, **314**, 821–825.

19 H. Dau, P. Liebisch and M. Haumann, The Structure of the Manganese Complex of Photosystem II in its Dark-stable S_1 State – EXAFS Results in Relation to Recent Crystallographic Data, *Phys. Chem. Chem. Phys.*, 2004, **6**, 4781–4792.

20 M. Haumann, C. Muller, P. Liebisch, L. Iuzzolino, J. Dittmer, M. Grabolle, T. Neisius, W. Meyer-Klaucke and H. Dau, Structural and Oxidation State Changes of the Photosystem II Manganese Complex in Four Transitions of the Water Oxidation Cycle ($S_0 \rightarrow S_1$, $S_1 \rightarrow S_2$, $S_2 \rightarrow S_3$, and S_3, $S_4 \rightarrow S_0$) Characterized by X-ray Absorption Spectroscopy at 20 K and Room Temperature, *Biochemistry*, 2005, **44**, 1894–1908.

21 R. D. Britt, K. A. Campbell, J. M. Peloquin, M. L. Gilchrist, C. P. Aznar, M. M. Dicus, J. Robblee and J. Messinger, Recent Pulsed EPR Studies of the Photosystem II Oxygen-evolving Complex: Implications as to Water Oxidation Mechanisms, *Biochim. Biophys. Acta, Bioenerg.*, 2004, **1655**, 158–171.

22 A. Haddy, EPR Spectroscopy of the Manganese Cluster of Photosystem II, *Photosynth. Res.*, 2007, **92**, 357–368.

23 A. Haddy, K. V. Lakshmi, G. W. Brudvig and H. A. Frank, Q-band EPR of the S_2 State of Photosystem II Confirms an $S = 5/2$ Origin of the X-band $g = 4.1$ Signal, *Biophys. J.*, 2004, **87**, 2885–2896.

24 N. Cox, M. Retegan, F. Neese, D. Pantazis, A. Boussac and W. Lubitz, Electronic Structure of the Oxygen-evolving Complex in Photosystem II Prior to O–O Bond Formation, *Science*, 2014, **345**, 804–808.

25 K. A. Campbell, J. M. Peloquin, D. P. Pham, R. J. Debus and R. D. Britt, Parallel Polarization EPR Detection of an S_1 State "Multiline" EPR Signal in Photosystem II Particles from *Synechocystis* sp. PCC 6803, *J. Am. Chem. Soc.*, 1998, **120**, 447–448.

26 K. V. Lakshmi, S. S. Eaton, G. R. Eaton, H. A. Frank and G. W. Brudvig, Analysis of Dipolar and Exchange Interactions Between Manganese and Tyrosine Z in the S_2Y_Z Center of Acetate-inhibited Photosystem II *via* EPR Spectral Simulations at X- and Q-Bands, *J. Phys. Chem. B*, 1998, **102**, 8327–8335.
27 K. V. Lakshmi, S. S. Eaton, G. R. Eaton and G. W. Brudvig, Orientation of the Tetranuclear Manganese Cluster and Tyrosine Z in the O_2-evolving Complex of Photosystem II: An EPR Study of the S_2Y_Z State in Oriented Acetate-inhibited Photosystem II Membranes, *Biochemistry*, 1999, **38**, 12758–12767.
28 J. Messinger, J. H. Robblee, W. O. Yu, K. Sauer, V. K. Yachandra and M. P. Klein, The S_0 State of the Oxygen-evolving Complex in Photosystem II is Paramagnetic: Detection of EPR Multiline Signal, *J. Am. Chem. Soc.*, 1997, **119**, 11349–11350.
29 L. V. Kulik, B. Epel, W. Lubitz and J. Messinger, ^{55}Mn Pulse ENDOR at 34 GHz of the S_0 and S_2 States of the Oxygen-evolving Complex in Photosystem II, *J. Am. Chem. Soc.*, 2005, **127**, 2392–2393.
30 H. A. Chu, W. Hillier, N. A. Law and G. T. Babcock, Vibrational Spectroscopy of the Oxygen-evolving Complex and of Manganese Model Compounds, *Biochim. Biophys. Acta, Bioenerg.*, 2001, **1503**, 69–82.
31 T. Noguchi and M. Sugiura, Flash-induced Fourier Transform Infrared Detection of the Structural Changes During the S-state Cycle of the Oxygen-evolving Complex in Photosystem II, *Biochemistry*, 2001, **40**, 1497–1502.
32 T. Noguchi, Fourier-transform Infrared Analysis of the Photosynthetic Oxygen-evolving Center, *Coord. Chem. Rev.*, 2008, **252**, 336–346.
33 R. J. Debus, FTIR Studies of Metal Ligands, Networks of Hydrogen Bonds, and Water Molecules Near the Active Site Mn_4CaO_5 Cluster in Photosystem II, *Biochim. Biophys. Acta, Bioenerg.*, 2015, **1847**, 19–34.
34 P. E. M. Siegbahn, Structures and Energetics for O_2 Formation in Photosystem II, *Acc. Chem. Res.*, 2009, **42**, 1871–1880.
35 E. M. Sproviero, J. A. Gascon, J. P. McEvoy, G. W. Brudvig and V. S. Batista, Computational Studies of the O_2-evolving Complex of Photosystem II and Biomimetic Oxomanganese Complexes, *Coord. Chem. Rev.*, 2008, **252**, 395–415.
36 E. M. Sproviero, J. A. Gascon, J. P. McEvoy, G. W. Brudvig and V. S. Batista, Quantum Mechanics/Molecular Mechanics Structural Models of the Oxygen-evolving Complex of Photosystem II, *Curr. Opin. Struct. Biol.*, 2007, **17**, 173–180.
37 M. Askerka, G. W. Brudvig and V. S. Batista, The O_2-Evolving Complex of Photosystem II: Recent Insights from Quantum Mechanics/Molecular Mechanics (QM/MM), Extended X-ray Absorption Fine Structure (EXAFS), and Femtosecond X-ray Crystallography Data, *Acc. Chem. Res.*, 2017, **50**, 41–48.
38 P. E. M. Siegbahn, Water Oxidation by PSII – A Quantum Chemical Approach, in *Mechanisms of Primary Energy Transduction in Biology*, ed. M. Wikström, The Royal Society of Chemistry, 2017.
39 S. Milikisiyants, R. Chatterjee, C. S. Coates, F. H. M. Koua, J. R. Shen and K. V. Lakshmi, The Structure and Activation of Substrate Water Molecules in the S_2 State of Photosystem II Studied by Hyperfine Sublevel Correlation Spectroscopy, *Energy Environ. Sci.*, 2012, **5**, 7747–7756.
40 H. Dau, T. Noguchi, J. Messinger, K. Moran and W. Hillier, FTIR Detection of Water Reactions in the Oxygen-evolving Centre of Photosystem II – Discussion, *Philos. Trans. R. Soc. London, Ser. B*, 2008, **363**, 1194–1195.

41 D. Shevela, K. Beckmann, J. Clausen, W. Junge and J. Messinger, Membrane-inlet Mass Spectrometry Reveals a High Driving Force for Oxygen Production by Photosystem II, *Proc. Natl. Acad. Sci. U. S. A.*, 2011, **108**, 3602–3607.
42 D. Shevela and J. Messinger, Studying the Oxidation of Water to Molecular Oxygen in Photosynthetic and Artificial Systems by Time-resolved Membrane-inlet Mass Spectrometry, *Front. Plant Sci.*, 2013, **4**, 473.
43 W. Hillier and T. T. Wydrzynski, The Affinities for the Two Substrate Water Binding Sites in the O_2-evolving Complex of Photosystem II Vary Independently During S-state Turnover, *Biochemistry*, 2000, **39**, 4399–4405.
44 H.-L. Huang and G. W. Brudvig, Kinetic Modeling of Substrate-Water Exchange in Photosystem II, *BBA Adv.*, 2021, **1**, 100014.
45 K. J. Young, L. A. Martini, R. L. Milot, R. C. Snoeberger, V. S. Batista, C. A. Schmuttenmaer, R. H. Crabtree and G. W. Brudvig, Light-driven Water Oxidation for Solar Fuels, *Coord. Chem. Rev.*, 2012, **256**, 2503–2520.
46 L. Hammarström, L. C. Sun, B. Akermark and S. Styring, Artificial Photosynthesis: Towards Functional Mimics of Photosystem II?, *Biochim. Biophys. Acta, Bioenerg.*, 1998, **1365**, 193–199.
47 S. Styring, Artificial Photosynthesis for Solar Fuels, *Faraday Discuss.*, 2012, **155**, 357–376.
48 I. Rivalta, G. W. Brudvig and V. S. Batista, Oxomanganese Complexes for Natural and Artificial Photosynthesis, *Curr. Opin. Chem. Biol.*, 2012, **16**, 11–18.
49 K. Linke and F. M. Ho, Water in Photosystem II: Structural, Functional and Mechanistic Considerations, *Biochim. Biophys. Acta, Bioenerg.*, 2014, **1837**, 14–32.
50 L. Vogt, D. J. Vinyard, S. Khan and G. W. Brudvig, Oxygen-evolving Complex of Photosystem II: An Analysis of Second-shell Residues and Hydrogen-bonding Networks, *Curr. Opin. Chem. Biol.*, 2015, **25**, 152–158.
51 H. Ishikita, W. Saenger, B. Loll, J. Biesiadka and E. W. Knapp, Energetics of a Possible Proton Exit Pathway for Water Oxidation in Photosystem II, *Biochemistry*, 2006, **45**, 2063–2071.
52 A. G. Gabdulkhakov, V. G. Kljashtorny and M. V. Dontsova, Molecular Dynamics Studies of Pathways of Water Movement in Cyanobacterial Photosystem II, *Crystallogr. Rep.*, 2015, **60**, 83–89.
53 C. J. Gisriel, J. Wang, J. Liu, D. A. Flesher, K. M. Reiss, H.-L. Huang, K. R. Yang, W. H. Armstrong, M. R. Gunner, V. S. Batista, R. J. Debus and G. W. Brudvig, High-resolution Cryo-EM Structure of Photosystem II from the Mesophilic Cyanobacterium, *Synechocystis* sp. PCC 6803, *Proc. Natl. Acad. Sci. U. S. A.*, 2022, **119**, e2116765118.
54 Z. Deak, S. Peterson, P. Geijer, K. A. Ahrling and S. Styring, Methanol Modification of the Electron Paramagnetic Resonance Signals from the S_0 and S_2 States of the Water-oxidizing Complex of Photosystem II, *Biochim. Biophys. Acta, Bioenerg.*, 1999, **1412**, 240–249.
55 R. D. Britt, J. L. Zimmermann, K. Sauer and M. P. Klein, The State of Manganese Iin the Photosynthetic Apparatus. 10. Ammonia Binds to the Catalystic Mn of the Oxygen-evolving Complex of Photosystem II – Evidence by Electron-spin-echo Envelope Modulation Spectroscopy, *J. Am. Chem. Soc.*, 1989, **111**, 3522–3532.

56 W. F. Beck, J. C. de Paula and G. W. Brudvig, Ammonia Binds to the Manganese Site of the O_2-evolving Complex of Photosystem II in the S_2 State, *J. Am. Chem. Soc.*, 1986, **108**, 4018–4022.

57 P. H. Oyala, T. A. Stich, R. J. Debus and R. D. Britt, Ammonia Binds to the Dangler Manganese of the Photosystem II Oxygen-evolving Complex, *J. Am. Chem. Soc.*, 2015, **137**, 8829–8837.

58 P. H. Oyala, T. A. Stich, J. A. Stull, F. Yu, V. L. Pecoraro and R. D. Britt, Pulse Electron Paramagnetic Resonance Studies of the Interaction of Methanol with the S_2 State of the Mn_4O_5Ca Cluster of Photosystem II, *Biochemistry*, 2014, **53**, 7914–7928.

59 D. A. Force, D. W. Randall, G. A. Lorigan, K. L. Clemens and R. D. Britt, ESEEM Studies of Alcohol Binding to the Manganese Cluster of the Oxygen Evolving Complex of Photosystem II, *J. Am. Chem. Soc.*, 1998, **120**, 13321–13333.

60 W. F. Beck and G. W. Brudvig, Reactions of Hydroxylamine with the Electron-Donor Side of Photosystem II, *Biochemistry*, 1987, **26**, 8285–8295.

61 D. A. Marchiori, P. H. Oyala, R. J. Debus, T. A. Stich and R. D. Britt, Structural Effects of Ammonia Binding to the Mn_4CaO_5 Cluster of Photosystem II, *J. Phys. Chem. B*, 2018, **122**, 1588.

62 H. Nagashima and H. Mino, Location of Methanol on the S_2 State Mn Cluster in Photosystem II Studied by Proton Matrix Electron Nuclear Double Resonance, *J. Phys. Chem. Lett.*, 2017, **8**, 621–625.

63 M. Retegan and D. A. Pantazis, Interaction of Methanol with the Oxygen-evolving Complex: Atomistic Models, Channel Identification, Species Dependence, and Mechanistic Implications, *Chem. Sci.*, 2016, **216**, 6463–6476.

64 I. Ghosh, G. Banerjee, K. Reiss, C. J. Kim, R. J. Debus, V. S. Batista and G. W. Brudvig, D1-S169A Substitution of Photosystem II Reveals a Novel S_2 state Structure, *Biochim. Biophys. Acta, Bioenerg.*, 2020, **1861**, 148301.

65 I. Ghosh, G. Banerjee, C. J. Kim, K. Reiss, V. S. Batista, R. J. Debus and G. W. Brudvig, D1-S169A Substitution of Photosystem II Perturbs Water Oxidation, *Biochemistry*, 2019, **58**, 1379–1387.

66 M. Retegan and D. A. Pantazis, Differences in the Active Site of Water Oxidation among Photosynthetic Organisms, *J. Am. Chem. Soc.*, 2017, **139**, 14340–14343.

67 G. Banerjee, I. Ghosh, C. J. Kim, R. J. Debus and G. W. Briudvig, Substitution of the D1-Asn87 Site in Photosystem II of Cyanobacteria Mimics the Chloride-binding Characteristics of Spinach Photosystem II, *J. Biol. Chem.*, 2018, **293**, 2487–2497.

68 P. Höfer, A. Grupp, H. Nebenführ and M. M. Mehring, Hyperfine Sublevel Correlation (HYSCORE) Spectroscopy: A 2D ESR Investigation of the Squaric Acid Radical, *Chem. Phys. Lett.*, 1986, **132**, 279–282.

69 G. W. Brudvig, J. L. Casey and K. Sauer, The Effect of Temperature on the Formation and Decay of the Multiline Electron Paramagnetic Resonance Signal Species Associated With Photosynthetic Oxygen Evolution, *Biochim. Biophys. Acta*, 1983, **723**, 366–371.

70 G. C. Dismukes and Y. Siderer, Intermediates of a Polynuclear Manganese Center Involved in Photosynthetic Oxidation of Water, *Proc. Natl. Acad. Sci. U. S. A.*, 1981, **78**, 274–278.

71 R. D. Britt, G. A. Lorigan, K. Sauer, M. P. Klein and J. L. Zimmermann, The $g = 2$ Multiline EPR Signal of the S_2 State of the Photosynthetic Oxygen-evolving

Complex Originates from a Ground Spin State, *Biochim. Biophys. Acta*, 1992, **1140**, 95–101.

72 T. A. Stich, G. J. Yeagle, R. J. Service, R. J. Debus and R. D. Britt, Ligation of D1-His332 and D1-Asp170 to the Manganese Cluster of Photosystem II from Synechocystis Assessed by Multifrequency Pulse EPR Spectroscopy, *Biochemistry*, 2011, **50**, 7390–7404.

73 K. Yang, K. V. Lakshmi, G. W. Brudvig and V. S. Batista, Is Deprotonation of the Oxygen-Evolving Complex of Photosystem II during the $S_1 \rightarrow S_2$ Transition Suppressed by Proton Quantum Delocalization?, *J. Am. Chem. Soc.*, 2021, **143**, 8324–8332.

74 D. A. Pantazis, W. Ames, N. Cox, W. Lubitz and F. Neese, Two Interconvertible Structures that Explain the Spectroscopic Properties of the Oxygen-Evolving Complex of Photosystem II in the S_2 State, *Angew. Chem., Int. Ed.*, 2012, **51**, 9935–9940.

75 D. A. Pantazis, M. Orio, T. Petrenko, S. Zein, W. Lubitz, J. Messinger and F. Neese, Structure of the Oxygen-evolving Complex of Photosystem II: Information on the S_2 State Through Quantum Chemical Calculation of Its Magnetic Properties, *Phys. Chem. Chem. Phys.*, 2009, **11**, 6788–6798.

76 J.-H. Su, N. Cox, W. Ames, D. A. Pantazis, L. Rapatskiy, T. Lohmiller, L. V. Kulik, P. Dorlet, A. W. Rutherford, F. Neese, A. Boussac, W. Lubitz and J. Messinger, The Electronic Structures of the S_2 States of the Oxygen-evolving Complexes of Photosystem II in Plants and Cyanobacteria in the Presence and Absence of Methanol, *Biochim. Biophys. Acta, Bioenerg.*, 2011, **1807**, 829–840.

77 J.-H. Su, K. G. V. Havelius, F. Mamedov, F. M. Ho and S. Styring, Split EPR Signals from Photosystem II Are Modified by Methanol, Reflecting S State-Dependent Binding and Alterations in the Magnetic Coupling in the $CaMn_4$ Cluster, *Biochemistry*, 2006, **45**, 7617–7627.

78 T. Yamauchi, H. Mino, T. Matsukawa, A. Kawamori and T.-a. Ono, Parallel Polarization Electron Paramagnetic Resonance Studies of the S_1 State Manganese Cluster in the Photosynthetic Oxygen-evolving System, *Biochemistry*, 1997, **36**, 7520–7526.

79 H. Yata and T. Noguchi, Mechanism of Methanol Inhibition of Photosynthetic Water Oxidation As Studied by Fourier Transform Infrared Difference and Time-resolved Infrared Spectroscopies, *Biochemistry*, 2018, **57**, 4803–4815.

80 A. Schweiger and G. Jeschke, *Principles of Pulse Electron Paramagnetic Resonance*, Oxford University Press, 2001.

81 J. A. Stull, T. A. Stich, R. J. Service, R. J. Debus, S. K. Mandal, W. H. Armstrong and R. D. Britt, ^{13}C ENDOR Reveals That the D1 Polypeptide C-terminus is Directly Bound to Mn in the Photosystem II Oxygen-evolving Complex, *J. Am. Chem. Soc.*, 2010, **132**, 446.

82 S. Schinzel, J. Schraut, A. V. Arbuznikov, P. E. M. Siegbahn and M. Kaupp, Density Functional Calculations of ^{55}Mn, ^{14}N and ^{13}C Electron Paramagnetic Resonance Parameters Support an Energetically Feasible Model System for the S_2 State of the Oxygen-evolving Complex of Photosystem II, *Chem.–Eur. J.*, 2010, **16**, 10424–10438.

83 N. J. Beal, T. A. Corry and P. J. O'Malley, Comparison between Experimental and Broken Symmetry DensityFunctional Theory (BS-DFT) Calculated Electron ParamagneticResonance (EPR) Parameters of the S_2 State of the

Oxygen-evolving Complex of Photosystem II in Its Native (Calcium) and Strontium-substituted Form, *J. Phys. Chem. B*, 2017, **121**, 11273–11283.

84 T. Lohmiller, V. Krewald, M. P. Navarro, M. Retegan, L. Rapatskiy, M. M. Nowaczyk, A. Boussac, F. Neese, W. Lubitz, D. A. Pantazis and N. Cox, Structure, Ligands and Substrate Coordination of the Oxygen-evolving Complex of Photosystem II in the S_2 State: A Combined EPR and DFT Study, *Phys. Chem. Chem. Phys.*, 2014, **16**, 11877–11892.

85 J. Schraut and M. Kaupp, On Ammonia Binding to the Oxygen-Evolving Complex of PhotosystemII:A Quantum Chemical Study, *Chem.–Eur. J.*, 2014, **20**, 7300–7308.

86 M. P. Navarro, W. M. Ames, H. Nilsson, T. Lohmiller, D. A. Pantazis, L. Rapatskiy, M. M. Nowaczyk, F. Neese, A. Boussac, J. Messinger, W. Lubitz and N. Cox, Ammonia Binding to the Oxygen-evolving Complex of Photosystem II Identifies the Solvent-Exchangeable Oxygen Bridge (μ-oxo) of the Manganese Tetramer, *Proc. Natl. Acad. Sci. U. S. A.*, 2013, **110**, 15561–15566.

87 F. M. Ho and S. Styring, Access Channels and Methanol Binding Site to the $CaMn_4$ Cluster in Photosystem II Based on Solvent Accessibility Simulations, With Implications for Substrate Water Access, *Biochim. Biophys. Acta, Bioenerg.*, 2008, **1777**, 140–153.

88 A. Gabdulkhakov, A. Guskov, M. Broser, J. Kern, F. Müh, W. Saenger and A. Zouni, Probing the Accessibility of the Mn_4Ca Cluster in Photosystem II: Channels Calculation, Noble Gas Derivatization, and Cocrystallization With DMSO, *Structure*, 2009, **17**, 1223–1234.

89 S. Vassiliev, T. Zaraiskaya and D. Bruce, Exploring the Energetics of Water permeation in Photosystem II by Multiple Steered Molecular Dynamics Simulations, *Biochim. Biophys. Acta*, 2012, **1817**, 1671–1678.

90 K. Ogata, T. Yuki, M. Hatakeyama, W. Uchida and S. Nakamura, All-atom Molecular Dynamics Simulation of Photosystem II Embedded in Thylakoid Membranes, *J. Am. Chem. Soc.*, 2013, **135**, 15670–15673.

Faraday Discussions

ROYAL SOCIETY OF CHEMISTRY

PAPER

Enzymatic X-ray absorption spectroelectrochemistry†

Karolina Cząstka,[a] Alaa A. Oughli,[b] Olaf Rüdiger *[a] and Serena DeBeer *[a]

Received 29th October 2021, Accepted 13th December 2021
DOI: 10.1039/d1fd00079a

The ability to observe the changes that occur at an enzyme active site during electrocatalysis can provide very valuable information for understanding the mechanism and ultimately aid in catalyst design. Herein, we discuss the development of X-ray absorption spectroscopy (XAS) in combination with electrochemistry for *operando* studies of enzymatic systems. XAS has had a long history of enabling geometric and electronic structural insights into the catalytic active sites of enzymes, however, XAS combined with electrochemistry (XA-SEC) has been exceedingly rare in bioinorganic applications. Herein, we discuss the challenges and opportunities of applying *operando* XAS to enzymatic electrocatalysts. The challenges due to the low concentration of the photoabsorber and the instability of the protein in the X-ray beam are discussed. Methods for immobilizing enzymes on the electrodes, while maintaining full redox control are highlighted. A case study of combined XAS and electrochemistry applied to a [NiFe] hydrogenase is presented. By entrapping the [NiFe] hydrogenase in a redox polymer, relatively high protein concentrations can be achieved on the electrode surface, while maintaining redox control. Overall, it is demonstrated that the experiments are feasible, but require precise redox control over the majority of the absorber atoms and careful controls to discriminate between electrochemically-driven changes and beam damage. Opportunities for future applications are discussed.

Introduction

Enzymatic X-ray absorption spectroscopy

X-ray absorption spectroscopy (XAS) has had a longstanding and an important impact on our understanding of metalloenzyme active sites.[3–14] In particular, the extended X-ray absorption fine structure (EXAFS) region of the spectrum is well known for having provided the first experimental measurements of the

[a]*Max Planck Institute for Chemical Energy Conversion, Mülheim an der Ruhr, DE, Germany. E-mail: serena.debeer@cec.mpg.de; olaf.ruediger@cec.mpg.de*

[b]*Technical University Munich, Campus Straubing for Biotechnology and Sustainability, Uferstraße 53, 94315 Straubing, Germany*

† Electronic supplementary information (ESI) available. See DOI: 10.1039/d1fd00079a

metal–ligand bond distances in the complex FeMo-cofactor of nitrogenase[11,14–16] and in the Mn_4CaO_5 oxygen evolving complex (OEC) of photosystem II (PSII).[17,18] While in both cases, the EXAFS data could not provide a definitive three-dimensional topological model, for both systems, EXAFS was able to provide average metal–ligand bond distances with high accuracy (~0.02 Å) and importantly, several decades before the first crystal structures of either enzyme was reported.[19,20]

Much of the power of XAS lies in its ability to provide element specific information about changes in the oxidation state and the metrical parameters of a given photoabsorber. For instance, XAS has been utilized at both the Mo and Fe K-edges in order to obtain geometric and electronic structural information about the FeMo-cofactor and correspondingly at the Mn and Ca K-edges for the OEC.[21–32] The resultant data provide information about the average oxidation state of each metal and the identity, coordination number and distances of the ligands coordinated to the selected metal, allowing for a relatively detailed picture of the active site geometric and electronic structure to be derived. As such, XAS is of particular utility for the characterization of reactive intermediates in an enzyme's catalytic cycle. The OEC of PSII is perhaps the most well-characterized example, in large part because laser flashing allows one to discretely advance from the resting state to the various intermediate "S states" of the catalytic cycle.[33] However, in addition to the OEC, the intermediates in a wide range of metalloproteins have been characterized by XAS. These include the Fe and Mo sites of nitrogenase intermediates,[34,35] the binuclear iron sites in soluble methane monooxygenase[36,37] and ribonucleotide reductases,[38–41] the trinuclear copper site in laccase,[42] and the Ni site in methyl coenzyme M reductase,[43] to name only a few. In almost all cases, the enzymes are prepared utilizing freeze quench methods to stop the reaction at a desired time point and enable the measurement of the XAS in the frozen state (most typically at LN_2 or LHe temperatures). Measurements of enzymes at cryogenic temperatures are particularly desirable as this helps limit the rate of X-ray beam induced radiation damage. Low temperatures also limit the thermal disorder contributions, thus simplifying the analysis of EXAFS data. While such an approach has led to the characterization of numerous enzyme intermediates, it is limited by the fact that it does not capture the dynamics of an operating catalyst. In this regard, XAS studies of enzymes under operating conditions are of great interest. While *operando* XAS of heterogeneous catalysts has seen increasing application in recent years,[44–46] applications to enzymes remain very limited.[47] One area of potentially great utility for *operando* XAS studies of enzymes is the application of spectroelectrochemistry (SEC). In the next sections, we provide a brief overview of the motivation for SEC, the advantages of combining SEC with XAS (XA-SEC) and the challenges associated with it. We then describe recent literature reports of XA-SEC to enzymes and molecular models and report our own recent application of XA-SEC to the enzyme hydrogenase. Finally, we provide a road map for future applications of XA-SEC to enzymes, with an emphasis on sample requirements and optimizing experimental design. We hope that this discussion will provide the scientific community with key principles to help guide the design of future XA-SEC experiments.

Spectroelectrochemistry: challenges and opportunities

Spectroelectrochemistry is a technique that combines two classical methods, electrochemistry and spectroscopy. The first concept of a spectroelectrochemical experiment was reported over 50 years ago,[48] and since that time a large variety of spectroscopic methods combined with electrochemistry have been reported.[49–53] Spectroscopic techniques include for example UV/Vis,[48,51,54] Raman,[51,55,56] IR,[55,57–66] and X-ray based spectroscopic techniques.[44,67–69] Less commonly developed spectroelectrochemical methods involve EPR,[51,56,60] and NMR.[70] Electrochemistry by itself is a technique that enables one to determine reaction kinetics, the catalytic activity, and to understand the stability of the system and its behavior towards substrates or inhibitors under the reaction conditions. Nevertheless, electrochemistry only provides activity measurements. Hence, in order to obtain information about the electronic and geometric changes that take place in an operating catalytic system, electrochemistry may be utilized in combination with spectroscopic techniques, as noted above. Accordingly, a spectroelectrochemical experiment can provide a collaborative outcome between spectroscopic and electrical signals that helps to explain the process occurring at the electrode. Therefore, spectroelectrochemical experiments can enable a more qualitative, and in some cases quantitative, picture of the thermodynamic, molecular, and mechanistic aspects of the system under study.

Despite these clear advantages, spectroelectrochemistry also presents several limitations. While the low sensitivity of electrochemical methods allows for fast or even extremely fast measurements, these low concentrations result, in most cases, in long spectroscopic measurement times. Another limitation at the moment of writing this manuscript, is that there are very few commercial spectroelectrochemical setups available. Hence, usually each spectroscopic technique requires the design of a specialized spectroelectrochemical cell, which might impose non-ideal electrode geometries, electrolyte composition, volume, *etc.* Moreover, the choice of solvents, supporting electrolytes, and electrode materials can be limited by the requirements of the spectroscopic methods. Another limitation to be considered in SEC, is that the incident electromagnetic radiation can cause physical and chemical changes to the sample, which are not related to pure electrochemical processes. This is particularly relevant for X-ray spectroscopy, where sample damage induced by the incident beam may cause the formation of the species which are unrelated to the actual mechanism, but may nevertheless affect the electrochemical results. This requires carefully designed experimental protocols, which will be discussed further in a subsequent section.

Given the many advantages of SEC and the ability of XAS to provide element specific access to changes in geometric and electronic structure, the appeal of XA-SEC is clear. However, for the study of enzymes, X-ray absorption SEC can be particularly challenging. For homogeneous catalysts in solution, one possible way of performing SEC is by bulk electrolysis. This works well with small molecules, but such experiments become very inefficient for large and slowly diffusing enzymes, making the use of redox mediators a requirement. If one is aiming to study a catalytic system under turnover conditions, having enzyme, substrate and mediators freely diffusing will result in most of the cases, in a very complex system to analyze, or require impractical amounts of enzyme. A possible solution is to immobilize the enzyme on the electrode surface in a configuration where direct electron transfer (DET) between enzyme and electrode can be established. This technique, called protein film electrochemistry (PFE), can be used in combination with spectroscopy, but it introduces additional challenges to the experiment. The main limitation of combining PFE with spectroscopy, especially when the spectroscopic technique aims at the metal center of the enzyme, is that the relatively large molecular weight of the protein backbone imposes a limitation on the maximum concentration of absorber that can be reached on the electrode surface. PFE on flat electrode surfaces usually are in the range of a few picomol cm^{-2} of enzyme,[71,72] well below the detection limit of a typical XAS experiment. Porous or nanostructured electrodes can be used to compensate for the low enzyme coverage. Another important aspect to consider when combining PFE with SEC, is that the signal response to the applied potential will be proportional to the amount of enzyme that is within a direct electron transfer regime.[73,74] This is not trivial for many enzymes, where the electrons will tunnel to the enzyme only when the enzyme is adequately oriented on the electrode surface. It is therefore desirable to efficiently wire the enzyme to the electrode to maximize the changes in the spectra induced by the applied potential. This is particularly relevant for XAS experiments, where the expected shifts on the absorption edge can, in some cases, be very subtle. Porous electrodes, carbon nanotubes, and conductive polymers can be used to favor electron transfer between enzymes and the electrode and should be considered when planning SEC. The final challenge imposed in XA-SEC experiments on enzymes is that, in order to achieve *operando* conditions, the measurements usually have to be carried in liquid water-based electrolyte, which typically means room temperature or minimally above freezing. Many enzymes, and in particular their active sites, are sensitive to X-ray photodamage, which can be minimized by measuring frozen samples at cryogenic temperatures, which is clearly not possible in *operando* SEC studies. As a result, very few XA-SEC experiments have been reported in bioinorganic chemistry. In the next sections, we briefly review the cases where XAS and electrochemistry experiments have been reported on enzymes and molecular catalysts.

XA-SEC in the literature

XAS is often used in combination with electrochemical measurements for studies on heterogeneous catalysts, and several recent reviews have summarized the field of *in situ* and *operando* XA-SEC.[44–46] In contrast, only relatively few reports of XA-SEC on molecular or enzymatic catalysts have been published to date. Here, we briefly review these reports.

One of the first XA-SEC studies on a molecular system came from Pushkar and co-workers,[75] where they measured a Ru(bda) catalyst (bda is 2,2′-bipyridine-6,6′-dicarboxylate) that is catalytically active for the oxygen evolution reaction. They immobilized the catalyst on a porous indium tin oxide (ITO). Their *in situ* Ru K-edge EXAFS analysis allowed for the detection and characterization of the key 7-coordinate Ru(V)=O intermediate state.

Recently, Crespilho and co-workers[47] reported an *operando* XAS study of bilirubin oxidase from *Myrothecium verrucaria* (*Mv*BOD). They illustrated the evolution of a redox reaction by probing the Cu K-edge of this copper-containing oxidoreductase. The protein was absorbed on a carbon nanoparticle mesoporous electrode in order to increase the protein coverage on the electrode. XA-

SEC measurements provided valuable information about the Cu ions within the enzymatic structure of *Mv*BOD and shed light on its role in the catalytic activity for the oxygen reduction reaction, highlighting the importance of doing these experiments under spectroelectrochemical *operando* conditions.

In another study, Bhattacharyya *et al.*,[76] reported *operando* XAS of heme-based biomolecules. They performed Fe K-edge XAS studies in homogenous conditions under electrochemical control simultaneously measuring cyclic voltammetry. They identified that the axial coordination and spin state of the iron active center are closely related to the redox response. Nevertheless, it is clear that these measurements posed experimental challenges. Close inspection of the reported data indicate that proper normalization of the Fe K-edge spectra could not be achieved and thus make it difficult to assess whether the observed differences are due to changes induced by the applied potential and/or differences in the background/normalization.

In another study, Liang and co-workers[77] published an *in situ/operando* XAS study of nickel phthalocyanime (NiPc) supported on carbon nanotubes as molecularly dispersed electrocatalysts (MDEs) that are active for CO_2 reduction to CO. Their Ni K-edge XAS analysis allowed them to understand how molecular engineering modulates the electrocatalytic properties of NiPc MDEs. Furthermore, they were able to identify a reduction site, where the first step reduction was located on the phthalocyanine ligand, whereas the second reduction was partially on the Ni center.

Lassalle-Kaiser *et al.*, designed a transmission cell for time-resolved XAS under electrochemical control.[78] Combined with a fast scanning monochromator, they could monitor the changes on a Fe solution, with good signal-to-noise spectra for measuring times as short as 5 s, but with the limitation of using a 20 mM Fe solution, a concentration that cannot be reached with most enzymes.[78]

We have also shown that *in situ* XAS can be used to identify reactive intermediates on a molecular Ru water oxidation complex, using a XA-SEC cell to titrate complexes in solution while measuring XAS.[79] The extremely reactive nature of the Ru(v) intermediate required low temperatures and rapid collecting methods. Additionally, given the broad nature of the spectra at the Ru K-edge, we complemented the results with EPR spectroscopy.[79]

The lack of more XA-SEC studies on molecular and enzymatic systems are indicative of how challenging these experiments are. Further development of this technique is necessary, so that in the future it can be utilized in a more routine and straightforward manner. This need motivated the XA-SEC studies on [NiFe] hydrogenase, which are reported in the section that follows.

Brief introduction to [NiFe] hydrogenases

Since [NiFe] hydrogenases are well studied using different spectroscopic, as well as spectroelectrochemical techniques,[80] we used this enzyme as a model system to see if we could observe electronic and geometric changes that occur during electrocatalysis by utilizing XAS spectra. The possibility of loading the enzyme into a redox polymer allows the enzyme surface concentration to be increased[81] in order to enable its detectability by XAS.

In this brief introduction to [NiFe] hydrogenases, we want to highlight the most relevant aspects for understanding the measured X-ray spectra.

Hydrogenases catalyze the H_2 oxidation in a reversible manner.[1,82] The active site of [NiFe] hydrogenases (Fig. 1A), consists of a sulfur bridged bimetallic mixed Ni–Fe active-center with an open coordination site on the Ni ion, where the substrates (H_2 and H^+), as well as inhibitors (O_2, CO) can bind. The Ni ion is coordinated by four cysteinyl thiolate ligands, where two of them are bridging to the iron ion and two others are bound to the Ni in a terminal fashion. The iron ion is coordinated by three diatomic inorganic ligands (CO and CN^-) forming octahedral site symmetry.[2] The bridging ligand coordinating to the Ni and the Fe changes identity depending on the redox state of the enzyme. [NiFe] hydrogenases contain chain iron sulfur clusters, which link the active site electronically with the surface of the protein, allowing for fast electron transfer (ET) during catalysis. The Fe ion remains formally in a divalent configuration throughout the catalytic cycle.

The active site changes in [NiFe] hydrogenases thus occur largely on the Ni ion (Fig. 1B), which changes its oxidation state and geometry during catalysis. The [NiFe] hydrogenase from *Dv*MF is oxygen sensitive, but in its inactive form (Ni-A and Ni-B) is oxygen resistant. The Ni in these oxidized states is in a trivalent Ni electronic configuration organized in distorted square-pyramidal (C_{4v}) geometry.[83,84] Upon one-electron reduction the activation occurs, resulting in an active Ni-Sl_a with a divalent Ni ion.[85] The first step of the catalytic cycle is the reaction of

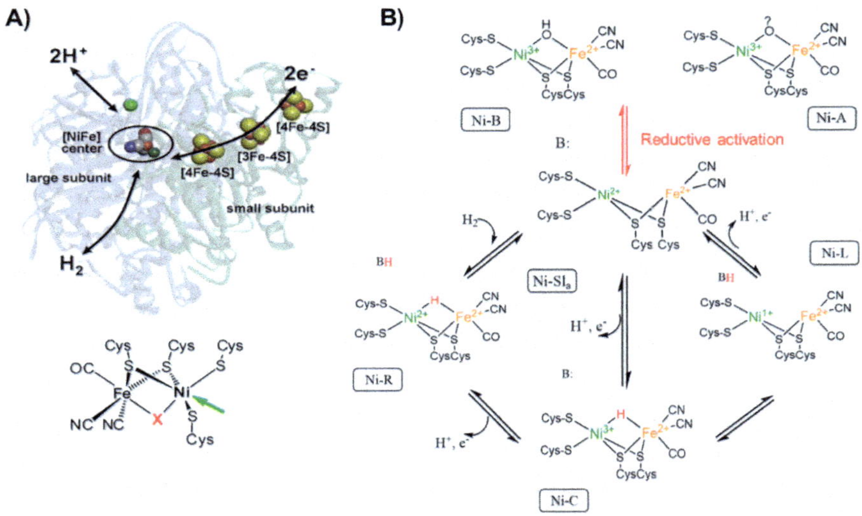

Fig. 1 (A) [NiFe] hydrogenase from *Dv*MF [1] with schematically indicated electron transfer chain (*via* iron-sulfur clusters), and pathways for the hydrogen and the proton transfer. The green sphere above the active center corresponds to the Mg ion. On the bottom, the chemical structure of the active site of [NiFe] hydrogenase. The arrow indicates the open metal coordination site, X corresponds to the bridging ligand that changes identity during catalytic activity. Adapted with permission from Lubitz, *et al.*, "Hydrogenases", *Chem. Rev.*, 2014, **114**(8), 4081. Copyright 2014 American Chemical Society.[2] (B) Proposed inactivation/reactivation process (red arrows) and catalytic cycle (black arrows) for [NiFe] hydrogenases. The activation of Ni-A and Ni-B goes through the Ni-SU and Ni-Si_r intermediates, which have been left out of the figure for clarity. The initial acceptor of the proton from heterolytic H_2 cleavage remains unclear, thus here a generic proton-acceptor is presented as B.

H_2 with Ni-SI_a state, which results in heterolytic H_2 cleavage to generate the Ni-R state with a bridging hydride in the [Ni(II)Fe(II)] active site, where Ni is organized in a distorted square-planar geometry (D_{4h}).[86] In the next step of the catalytic cycle, one proton and one electron is removed, forming the Ni-C intermediate, which retains a bridging hydride in the [Ni(III)Fe(II)] active site and consists of an $[4Fe4S]^{1+}$ cluster.[2] The final step of the catalytic cycle is believed to be a proton and ET from the bridging hydride in the Ni-C resulting in the Ni-L intermediate formation (with [Ni(I)Fe(II)] active site configuration with an open bridge position)[87] and finally regenerating the Ni-SI_a state.[88]

Results and discussion

Operando XAS study of *Desulfovibrio vulgaris* Miyazaki F [NiFe] hydrogenase in a redox polymer matrix

Before performing the *operando* measurements, it is important to first establish the changes that may be expected during the catalytic cycle by measuring reference samples. In order to do this the reference samples **1**, **2**, and **3** were first characterized by FTIR. FTIR characterization of the reference frozen solution of [NiFe] hydrogenase was carried out to quantify the relative distribution of redox states contributing to each of the three samples. Through using FTIR, CO and CN^- ligands coordinated directly to the iron ion at the active center of [NiFe] hydrogenase were probed. These ligands are very sensitive to the electronic changes at the active center of [NiFe] hydrogenase due to their strong π-accepting character (with decreasing oxidation state at the [NiFe] hydrogenase active center greater back-donation of the charge from metal to ligand occurs, hence causing the CN and CO stretching modes to shift to lower frequencies). Thus, by FTIR spectra indirect information about the oxidation state of the metals at the active center can be derived. The band positions for the different redox states of the *Desulfovibrio vulgaris* Miyazaki F (*Dv*MF) [NiFe] hydrogenase have been identified and assigned previously.[89]

The area under the FTIR spectral peaks has been used to estimate the sample composition. For the as isolated sample **1** (Fig. 2), it was found that two intermediate states were present, with 81% Ni-B and 19% Ni-Sl_r (Table 1). Upon enzyme incubation in H_2 (sample **2**, Fig. 2) 70% of Ni-R, 25% of Ni-C, and 5% of Ni-Sl_r were present. [NiFe] hydrogenase reduced with viologen ($E = -300$ mV *vs.* SHE) and NaDT (with the redox couple potential of −420 mV *vs.* SHE at pH 7, sample **3**) yields Ni-C (∼70%), and Ni-R (∼25%), and very little (5%) Ni-Sl_r. Furthermore, based on this FTIR-derived sample composition, the average oxidation state at the Ni center is 2.78, 2.25, and 2.70 (Table 1) for sample **1**, **2**, and **3**, respectively. This analysis is based on assuming formal trivalent oxidation states for Ni-B and Ni-C, and formal divalent oxidation states for Ni-R and Ni-SI_r, as summarized in Table 1. Based on the sample compositions, we thus expect changes in both the Ni oxidation state, coordination number, and local site symmetry, which should be observable by Ni K-edge XAS.

Fig. 3 shows the Ni K-edge XAS of **1**, **2** and **3**. While differences are observed, the Ni(II) rich sample **2** actually has the highest energy pre-edge and rising edge feature, while the Ni(III) rich complexes **1** and **3** have lower energy pre-edge and rising edge features (Table 2). This is due to the cancelling effects of changes in oxidation and site symmetry (as discussed in the ESI†) and highlights the

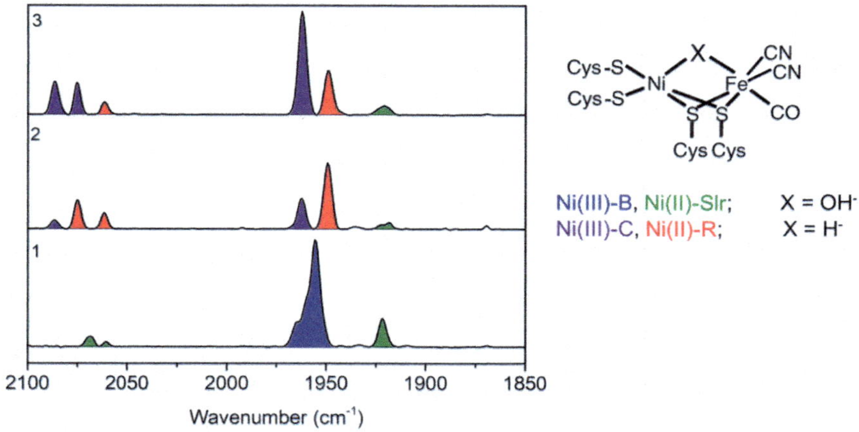

Fig. 2 (Left) Baseline corrected FTIR spectra of a *Dv*MF [NiFe] hydrogenase as isolated **1**, incubated 4 hours in H_2 atmosphere **2**, and in 15 μl of viologen and 40 μl NaDT **3**. Band positions for the different state are as follows: Ni(III)-B (1955, 2081, and 2090 cm^{-1}), Ni(II)-Sl_r (1922, 2061, 2070 cm^{-1}), Ni(III)-C (1961, 2074, 2085 cm^{-1}), Ni(II)-R (1948, 2061, 2074 cm^{-1}). (Right) Schematic representation of *Dv*MF [NiFe] hydrogenase active site, where X is identified for each redox state.

Table 1 Percentage of reference sample composition with [NiFe] hydrogenase intermediates and average of the oxidation state at the Ni center based on the FTIR spectra presented in Fig. 2

Sample	$Ni^{(III)}$-B C_{4v}	$Ni^{(III)}$-C D_{2d}	$Ni^{(II)}$-R D_{4h}	$Ni^{(II)}$-Sl_r C_{4v}	Average oxidation state at Ni[a]
1	81%			19%	2.78
2		25%	70%	5%	2.25
3		70%	25%	5%	2.70

[a] The average oxidation state at Ni was derived using the following formula: $x \approx \frac{(n\%\ Ni^{3+} \times 3) + (n\%\ Ni^{3+} \times 2)}{100\%}$, where n is the sum of the percentage of quantified [NiFe] hydrogenase intermediates where Ni is in the trivalent or divalent electronic configuration.

importance of measuring reference samples to highlight the expected limits that may be observed during XA-SEC.

Operando XAS spectroelectrochemistry

[NiFe] hydrogenase entrapped in the redox polymer on a GC electrode (Fig. 4, pink line) exhibits one pre-edge feature at 8333.0 eV. The Ni K-edge arises at 8342.8 eV (Table 3). The observed changes in shape and the intensity of the Ni K-edge of hydrogel film indicate that the local geometry of the Ni active center is modulated upon electrode preparation, as compared to reference sample **1**, suggesting a more distorted tetrahedral configuration (T_d).

The Ni K-edge XAS spectrum of the *Dv*MF [NiFe] hydrogenase-polymer-modified electrode at open circuit potential (OCP), in phosphate buffer pH 7

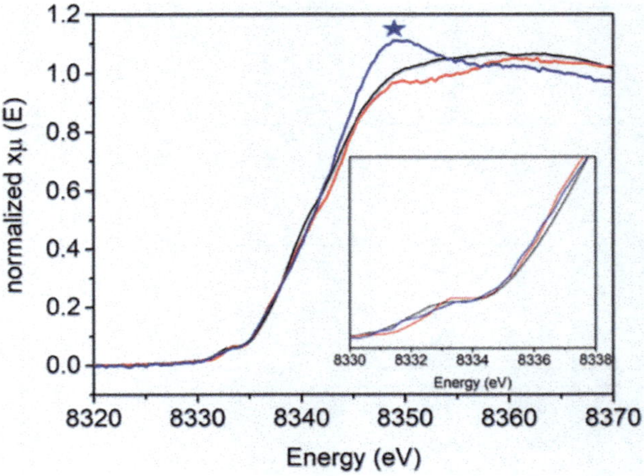

Fig. 3 Ni K-edge XAS spectra of [NiFe] hydrogenase from *Dv*MF as isolated **1** (black line), incubated 4 hours with H_2 **2** (red line) and reduced with NaDT/viologen **3** (blue line) measured at 10 K at SSRL 9-3 beamline. Blue asterisk corresponds to the energy edge at 8339 eV energy.

Table 2 Ni K-edge XAS features of the [NiFe] hydrogenase reference frozen solutions presented in Fig. 3

Sample	1s → 3d pre-edge energy (eV)	Edge energy (E_0, eV)	Edge energy at half-height (eV)
1	8331.9	8338.7	8340.5
2	8332.3	8339.1	8340.9
3	8332.0	8341.7	8341.8

saturated with H_2 is shown in Fig. 5. The spectra were measured after 1 hour of H_2 saturated electrolyte flow to activate and let the polymer equilibrate with the buffer. Ni K-edge XAS collected at OCP and H_2 flow exhibit a pre-edge feature at 8333.0 eV. The pre-edge intensity subtly increases and the edge shifts 1.0 eV toward lower energy (Fig. 5, Table 3 and ESI, Fig. S6†), and its intensity slightly decreases indicating a similar trend observed for the frozen solution of [NiFe] hydrogenase incubated for 4 hours with H_2 (Fig. 3). However, the shape of the edge line is still sharper than those observed for the reference sample. The observed Ni K-edge may be a result of a shorter enzyme contact time with H_2, or that the enzyme in the presence of the viologen polymer experiences a different redox potential and therefore the ratio of states is different than under H_2 without the polymer. The spectroelectrochemical experiment started at +200 mV where catalytic current for H_2 oxidation was observed (Fig. 5B and C). The Ni K-edge spectrum at this high potential shows a subtle shift (~0.2 eV) towards higher energy on the Ni K-edge line (Fig. 5, Table 3 and ESI Fig. S7†), suggesting either subtle sample oxidation or subtle geometry changes at the active center. At −400 mV, the pre-edge feature gains intensity and the edge shifts 1.1 eV toward lower energy, and loses intensity (Fig. 5, Table 3 and ESI, Fig. S8†). This suggests

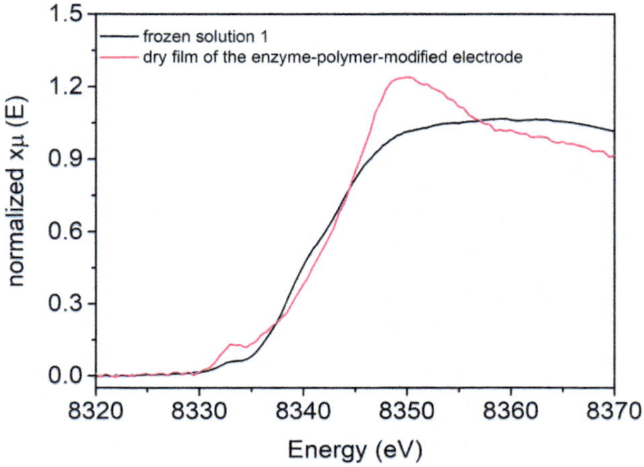

Fig. 4 Ni K-edge XAS spectra collected at 298 K at SAMBA Soleil beamline for a hydrogel film with [NiFe] hydrogenase from *Dv*MF on GC electrode (pink line), and reference frozen solution of **1** (black line) measured at 10 K at SSRL 9-3 beamline.

Table 3 Ni K-edge XAS features of the [NiFe] hydrogenase hydrogel film in the presence/absence of hydrogen

Conditions of X-ray studies	Edge energy (E_0, eV)	Edge energy at half-height (eV)	Edge intensity at 8349 eV
Dry hydrogel film	8342.8	8343.0	1.24
OCP, H_2	8341.8	8342.5	1.22
E = +200 mV *vs.* SHE, H_2	8342.0	8342.3	1.20
E = −400 mV *vs.* SHE, H_2	8340.9	8341.1	1.12
E = −150 mV *vs.* SHE, H_2	8341.4	8342.5	1.23
E = +200 mV *vs.* SHE, H_2	8342.0	8342.3	1.20

reduction of the Ni center and a change of the symmetry around the Ni ion. The resulting spectrum is similar to sample **2**, which contained mostly Ni-R. Increasing the potential back to +200 mV leaves spectra very similar to the previous one, showing reversibility of the process, something that would be incompatible with beam damage (Table 3). These spectra are also similar to the Ni-C rich spectrum measured at low temperature (sample **3**, Fig. 3 and S13†). The pre-edge feature remains at the same position for all measured conditions. Only subtle changes in its intensity have been observed, indicating minor changes in symmetry of Ni at the active site. These results are in agreement with what was observed using a transmission optically transparent thin layer (OTTLE) FTIR cell, where it was found that when [NiFe] hydrogenase is activated and trapped in the polymer, it can access just the catalytically active reduced state such as Ni^{II}-SI_a, Ni^{III}-C, and Ni^{II}-R, since the redox potential of the electron carrier, the viologen, is not positive enough to oxidize the enzyme back to Ni-B.[90] The previously reported FTIR SEC experiments were done in the absence of H_2. Under H_2 oxidation turnover conditions, the ratio of these states will be different. Additionally, kinetic modelling of the diffusion-reaction processes of a thick polymer-enzyme film

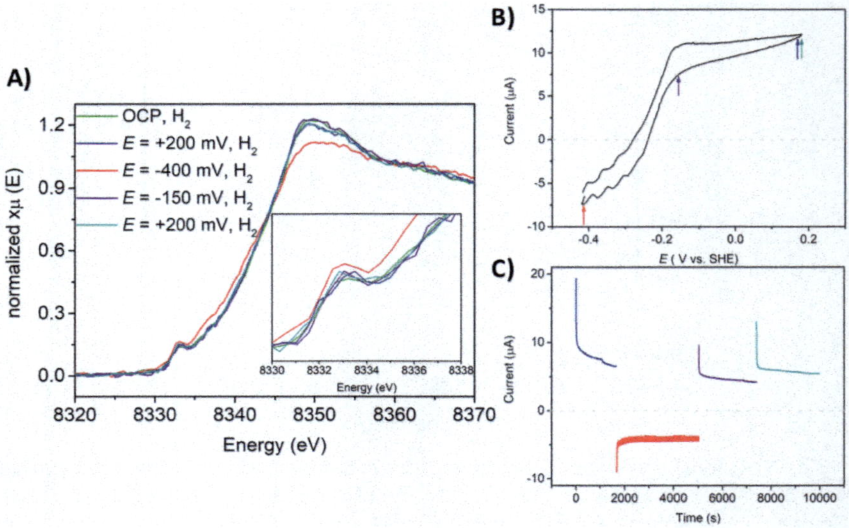

Fig. 5 (A) Ni K-edge XAS spectra collected at 298 K at SAMBA Soleil beamline for a hydrogel film with [NiFe] hydrogenase from *Dv*MF on GC electrode, measured upon the flow of phosphate electrolyte pH = 7 saturated with H_2 and OCP (green line), at $E = +200$ mV (blue line), at $E = -400$ mV (red line), at $E = -150$ mV (violet line), at $E = +200$ mV *vs.* SHE (dark cyan line). (B) Cyclic voltammogram (CV) of a GC electrode with a [NiFe] hydrogenase hydrogel in phosphate buffer at pH = 7 and H_2 flow. Measured at room temperature and 20 mV s^{-1} of scan rate. The colored arrows indicate at what potential X-ray *operando* measurements have been conducted. (C) Chronoamperometry (CA) experiments showing H_2 evolution and H_2 oxidation currents, measured at $E = +200$ mV (blue line), at $E = -400$ mV (red line), at $E = -150$ mV (violet line), and at $E = +200$ mV (dark cyan line). The horizontal dashed lines on CV and CA indicates a zero line current.

under turnover conditions has predicted the simultaneous formation of catalytic layers with non-catalytic layers within the film, meaning that only a fraction of the enzyme is involved in catalysis due to the diffusion limitation of electrons and H_2 through the polymer film.[81] Since the XAS probes the whole film, the data analysis of such a system becomes very complex. Taking all the results into account, it appears that a mixture of Ni-R and Ni-C is present at −400 mV, where the Ni-R intermediate dominates the spectra.

The catalytic H_2 oxidation current measured during multiple-steps chronoamperometry experiments display stable catalytic currents of ~21 μA cm^{-2} for H_2 oxidation, lower than what has been reported on a rotating disk electrode (RDE).[90] This may be a result of limiting the H_2 supply in the flow cell, compared to the RDE setup. The CV measured after *operando* XAS studies indicates ~34% loss in the H_2 oxidation activity (ESI, Fig. S14†). This can be explained by the fact that during the whole course of the experiment the enzyme-polymer film is slowly washed off the electrode surface while flowing electrolyte through the spectroelectrochemical flow-cell.

These experiments demonstrate that *operando* electrochemical X-ray experiments are possible, although extremely challenging. The use of polymer films yields an adequate protein loading. We note that our present data suggests that measurements of a monolayer are not currently possible (see ESI†). While the polymer films circumvent this issue, the formation of catalytic and non-catalytic layers associated with H_2 and redox gradients, and the complexity of the changes around the Ni ion in the [NiFe] hydrogenase during the catalytic cycle, complicate the data analysis. Furthermore, these investigations indicate that the XA-SEC spectroelectrochemical experiments require precise redox control over the majority of the studied metal centers, to maximize the response of the spectra to the applied potential. Additionally, careful controls are required to distinguish between electrochemically driven changes and enzyme degradation, either just as a result of enzyme instability or induced by the X-ray beam.

Recommendations and concluding remarks

In this section, we want to offer the reader a step-by-step guide on how to plan and execute XA-SEC experiments on enzymes.

As we have stated already in this article, measuring XA-SEC on an enzyme immobilized on an electrode surface is challenging, so it is advisable to gather some relevant information before collecting SEC data. Finding existing literature XAS/EXAFS data, and/or measuring the XAS/EXAFS of the studied protein in solution and at low temperature (as well as chemically reduced/oxidized, if experimentally possible) will on the one hand serve as a good reference of the undamaged protein and on the other hand indicate the expected range of changes that could be observed during the SEC experiment.

The next step is to design the SEC cell. As stated above, the experiment will be carried out in fluorescence mode. There are several factors that can be used to maximize the signal that reaches the detector in such a configuration. Defocusing the beam, or the use of rather large beam sizes has the advantage of exposing an increased electrode area, and therefore the amount of absorber. In this sense, bending magnet or wiggler beamlines are better suited to these experiments than beamlines using an undulator insertion device. An advantage of a defocused beam is that the dose per area unit is lowered, which may, to a certain extent, help to prevent beam damage on the sample. Another advantage is that a large beam averages the signal over the electrode surface and minimizes the effect of enzyme film inhomogeneity. On the other hand, the use of large beam spots requires the use of large electrodes. Employing a large working electrode requires the use of an even larger counter electrode. Building such electrochemical cells has to be done with extreme care, and the cell geometry has to be designed so that the applied electrode potential is homogeneous over the whole working electrode surface, to ensure that the whole sample is measured at the same applied electrochemical conditions.

While ideally we would place the electrode at 45° to the beam, decreasing the angle between electrode and beam will result in an increase of effective area of the electrode being probed. Grazing incidence XAS may be particularly useful for measuring electrodes that have been modified with thick enzyme-modified layers (*e.g.* porous materials, redox polymers and in general composite electrodes). Such configurations could yield layered structures, resulting from electron or substrate/products diffusion limitations through the film, where the enzyme could assume different electronic structures depending on the position of the enzyme in the film, which could be studied by varying the grazing angle.[91]

The shape of the window should be carefully designed based on the aforementioned criteria, considering the beam path and the projected size of the beam on the electrode. The materials used to construct the cell have to be free of interfering metals that could contribute to the spectra and be compatible with the used electrolytes.

For air-sensitive proteins, the cell has to be designed in such a way that it can be loaded and operated at the facilities available in the synchrotron. Gloveboxes are usually available at most synchrotrons, but the beamline is usually open to air and may be located far from the glovebox. Consider designing a cell that is gas-tight, that can be assembled in the glovebox and then can be anaerobically sealed for transfer to the beamline. If it is a flow cell, the tubes and pump system should be gas-tight and leak-safe to be operated on the beamline.

At the beamline, the first step is to align the cell with the beam and optimize detector positioning. Since in most cases one will be working with a very weak fluorescence signal, a large solid state detector with reasonable energy resolution for rejection of background scattering signals is very helpful. Low-pass Z-1 filters in combination with Soller slits may be utilized to further reject scattering contributions. The distance between sample and detector has to be carefully selected to optimize the counts reaching the detector. Unless working at high energies, the SEC cell will block the beam almost completely, making it difficult to measure a reference foil at the same time as the sample. In such cases, one should remember to measure a reference foil regularly during the beam time, or alternatively, one can utilize a scattering reference upstream in which a portion of the incident beam is deflected to a reference foil just outside the primary beam path.

The next step is to scan the cell surface (preferably using a filter to attenuate the beam and prevent any beam damage during these steps) in order to map the location of the sample on the electrode. Then at a given spot, one measures quick XAS scans, without any attenuating beam filters in order to evaluate how quickly the sample degrades. This can be judged by comparing the edge with reference samples measured at low temperature, and by observing how fast the XAS spectrum shape changes. Usually, the beam will induce metal reduction, which manifests by a shift of the edge to lower energies, but depending on the nature of the damage, changes in the pre-edge region or even in the white line intensity can also be observed. If the spectra are not changing over time, but the shape is different than what was measured for the reference sample at low temperature, this may indicate that the first measured spectra is already damaged. In that case, one should attenuate the beam with carbon or aluminum filters, and/or defocus the beam, and repeat the damage scans to find the exposure limit for a given position of the beam on the electrode. Then the XA-SEC experiment can begin. During the measurements it is advisable to keep track of the beam positions on the electrode that have been irradiated.

Regarding the system to be studied, it is important to have an optimized protein attachment protocol to the electrode surface that yields not only satisfactory electrode coverage, but also great stability of the protein film, since the SEC experiment will require long measuring times. When using redox active matrices (such as polymer films) for increasing the enzyme loading, the film thickness should be adjusted to avoid the concomitant formation of catalytic and non-catalytic layers. Homogeneous distribution of reaction rates, of concentrations and of redox states within the film are desired and obtained when catalysis is limiting the overall current output. This can be verified by comparing the current for different film thicknesses[92] and is typically attainable for film thicknesses below 10 μm for hydrogenase modified dendrimer films,[93] which are more homogeneous than polymer films at low thicknesses.[94,95]

Performing experiments with dynamic systems and under turnover conditions, means that the spectra will represent steady state conditions. These types of experiments are useful to identify rate-limiting reaction steps and could be also used to investigate inactivation processes or enzyme degradation pathways under turnover conditions.

Summary

Herein, we have briefly examined the possibilities of XA-SEC and reviewed the very few reports where this technique has been utilized on enzymes and molecular catalysts. We identified the challenges associated with XA-SEC experiments, such as low concentration of the photoabsorber and the instability of the protein in the X-ray beam. Further, we have highlighted the importance of the methods of enzyme immobilization, in order to obtain satisfactory electrode coverage and stability, while maintaining full redox control. Overall, we have demonstrated that the XA-SEC experiments are feasible, but require careful controls to discriminate between electrochemically-driven changes and enzyme degradation as a result of either enzyme instability and/or beam induced damage. Finally, we provided a road map for future applications of XA-SEC to enzymes, as well as the key principles to provide guidance in the experimental design. Further developments of this technique are still necessary so that in the future it can be utilized in a more routine and straightforward manner.

Author contributions

SD and OR planned and designed the study. The beamtime data collection and analysis was carried on by KC, OR and SD. AO contributed to the polymer synthesis. The manuscript has been written by KC, OR and SD and checked by all authors.

Conflicts of interest

There are no conflicts to declare.

Acknowledgements

The authors acknowledge the Max Planck Society for funding. SSRL and SOLEIL are acknowledged for allocating beamtime. Use of the SSRL is supported by the U.S. Department of Energy (DOE), Office of Science, Office of Basic Energy Sciences, under Contract DE-AC02-76SF00515. The authors gratefully acknowledge the Proposal Review Committee of SOLEIL for provision of their synchrotron radiation facilities (Proposal No. 20171380). Matthew Latimer at beamline 9-3 (SSRL), Menlo Park, CA, USA Dr Andrea Zitolo, Dr Gautier Landort and Guillaume Alizon at beamline SAMBA (SOLEIL), Saint-Aubin, France, are gratefully acknowledged for their technical assistance. The authors thank Prof. Dr Nicolas Plumeré for his contribution to polymer synthesis and helpful discussions. Birgit

Nöring, Christoph Laurich, and the mechanic workshop at the MPI CEC are thanked for manufacturing the SEC cells. Open Access funding was provided by the Max Planck Society.

Notes and references

1 H. Ogata, S. Hirota, A. Nakahara, H. Komori, N. Shibata, T. Kato, K. Kano and Y. Higuchi, *Structure*, 2005, **13**, 1635–1642.

2 W. Lubitz, H. Ogata, O. Rüdiger and E. Reijerse, *Chem. Rev.*, 2014, **114**, 4081.

3 K. B. Musgrave, H. I. Liu, L. Ma, B. K. Burgess, G. Watt, B. Hedman and K. O. Hodgson, *JBIC, J. Biol. Inorg. Chem.*, 1998, **3**, 344–352.

4 S. J. George, B. M. Barney, D. Mitra, R. Y. Igarashi, Y. Guo, D. R. Dean, S. P. Cramer and L. C. Seefeldt, *J. Inorg. Biochem.*, 2012, **112**, 85–92.

5 J. Kowalska and S. DeBeer, *Biochim. Biophys. Acta, Mol. Cell Res.*, 2015, **1853**, 1406–1415.

6 J. K. Kowalska, A. W. Hahn, A. Albers, C. E. Schiewer, R. Bjornsson, F. A. Lima, F. Meyer and S. DeBeer, *Inorg. Chem.*, 2016, **55**, 4485–4497.

7 K. M. Lancaster, M. Roemelt, P. Ettenhuber, Y. Hu, M. W. Ribbe, F. Neese, U. Bergmann and S. DeBeer, *Science*, 2011, **334**, 974–977.

8 C. Van Stappen, R. Davydov, Z.-Y. Yang, R. Fan, Y. Guo, E. Bill, L. C. Seefeldt, B. M. Hoffman and S. DeBeer, *Inorg. Chem.*, 2019, **58**, 12365–12376.

9 P. Rodríguez-Maciá, L. M. Galle, R. Bjornsson, C. Lorent, I. Zebger, Y. Yoda, S. P. Cramer, S. DeBeer, I. Span and J. A. Birrell, *Angew. Chem., Int. Ed.*, 2020, **59**, 16786–16794.

10 M. C. Corbett, Y. Hu, A. W. Fay, M. W. Ribbe, B. Hedman and K. O. Hodgson, *Proc. Natl. Acad. Sci. U. S. A.*, 2006, **103**, 1238–1243.

11 S. P. Cramer, K. O. Hodgson, W. O. Gillum and L. E. Mortenson, *J. Am. Chem. Soc.*, 1978, **100**, 3398–3407.

12 R. G. Shulman, P. Eisenberger, B. K. Teo, B. M. Kincaid and G. S. Brown, *J. Mol. Biol.*, 1978, **124**, 305–321.

13 R. G. Shulman, P. Eisenberger and B. M. Kincaid, *Annu. Rev. Biophys. Bioeng.*, 1978, **7**, 559–578.

14 C. Van Stappen, A. T. Thorhallsson, L. Decamps, R. Bjornsson and S. DeBeer, *Chem. Sci.*, 2019, **10**, 9807–9821.

15 S. P. Cramer, W. O. Gillum, K. O. Hodgson, L. E. Mortenson, E. I. Stiefel, J. R. Chisnell, W. J. Brill and V. K. Shah, *J. Am. Chem. Soc.*, 1978, **100**, 3814–3819.

16 M. R. Antonio, B. K. Teo, W. Orme-Johnson, M. J. Nelson, S. E. Groh, P. A. Lindahl, S. M. Kauzlarich and B. A. Averill, *J. Am. Chem. Soc.*, 1982, **104**, 4703–4705.

17 V. K. Yachandra, R. Guiles, A. E. McDermott, J. L. Cole, R. D. Britt, S. Dexheimer, K. Sauer and M. P. Klein, *Biochemistry*, 1987, **26**, 5974–5981.

18 D. B. Goodin, V. K. Yachandra, R. D. Britt, K. Sauer and M. P. Klein, *Biochim. Biophys. Acta, Bioenerg.*, 1984, **767**, 209–216.

19 J. Kim, D. Woo and D. Rees, *Biochemistry*, 1993, **32**, 7104–7115.

20 A. Zouni, H.-T. Witt, J. Kern, P. Fromme, N. Krauss, W. Saenger and P. Orth, *Nature*, 2001, **409**, 739–743.

21 C. Mller, P. Liebisch, M. Barra, H. Dau and M. Haumann, *Phys. Scr.*, 2005, 847, DOI: 10.1238/physica.topical.115a00847.

22 P. Gatt, R. Stranger and R. J. Pace, *J. Photochem. Photobiol., B*, 2011, **104**, 80–93.
23 C. Glöckner, J. Kern, M. Broser, A. Zouni, V. Yachandra and J. Yano, *J. Biol. Chem.*, 2013, **288**, 22607–22620.
24 M. Risch, K. Klingan, F. Ringleb, P. Chernev, I. Zaharieva, A. Fischer and H. Dau, *ChemSusChem*, 2012, **5**, 542–549.
25 I. Zaharieva, M. M. Najafpour, M. Wiechen, M. Haumann, P. Kurz and H. Dau, *Energy Environ. Sci.*, 2011, **4**, 2400–2408.
26 M. C. Corbett, Y. Hu, A. W. Fay, M. W. Ribbe, B. Hedman and K. O. Hodgson, *Proc. Natl. Acad. Sci.*, 2006, **103**, 1238–1243.
27 K. B. Musgrave, H. I. Liu, L. Ma, B. K. Burgess, G. Watt, B. Hedman and K. O. Hodgson, *JBIC, J. Biol. Inorg. Chem.*, 1998, **3**, 344–352.
28 A. W. Fay, M. A. Blank, C. C. Lee, Y. Hu, K. O. Hodgson, B. Hedman and M. W. Ribbe, *Angew. Chem., Int. Ed.*, 2011, **50**, 7787–7790.
29 B. Hedman, P. Frank, S. F. Gheller, A. L. Roe, W. E. Newton and K. O. Hodgson, *J. Am. Chem. Soc.*, 1988, **110**, 3798–3805.
30 F. A. Lima, R. Bjornsson, T. Weyhermüller, P. Chandrasekaran, P. Glatzel, F. Neese and S. DeBeer, *Phys. Chem. Chem. Phys.*, 2013, **15**, 20911–20920.
31 R. Bjornsson, F. A. Lima, T. Spatzal, T. Weyhermueller, P. Glatzel, E. Bill, O. Einsle, F. Neese and S. DeBeer, *Chem. Sci.*, 2014, **5**, 3096–3103.
32 M. A. Beckwith, W. Ames, F. D. Vila, V. Krewald, D. A. Pantazis, C. Mantel, J. Pécaut, M. Gennari, C. Duboc, M.-N. Collomb, J. Yano, J. J. Rehr, F. Neese and S. DeBeer, *J. Am. Chem. Soc.*, 2015, **137**, 12815–12834.
33 J. Messinger, J. H. Robblee, U. Bergmann, C. Fernandez, P. Glatzel, H. Visser, R. M. Cinco, K. L. McFarlane, E. Bellacchio, S. A. Pizarro, S. P. Cramer, K. Sauer, M. P. Klein and V. K. Yachandra, *J. Am. Chem. Soc.*, 2001, **123**, 7804–7820.
34 C. Van Stappen, L. Decamps, G. E. Cutsail, R. Bjornsson, J. T. Henthorn, J. A. Birrell and S. DeBeer, *Chem. Rev.*, 2020, **120**, 5005–5081.
35 C. Van Stappen, A. T. Thorhallsson, L. Decamps, R. Bjornsson and S. DeBeer, *Chem. Sci.*, 2019, **10**, 9807–9821.
36 R. G. Castillo, R. Banerjee, C. J. Allpress, G. T. Rohde, E. Bill, L. Que, J. D. Lipscomb and S. DeBeer, *J. Am. Chem. Soc.*, 2017, **139**, 18024–18033.
37 G. E. Cutsail, R. Banerjee, A. Zhou, L. Que, J. D. Lipscomb and S. DeBeer, *J. Am. Chem. Soc.*, 2018, **140**, 16807–16820.
38 N. Voevodskaya, F. Lendzian, O. Sanganas, A. Grundmeier, A. Gräslund and M. Haumann, *J. Biol. Chem.*, 2009, **284**, 4555–4566.
39 J. M. Younker, C. M. Krest, W. Jiang, C. Krebs, J. M. Bollinger and M. T. Green, *J. Am. Chem. Soc.*, 2008, **130**, 15022–15027.
40 E. M. Sproviero, *Dalton Trans.*, 2017, **46**, 4724–4736.
41 L. M. K. Dassama, A. Silakov, C. M. Krest, J. C. Calixto, C. Krebs, J. M. Bollinger and M. T. Green, *J. Am. Chem. Soc.*, 2013, **135**, 16758–16761.
42 S.-K. Lee, S. D. George, W. E. Antholine, B. Hedman, K. O. Hodgson and E. I. Solomon, *J. Am. Chem. Soc.*, 2002, **124**, 6180–6193.
43 R. Sarangi, M. Dey and S. W. Ragsdale, *Biochemistry*, 2009, **48**, 3146–3156.
44 J. Timoshenko and B. Roldan Cuenya, *Chem. Rev.*, 2021, **121**, 882–961.
45 Y. Zhai, Z. Zhu, S. Zhou, C. Zhu and S. Dong, *Nanoscale*, 2018, **10**, 3089–3111.
46 C. H. M. van Oversteeg, H. Q. Doan, F. M. F. de Groot and T. Cuk, *Chem. Soc. Rev.*, 2017, **46**, 102–125.

47 L. J. A. Macedo, A. Hassan, G. C. Sedenho and F. N. Crespilho, *Nat. Commun.*, 2020, **11**, 316.
48 T. Kuwana, R. K. Darlington and D. W. Leedy, *Anal. Chem.*, 1964, **36**, 2023–2025.
49 J. J. A. Lozeman, P. Führer, W. Olthuis and M. Odijk, *Analyst*, 2020, **145**, 2482–2509.
50 H. Oberacher, F. Pitterl, R. Erb and S. Plattner, *Mass Spectrom. Rev.*, 2015, **34**, 64–92.
51 L. Dunsch, *J. Solid State Electrochem.*, 2011, **15**, 1631–1646.
52 A. J. Wain and M. A. O'Connell, *Adv. Phys.: X*, 2017, **2**, 188–209.
53 Y. J. Tong, *Curr. Opin. Electrochem.*, 2017, **4**, 60–68.
54 K. J. Lee, N. Elgrishi, B. Kandemir and J. L. Dempsey, *Nat. Rev. Chem.*, 2017, **1**, 0039.
55 L. León and J. D. Mozo, *TrAC, Trends Anal. Chem.*, 2018, **102**, 147–169.
56 T. M. Cotton and R. P. Van Duyne, *Biochem. Biophys. Res. Commun.*, 1978, **82**, 424–433.
57 N. Chongdar, J. A. Birrell, K. Pawlak, C. Sommer, E. J. Reijerse, O. Rüdiger, W. Lubitz and H. Ogata, *J. Am. Chem. Soc.*, 2018, **140**, 1057–1068.
58 C. Sommer, A. Adamska-Venkatesh, K. Pawlak, J. A. Birrell, O. Rüdiger, E. J. Reijerse and W. Lubitz, *J. Am. Chem. Soc.*, 2017, **139**, 1440–1443.
59 P. Rodríguez-Maciá, K. Pawlak, O. Rüdiger, E. J. Reijerse, W. Lubitz and J. A. Birrell, *J. Am. Chem. Soc.*, 2017, **139**, 15122–15134.
60 A. Adamska, A. Silakov, C. Lambertz, O. Rüdiger, T. Happe, E. Reijerse and W. Lubitz, *Angew. Chem., Int. Ed.*, 2012, **51**, 11458–11462.
61 N. Chongdar, K. Pawlak, O. Rüdiger, E. J. Reijerse, P. Rodríguez-Maciá, W. Lubitz, J. A. Birrell and H. Ogata, *JBIC, J. Biol. Inorg. Chem.*, 2020, **25**, 135–149.
62 P. Rodríguez-Maciá, L. Kertess, J. Burnik, J. A. Birrell, E. Hofmann, W. Lubitz, T. Happe and O. Rüdiger, *J. Am. Chem. Soc.*, 2019, **141**, 472–481.
63 M. Osawa, *Surface-Enhanced Infrared Absorption*, Springer Berlin Heidelberg, 2001, pp. 163–187, DOI: 10.1007/3-540-44552-8_9.
64 M. Osawa, *Bull. Chem. Soc. Jpn.*, 1997, **70**, 2861–2880.
65 P. A. Ash, S. E. T. Kendall-Price and K. A. Vincent, *Acc. Chem. Res.*, 2019, **52**, 3120–3131.
66 P. A. Au-Ash, R. Au-Hidalgo and K. A. Au-Vincent, *JoVE*, 2017, e55858, DOI: 10.3791/55858.
67 J. Velasco Vélez, L. Falling, D. Bernsmeier, M. Sear, P. Clark, T. Chan, E. Stotz, M. Hävecker, R. Kraehnert, A. Knop-Gericke, C. Chuang, D. Starr, M. Favaro and R. Mom, *J. Phys. D: Appl. Phys.*, 2020, DOI: 10.1088/1361-6463/abd2ed.
68 L. R. Sharpe, W. R. Heineman and R. C. Elder, *Chem. Rev.*, 1990, **90**, 705–722.
69 Y. Yang, X. Liu, Z. Dai, F. Yuan, Y. Bando, D. Golberg and X. Wang, *Adv. Mater.*, 2017, **29**, 1606922.
70 R. Boisseau, U. Bussy, P. Giraudeau and M. Boujtita, *Anal. Chem.*, 2015, **87**, 372–375.
71 H. R. Pershad, J. L. C. Duff, H. A. Heering, E. C. Duin, S. P. J. Albracht and F. A. Armstrong, *Biochemistry*, 1999, **38**, 8992.
72 A. Parkin, G. Goldet, C. Cavazza, J. C. Fontecilla-Camps and F. A. Armstrong, *J. Am. Chem. Soc.*, 2008, **130**, 13410.
73 K. A. Vincent, A. Parkin and F. A. Armstrong, *Chem. Rev.*, 2007, **107**, 4366–4413.

74 H. R. Pershad, J. L. Duff, H. A. Heering, E. C. Duin, S. P. Albracht and F. A. Armstrong, *Biochemistry*, 1999, **38**, 8992–8999.

75 D. Lebedev, Y. Pineda-Galvan, Y. Tokimaru, A. Fedorov, N. Kaeffer, C. Copéret and Y. Pushkar, *J. Am. Chem. Soc.*, 2018, **140**, 451–458.

76 R. N. Samajdar and A. J. Bhattacharyya, *J. Phys. Chem. B*, 2021, **125**, 5258–5264.

77 X. Zhang, Y. Wang, M. Gu, M. Wang, Z. Zhang, W. Pan, Z. Jiang, H. Zheng, M. Lucero, H. Wang, G. E. Sterbinsky, Q. Ma, Y.-G. Wang, Z. Feng, J. Li, H. Dai and Y. Liang, *Nat. Energy*, 2020, **5**, 684–692.

78 K. Cheaib, B. Maurice, T. Mateo, Z. Halime and B. Lassalle-Kaiser, *J. Synchrotron Radiat.*, 2019, **26**, 1980–1985.

79 N. Levin, C. Casadevall, G. E. Cutsail III, J. Lloret-Fillol, S. DeBeer and O. Rüdiger, *ChemElectroChem*, 2022, **9**, e202101271.

80 H. Ogata, K. Nishikawa and W. Lubitz, *Nature*, 2015, **520**, 571.

81 V. Fourmond, S. Stapf, H. Li, D. Buesen, J. Birrell, O. Rüdiger, W. Lubitz, W. Schuhmann, N. Plumeré and C. Léger, *J. Am. Chem. Soc.*, 2015, **137**(16), 5494–5505.

82 A. K. Jones, E. Sillery, S. P. J. Albracht and F. A. Armstrong, *Chem. Commun.*, 2002, 866–867.

83 K. A. Bagley, E. C. Duin, W. Roseboom, S. P. J. Albracht and W. H. Woodruff, *Biochemistry*, 1995, **34**, 5527–5535.

84 K. A. Vincent, A. Parkin, O. Lenz, S. P. J. Albracht, J. C. Fontecilla-Camps, R. Cammack, B. Friedrich and F. A. Armstrong, *J. Am. Chem. Soc.*, 2005, **127**, 18179–18189.

85 H. Ogata, P. Kellers and W. Lubitz, *J. Mol. Biol.*, 2010, **402**, 428–444.

86 M. Van Gastel, M. Stein, M. Brecht, O. Schröder, F. Lendzian, R. Bittl, H. Ogata, Y. Higuchi and W. Lubitz, *J. Biol. Inorg. Chem.*, 2006, **11**, 41–51.

87 T. Krämer, M. Kampa, W. Lubitz, M. van Gastel and F. Neese, *ChemBioChem*, 2013, **14**, 1898–1905.

88 J. L. Barilone, H. Ogata, W. Lubitz and M. van Gastel, *Phys. Chem. Chem. Phys.*, 2015, **17**, 16204–16212.

89 D. Millo, M. E. Pandelia, T. Utesch, N. Wisitruangsakul, M. A. Mroginski, W. Lubitz, P. Hildebrandt and I. Zebger, *J. Phys. Chem. B*, 2009, **113**, 15344–15351.

90 N. Plumeré, O. Rüdiger, A. A. Oughli, R. Williams, J. Vivekananthan, S. Pöller, W. Schuhmann and W. Lubitz, *Nat. Chem.*, 2014, **6**, 822–827.

91 C. Maurizio, M. Rovezzi, F. Bardelli, H. G. Pais and F. D'Acapito, *Rev. Sci. Instrum.*, 2009, **80**, 063904.

92 H. Li, D. Buesen, S. Dementin, C. Léger, V. Fourmond and N. Plumeré, *J. Am. Chem. Soc.*, 2019, **141**, 16734–16742.

93 S. Hardt, S. Stapf, D. T. Filmon, J. A. Birrell, O. Rüdiger, V. Fourmond, C. Léger and N. Plumeré, *Nat. Catal.*, 2021, **4**, 251–258.

94 D. Buesen, H. Li and N. Plumeré, *Chem. Sci.*, 2020, **11**, 937–946.

95 H. Li, D. Buesen, R. Williams, J. Henig, S. Stapf, K. Mukherjee, E. Freier, W. Lubitz, M. Winkler and T. Happe, *Chem. Sci.*, 2018, **9**, 7596–7605.

Faraday Discussions

ROYAL SOCIETY OF CHEMISTRY

PAPER

Intermediate-spin iron(IV)-oxido species with record reactivity†

Peter Comba, [*ab] George Nunn,[c] Frederik Scherz[a] and Paul H. Walton [c]

Received 18th October 2021, Accepted 8th November 2021

DOI: 10.1039/d1fd00073j

The nonheme iron(IV)-oxido complex *trans*-N3-$[(L^1)Fe^{IV}{=}O(Cl)]^+$, where L^1 is a derivative of the tetradentate bispidine 2,4-di(pyridine-2-yl)-3,7-diazabicyclo[3.3.1]nonane-1-one, has an $S = 1$ electronic ground state and is the most reactive nonheme iron model system known so far, of a similar order of reactivity as nonheme iron enzymes (C–H abstraction of cyclohexane, −90 °C (propionitrile), $t_{1/2} = 3.5$ s). The reaction with cyclohexane selectively leads to chlorocyclohexane, but "cage escape" at the $[(L^1)Fe^{III}(OH)(Cl)]^+$/cyclohexyl radical intermediate lowers the productivity. Ligand field theory is used herein to analyze the d–d transitions of $[(L^1)Fe^{IV}{=}O(X)]^{n+}$ ($X = Cl^-$, Br^-, MeCN) in comparison with the thoroughly characterized ferryl complex of tetramethylcyclam (TMC = L^2; $[(L^2)Fe^{IV}{=}O(MeCN)]^{2+}$). The ligand field parameters and d–d transition energies are shown to provide important information on the triplet–quintet gap and its correlation with oxidation reactivity.

Introduction

The functionalization of hydrocarbon substrates with inert C–H bonds is among the more challenging and important reactions in synthetic organic chemistry.[1–3] Nature frequently uses iron based enzymes for these and other metabolically important oxidative transformations, and nonheme iron(IV)-oxido species have been identified as ubiquitous key intermediates in oxidative transformations of unactivated C–H bonds. The last two decades have seen considerable development in this area, involving enzymatic processes, bioinspired models and technical applications, based on structural data, spectroscopy and mechanistic work, involving experimental as well as computational studies.[4–11] Despite these advances in our understanding, important and still disputed key questions are

[a]*Universität Heidelberg, Anorganisch-Chemisches Institut, INF 270, D-69120 Heidelberg, Germany. E-mail: peter.comba@aci.uni-heidelberg.de*

[b]*Universität Heidelberg, Interdisziplinäres Zentrum für Wissenschaftliches Rechnen (IWR), Germany*

[c]*Department of Chemistry, University of York, Heslington, YORK, YO10 5DD, UK*

† Electronic supplementary information (ESI) available: Ligand field parameterization, as well as schemes and tables with computed ligand field transitions. See DOI: 10.1039/d1fd00073j

related to the importance of the driving force (*i.e.* the $Fe^{IV/III}$-oxido redox potential), where accurate and reliable experimental and computational data are still needed,[12–17] and also to the relevance of the spin ground state of the Fe^{IV}-oxido oxidant.

All enzymes have an $S = 2$ (high-spin) ground state, while many synthetic model systems are $S = 1$, and those with a quintet ground state are not necessarily the most reactive.[18–23] An important aspect is that, based on ligand field theory, the high-spin ground state is favored for trigonal bipyramidal geometries, as generally observed in enzymes, while many model systems have pseudo-tetragonal coordination geometries, such as those discussed in the present communication. DFT (density functional theory) analysis suggests that C–H abstraction by Fe^{IV}-oxido species generally proceeds on the quintet surface with a preference for the σ channel with a linear [Fe–O···H···C–R] transition state and the electron transferred into the d_{z^2} orbital,[24–27] and $Fe^{IV}{=}O$ oxidants with a triplet ground state generally cross the spin surface to react over the lower energy quintet transition state. The underlying concept of two-state-reactivity (TSR) indicates that the spin ground state of $Fe^{IV}{=}O$ oxidants might be less important than sometimes assumed, as long as the triplet–quintet energy gap is relatively small.[28–31]

We have recently shown that the $Fe^{IV}{=}O$ species with the tetradentate bispidine ligand L^1 (bispidine = 2,4-di(pyridine-2-yl)-3,7-diazabicyclo[3.3.1]nonane-1-one, see Scheme 1 for ligands discussed in this manuscript) has an $S = 1$ ground state and is the most reactive $Fe^{IV}{=}O$ oxidant known so far, with C–H abstraction rates probably as fast as those of enzymes.[21–23] We admit that this assignment is only based on the d–d transitions of this extremely reactive $Fe^{IV}{=}O$ species, and characterization by other spectroscopies (Mössbauer, resonance Raman, MCD *etc.*) are not yet available due to the difficulty to trap this oxidant that decays at −90 °C in seconds. Moreover, the ligand field analysis presented

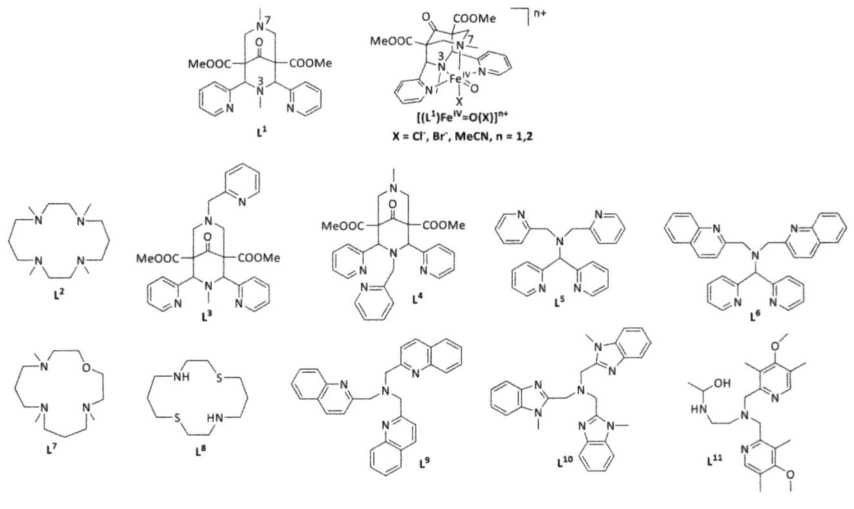

Scheme 1 Ligands discussed in this manuscript and structural drawing of $[(L^1)Fe^{IV}{=}O(X)]^{n+}$.

here supports our assignment. Note that a comparison of reactivities has to be made with caution because published rates have been measured under a variety of conditions, (i) using various substrates with differing C–H bond dissociation energies (BDE), (ii) determined in different solvents, and (iii) measured at different temperatures and generally without experimentally determined temperature dependences (a relevant collection of rate data is given in Table 1). An important attribute related to the reactivity of $S = 1$ $[(L^1)Fe^{IV}{=}O(Cl)]^+$ is that the spin crossover, *i.e.* the efficiency of TSR and hence the reactivity of a triplet $Fe^{IV}{=}O$ oxidant, depends on the triplet–quintet gap.[32–34] This follows from the mixing of the triplet ground state with the low lying quintet state *via* spin–orbit coupling, and a consequence is that the transmission coefficient in the Eyring equation depends on the spin–orbit coupling matrix elements. For tetragonal systems, the triplet–quintet gap primarily depends on the in-plane ligand field (see Scheme 2 for a simplified visualization of ligand field effects). On the basis of the assignment of the d–d transitions in analogy to thorough spectroscopic analyses of well-characterized tetragonal ferryl complexes,[35,36] the relative energy of the in-plane ligand field can be determined experimentally as a measure of the quintet–triplet gap, which then can be correlated to the reactivity, and this is the main contribution of the present communication.

Table 1 Selected d–d transitions, pseudo-first-order [s^{-1}] and second order [M^{-1} s^{-1}] rates involving mononuclear $Fe^{IV}{=}O$ species (note that standard deviations are not available for all rates reported and are therefore omitted; note also that the tabulated rates are for various substrates with differing bond dissociation energies, that they were measured in different solvents and at different temperatures – all given in the footnotes; therefore, the comparison of the rates needs to be made very cautiously)

L/X/spin	d–d transition/ nm		k/s^{-1}	$k/M^{-1}\ s^{-1}$	Ref.
L^1/MeCN/$S = 1$	565	786		2.9×10^{-1a}	15 and 22
L^1/Cl/$S = 1$	581	831	2×10^{-1b}	7.6×10^{2b}	22
L^1/Br/$S = 1$	607	(900)			
L^2/MeCN/$S = 2$	568	775		n.r.[c]	35
L^3/$S = 1$	500	730		4.9×10^{-3d}	15 and 37
L^4/$S = 1$	500	730		1.3×10^{-4d}	15 and 37
L^5/$S = 1$		695		8.3×10^{-2e}	38
L^6/$S = 1$		770		5.1×10^{-1e}	38
L^7/MeCN/$S = 1$	585	848, 992		1.0×10^{-2f}	39
L^8/MeCN/$S = 1$	596	815		3.7×10^{2g}	40
L^9/MeCN/$S = 2$				3.7×10^{-1h}	19
L^9/Cl/$S = 2$				5×10^{-2i}	41
L^9/Br/$S = 2$				7×10^{-2i}	41
L^{10}/—/$S = 1$	600	770		2.5×10^{-1j}	42
L^{11}/dinuclear/$S = 1$				8.1×10^{-2k}	43
TauD			13[l]		44
MMO				3.4×10^{4m}	45 and 46

[a] bzOH, MeCN, −35 °C. [b] Cyclohexane (chlorocyclohexane), EtCN, −90 °C. [c] No reaction with cyclohexane. [d] Cyclohexane, MeCN, 25 °C. [e] bzOH, MeCN, 25 °C. [f] Cyclohexane, CH_2Cl_2, −40 °C. [g] 1,4-CHD, MeCN, 20 °C. [h] Cyclohexane (cyclohexanone), MeCN, −40 °C. [i] Toluene, MeCN, −40 °C. [j] Cyclohexane, MeCN, −40 °C. [k] Cyclohexane, MeCN, 10 °C. [l] H_2O, 5 °C. [m] H_2O, 25 °C.

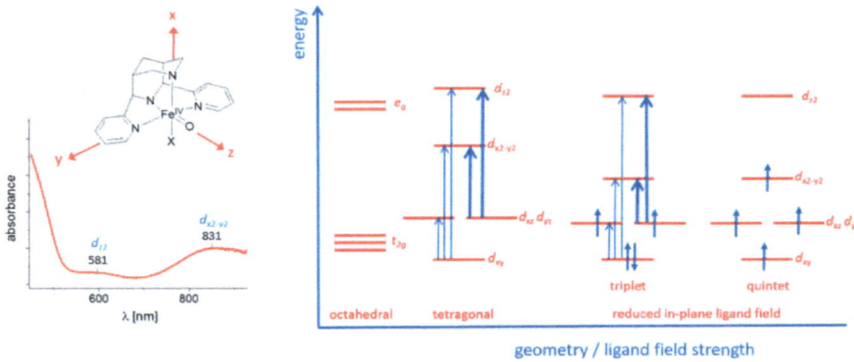

Scheme 2 Structure of $[(L^1)Fe^{IV}{=}O(X)]^{n+}$ with the coordinate system used in the ligand field analysis together with the d–d spectrum of the chlorido complex $[(L^1)Fe^{IV}{=}O(Cl)]^+$, and a simplified representation of the ligand field effects in pseudo-tetragonal $Fe^{IV}{=}O$ complexes (transitions in bold are those observed in the spectra of $[(L^1)Fe^{IV}{=}O(X)]^{n+}$ (X = MeCN, Cl^-, Br^-); all transitions have been assigned for $[(L^2)Fe^{IV}{=}O(MeCN)]^{2+}$, see below[36]).

While the effect of axial ligands on the reactivities of nonheme $Fe^{IV}{=}O$ centers has been studied in detail (in heme systems the tuning of reactivities by axial thiolate and selenolate has been shown to be of central importance[34,47–50]), in contrast there are only a few studies on in-plane ligand field effects with respect to reactivities of nonheme ferryl complexes.[21–23,38–40,51,52] Here we report a thorough analysis of the ligand field properties and possible correlations with the oxidation reactivities of a series of $[(L^1)Fe^{IV}{=}O(X)]^{n+}$ complexes, based on experimental electronic spectroscopic data and ligand field calculations.

Results and discussion

Experimental spectroscopic and reactivity data

The high reactivity of $[(L^1)Fe^{IV}{=}O(Cl)]^+$ and derivatives with other co-ligands and to some extent also of ferryl complexes with the pentadentate bispidine ligands L^3 and L^4, carries the problem that $Fe^{IV}{=}O$ species are difficult to trap and spectroscopically fully characterize – the half-life of $[(L^1)Fe^{IV}{=}O(Cl)]^+$ at −90 °C is only seconds[22] – and no crystal structure is available of any bispidine–$Fe^{IV}{=}O$ complex. Also, Mössbauer and resonance Raman spectra are not yet available from the complexes with the tetradentate bispidine L^1, and the electronic spectra discussed here are limited to solution UV-vis-NIR measurements obtained a few milliseconds after injecting the oxidant to the iron(II) precursor at low temperature (see Table 1 and Fig. 1). Thus, given the paucity of spectroscopic data, the assignment of the d–d transitions largely follows those of the fully characterized tetragonal $S = 1$ ferryl complex $[(L^2)Fe^{IV}{=}O(MeCN)]^{2+}$ based on vis-NIR and MCD spectra supported by experimentally calibrated DFT calculations (see Scheme 2).[36] Reasons for the low in-plane ligand field (low energy of the transition to the $d_{x^2-y^2}$ orbital) in bispidine complexes are that the bispidine cavity is very rigid and relatively large, and that particularly the Fe–N7 bond distance is more flexible and generally longer then the Fe–N3 bond (see Scheme 1 for the numbering).[53–55]

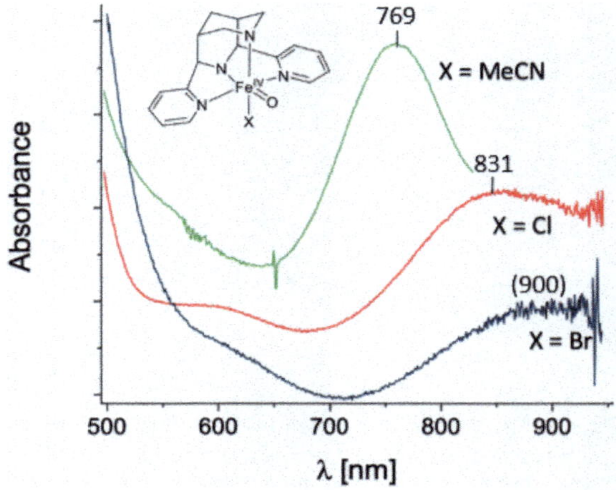

Fig. 1 Optical spectra (d–d transitions in the vis-NIR region) of $S = 1$ bispidine-iron(IV)-oxido complexes.

Importantly, the co-ligand X offers an additional simple possibility for tuning the in-plane ligand field.

The second problem in terms of data availability is that kinetic studies are hampered by the fact that there are two very fast self-decay processes in addition to reaction with a substrate:[22] the first involves the reaction of the precursor with the product to form an oxido-bridged diiron(III) resting state, *i.e.* an oxo-transfer reaction (OAT) involving the ferryl complex and its precursor. The other is an intramolecular C–H activation (HAT), involving the N7 methyl group and leading to demethylation. From a kinetic analysis of $[(L^1)Fe^{IV}{=}O(Cl)]^+$ in MeCN at −90 °C it follows that even with a 10-fold excess of iodosylbenzene as oxidant, the pseudo-first-order formation rate of the ferryl oxidant is only just over 5 times larger than HAT with cyclohexane, and the self-decay processes are only 5 (OAT) to 100 (HAT) times slower under these conditions than the reaction with substrate.[22] Preliminary data suggest that the reactivity of the corresponding bromido complex $[(L^1)Fe^{IV}{=}O(Br)]^+$, for which no full kinetic analysis is available, is about one order of magnitude larger than that of the chlorido complex, that of the MeCN complex under similar conditions is around one order of magnitude slower,[21] and this correlates qualitatively with the main d–d transitions (see Fig. 1 and 2).

Ligand field analysis

The ligand field computer program, *Kestrel*,[56] was used for the analysis of the optical spectra of $[(L^1)Fe^{IV}{=}O(X)]^{n+}$ (X = MeCN, Cl^-, Br^-). *Kestrel* uses a configurational interaction method to calculate the ligand field energies of d electron Russell–Saunders states, using the traditional ligand field parameters (Racah B and C, spin–orbit coupling) and angular overlap model (AOM) *e*-values as inputs. The differences in calculated energies from *Kestrel* can be fitted to experimental transition energies to determine the ligand field parameters. In the present case, there are two possible strategies to reduce the complexity of the fitting procedure, *i.e.* to either use pseudo-D_{4h} or pseudo-C_{4h} symmetry of the coordination sphere.

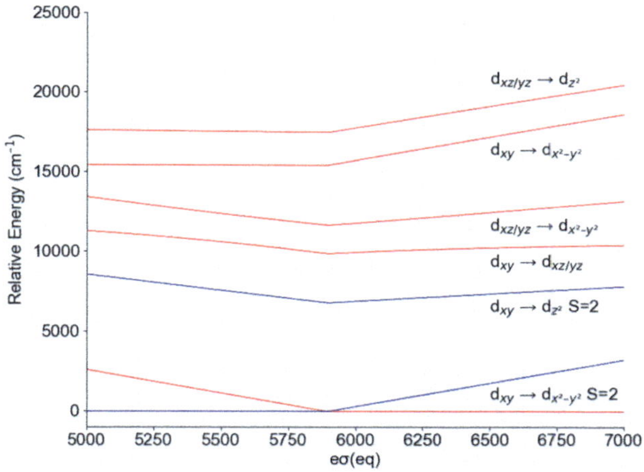

Fig. 2 Ligand field calculated energies of d electron states for octahedral geometry with the transferred axial parameters (see text), as a function of the average e_σ(equatorial) value. Red lines are triplet d-states, the blue lines are the low-lying quintet excited states. Racah B = 600 cm^{-1}, C/B = 5, in the absence of spin–orbit coupling. The triplet $d_{xz/yz} \rightarrow d_{x^2-y^2}$ and $d_{xz/yz} \rightarrow d_{z^2}$ transitions can mix *via* static correlation (interelectronic repulsion), which means that the transitions cannot be taken as pure electronic states and their relative ordering can be swapped. We have chosen the labeling to be consistent with ref. 35 and 36.

The former is less ambiguous and is used here. With assumed D_{4h} symmetry, one obtains an axial ligand field for the Fe^{IV}=O/Fe^{IV}–NCMe axis of $[(L^2)Fe^{IV}$=O(MeCN)$]^{2+}$ or the Fe^{IV}=O/Fe^{IV}–N3 axis of $[(L^1)Fe^{IV}$=O(X)$]^{n+}$ (X = MeCN, Cl^-, Br^-), respectively, and an in-plane ligand field for the four tertiary amines of $[(L^2)Fe^{IV}$=O(MeCN)$]^{2+}$ or N7, the two pyridines and X for $[(L^1)Fe^{IV}$=O(X)$]^{n+}$ (X = MeCN, Cl^-, Br^-), respectively. Note here, therefore, that the calculated e_σ values for the ligating atoms in either axial or equatorial directions represent the *average* e_σ value for each ligating atom in that direction. The parameterization must be reduced to include as few parameters as possible. Accordingly, the strategy herein was to determine axial e_σ and e_π parameters and a C/B ratio from analysis of $[(L^2)Fe^{IV}$=O(MeCN)$]^{2+}$, allowing the equatorial e_σ and Racah B parameters to vary freely. All other π interactions were fixed to 0 cm^{-1}.

In the first instance, the thoroughly analyzed spectra (vis-NIR and MCD) of the pseudo-C_{4v} symmetrical $[(L^2)Fe^{IV}$=O(MeCN)$]^{2+}$ model complex,[35,36] for which also an X-ray structure is available,[57] was modeled with *Kestrel*. This analysis established *e* parameters for the axial ligands and the equatorial amines, along with a C/B ratio and a value for Racah B (Table 2). While the analysis largely confirmed the original assignment of spectroscopic bands in the vis-NIR and MCD spectra, it was not possible to recreate with *Kestrel* the assignment of the band at 10 400 cm^{-1} as a spin-allowed d_{xy} to $d_{x^2-y^2}$ transition. Instead, we assign this band to the formally spin-forbidden d_{xy} to d_{z^2} transition, which gains intensity through large triplet character that arises from spin–orbit coupling (see discussion below and ESI†).

The ligand field parameters for the $[(L^1)Fe^{IV}$=O(X)$]^{n+}$ (X = MeCN, Cl^-, Br^-) complexes were then fitted using the assignments of the visible spectrum of $[(L^2)$

Table 2 AOM parameters [cm^{-1}] for $[(L^1)Fe^{IV}{=}O(X)]^+$ (X = MeCN, Cl, Br) and $[(L^2)Fe^{IV}{=}O(MeCN)]^+$.[a]

[cm^{-1}]	L^2/MeCN[b]	L^1/MeCN	L^1/Cl	L^1/Br
e_σ(axial)[c]	9288	9288	9288	9288
e_π(axial)[c]	5584	5584	5584	5584
e_σ(equatorial)[c]	6770	6893	6201	5782
e_π(equatorial)[c]	0	0	0	0
Racah B[d]	500	617	594	588

[a] As obtained from an LFT analysis of the vis-NIR spectra assuming an angular octahedral geometry. [b] Adapted from an LFT analysis of the spectroscopic data of $[(L^2)Fe^{IV}{=}O(NCCH_3)]^{2+}$. [c] The complexes are assumed to have C_{4v} symmetry, 'axial' refers to the Fe–O axis including the ligand *trans* to the oxido group (L^2/MeCN: MeCN; L^1/X: R_3N3); 'equatorial' denotes the average ligand field strength of the in-plane ligands (L^2/MeCN: NR_3; L^1/X: N7, py, py, X). [d] C/B = 5.0.

$Fe^{IV}{=}O(MeCN)]^+$ as a basis. As such, the axial parameters were transferred from $[(L^2)Fe^{IV}{=}O(MeCN)]^+$ and the C/B ratio of 5 was held constant. The final ligand field parameters in Table 2 were then determined by adjusting empirically to give fits to the experimental spectra of ($[(L^1)Fe^{IV}{=}O(X)]^{n+}$ (X = MeCN, Cl^-, Br^-) and $[(L^2)Fe^{IV}{=}O(MeCN)]^{2+}$. The electronic parameters were also checked against preliminary AILFT calculations of the complexes and, importantly, with parameters used for AOM calculations of various other coordination compounds.[58–60] Table 3 summarizes the experimentally observed d–d transitions of $[(L^2)Fe^{IV}{=}O(MeCN)]^{2+}$ and $[(L^1)Fe^{IV}{=}O(X)]^{n+}$ (X = MeCN, Cl^-, Br^-) together with those computed with the parameters of Table 2. As expected, there is excellent agreement between all observed and computed transitions.

Table 3 Experimental and AOM-computed[a] electronic transitions of $[(L^1)Fe^{IV}{=}O(X)]^{n+}$ (X = MeCN, Cl^-, Br^-) and $[(L^2)Fe^{IV}{=}O(MeCN)]^{2+}$ in cm^{-1}

	L^2/MeCN		L^1/MeCN		L^1/Cl		L^1/Br	
Ligands	Exp.	AOM	Exp.	AOM	Exp.	AOM	Exp.	AOM
^{3}H to ^{5}D gap	4500[b]	5436		2438		1069		<1000[d]
$d_{xz},d_{yz} \rightarrow d_{x^2-y^2}$[e]	12 900	12 812	13 004	13 004	12 034	12 034	(11 110)	11 497
$d_{xy} \rightarrow d_{xz},d_{yz}$	10 600	10 488		10 373		10 102		9775
$d_{xy} \rightarrow d_{x^2-y^2}$	*10 400*	*10 472*[c]		18 266		16 326		15 123
$d_{xz},d_{yz} \rightarrow d_{z^2}$[e]	17 600	17 568	19 011	19 011	17 211	17 212	16 474	16 289
$d_{xy} \rightarrow d_{z^2}$	24 900	25 264		25 717		24 479		23 945

[a] Parameters from Table 2, C/B = 5, no SOC. [b] Computed value, see ref. 36. [c] For $[(L^2)Fe^{IV}{=}O(MeCN)]^{2+}$ this is the triplet–quintet transition $d_{xy} \rightarrow d_{z^2}$, for $[(L^1)Fe^{IV}{=}O(X)]^{n+}$ the transition is assigned to $d_{xy} \rightarrow d_{x^2-y^2}$ in the triplet state, see text and ESI. [d] This value is heavily dependent on the C/B ratio. C/B = 5 gives a quintet–triplet spin gap of 1.2 cm^{-1}. A value of C/B > 4.65 places this energy below 1000 cm^{-1}. The other computed energies are assuming C/B = 5. [e] These transitions can mix *via* static correlation (interelectronic repulsion), which means that the transitions cannot be taken as pure electronic states and their relative ordering can be swapped. We have chosen the labeling to be consistent with that in ref. 35 and 36.

The resulting parameter set (Table 2) reveals a decrease in the equatorial bonding strength across the series of bispidine complexes, as expected from the position of the exogenous ligand in the spectrochemical series. A consistent value of Racah B is also seen across the series but one which is higher than the Racah B value for $[(L^2)Fe^{IV}{=}O(MeCN)]^+$, suggesting a comparatively low degree of covalency in the FeL^1 series of complexes.

A particular observation, which is germane to the wider discussion about the reactivity of these complexes, is that the calculated triplet–quintet energy gap decreases across the series and is, as expected, correlated with the weakening of the equatorial field strength. However, the decrease of the quintet–triplet gap going from $[(L^2)Fe^{IV}{=}O(MeCN)]^+$ to $[(L^1)Fe^{IV}{=}O(X)]^+$ is also related to a change in the magnitudes of the interelectronic repulsion parameters Racah B and C. Thus, the triplet–quintet gap can be seen to be a combined factor of equatorial ligand field strength and the overall covalency in the complex.

As described above, it is instructive to re-examine the very low energy transition in $[(L^2)Fe^{IV}{=}O(MeCN)]^{2+}$ at 10 400 cm^{-1}, which was originally assigned to a spin-allowed d_{xy} to $d_{x^2-y^2}$ transition. Our ligand field analysis with *Kestrel* places this transition at a higher energy, thus requiring a different assignment to the 10 400 cm^{-1} transition. We note in this regard that the formally spin-forbidden d_{xy} to d_{z^2} transition energy (upper blue line in Fig. 2) is a close match to the experimental transition energy, and that such a transition can gain significant intensity through spin–orbit coupling of the triplet and quintet states. We thus assign this band as a "triplet to quintet" d_{xy} to d_{z^2} transition. In accord with this assignment, we note that the low energy band in the MCD spectrum of $[(L^2)Fe^{IV}{=}O(MeCN)]^{2+}$ is *z*-polarized as expected for a d_{xy} to d_{z^2} transition, and that $[(L^2)Fe^{IV}{=}O(MeCN)]^{2+}$ is a very sluggish oxidant, suggesting that the triplet–quintet gap in this complex is relatively high, incommensurate with a low ligand field strength that would otherwise be signaled by assigning the 10 400 cm^{-1} band to a d_{xy} to $d_{x^2-y^2}$ transition.

A feature of *Kestrel* is that the variation of d state energies can be plotted as a function of a ligand field variable, akin to a Tanabe–Sugano diagram. Such a variation plot for $[(L^1)Fe^{IV}{=}O(X)]^{n+}$ is shown in Fig. 2, where the average equatorial e_σ(X) value is varied. Qualitatively, the energies agree well with the observed trends in reactivity and also with the computed quintet–triplet energy gaps based on various quantum chemical methods, including coupled cluster calculations.[21] The plot also shows that in the case of $[(L^1)Fe^{IV}{=}O(Br)]^+$ (e_σ = 5782 cm^{-1}) the molecule is near the spin-crossover limit and although its triplet–quintet energy gap is not precisely established, it appears to be small, in line with its high reactivity. Note however that some caution needs to be applied to precisely interpreting the calculated quintet–triplet energy gap, which depends greatly on the value of Racah C.

A further observation from this analysis is that the position of the major d–d transition at 800 nm, 768 nm, 850 nm, 900 nm, for $[(L^2)Fe^{IV}{=}O(MeCN)]^{2+}$, $[(L^1)Fe^{IV}{=}O(MeCN)]^{2+}$, $[(L^1)Fe^{IV}{=}O(Cl)]^+$ and $[(L^1)Fe^{IV}{=}O(Br)]^+$, is not necessarily a useful proxy for the triplet–quintet energy separation (as per Scheme 2). As Fig. 2 shows, the lowest energy band usually observable in the visible spectrum (*i.e.* >10 000 cm^{-1}) is a d_{xy} to $d_{xz/yz}$ transition, the energy of which does not track well the variation of the triplet–quintet gap with e_σ ligand field strength (lower blue line in Fig. 2). On the other hand, the $d_{xz/yz}$ to $d_{x^2-y^2}$ transition, which is almost certainly always observable in experiment, does act as a useful proxy for the triplet–quintet gap, and is one that could usefully be employed to estimate the triplet–quintet gap.

The variation of the transition energies with Racah B is shown in Fig. 3. In the context of the discussion above about the origin of the low energy band (10 400 cm^{-1} for $[(L^2)Fe^{IV}{=}O(MeCN)]^{2+}$), Fig. 3 shows that the second quintet excited state crosses the low energy triplet manifold at lower values of B and thus can mix *via* spin–orbit coupling, thus lending intensity to any "triplet to quintet" d_{xy} to d_{z^2} transition, which then may appear as a very low energy band in the visible spectrum for a complex with low B values (*i.e.* higher degrees of covalency). Indeed, the wider correlation of Racah B with the quintet–triplet energy is clearly evident from Fig. 3, suggesting a strategy for future ligand design that balances covalency and ligand field strengths, to minimize the triplet–quintet gap and thus increase reactivity.

The ligand field analysis also allowed calculation of the zero-field splitting parameter D. For $[(L^2)Fe^{IV}{=}O(MeCN)]^{+}$ this could be compared to experimental data (field Mössbauer: 28 cm^{-1}, HF-EPR: 26.95 cm^{-1}).[61] The value computed with the parameters in Table 2 and a spin–orbit coupling parameter of 415 cm^{-1} is D = 28.1 cm^{-1} (see ESI, Table S11†), in excellent agreement with experimental, lending further confidence in the fitted ligand field parameters. Across the series, with the decrease of the triplet–quintet gap, the computed value of D increases ($[(L^2)Fe^{IV}{=}O(MeCN)]^{+}$: 28.1 cm^{-1}, $[(L^1)Fe^{IV}{=}O(MeCN)]^{2+}$: 36.5 cm^{-1}, $[(L^1)Fe^{IV}{=}O(Cl)]^{+}$: 56.2 cm^{-1}, $[(L^1)Fe^{IV}{=}O(Br)]^{+}$: 73.9 cm^{-1}). These computed D values again have to be treated with caution, as they strongly depend on the values of Racah

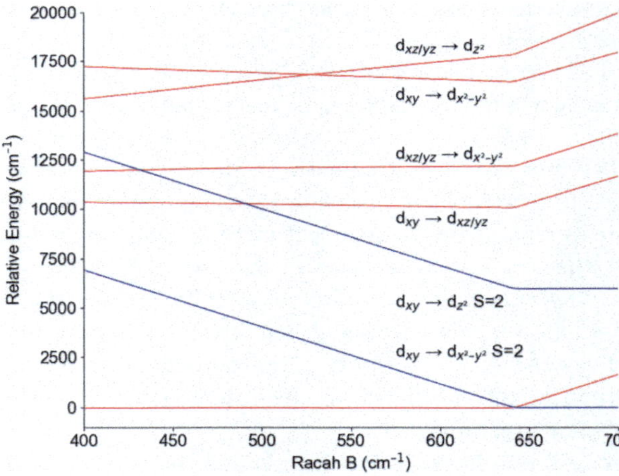

Fig. 3 Ligand field calculated energies of d electron states for octahedral geometry with the transferred axial parameters (see text), as a function of the Racah B parameter. Red lines are triplet d-states, the blue lines are the low-lying quintet excited states. Equatorial e_σ = 6300 and C/B = 5, in the absence of spin–orbit coupling. The triplet $d_{xz/yz} \rightarrow d_{x^2-y^2}$ and $d_{xz/yz} \rightarrow d_{z^2}$ transitions can mix *via* static correlation (interelectronic repulsion), which means that the transitions cannot be taken as pure electronic states and their relative ordering can be swapped. We have chosen the labeling to be consistent with ref. 35 and 36.

parameters and spin–orbit coupling. Notwithstanding this caution however, it appears that within the group of complexes studied here, specifically in the group of the three bispidine complexes, the variation of the electronic parameters should be minimal. The fact that the zero-field splitting increases with decreasing quintet–triplet gap, and that there is a significant increase from the TMC to the bispidine system is expected, and it will be of interest to probe this experimentally. In the case of $[(L^1)Fe^{IV}{=}O(Br)]^+$ the ground state is heavily mixed, such that it has a large quintet component. This is of particular importance with respect to an assumed correlation of the triplet–quintet gap with the oxidation reactivity because the spin–orbit coupling matrix elements are known to be related to the activation barrier.[21,32]

Conclusion

$[(L^1)Fe^{IV}{=}O(Cl)]^+$ as an efficient oxidant for organic substrates, in particular also for the difficult to oxidize substrate cyclohexane with less than 10 kcal mol^{-1} more activated C–H bonds as the "dream substrate" methane. This nonheme model system has been studied in detail in terms of oxidation power (driving force), oxidation efficiency (activation barrier of the rate determining step) and product selectivity,[15,22,23] and has been shown to be the most reactive ferryl oxidant known although it has, in contrast to all enzymes "the wrong spin ground state" with respect to the pathway with the most efficient transition state. The probable reason for this interesting behavior is a very low energy gap between the triplet ground and the quintet excited state, and this has been suspected on the basis of *ab initio* quantum chemical calculations.[21] In the current communication we discuss a ligand field analysis of $[(L^1)Fe^{IV}{=}O(X)]^{n+}$ with X = MeCN, Cl^-, Br^- as in-plane ligand with varying donor strength *cis* to the oxido group. The ligand field analysis indicates that with decreasing ligand field strength (MeCN > Cl > Br) there is a decrease of the triplet–quintet gap concomitant with a shift of the main d–d transition towards lower energy, an increase of the zero-field splitting and an increase of the oxidation reactivity. The present analysis is an excellent qualitative and semi-quantitative basis for a quantitative analysis that most importantly should involve additional spectroscopic techniques and temperature-dependent kinetic analyses, which both are not trivial with an $Fe^{IV}{=}O$ complex that decays already in the absence of external substrates in seconds at −90 °C, as well as quantum chemical studies including CASSCF/NEVPT2 optimized wavefunctions and AILFT, which also have limits – not least because the generally DFT-optimized structure might not be in the minimum of the *ab initio* wavefunction.

Experimental section

Details of the computational methods

All ligand field calculations were carried out with the development version of the *Kestrel* program, a ligand field program for d-block transition metal compounds. All calculations used a complete d^4 Russell–Saunders basis set. The AOM parameters for $[(L^2)Fe^{IV}{=}O(MeCN)]^{2+}$ were obtained by a variation of the equatorial amine e_σ parameters and the axial e_σ and e_π parameters, Racah B and C (see ESI† for more information). The zero-field splitting was recreated by variation of the spin–orbit coupling parameter between 0 and the free-ion value of 515 cm^{-1}. The variation diagrams were generated from Kestrel's calculations.

Conflict of interest

The authors declare no conflict of interest.

Acknowledgements

Financial support from the University of Heidelberg and the German Science Foundation (DFG) is gratefully acknowledged. This study was conducted within the Max Planck School Matter to Life, supported by the BMBF in collaboration with the Max Planck Society. We are grateful for computational resources provided by the bwForCluster JUSTUS, funded by the Ministry of Science, Research and Arts and the Universities of the State of Baden-Württemberg, Germany, within the framework program bwHPC-C5.

References

1 J.-E. Bäckvall, *Modern Oxidation Methods*, Wiley-VCH, 2004.
2 L. Vicens, G. Olivo and M. Costas, *ACS Catal.*, 2020, **10**, 8611–8631.
3 S. Rana, J. P. Biswas, S. Paul, A. Paik and D. Maiti, *Chem. Soc. Rev.*, 2021, **50**, 243–472.
4 C. Krebs, D. Galonici Fujimori, C. T. Walsh and J. M. Bollinger Jr, *Acc. Chem. Res.*, 2007, **40**, 484–492.
5 E. I. Solomon, T. C. Brunold, M. I. Davis, J. N. Kensley, S.-K. Lee, N. Lehnert, F. Neese, A. J. Skulan, Y.-S. Yang and J. Zhou, *Chem. Rev.*, 2000, **100**, 235.
6 M. M. Abu-Omar, A. Loaiza and N. Hontzeas, *Chem. Rev.*, 2005, **105**, 2227–2252.
7 A. R. McDonald and L. Que Jr, *Coord. Chem. Rev.*, 2013, **257**(2), 414–428.
8 W. Nam, *Acc. Chem. Res.*, 2015, **48**, 2415–2423.
9 W. Nam, Y.-M. Lee and S. Fukuzumi, *Acc. Chem. Res.*, 2014, **47**, 1146–1154.
10 P. Comba, M. Kerscher, M. Krause and H. F. Schöler, *Environ. Chem.*, 2015, **12**, 381–395.
11 X. Engelmann, I. Monte-Perez and K. Ray, *Angew. Chem., Int. Ed.*, 2016, **55**, 7632–7649.
12 J. M. Mayer, *Acc. Chem. Res.*, 2011, **44**, 36–46.
13 D. Wang, K. Ray, M. J. Collins, E. R. Farquhar, J. R. Frisch, L. Gomez, T. A. Jackson, M. Kerscher, A. Waleska, P. Comba, M. Costas, E. Münck and L. Que Jr, *Chem. Sci.*, 2013, **4**, 282–291.
14 P. Comba, H. Wadepohl and A. Waleska, *Aust. J. Chem.*, 2014, **67**, 398–404, Heron 6 Special Issue.
15 P. Comba, S. Fukuzumi, C. Koke, A. M. Löhr and J. Straub, *Angew. Chem., Int. Ed.*, 2016, **55**, 11129–11133.
16 P. Comba, A.-M. Löhr, F. Pfaff and K. Ray, *Isr. J. Chem.*, 2020, **60**, 957–962, Metal-oxido Special Issue.
17 P. Comba, D. Faltermeier and B. Martin, *Z. Anorg. Allg. Chem.*, 2020, **646**, 1839–1845, Special Issue for H. W. Roesky.

18 J. England, Y. Guo, K. M. V. Heuvelen, M. A. Cranswick, G. T. Rohde, E. L. Bominaar, E. Münck and L. Que Jr, *J. Am. Chem. Soc.*, 2011, **133**, 11880–11883.
19 A. N. Biswas, M. Puri, K. K. Meier, W. N. Oloo, G. T. Rohde, E. L. Bominaar, E. Münck and L. Que Jr, *J. Am. Chem. Soc.*, 2015, **137**, 2428–2431.
20 C. Kupper, B. Mondal, J. Serrano-Plana, I. Klawitter, F. Neese, M. Costas, S. Ye and F. Meyer, *J. Am. Chem. Soc.*, 2017, **139**, 8939–8949.
21 P. Comba, D. Faltermeier, S. Krieg, B. Martin and G. Rajaraman, *Dalton*, 2020, **49**, 2888–2894.
22 M. Abu-Odeh, K. Bleher, N. J. Britto, P. Comba, M. Gast, M. Jaccob, M. Kerscher, S. Krieg and M. Kurth, *Chem.–Eur. J.*, 2021, **27**, 11377–11390.
23 K. Bleher, P. Comba, D. Faltermeier, A. Gupta, M. Kerscher, S. Krieg, B. Martin, G. Velmurugan and S. Yang, *Chem.–Eur. J.*, 2022, **28**, e202103452.
24 S. Ye and F. Neese, *Curr. Opin. Chem. Biol.*, 2009, **12**, 89–98.
25 C. Geng, S. Ye and F. Neese, *Angew. Chem., Int. Ed.*, 2010, **49**, 5717–5720.
26 L. Roy, *ChemPlusChem*, 2019, **84**, 893–906.
27 D. Mandal, D. Mallick and S. Shaik, *Acc. Chem. Res.*, 2018, **51**, 107–117.
28 S. Shaik, D. Danovich, A. Fiedler, D. Schröder and H. Schwarz, *Helv. Chim. Acta*, 1995, **78**, 1393–1407.
29 D. Schröder, S. Shaik and H. Schwarz, *Acc. Chem. Res.*, 2000, **33**, 139–145.
30 D. Usharani, D. Janardanan, C. Li and S. Shaik, *Acc. Chem. Res.*, 2012, **46**, 471–482.
31 S. Shaik, H. Hirao and D. Kumar, *Acc. Chem. Res.*, 2007, **40**, 532–542.
32 D. Danovich and S. Shaik, *J. Am. Chem. Soc.*, 1997, **119**, 1773–1789.
33 J. Harvey, *Phys. Chem. Chem. Phys.*, 2007, **9**, 331–349.
34 H. Hirao, L. Que Jr, W. Nam and S. Shaik, *Chem.–Eur. J.*, 2008, **14**, 1740–1756.
35 A. Decker, J.-U. Rohde, E. J. Klinker, S. D. Wong, L. Que Jr and E. I. Solomon, *J. Am. Chem. Soc.*, 2007, **129**, 15983–15996.
36 A. Decker, J. U. Rohde, L. Que Jr and E. I. Solomon, *J. Am. Chem. Soc.*, 2004, **126**, 5378–5379.
37 M. R. Bukowski, P. Comba, A. Lienke, C. Limberg, C. Lopez de Laorden, R. Mas-Balleste, M. Merz and L. Que Jr, *Angew. Chem., Int. Ed.*, 2006, **45**, 3446.
38 G. Mukherjee, C. W. Z. Lee, S. S. Nag, A. Alili, F. G. Cantu Reinhard, D. Kumar, C. V. Sastri and S. P. de Visser, *Dalton Trans.*, 2018, **47**, 14945–14957.
39 I. M. Perez, X. Engelmann, Y.-M. Lee, M. Yoo, E. Kumaran, E. R. Farquhar, E. Bill, J. England, W. Nam, M. Swart and K. Ray, *Angew. Chem., Int. Ed.*, 2017, **56**, 14384–14388.
40 J. Deutscher, P. Gerschel, K. Warm, U. Kuhlmann, S. Mebs, M. Haumann, H. Dau, P. Hildebrandt, U.-P. Apfel and K. Ray, *Chem. Commun.*, 2021, **57**, 2947–2950.
41 M. Puri, A. N. Biswas, R. Fan, Y. Guo and L. Que, *J. Am. Chem. Soc.*, 2016, **138**, 2484–2487.
42 M. S. Seo, N. H. Kim, K.-B. Cho, J. E. So, S. K. Park, M. Clémancey, R. Garcia-Serres, J.-M. Latour, S. Shaik and W. Nam, *Chem. Sci.*, 2011, **2**, 1039–1045.
43 D. Wang, E. R. Farquhar, A. Stubna, E. Münck and L. Que Jr, *J. Am. Chem. Soc.*, 2009, **1**, 145–150.
44 J. C. Price, E. W. Barr, L. M. Hoffart, C. Krebs and J. M. Bollinger Jr, *Biochem*, 2005, **44**, 8138–8147.

45 A. M. Valentine, S. S. Stahl and S. J. Lippard, *J. Am. Chem. Soc.*, 1999, **121**, 3876–3887.

46 B. J. Brazeau and J. D. Lipscomb, *Biochem*, 2000, **39**, 13503–13515.

47 T. H. Yosca, J. Rittle, C. M. Krest, E. L. Onderko, A. Silakov, J. C. Calixto, R. K. Behan and M. T. Green, *Science*, 2013, **342**, 825–829.

48 K. Mittra and M. T. Green, *J. Am. Chem. Soc.*, 2019, **141**, 5504–5510.

49 C. V. Sastri, J. Lee, K. Oh, Y. J. Lee, T. A. Jackson, K. Ray, H. Hirao, W. Shin and J. A. Halfen, *Proc. Natl. Acad. Sci. U. S. A.*, 2007, **104**(49), 19181–19186.

50 R. Kumar, A. Ansari and G. Rajaraman, *Chem.–Eur. J.*, 2018, **24**, 6818–6827.

51 A. Anastasi, P. Comba, J. McGrady, A. Lienke and H. Rohwer, *Inorg. Chem.*, 2007, **46**, 6420–6426.

52 J. Bautz, P. Comba, C. Lopez de Laorden, M. Menzel and G. Rajaraman, *Angew. Chem., Int. Ed.*, 2007, **46**, 8067.

53 P. Comba, M. Kerscher, M. Merz, V. Müller, H. Pritzkow, R. Remenyi, W. Schiek and Y. Xiong, *Chem.–Eur. J.*, 2002, **8**, 5750–5760.

54 C. Bleiholder, H. Börzel, P. Comba, R. Ferrari, A. Heydt, M. Kerscher, S. Kuwata, G. Laurenczy, G. A. Lawrance, A. Lienke, B. Martin, M. Merz, B. Nuber and H. Pritzkow, *Inorg. Chem.*, 2005, **44**, 8145–8155.

55 P. Comba, M. Kerscher and W. Schiek, *Prog. Inorg. Chem.*, 2007, **55**, 613–704.

56 G. Nunn, M. A. Bates, M. C. R. Cockett and P. H. Walton, manuscript in preparation.

57 J.-U. Rohde, J.-H. In, M. H. Lim, W. W. Brennessel, M. R. Bukowski, A. Stubna, E. Münck, W. Nam and L. Que Jr, *Science*, 2003, **299**, 1037–1039.

58 H. Weihe and H. U. Güdel, *J. Am. Chem. Soc.*, 1998, **120**, 2870–2879.

59 P. Comba, *Coord. Chem. Rev.*, 1999, **182**, 343–371.

60 M. Atanasov, P. Comba, C. A. Daul and F. Neese, in *Models, Mystery and Magic of Molecules*, ed. J. C. A. Boeyens and J. Ogilvie, Springer, 2008, p. 411.

61 J. Krzystek, J. England, K. Ray, A. Ozarowski, D. Smirnov, L. Que Jr and J. Telser, *Inorg. Chem.*, 2008, **47**, 3483–3485.

Faraday Discussions

ROYAL SOCIETY OF CHEMISTRY

DISCUSSIONS

Electron transfer, spectroscopy and theory: general discussion

Anjali Anilkumar, Philip Ash, Akhil R. Chakravarty, Peter Comba, Serena DeBeer, Abhishek Dey, Apparao Draksharapu, Debabrata Goswami, Shinobu Itoh, Kenneth Karlin, K. V. Lakshmi, Shyamalava Mazumdar, Dimitrios Pantazis, David Parker, Lawrence Que, Gopalan Rajaraman, Sankar Prasad Rath, Chivukula Sastri, Sayam Sen Gupta and Edward I. Solomon

DOI: 10.1039/d2fd90013k

Shyamalava Mazumdar opened the discussion of the paper by Abhishek Dey: Thank you for the excellent talk. You talked about the possibilities of different mechanisms in model systems. What is the mechanism in the case of real proteins, such as peroxidase or oxygenases? How does the asymmetric protein environment with very well defined channels *etc.* affect the pathways?

Abhishek Dey answered: The whole point of including the second sphere residues is not only to mimic the function of the enzymes but also to mimic the mechanism of the enzyme. We have reported iron porphyrins with pendant pyridine and amine groups which form compound **I** from peroxides at rates faster than that of HRPs (see ref. 1). Accordingly, these were found to be functional models of HRP and in some cases show a ping-pong mechanism with comparable K_m and K_{cal} (see ref. 2). Apart from the second sphere residues, hydrophobic residues installed in the distal site mimic the selectivity of monoxygenases. Preferential oxidation of primary C–H over benzylic or tertiary C–H to alcohol without any overoxidation to ketones is reported in ref. 3.

1 S. Bhunia, A. Ranu, S. G. Dey, A. Ivancich and A. Dey, *Chem. Sci.*, 2020, **11**(10), 2681–2695.
2 S. Bhakta, A. Nayek, B. Roy and A. Dey, *Inorg. Chem.*, 2019, **58**(5), 2954–2964.
3 M. Mukherjee and A. Dey, *ACS Cent. Sci.*, 2019, **5**(4), 671–682.

Chivukula Sastri asked: By comparing porphyrin and thiomethyl porphyrin, we found that these are more reactive compared to *ortho*-methyl, would you expect something similar? If I compare F 20 porphyrin with octa methyl porphyrin, it is known that F20 is more reactive but less efficient and the opposite is true for octa methyl. So, for O_2 reduction, would you expect the same trend?

Abhishek Dey replied: Yes, we might.

Sayam Sen Gupta commented: When you described the formation of iron(III) superoxo, the EPR spectra were fitted to two species, one low spin Fe(III) and one high spin Fe(III). What is the origin of the high spin Fe(III) species?

Abhishek Dey replied: It is the ferric porphyrin formed after the auto-oxidation of ferrous porphyrin.

Sayam Sen Gupta asked: Can you generate the Fe(III) superoxo by starting from the Fe(III) complex and KO_2?

Abhishek Dey answered: Yes, you can. You have to add an aza crown to stabilize the K^+.

Akhil R. Chakravarty queried: Is histidine or free amine binding involved?

Abhishek Dey replied: No. The crystal structure shows that the amine is not bound to the metal due to its distance from the porphyrin ring.

Chivukula Sastri questioned: Does the electron potential tell you about the rate or efficiency?

Abhishek Dey responded: Not really. The rate is controlled by the one of the several $4e/4H^+$ steps involved in O_2 reduction. This depends strongly on the electronic structure of the iron porphyrin. The selectivity, on the other hand, is mostly derived from the stability of the Fe(III)–OOH intermediate from hydrolysis.

Philip Ash commented: The Fe(III/II) reduction potential seems to be a key factor in determining the rate determining step of these reactions. This is in many ways analogous to natural heme protein systems, where axial ligation and secondary coordination sphere effects tune the Fe(III/II) potential dramatically and, hence, affect the reactivity. Have you studied the effect of axial ligation directly using electrode modification strategies to immobilise heme to electrode surfaces? Does the variation in the Fe(III/II) potential, and the rate determining step for ORR, follow the same trends as those you have reported here?

Abhishek Dey responded: Yes, we have. I would recommend that you see the latest reviews we have published on the subject in ref. 1 and 2.

1 S. Dey, B. Mondal, S. Chatterjee, A. Ranu, S. Amanullah and A. Dey, *Nat. Rev. Chem.*, 2017, **1**, 0098.
2 M. Mukherjee and A. Dey, *JACS Au.*, 2021, **1**(9), 1296–1311.

Philip Ash said: Steady-state kinetic isotope effects can help in identifying a rate determining step, but in terms of the microscopic mechanism the detail can be limited. Especially for systems where there is a nearby pendant base, such as those in Scheme 1 of the paper (DOI: 10.1039/d1fd00076d), is it possible that a transient-state kinetic isotope effect is more diagnostic in terms of distinguishing between sequential and concerted proton-coupled electron transfer mechanisms, such as pathway-I, pathway-II, and PCET in Scheme 2 of the paper?

Abhishek Dey replied: You are absolutely right. Such information will be crucial for a more critical evaluation of the roles played by these second sphere residues. The nature of the chemical steps are different along the ORR reaction co-ordinate and it is only natural that the second sphere residues will affect them differently.

Gopalan Rajaraman remarked: The gold surface is known to be non-innocent towards the bound substrates. We have seen that oxidation/reduction is very common. Do you see any such surface interference in the catalysis? Also, molecules deposited on gold are known to interact *via* surface (exchange coupling *via* the surface between two or more molecules), is there such a possibility here? If so, what is your take on the reactivity keeping this in mind?

Abhishek Dey responded: The Au surface is covered with SAM, which suppresses the catalytic activity of Au. These molecules are far from the Au surface due to the octanethiol SAM. This removes any interaction involving the substrate.

Shyamalava Mazumdar opened the discussion of the paper by David Parker: Thank you for the excellent talk. You have done both steady state and time-resolved experiments – do you see any major differences in the values of the FRET efficiency with the two techniques? It is known that there are interferences in steady-state fluorescence techniques, which may be avoided using time-resolved methods. What is the value of the orientation factor (kappa squared) used in the analyses of the FRET results and how do you justify it? Is there any structural evidence for a random orientation or there is certain restricted motion of the chromophore(s)?

David Parker replied: The orientation factor (kappa squared) used in the FRET calculations was two thirds, in common with the traditional estimate used. Such an assumption accords with the fact that both the Eu donor and the acceptor are not constrained in their motion, and so can be assumed to adopt random orientational motions as if freely diffusing. We saw no particular differences in the FRET efficiency as the time gate was varied.

Debabrata Goswami asked: Has there been any work to quantify the ultrafast timescales of the lanthanide ion relaxation process? I understand that you have not gone into the ultrafast domain – do you have any evidence, besides FRET circumstantial evidence, of extremely fast transfer?

David Parker said: The lanthanide excited state in these Eu complexes has a lifetime of the order of a millisecond normally. It is subject to non-radiative deactivation by vibrational quenching by electron transfer and by energy transfer. We estimated that the second order rate constant that defines the energy transfer process in the case of efficient eT (*e.g.* to near-IR cyanine dyes) was of the order of 10^9, whereas when the spectral overlap integral was much poorer, (*e.g.* with Mn or Cu^{2+} ions, this value decreased by a factor of a thousand. There has been very little work using modern 'ultra-fast' techniques to monitor any of these processes, and I would say that there is considerable scope for new experimental studies to be undertaken that may allow a deeper understanding of these competing non-radiative deactivation pathways.

Gopalan Rajaraman enquired: Did you look at any other lanthanides, other than Eu? How important is the $J = 0$ ground state of Eu(III) in the observance of the luminescence?

David Parker responded: The work described (DOI: 10.1039/d1fd00059d) has focused on Eu complexes, for good reason. First, they have the best defined emission spectral fingerprints arising from the absence of degeneracy of the 5-D-0 ($J = 0$) emissive excited state. Transitions occur to the ground state 7-F-*n* manifest ($n = J = 0$ to 6 are visible) in the red region of the visible spectrum, *i.e.* 580 to 750 nm. Second, they are less prone to vibrational quenching of the excited state by energy-matched XH oscillators (X = O, N or C), so that the quenching processes by electron or energy transfer can be studied with more confidence.

We and others have examined terbium and Dy analogues (see ref. 1 and 2) and the near-IR emitting systems of Yb and Nd can also be studied, although they have much lower solution quantum yields and emission lifetimes that are in the microsecond rather than the millisecond regime.

1 J. W. Walton, R. Carr, N. H. Evans, A. M. Funk, A. M. Kenwright, D. Parker, D. S. Yufit, M. Botta, S. De Pinto and K.-L. Wong, *Inorg. Chem.*, 2012, **51**, 8042–8056.
2 D. Parker, J. D. Fradgley and K.-L. Wong, *Chem. Soc. Rev.*, 2021, **50**, 8193–8213.

Anjali Anilkumar asked: Have you done any experiments with SNAP tags and cancer cell lines?

David Parker replied: We have not done such work, but many others have. The SNAP TAG methodology was introduced by Horst Vogel in 2005 (ref. 1) and nowadays such kits are available commercially, *e.g.* New England Biolabs.

1 N. George, H. Pick, H. Vogel, N. Johnsson and K. Johnsson, *J. Am. Chem. Soc.*, 2004, **126**, 8896–8897.

Anjali Anilkumar asked: You mentioned about directing to different organelles, how was it directed?

David Parker replied: The luminescent Eu complexes discussed in the paper (DOI: 10.1039/d1fd00059d) are able to enter cells when conjugated to a targeting vector, such as a small peptide. Depending on the nature of the peptide, the conjugate localises in different organelles, such as the ER or the Golgi apparatus. Examples of this approach have been published by many others and by ourselves, *e.g.* ref. 1. In addition, the parent Eu complexes themselves have a tendency to localise to different organelles, in a manner which is a function of their structure (primarily their aromatic/heterocyclic chromophores) and charge and their chirality. For a striking example of the role of the chirality of the Eu probe, see ref. 2.

1 M. Starck, J. D. Fradgley, S. Di Vita, J. A. Mosely, R. Pal and D. Parker, *Bioconjugate Chem.*, 2020, **31**, 229–240.
2 P. Stachelek, L. Mackenzie, D. Parker and R. Pal, *Nat. Commun.*, 2022, **13**, 553–560.

Anjali Anilkumar remarked: What kinds of cell lines have you used for these endosomal experiments?

David Parker replied: The work discussed in the paper (DOI: 10.1039/d1fd00059d) used mouse skin fibroblast cells (NIH-3T3). Similar behaviour was observed with CHO cells (Chinese hamster ovarian) and with MCF7 (breast cancer) cells.

Anjali Anilkumar asked: I've heard about cancer cell lines having a lower pH. SNAP tags are pH sensitive – can they be used for targeting cancer cells?

David Parker replied: Different organelles in all cell types have different mean pH values, varying from the acidic nature of lysosomes (pH 4.5) to the more basic pH associated with, for example, the inner compartments of mitochondria. Our pH sensitive Eu probes are designed to be used to monitor receptor internalisation into endosomes, and the lowering of pH that occurs as endosomes age and slowly develop and mature into lysosomes. In normal cells, the intracellular pH is lower than the extracellular pH (pHe), with pHi and pHe values typically in the ranges of 7.0–7.2 and 7.3–7.4, respectively. Cancer cells are generally associated with slightly higher values of pHi (7.12–7.65) and a lower pHe value (6.4–6.9), although the extent of the small reduction in extracellular pH is very dependent on the system being examined. It is this slight acidification in pH that has led some workers to pursue pH dependent imaging agents to map pH *in vivo*, thereby aiding the identification of tumour tissue in principle.

Shinobu Itoh opened the discussion of the paper by Gopalan Rajaraman: Can you estimate the reactivity difference between the oxo and oxyl in the same ligand environment? We synthetic chemists are always interested in the activity and also the physical chemistry properties of mononuclear species. The calculations you have done are mostly on the spin state, but I am interested in the reactivity difference between the oxo and oxyl.

Gopalan Rajaraman responded: Although in this manuscript (DOI: 10.1039/d1fd00072a), we have not performed calculations on the reactivity of oxo *vs.* oxyl species, we agree that this is of interest. It is anticipated that during the reactivity, as the substrate approaches the species at the transition state, an electron transfer or a proton transfer followed by the electron transfer process is expected to occur and, at this stage, we believe that the dominant species is likely to be $Co^{III}–O^{\bullet}$ rather than a $Co^{IV}{=}O$ one. How the variation in the % of oxyl *vs.* oxo character alters the reactivity is something that we are exploring at the moment, and we hope to come up with some answers in line with your interests soon.

Shinobu Itoh enquired: This has been calculated for the Co system – what about Ni or Cu? What happens when you change Co to Ni or Cu?

Gopalan Rajaraman answered: We have not performed these calculations at the moment. However, we believe that the oxyl character will increase if we move to Ni and Cu, starting even with a +3 oxidation state. We have studied the formation of Ni(III)–oxo species in a trigonal bipyramidal ligand architecture earlier and observed a significant oxyl radical character on the oxygen in any spin state (see ref. 1). Therefore, Ni(III)–oxo is not true Ni(III)=O but exists in the Ni(II)–O˙ form. Although the observation was based completely on a single reference DFT study, a closer look using CASSCF/NEVPT2 will be more helpful in this aspect.

1 B. Pandey, K. Ray and G. Rajaraman, *Z. Anorg. Allg. Chem.*, 2018, **644**, 790–800.

Shinobu Itoh asked: If you go to group 10 or 11, maybe there will be oxyl character?

Gopalan Rajaraman replied: Based on our previous observation on Co and Ni systems, we believe that a significant oxyl character is expected for groups 10 and 11, even with a lower oxidation state, such as +3.

Abhishek Dey said: Spectroscopy can be highly contentious in Co(IV)=O. If you reconsider the data, maybe it can rationalized by quantum mixing of states rather than being one or the other?

Gopalan Rajaraman answered: We agree that this is a good proposal which is certainly of interest to us. We will explore the possibility of looking at various spectral features that could provide concrete evidence for such an admixed state. Among all, we believe that EPR *g*-tensors could be very helpful; however, only limited experimental data is available on this front. Vibrational frequencies are derived from single reference DFT methods, and though they offer some qualitative understanding of the nature of the species that lies at the extremes, they can not be taken at their face value for the admixed state. We are currently looking at various other possibilities.

Peter Comba remarked: These are very interesting observations. The interesting question, of course, is how the reactivity changes along the metal(IV)–oxo metal(III)–oxyl continuum. I realize that this is not the focus of your paper (DOI: 10.1039/d1fd00072a) but wonder whether you have some ideas or results to suggest a possible answer to this question. A related question of interest is how this translates to iron(IV)–oxo (and manganese(IV)–oxo). Obviously, there is an additional complication with spin states, but my question would be how the reactivity can be modulated by increasing/decreasing the spin density on the oxygen.

Gopalan Rajaraman replied: We agree that this is an interesting aspect that we have not focused on in this paper (DOI: 10.1039/d1fd00072a). Many examples that we have studied previously indicate that a strong oxyl radical character is developed at the transition state due to the transfer (or partial transfer) of an electron from the substrate to the catalyst. However, this is strongly dependent on the ligand framework. Whether the catalyst undergoes valence tautomerism to oxyl before reactivity or the oxo itself activates the C–H bond is an intriguing question that has not been answered. As you have stated, iron(IV)–oxo and Mn(IV)–oxo are also of interest, but our calculations suggest that in the ligand architecture studied, these are mostly oxo at their equilibrium geometry with a very less or negligible oxyl character. We believe that increasing the radical character on oxygen is likely to enhance the reactivity. This could, in principle, be carried out also using external stimuli such as an oriented electric field – an area that is quickly developing at the moment.

Dimitrios Pantazis asked: You performed state-averaged CASSCF calculations, so that the orbitals are optimized simultaneously for different states and electron configurations. This means that there is a common set of orbitals describing all states, as opposed to a state-specific orbital optimization. I wonder whether in the present case, this approach may directly affect the predicted relative energies and electromer ratios, perhaps artificially favoring higher-energy states?

Gopalan Rajaraman responded: We have performed several sets of calculations to assess and analyse the variation of % of electromeric species computed using the CASSCF method. In particular, we have performed CAS(11,8) using 11 electrons in five metal d-orbitals and three oxo p-orbitals, including two oxo-p_π (π_x and π_y) orbitals and one oxo-p_σ (σ_z) orbital. Furthermore, we have also extended the reference space to the ligated nitrogen-orbitals, N_{axial} and $N_{equatorial}$ – a CAS(15,10). In all these cases where even the reference space is varied, which is also expected to influence the % of electromeric species, the variation was found to be minimal (±3%). With the reference space utilised for species 1 to 3, the CASSCF calculations are able to correctly reproduce $S = 3/2$ as the ground state for 1 and $S = 1/2$ as the ground state for 2 and 3. This is in agreement with experiments. This suggests that the energy ordering is maintained in the method employed. Furthermore, when such calculations were performed in $Fe^{IV}{=}O$ and $Mn^{IV}{=}O$ species (species 6 and 5), we see only very little mixing (5–8%). For $Mn^{IV}{=}O$ species, this is also in agreement with the experiments performed with slightly different ligand architecture. This clearly suggests that, despite the limitations, these SA-CASSCF calculations yield the % electromeric species correctly. However, a note of caution is due, which has now been added to the paper (DOI: 10.1039/d1fd00072a)

Shyamalava Mazumdar said: Thank you for the excellent talk. Do the oxo- and oxyl-forms exist in equilibrium or in some type of quantum mechanical mixture at the ground state? How is the temperature dependence expected?

Gopalan Rajaraman replied: Our results suggest that this is a quantum mechanical mixture of two tautomeric valence structures contributing a different per cent to the overall ground state wavefunction. A parallel to this proposal can be drawn from the spin admixed state proposal of Maltempo in heme proteins (see ref. 1.) As this is a quantum mechanical mixture, we do not expect a temperature dependency. However, due to such mixing, the *g*-anisotropies are expected to be different from the oxo and oxyl counterparts, which could be utilized to probe such a mixture.

1 M. M. Maltempo and T. H. Moss, *Q. Rev. Biophys.*, 1976, **9**, 181–215.

Sankar Prasad Rath asked: Has there been any experimental evidence for the mixing of M(IV)O with M(III)–O·? Has the same mixing also been observed in the case of iron?

Gopalan Rajaraman replied: Our calculations indicate very small or negligible mixing of these two species for both Mn as well as iron(IV)–oxo analogues. Earlier experimental studies performed on different ligand architectures with Mn(II/IV/V)=O species by Borovik and co-workers suggest that Mn–O• is absent (see ref. 1) and this supports our data. However, we would like to note here that ligand architecture plays a crucial role in this valence tautomerism and without explicit calculations, it is not always possible to extend the interpretation to other geometries.

1 R. Gupta, T. Taguchi, B. Lassalle-Kaiser, E. L. Bominaar, J. Yano, M. P. Hendrich and A. Borovik, *Proc. Natl. Acad. Sci. U. S. A.*, 2015, **112**, 5319–5324.

Abhishek Dey asked: If your ground-state wavefunction is mixed not because you are actively mixing low lying excited states but quantum mixed, would your spectroscopy show this? *G* values of EPR depend on the quantum mixing – can this be used to quantify it in this case?

Gopalan Rajaraman answered: Yes, it is possible. We believe that spectral characteristics such as EPR could be very helpful. The mixing is expected to yield a pseudo doublet state and, if this is isolated enough from other excited states, it will give *g*-values that are different to the individual contributing configurations/states. We have looked at such states very recently for a square planar Fe(III) complex[1] and would be happy to look at this for Co(IV)=oxo/oxyl species, if available.

1 K. Uddin Ansari, P. Kumar, L. Spillecke, P. Shukla, S. Tripathi, S. Dey, R. Klingeler, G. Blondin, G. Rajaraman and M. Shanmugam, manuscript submitted.

Abhishek Dey said: The correlation is very interesting – did you check for any correlation for oxos in the trigonal ligand field? These have different mixing between the d-orbitals involved in binding oxo and the auxiliary ligand.

Also, you mentioned that you are going to be able to predict the optical/vibrational spectrum of Co(III) oxyl or oxo, but with the ground state being a mixture, how will you do that? Those are all excited states.

Gopalan Rajaraman responded: We have not looked at any other ligand field other than octahedral and square pyramidal. The trigonal bipyramidal is certainly interesting, and we plan to explore this in the future. The vibrational features were already calculated based on the DFT structures, and our calculations yield 886 cm^{-1}, 621 cm^{-1} and 615 cm^{-1} for complexes 1–3, respectively. Although rare, the Co–O bond stretching frequency for a true Co^{IV}=O is reported to be 840 cm^{-1}. This reported data is not for a coordination complex but for a cobalt oxide nanoparticle observed during the water oxidation process (see ref. 1). If we compare this value to the computed one, we suggest that complex 1 is closer to an ideal Co^{IV}=O while species 2 and 3 are not. At the moment, we have not calculated the optical features, though this is readily available in our CASSCF calculations. The vibrational features of the mixed states are not straightforward, as this is coming from DFT. There are new developments, including multiconfiguration pair-density functional theory (MC-PDFT) which could be helpful, but we have not looked into this aspect yet.[2]

1 M. Zhang, M. De Respinis and H. Frei, *Nat. Chem.*, 2014, **6**, 362–367.
2 C. A. Gaggioli, S. J. Stoneburner, C. J. Cramer and L. Gagliardi, *ACS Catal.*, 2019, **9**(9), 8481–8502.

Dimitrios Pantazis opened a general discussion of the paper by K. V. Lakshmi: Congratulations on your work (DOI: 10.1039/d1fd00094b), which showcases yet again the central role of magnetic resonance spectroscopies in OEC research. As you mentioned, in my group, we have modelled the diverse channel architecture between cyanobacterial and higher plant PSII, described the N87A difference and anticipated the change in permeability to methanol upon this mutation,[1] so it is fantastic that our predictions have received this experimental validation. I note that we recently used this knowledge to design methanol inhibition experiments that allowed us to arrest water binding in the S3 state of spinach PSII.[2] As you also noted, we have common computational models of methanol binding that match the ^{13}C hyperfine parameters of those you present here,[3] so I am convinced that the structural interpretation you suggest of methoxy binding to Mn^4 is the most plausible scenario. I wonder though how we should interpret the two orders of magnitude difference between the weak dipolar hyperfine interaction reported by Oyala *et al.* for spinach PSII[4] and your results? Is it a genuine difference in binding in the case of the cyanobacterial N87A mutant, which would imply that the local mutation is insufficient to render the cyanobacterial OEC site "plant-like"? Or is the mutant structurally similar locally to spinach but there are other issues related to the methodology, sample preparation, or experimental conditions?

1 M. Retegan and D. A. Pantazis, *J. Am. Chem. Soc.*, 2017, **139**, 14340–14343.
2 G. Zahariou, N. Ioannidis, Y. Sanakis and D. A. Pantazis, *Angew. Chem., Int. Ed.*, 2021, **60**, 3156–3162.
3 M. Retegan and D. A. Pantazis, *Chem. Sci.*, 2016, **7**, 6463–6476.
4 P. H. Oyala, T. A. Sitch, J. A. Stull, F. Yu, V. L. Pecoraro and R. D. Britt, *Biochemistry*, 2014, **53**, 7914–7928.

K. V. Lakshmi answered: Continuous-wave and pulse EPR spectroscopy have provided valuable insights into the geometric and electronic structure of the higher oxidation state intermediates of the oxygen-evolving complex (OEC) of photosystem II (PSII). In the present study on the binding of methanol to the S2 state of the OEC of D1-N87A PSII (DOI: 10.1039/d1fd00094b), the ^{13}C hyperfine coupling constants calculated from the QM/MM model are in agreement with the experimental couplings determined by 2D ^{13}C HYSCORE spectroscopy, as well as previous computational couplings published by your research group.[1] Hence, the replacement of the water-derived ligand, W2, at the Mn^4 ion by a methoxy group is the most plausible model for methanol binding in the S2 state of the OEC of PSII. Regarding the difference between spinach and D1-N87A PSII, as you mention, the ^{13}C hyperfine coupling parameters for methanol binding in the S2 state of spinach PSII determined by Oyala *et al.*[2] were much weaker than the ^{13}C hyperfine couplings constants of methanol binding to the S2 state of cyanobacterial D1-N87A PSII. The pulse EPR methods, sample preparation and experimental conditions are similar in both studies. There is a genuine difference in the binding mode in D1-N87A PSII in comparison with spinach PSII. As interpreted by both Oyala *et al.* and your research group, methanol binds in the vicinity of the Mn^4Ca–oxo cluster or at the Ca^{2+} ion in the S2 state of spinach PSII. But in the case of D1-N87A PSII, the methoxy group is directly coordinated to the paramagnetic

Mn^4 ion in the S2 state, which results in larger ^{13}C isotropic and anisotropic hyperfine coupling constants. It appears that the D1-N87A mutation in cyanobacterial PSII results in rearrangement of the extensive hydrogen-bonding network in the narrow/broad channel and the OEC, which allows for coordination of the methoxy group at the Mn^4 ion. This is in agreement with previous observations by Brudvig and coworkers[3] that the D1-N87A mutation perturbs the Cl^- binding site in the broad channel. This is interesting as it points to a possible role of Cl^- in substrate binding in the OEC.

1 M. Retegan and D. A. Pantazis, *Chem. Sci.*, 2016, **7**, 6463–6476.
2 P. H. Oyala, T. A. Sitch, J. A. Stull, F. Yu, V. L. Pecoraro and R. D. Britt, *Biochemistry*, 2014, **53**, 7914–7928.
3 G. Banerjee, I. Ghosh, C. J. Kim, R. J. Debus and G. W. Brudvig, *J. Biol. Chem.*, 2018, **293**, 2487–2497.

Dimitrios Pantazis asked: Have you quantified the population of methanol-bound OEC in the S2 state of the N87A mutant, and can you confirm whether it can advance to the S3 state? The direct coordination of methoxy to Mn^4 seems more likely to be inhibitory compared to a more remote second-sphere interaction of methanol, as proposed for spinach PSII.

K. V. Lakshmi replied: Previous studies of spinach PSII have shown that methanol binding in the OEC does not inhibit oxygen evolution, but this was a second sphere interaction with methanol. We have not conducted studies on the S-state advancement of methanol-bound D1-N87A PSII. I agree that it would be interesting to probe the advancement to the S3 state as direct coordination of a methoxy group to the Mn^4 ion could influence oxygen evolution in D1-N87A PSII.

Sankar Prasad Rath remarked: You have presented a X-ray structure, have you seen any gross structural differences with other reported structure of cyanobacteria?

K. V. Lakshmi responded: The methanol-bound structures of the oxygen-evolving complex that I presented in my talk are QM/MM models. We don't observe gross structural differences between the methanol-bound QM/MM models of the S2 state and the X-ray crystal structures of wild-type cyanobacterial photosystem II except for the replacement of the water-derived and amino acid ligands at the Mn^4 or Mn^1 ion by a methoxy group or methanol, respectively.

Philip Ash opened a general discussion of the paper by Serena DeBeer: I appreciate the fact that you have presented *in situ* electrocatalytic data of your enzyme/polymer films in this paper, as it is a very important but sadly often overlooked part of enzymatic/*in situ*/*in operando* spectroelectrochemical measurements, especially in the biological catalysis community.

In similar steady-state infrared measurements on [NiFe] hydrogenase using protein film infrared electrochemistry (PFIRE), we tend to find virtually no change in the distribution of active site redox states as a function of applied potential under turnover conditions, and in the absence of any inactivation or inhibition processes.[1,2] These observations don't seem to be reflected in the XAS changes you observe here. Does this difference reinforce the critical importance of good electrode characterisation for each specific *in situ* spectroelectrochemical measurement?

1 R. Hidalgo, P. A. Ash, A. J. Healy and K. A. Vincent, *Angew. Chem., Int. Ed.*, 2015, **54**, 7110.
2 P. A. Ash, S. E. T. Kendall-Price and K. A. Vincent, *Acc. Chem. Res.*, 2019, **52**, 3120.

Serena DeBeer responded: The changes we are seeing in the Ni K-edge XAS spectra are also relatively modest. But, yes, I certainly agree that good electrode preparation and characterization is really important. When we are preparing the polymer films, for instance, it is really important that the majority of the sample is actually under electrochemical control in order to see changes. Also, diffusion through the polymer film will play a role, such that for thick polymer films only a fraction of the enzyme is involved in catalysis due to the diffusion limitation of electrons and H_2. All of these factors will limit our ability to see potential changes at the active site.

Philip Ash asked: Have you, or anyone else, measured steady-state infrared spectra under turnover conditions for these polymer/hydrogenase films? In the case of the hydrogel films, SEIRA will presumably only give information on the enzyme closest to the electrode and so might not be sensitive to dispersion effects throughout the film.

Serena DeBeer replied: My colleague Olaf Rüdiger has tried to measure these polymers using SEIRA, but unfortunately, since SEIRA only measures within ~10 nm of the nanostructured gold film, he was not able to get any signal. The polymer encapsulates the enzyme very efficiently and it covers the electrode surface, not leaving much enzyme near the Au. Additionally, these are hydrogels, and so when in contact with the electrolyte they swell, displacing the enzyme further away from the electrode.

Philip Ash remarked: What do you think the possibilities are for applying advanced spectroscopic methods, such as HERF-XAS, RIXS, pink-beam XAS *etc.*, given the necessarily longer acquisition times or higher photon flux?

Serena DeBeer responded: Conventional XAS combined with electrochemistry on immobilizied enzymes is already really challenging, and so at least at the moment, I am hesitant about applying more advanced X-ray spectroscopic approaches. In order to optimize the amount of enzyme that is under electrochemical control, ideally one wants a monolayer on the electrode. However, for enzymes, this means that the coverage of the photoabsorbing metal atoms is in the order of picomol/cm^2. This is really beyond the detection limit even for standard XAS and, hence, we need to go to the polymer films to improve the effective enzyme concentration that we are probing. With the polymer films, we are able to obtain effective Ni concentrations in the micromolar range. This is, however, still too low for HERFD or RIXS experiments. Furthermore, as HERFD and RIXS measurements generally require focused beams to realize good resolution, the damage issues become even more problematic. I think for HERFD (or RIXS) XA-SEC on enzymes, we will really need both fast scanning incident beam monochromators and improvements in the solid angle of detection to make these experiments feasible. For molecular systems, where one intrinsically can obtain better coverage, I am more optimistic about utilizing advanced X-ray spectroscopic approaches.

Abhishek Dey said: I wonder if one can conceive of a 2D X-ray method where the electrode material is used as a secondary source of low energy X-rays by ionizing it with the primary X-ray? In that case, the molecules near the electrode will be locally excited and may resolve the bulk of non-reactive material in the beam path in the X-ray spectro-electrochemistry measurements. Such a set-up, in many ways, would provide the advantages analogous to surface enhancement.

Serena DeBeer answered: It is certainly an intriguing idea, and I am not aware of anyone having tried something like this. I would only caution that generating low energy X-rays locally will likely result in even greater issues with damage processes.

Apparao Draksharapu remarked: The majority of studies have focused on the edge values – is it possible to get the XAS information/data also? The edge value of a metal complex highly depends on the oxidation state and attached ligands. If the coordinated ligands change, then there is the possibility of having a different edge. How can we be sure of the oxidation state assignment without EXAFS data?

Serena DeBeer responded: Yes, the changes in the edge will be coupled to changes in oxidation state, spin state and coordination number. These changes have a particularly pronounced effect on the pre-edge region. However, we are also able to fairly reliably calculate this spectral region and use time-dependent DFT to limit the possible structures that may be consistent with the observed experimental data. Of course, it is also great if we have additional data from EXAFS to corroborate this. This is possible but will simply depend on sample concentration in terms of the feasibility. Also, I would like to note that in addition to XAS and EXAFS, we always try to use as many other spectroscopic probes as possible (IR, EPR, Mössbauer, *etc.*) in order to have a more holistic picture of the electronic structure.

Apparao Draksharapu said: Reaching high concentrations of enzymatic samples is challenging. Could you use model compounds/molecular systems to achieve Fe(V)/Fe(IV) by XAS spectroelectrochemistry? Fe(V) is not an easy molecule to synthesize but there is a possibility of achieve this by electrochemistry – do you have any thoughts on this?

Serena DeBeer responded: Yes, it is certainly possible to use XA-SEC to study molecular models. We recently applied this approach to identify a Ru(V)–oxo intermediate in a molecular water oxidation complex (see ref. 1), and this could be equally well applied to molecular iron complexes.

1 N. Levin, C. Casadevall, G. E. Cutsail III, J. Lloret-Fillol, S. DeBeer and O. Rüdiger, *ChemElectroChem*, 2021, **9**, e202101271.

Apparao Draksharapu opened a general discussion of the paper by Peter Comba: The UV-vis spectrum of the Fe(IV)=O complex discussed in the talk looks similar to that of the $S = 1$ Fe(IV)–oxo compound. Is there any marker band (or epsilon value) to realize the existence of a high spin Fe(IV)=O complex by UV-vis absorption spectroscopy?

Peter Comba replied: The UV-vis-NIR spectra of the three Fe(IV)=O complexes with MeCN, Cl^- and Br^- co-ligands indeed have d–d transitions, as expected for an $S = 1$ ground state. For the MeCN complex, we have published Mössbauer data that confirm this assignment, and for the chlorido complex, we have unpublished preliminary Mössbauer data, also confirming an $S = 1$ ground state. The ligand field analysis presented allows us to fully understand the spectroscopic changes upon variation of the ligand. For tetragonal $S = 2$ systems, there are so far no thorough assignments of d–d transitions that would allow a similar analysis.

Apparao Draksharapu said: What do you think one needs to have in a ligand to stabilize this Fe(IV)=O high spin compound?

Peter Comba responded: Nature does it with trigonal symmetry; ligand field theory tells us that this leads to a preference for high-spin electronic configuration. With tetragonal symmetry, one needs to decrease the in-plane ligand field. With our tetradentate bispidine systems, we have done this by variation of the monodentate co-ligand *cis* to the oxido group, *trans* to N7 (MeCN *vs.* Cl^- *vs.* Br^-). Another possibility is to substitute the amine N7 *trans* to the monodentate ligand (R_2NH *vs.* R_2NCH_3 *vs.* R_2NCF_3 *vs.* R_2NF – only the first two have been studied so far). Moreover, the two other donors (pyridine) in the *xy*-plane can be varied, and we have done some work in other areas with *para*-substituted pyridine groups (H *vs.* OMe *vs.* Cl *vs.* NO_2 (ref. 1)). The most simple system with a weak in-plane ligand field of course is the aqua ion, and this has been shown to be high-spin.[2]

1 P. Comba, M. Morgen and H. Wadepohl, *Inorg. Chem.*, 2013, **52**, 6481–6501.
2 O. Pestovsky, S. Stoian, E. L. Bominaar, X. Shan, E. Münck, L. Que Jr. and A. Bakac, *Angew. Chem., Int. Ed.*, 2005, **44**, 6871–6874.

Apparao Draksharapu said: The ligands discussed in the talk are tetradentate in nature. The oxido ligand acts as a fifth ligand, and the solvent MeCN likely binds to the metal as the sixth ligand. Is it possible to stabilize the Fe(IV) high spin state by playing with the sixth ligand instead of the chelating tetradentate one?

Peter Comba replied: This indeed is the main idea behind this system and the data discussed; the tetradentate bispidine ligand enforces a tetragonal structure with two open sites, *cis* to each other, one occupied by the oxido group. Therefore, a monodentate co-ligand in the *xy* plane can be used to tune the energy of the $d_{x^2-y^2}$ orbital, *i.e.*, the triplet–quintet gap, the reactivity and the spectroscopic properties. Starting with the iron(II) triflate, chloride and bromide precursors, we were able to obtain the MeCN, Cl^- and Br^- complexes, respectively, and their ligand field spectra behave as one would expect.

Edward I. Solomon remarked: I agree with Peter's idea that as the ligand field transition energy he observes in absorption decreases, the $S = 2$ state will come down in energy relative to the $S = 1$ ground state of the Fe(IV)=O. However, there is an assignment in Peter's paper (DOI: 10.1039/d1fd00073j) of an electronic transition in our MCD spectrum of the TMCFe(IV)=O complex of Larry Que that I do not agree with. The correct assignment of a transition is based on detailed spectroscopic insight and not just a parameterized angular overlap model. This has to do with the assignment of the ligand field transition at 10 400 cm^{-1} (band I in Fig. 4a of ref. 1). We assigned that to the spin allowed xy to $x^2 - y^2$ ligand field transition, while Peter's paper assigns this to the spin forbidden xy to z^2 transition. From the temperature dependence of the MCD spectrum at high magnetic field in Fig. 4a of ref. 1, this is the lowest energy, spin allowed, z polarized transition which is xy to $x^2 - y^2$ and, thus, reflects the strength of the equatorial ligand field. This is further confirmed by correlating the TMCFe(IV)=O to Larry Que's N4PyFe(IV)=O complex. From our NRVS data in Fig. 1A and C in ref. 2, the N4Py complex has a stronger equatorial ligand field relative to that of TMC and from Fig. 4A and C in ref. 1: this band I shifts up in energy to ~14 000 cm^{-1}, as expected for the xy to $x^2 - y^2$ transition. Note that the spin forbidden xy to z^2 transition assignment proposed in Peter's presentation would go down in energy. From our correlation of the MCD to the absorption data in Fig. 3 of ref. 1, band III is the $xz{\cdot}yz$ to $x^2 - y^2$ transition and has most of the intensity in the absorption spectrum. Its energy does decrease as the strength of the equatorial ligand field decreases, which would correlate with a decrease in the energy of the quintet relative to that of the triplet ground state, as proposed in this paper (DOI: 10.1039/d1fd00073j).

In addition, the spin forbidden transition from xy to z^2 would gain intensity by spin–orbit coupling to the nearby spin allowed d–d transitions. To get the selection rules (*i.e.* polarizations), you need to use the double group to include the spin. This results in the xy to z^2 spin forbidden transition being x,y as well as z polarized. And since band III has most of the absorption intensity and is x,y polarized there would be a significant x,y contribution to the VT MCD of band I if it were this transition, which is not the case.

1 E. Decker, J.-U. Rohde, E. J. Klinker, S. D. Wong, L. Que and E. I. Solomon, *J. Am. Chem. Soc.*, 2007, **129**, 15983–15996.
2 C. B. Bell III, S. D. Wong, Y. Diao, E. J. Klinker, A. L. Tenderholt, M. C. Smith, J.-U. Rohde, L. Que Jr., S. P. Cramer and E. I Solomon, *Angew. Chem.*, 2008, **120**, 9211–9214.

Peter Comba answered: We very much appreciate that Ed Solomon agrees with our general conclusion based on the ligand field analysis, *i.e.* that the high reactivity of our $S = 1$ iron(IV)oxido systems is related to relatively small quintet–triplet energy gaps, that this can be tuned by modifying the in-plane ligand field and that this can be probed by ligand field spectroscopy. In our contribution (DOI: 10.1039/d1fd00073j), we report the absorption spectroscopy of three of our complexes in comparison with Larry Que/Wonwoo Nam's TMC system,[1] for which Ed Solomon has published a detailed spectroscopic study based on well-resolved MCD spectra.[2] An important feature in any spectroscopic analysis is the correct assignment of the electronic transitions. We propose an assignment of the TMC system that differs in one of the 5 observed transitions, and this leads to a fully consistent interpretation of all experimental data available, *i.e.* the absorption spectra of the four ferryl complexes considered, the zero-field splitting of the TMC complex, as well as the qualitative order of the zero-field splittings of our complexes (experimentally determined values are not yet available), and the relative order of triplet–quintet gaps (related to the observed reactivities). In a direct reply to some of the specific comments raised by Ed Solomon, we note:

(i) Kestrel, the ligand field program used in our analysis, allows for a full CI treatment, in which spin–orbit coupling is also included (*cf.* ref. 3). In this context, a thorough LFT analysis offers advantages over DFT, as the latter does not usually allow for multiconfigurational transitions.

(ii) Based on the analysis with Kestrel, we assign band I to an *xy* based transition, where the excited state is comprised of a heavily spin–orbit coupled CI state of the z^2 quintet and *xz*/*yz* triplet.

(iii) The potential covalency between the ligand orbitals and the metal d-orbitals could indeed compromise the ligand field effective-Hamiltonian upon which Kestrel depends. In this regard, however, we note that Kestrel successfully fits values for all other transitions and the zero field splitting, and that the observed d–d transitions of $S = 1$ ferryl complexes generally have ε values in the range of 300–400 L mol^{-1} cm^{-1}, indicative of the largely intact *g* symmetry of the d-orbitals, commensurate with only a small degree of any d-orbital–ligand orbital covalent interaction. We further note here that covalent interactions between the ligand orbitals and the 4s and 4p virtual orbitals of the metal center do not compromise the ligand field effective-Hamiltonian.

(iv) In terms of the polarizations observed in the MCD spectra, we note (based on double group treatment) that the observed polarizations and the temperature dependence of the MCD bands are in agreement with our assignment. In terms of the 'residual' intensity to band I at higher temperatures that might be expected to arise from the part *xy* polarization of band I required by our assignment, but is ostensibly absent in the spectra, we note that bands I and II in the MCD spectra are overlapping (as published with TMC and N4Py as supporting ligands). Therefore, their deconvolution and temperature variation is not unambiguous, especially since band II is strongly negative in MCD intensity, and, furthermore, that the temperature variation of band I had to be estimated from that of band V in the original work (Fig. S1 in ref. 2). Therefore, we do not see any strong arguments against our present ligand-field multiconfigurational interpretation. We further add that circumstantial support for our assignment arises from the observed reactivities as well as the shift of the main d–d transition, observed at around 750 to 900 nm – aspects with which Ed Solomon agrees.

Experiments currently in process, which might help to resolve the controversy, include the following: (i) we have recently been able to obtain freeze-quenched samples of two of our iron(IV)oxido complexes (others may follow), and these may help to obtain additional experimental parameters from field Mössbauer, HF-EPR and synchrotron experiments (*e.g.* zero-field splitting parameters). Whether or not well enough resolved MCD data are possible to obtain needs to be seen. A problem is that the samples freeze-quenched at −80 °C in milliseconds to seconds are not pure ferryl complexes, *i.e.* they contain the Fe^{II} precursor as well as oxido-bridged diiron(III) species. (ii) The reported MCD spectra with TMC and N4Py as supporting ligands should allow us to obtain an experimental value of the zero-field splitting parameter *D*. This has not been reported, and it would be interesting to see how the values obtained compare with those published from field Mössbauer (and HF-EPR) experiments and specifically, how the thus obtained values differ based on the two assignments by Ed Solomon and by us. (iii) Our ligand field treatment may be extended with a range of additional systems, including N4Py (where well-resolved MCD spectra are also available) and also including our ferryl complexes with the pentadentate bispidines and with other tetradentate bispidines with slightly different ligand fields. (iv) It would of course be interesting to obtain further information on the energy of the quintet–triplet gap. With quantum mechanics, this is a difficult task, and whether experimentally there is a chance to get a system with a small enough gap to observe spin-crossover experimentally (temperature-dependent spectroscopy) is nothing more than a dream at present. (v) Temperature-dependent kinetics may give some experimental information on the spin–orbit coupling matrix elements *via* the transmission coefficient in the Eyring treatment, and this is also part of our current studies.

1 J.-U. Rohde, J.-H. In, M. H. Lim, W. W. Brennessel, M. R. Bukowski, A. Stubna, E. Münck, W. Nam and L. Que Jr., *Science*, 2003, **299**, 1037.
2 A. Decker, J.-U. Rohde, L. Que and E. I. Solomon, *J. Am. Chem. Soc.*, 2004, **126**, 5378.
3 M. Gerloch, *Magnetism and ligand field analysis*, Cambridge University Press, Cambridge, 1983.

Kenneth Karlin remarked: I wonder if you can please explain the thermodynamics (BDEs involved) and/or kinetics that allow you to specifically observe cyclohexane oxidation in "normal" MeCN or EtCN (propionitrile) solvents; is there not significant solvent oxidation even with excesses of substrate and oxidant (*e.g.*, PhIO or other)? I guess perhaps this question may also apply to Larry Que's work. I would have thought you would get a really large amount of solvent oxidation. Can you measure solvent oxidation?

In addition, if you don't add in any substrate to acetonitrile or propionitrile (the solvent), is your intermediate stable? Does it self-decay?

Peter Comba responded: These are interesting and important questions asked by Ken Karlin, and I was first concerned about the possibility of C–H abstraction at the solvent more than 15 years ago when we first started to work with iron(IV) oxido compounds. I have discussed this with many colleagues and many times with my group, and I am still in the uncomfortable situation to have to answer: I do not know, I do not have a satisfactory and fundamentally correct answer. What I can say from our experimental observations is that: (i) our most reactive complexes (*i.e.* the bispidine–iron(IV)oxido–halogenido complexes discussed in our contribution) are not stable in acetonitrile or proprionitrile without substrate (cyclohexane or cyclopropane), even at −90 °C.[1] We have carefully analyzed the self-decay pathways, and these are the formation of the oxido-bridged diiron(III) complex (thermodynamic sink, resting state) and an intramolecular C–H activation, leading to demethylation of the ligand (oxidation of free chloride also leads to decay of the iron(IV)oxido species). With slightly less reactive derivatives (MeCN instead of Cl^- as the coligand or removal of the intramolecularly active methyl group in the ligand backbone *etc.*), we can get ferryl complexes that are more stable in solution when no substrate is added. (ii) Whenever we have tested for solvent oxidation, we only found traces at most. (iii) Most importantly, with our bispidine–iron(IV)oxido–chlorido complex discussed in this paper (DOI: 10.1039/

d1fd00073j), we get up to close to 80% yield of chlorocyclohexane (this is with a derivative, with the ferryl complex discussed in our contribution, it is approx. 50%) – admittedly the mechanism is complex (this has been studied in detail and published, see above) and the reaction is run with an excess of oxidant (iodosylbenzene), but we can show that the Cl in the organic product is transferred from an Fe–Cl unit (rebound), *i.e.* with respect to the iron(IV)oxido–chlorido complex the reaction is stoichiometric, and 80% yield with such a reactive species is respectably large, *i.e.* the side reactions (*e.g.* with MeCN, which as the solvent is in large excess) are not of major importance.

1 M. Abu-Odeh, K. Bleher, N. J. Britto, P. Comba, M. Gast, M. Jaccob, S. Krieg and M. Kurth, *Chem.–Eur. J.*, 2021, **27**, 11377–11390.

Kenneth Karlin commented: I did look at Jim Mayer's *Chem. Rev.*[1] and it has *t*-butoxyl radical which does not react with acetonitrile, even though the BDE says it might, but with propionitrile, you would have thought terminal methyl group would react – I am glad to hear that you have asked yourself this question.

1 J. J. Warren, T. A. Tronic and J. M. Mayer, *Chem. Rev.*, 2010, **110**, 6961–7001.

Peter Comba answered: There is nothing further to add.

Kenneth Karlin asked: In one or more of your previous publications, you note that excess oxidant and substrate are used in nitrile solvents for cyclohexane oxidation to avoid large difficulties with competing solvent oxidation, but I do not think that this can explain the lack of solvent oxidation.

Peter Comba replied: Apart from the general answer given to your question above, we note that as the oxidant, we generally use sPhIO, the more soluble version of iodosylbenzene. For the kinetics, we use an excess of the substrate (cyclohexane) to get pseudo-first order conditions. In addition, we have to use an excess of sPhIO (10-fold, limited by the solubility) to suppress as much as possible the self-decay to the oxido-bridged diiron(III) species; this is a second order reaction (Fe(II) precursor, Fe(IV)=O), which is very fast due to the very high reactivity of our Fe(IV)=O species. That is, we need to accelerate the formation rate of the Fe(IV)=O species (second order reaction, Fe(II) precursor, sPhIO). The other important point is that we have performed a blank (identical conditions as for the oxidation of cyclohexane by our Fe(IV)=O but without our Fe(II) precursor), and no cyclohexane is then oxidized, *i.e.* our Fe(IV)=O is needed for the C–H abstraction from cyclohexane.

Lawrence Que enquired: What is the substrate oxidation rate difference between the MeCN adduct and the Cl adduct? Which complex is more reactive?

Peter Comba answered: The Cl^- complex is much more reactive than the MeCN complex, and the Br^- species is again somewhat more reactive than the Cl^- complex. Part of these data are given in Table 1 of our contribution (DOI: 10.1039/d1fd00073j), but it is difficult to directly compare these rates since not all have been studied in detail so far and not all have been measured under identical conditions – more data are being measured at present (of importance specifically are also temperature-dependent kinetics in order to be able to directly compare the various systems mentioned in Table 1 of the paper). The current picture is that the second order rate of our Cl^- ferryl complex with cyclohexane at −90 °C is very high at 7.6×10^2 M^{-1} s^{-1}; the rate of the corresponding MeCN complex has only been measured with bzOH (BDE 82 *vs.* 96 kcal mol^{-1}), and at −35 °C, the second order rate is 2.9×10^{-1} M^{-1} s^{-1} (*i.e.* this still is a very reactive oxidant and one of the more reactive ferryl compounds known but is significantly slower than the Cl^- species). For the Br^- complex, we only have preliminary rates of 'self-decay' processes, and these are about one order of magnitude faster than the same reactions with the Cl^- oxidant.

Abhishek Dey queried: Do you think one (either σ or π) pathway being more active than the other is somehow related to the reactivity difference between the CH_3CN and Cl^- bound oxidants?

Peter Comba said: This question is very interesting, and we have considered this and similar issues ourselves. Computational studies were done along these lines but not such that we could at present directly compare the Cl^- and MeCN systems. The question is also related to the discussion with Gopalan Rajaraman after his lecture on 'the oxo-wall'. Moreover, it relates to a feature that was mentioned in my presentation, *i.e.* the relevance of the driving force of the C–H abstraction *vs.* that of the energy barrier. An important observation related to this question is that our tetradentate bispidine systems are very fast in general (even the slowest among them, *i.e.* that with MeCN). Qualitatively, and in agreement with Ed Solomon's comment, I assume that within our bispidine series (MeCN, Cl^- and Br^-), the variation is mainly related to the in-plane ligand field (triplet–quintet gap). If this simplistic view is appropriate, the important question arises of what then is responsible for the much higher reactivity of the tetradentate bispidine systems in comparison to ferryl species with different ligand platforms? The possibilities are: (i) the driving force plays an important role – we know that our $Fe^{IV}{=}O$ systems have relatively high redox potentials – possibly the highest among known ferryl complexes. (ii) As mentioned by Abhishek Dey, the reaction channel (σ *vs.* π) is an important factor. (iii) What I also see as a possibly relevant reason for enhancing the reactivity is the spin density on the oxygen atom (related to the $Fe^{IV}{=}O$ *vs.* Fe^{III}–oxyl continuum), *i.e.* spin polarization due to increasing spin density on the oxygen may influence the reactivity (barrier height). For all these possible interpretations, there are no unambiguous answers so far.

Edward I. Solomon remarked: In ref. 1, we showed that for an $S = 2$ Fe(IV)=O, the π and σ channels (*i.e.* frontier molecular orbitals, FMOs) have very similar transition state energies and, hence, similar H-atom abstraction (HAA) rates. The σ channel abstracts along the Fe–O bond (likely favored in chelate complexes due to equatorial ligand steric issues for the π channel), but can only rebound hydroxylate. Alternatively, the π channel is active in HAA perpendicular to the Fe–O bond. This is important in enabling selectivity in the reaction outcome. In ref. 2, we showed that for a non-heme Fe(IV)=O halogenase, HAA *via* the π channel enables halogenation of the H-atom abstracted substrate radical over the thermodynamically favored hydroxylation. The other point is that it is sometimes possible to perform absorption/MCD on a RFQ powder by making a mull or glass at low temperature. Alternatively, methods like NRVS (nuclear resonance vibrational spectroscopy) provide detailed structural insights directly on the RFQ powder.

1 M. Srnec, S. R. Iyer, L. M. K. Dassama, K. Park, S. D. Wong, K. D. Sutherlin, Y. Yoda, Y. Kobayashi, M. Kurokuzu, M. Saito, M. Seto, C. Krebs, J. M. Bollinger Jr. and E. I. Solomon, *J. Am. Chem. Soc.*, 2020, **142**, 18886–18896.
2 M. Srnec and E. I. Solomon, *J. Am. Chem. Soc.*, 2017, **139**, 2396–2407.

Peter Comba responded: We have studied (and published[1]) a study on the halogenation with the tetradentate iron(IV)oxido–chlorido complex discussed in our contribution. The chloride complex discussed is remarkable because it is not only probably the most efficient iron(IV)oxido system known to date but it also leads within an error limit of about 5% with full selectivity to chlorocyclohexane, *i.e.* 100% halogenation, no alcohol and no ketone (and, as mentioned by Ed Solomon, against the thermodynamic preference for the alcohol product). We have studied this in comparison with similar systems from Larry Que, Miquel Costas and others, and the qualitative result is that the Cl *vs.* OH rebound is imposed sterically. This is based on a DFT analysis, with all its problems. This analysis agrees to some extent with the conclusions mentioned and published by Ed Solomon, although one has to realize that the constraints in an enzyme are different from those in small molecule model systems.

1 K. Bleher, P. Comba, D. Faltermeier, A. Gupta, M. Kerscher, S. Krieg, B. Martin, G. Velmurugan and S. Yang, *Chem.–Eur. J.*, 2022, **28**, e202103452.

Conflicts of interest

There are no conflicts to declare.

Faraday Discussions

PAPER

Density functional theory investigation of Ru(II) and Os(II) asymmetric transfer hydrogenation catalysts†

Elizabeth M. Bolitho, [a] James P. C. Coverdale, [b] Juliusz A. Wolny, *[c] Volker Schünemann [c] and Peter J. Sadler *[a]

Received 20th October 2021, Accepted 30th November 2021
DOI: 10.1039/d1fd00075f

Transition metal ions have a unique ability to organise and control the steric and electronic effects around a substrate in the active site of a catalyst. We consider half-sandwich Ru(II) (Noyori-type) and Os(II) sulfonyldiamine 16-electron active catalysts [Ru/Os(η^6-*p*-cymene)(TsDPEN-H_2)], where TsDPEN is *N*-tosyl-1,2-diphenylethylenediamine containing *S,S* or *R,R* chiral centres, which catalyse the highly efficient asymmetric transfer hydrogenation of aromatic ketones to chiral alcohols using formic acid as a hydride source. We discuss the recognition of the prochiral ketone acetophenone by the catalyst, the protonation of a ligand NH and transfer of hydride from formate to the metal, subsequent transfer of hydride to one enantiotopic face of the ketone, followed by proton transfer from metal-bound NH_2, and regeneration of the catalyst. Our DFT calculations illustrate the role of the two chiral carbons on the *N,N*-chelated sulfonyldiamine ligand, the axial chirality of the π-bonded *p*-cymene arene, and the chirality of the metal centre. We discuss new features of the mechanism, including how a change in metal chirality of the hydride intermediate dramatically switches *p*-cymene coordination from η^6 to η^2. Moreover, the calculations suggest a step-wise mechanism involving substrate docking to the bound amine NH_2 followed by hydride transfer prior to protonation of the O-atom of acetophenone and release of the enantio-pure alcohol. This implies that formation and stability of the M–H hydride intermediate is highly dependent on the presence of the protonated amine ligand. The Os(II) catalyst is more stable than the Ru(II) analogue, and these studies illustrate the subtle differences in mechanistic behaviour between these $4d^6$ and $5d^6$ second-row and third-row transition metal congeners in group 8 of the periodic table.

[a]*Department of Chemistry, University of Warwick, Coventry, CV4 7AL, UK. E-mail: p.j.sadler@warwick.ac.uk*
[b]*School of Pharmacy, Institute of Clinical Sciences, University of Birmingham, Edgbaston, B15 2TT, UK*
[c]*Fachbereich Physik, Technische Universität Kaiserslautern, Kaiserslautern, Germany. E-mail: wolny@physik.uni-kl.de*

† Electronic supplementary information (ESI) available. See DOI: 10.1039/d1fd00075f

Introduction

Catalysts for transfer hydrogenation reactions are important in natural biological systems and in synthetic chemistry. In both cases, the common source of reducing equivalents is hydride, in biology originating from nicotinamide adenine dinucleotide (phosphate), NAD(P)H, or flavin mono- or di-nucleotides, $FMNH_2/FADH_2$, and in chemical transformations from formic acid or propanol. Examples are in Table 1. A range of boron, nitrogen, and silicon-based hydride donors are also used in organic transfer hydrogenations.[1] Typically substrates are aldehydes, ketones or imines. When the catalyst is chiral (natural or synthetic), transfer hydrogenation is often enantioselective (asymmetric transfer hydrogenation, ATH).

Although in some bacteria and archaea. Hydrogenases contain metals (Fe, Ni), in mammals hydrogenases appear to be only non-metalloenzymes. In synthetic chemistry, transition metal catalysts, especially complexes containing Ru(II), Os(II), Rh(III), or Ir(III), are often very efficient and offer a wide range of transformations. Here we concentrate on organometallic half-sandwich Ru(II) and Os(II) ATH catalysts for the asymmetric reduction of the ketone acetophenone, which is often used as a reference substrate. Such complexes are of interest not only for *in vitro* chemical transformations, but also for their possible activity in living biological cells and potential role in metal-based therapies.[2]

Ruthenium-catalysed ATH of ketones to afford optically pure alcohols was first described by Noyori *et al.* in the 1990s. The Noyori complex [Ru(*p*-cymene)(TsDPEN-H)Cl] **1** (where TsDPEN = *N*-tosyl-diphenylethylenediamine) is

Table 1 Examples of transfer hydrogenation catalysts, substrates, and reaction conditions

Catalyst	Substrate	Hydride	Conditions	Ref.
$[Ru^{II}(diphosphine)(diamine)Cl_2]$	Ketones and imines	Formic acid	Acidic	3 and 4
$[Ru^{II}(\eta^6\text{-arene})(TsDPEN)Cl]$	Ketones and imines	Formic acid	Acidic	5–11
$[(\eta^6\text{-}p\text{-cym})Ru(N,N')Cl]$	NAD^+	Formate	Aqueous	12
$[Ru^{II}(\eta^6\text{-}C_6H_5(CH_2)_3NH\text{-}DPEN\text{-}R)Cl]$	Ketones	Formic acid	Aqueous	13
$[Ru^{II}(\eta^6\text{-}Ph(CH_2)_3\text{-}(en)\text{-}N\text{-}R)Cl]$	NAD^+	Formate	Aqueous	14
$[Ru^{II}$(phenylpropylamino-TsDPEN) Cl]	Ketones and imines	Formic acid	Cancer cells	15
$[Os^{II}(\eta^6\text{-arene})(TsDPEN)]$	Acetophenone	Formic acid	Acidic	16 and 17
$[Os^{II}(\eta^6\text{-arene})(TsDPEN)]$	Pyruvate	Formate	Cancer cells	18
$[Os^{II}(\eta^6\text{-arene})(TsDPEN)]$	Quinones	Formic acid	Acidic	19
$[Os(\eta^6{:}\kappa^1\text{-}C_6H_5(CH_2)_3OH/O)(XY)]^+$	Pyruvate	Formate	Cancer cells	20
$[Ir^{III}(Cp^*)(N,N)(OH_2)]$	Quinones	NADH	Aqueous	21
$[(Cp^XRh^{III}(N,N')Cl)]^{n+}$	NAD^+	Formate	Aqueous	22
$[(Cp^XRh^{III}(N,N')Cl)]^{n+}$	Pyruvate	Formate	Aqueous	22
Lactate dehydrogenase (LDH)	Pyruvate	NADH	Cells	23
NAD(P)H oxidoreductases	NAD^+	Substrate	Cells	24

a pre-catalyst, which first requires treatment with a base, eliminating HCl to form a coordinatively-unsaturated 16-electron amido active catalyst [Ru(*p*-cymene)(TsDPEN-H_2)], a neutral complex deprotonated at each N of the chelate ring. The role of the sulfonyl substituent in stabilising the doubly-deprotonated chelated diamine ligand is notable.

The commonly accepted catalytic cycle (Fig. 1) requires hydride and proton sources, which facilitate the *in situ* generation of a neutral 18-electron hydride complex [Ru(*p*-cymene)(TsDPEN-H)H]. Once formed, the hydride complex can transfer hydride to ketones or imines enantioselectively *via* an outer-sphere ligand-assisted process.[5,25,26] The stereoselectivity of reduction is governed by the chirality of the phenylated carbon atoms of the diamine ligand, (*S*,*S*)-TsDPEN-H_2 typically yielding the *S*-alcohol reduction product, while (*R*,*R*)-TsDPEN-H_2 typically yields the *R*-alcohol.[27] The process regenerates the 16-electron active catalyst, which can re-accept hydride from the donor source, thus completing the catalytic cycle.

The classical transfer hydrogenation (TH) mechanism involves a 6-membered cyclic transition state (Fig. 1).[27] Such a mechanism allows simultaneous hydrogenation of O and C of the substrate carbonyl group by the concerted transfer of catalyst hydridic Ru–H and protic N–H hydrogens. Theoretical calculations have suggested that an alternative β-elimination/insertion mechanism is unlikely, as it requires unfavourable partial de-coordination of the η^6-arene ligand from the metal centre.[27] A mechanistic study of ATH in DMF/water using [Ru(*p*-cymene)(TsDPEN-H)Cl] showed that water can hydrogen bond with the substrate

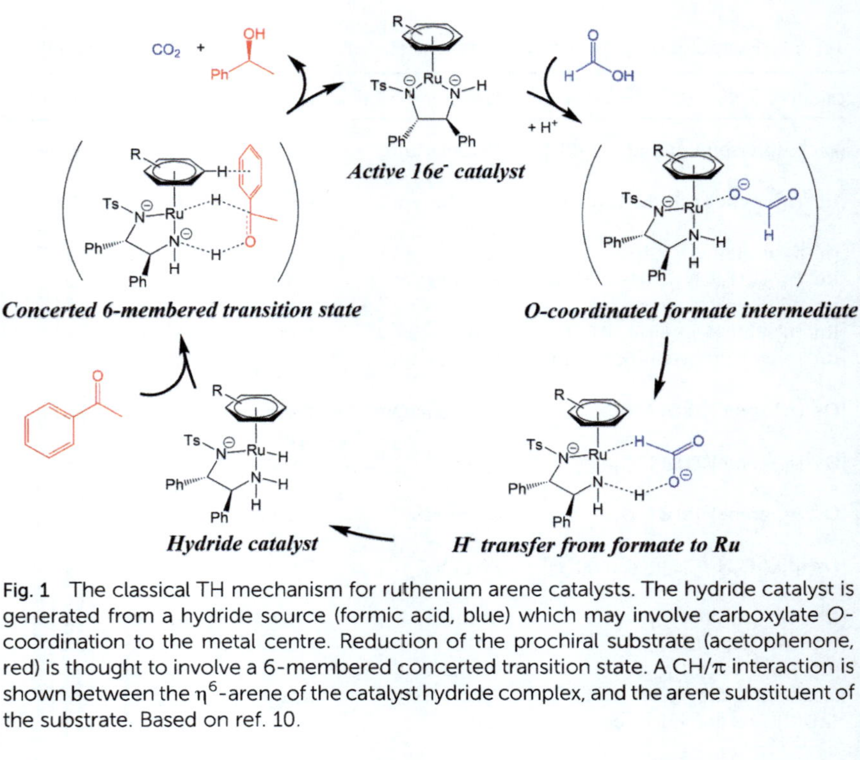

Fig. 1 The classical TH mechanism for ruthenium arene catalysts. The hydride catalyst is generated from a hydride source (formic acid, blue) which may involve carboxylate *O*-coordination to the metal centre. Reduction of the prochiral substrate (acetophenone, red) is thought to involve a 6-membered concerted transition state. A CH/π interaction is shown between the η^6-arene of the catalyst hydride complex, and the arene substituent of the substrate. Based on ref. 10.

during the transition state for hydride transfer, and that hydride transfer becomes more step-wise in nature, as opposed to being truly concerted.[28]

It is well accepted that observed enantioselectivities in ATH reductions are the result of various steric and electronic factors.[29] Interestingly, experiments have indicated that the hydride catalyst interacts with the sterically-congested face of the substrate, rather than the uncrowded face. It has been suggested that a favourable CH/π interaction occurs between the η^6-arene ligand (*p*-cymene) and the aromatic substituent of the carbonyl substrate (Fig. 1) stabilising the more sterically-crowded over the less-crowded transition state.[10,27] While the catalyst amine group directs the stereochemistry during the reduction, it is thought that the CH/π interaction stabilizes one of the diastereomeric transition states over the other. This theory has been supported by observations that electron-donating substituents on the substrate increase enantioselectivity, while those bearing electron-withdrawing substituents decrease enantioselectivity. These substituents modulate the ability of the substrate to partake in CH/π interactions,[27] thus influencing the stability of the transition state.

However, the theoretical calculations reported to date mostly use simplified models, lacking the pendant phenyl groups of the diamine ligand, as well as the tosylate functional group.[10] Similar CH/π interactions have been described between the substrate and the *p*-cymene ring in other ruthenium ATH catalysts, but again were not sterically hindered around the Ru–NH_2 active catalytic site, and did not bear pendant phenyl groups on the diamine ligand.[30,31] More recently, a DFT study of the asymmetric transfer hydrogenation of acetophenone *N*-benzylimine by [Ru(*p*-cymene)(TsDPEN-H)Cl] identified not only a CH/π interaction between one of the substrate arenes and the *p*-cymene ligand, but also a second interaction between the other arene ring on the substrate and the proximal pendant phenyl group of the diamine.[32]

The Os(II) analogues of Ru(II) ATH catalysts have been little studied.[33] Though typically considered more inert than Ru(II), and exhibiting slower ligand exchange kinetics, Os(II) complexes have been shown to catalyse TH reactions at comparable, or sometimes greater, rates than their Ru(II) analogues.[16] Osmium complexes bearing the chiral ligand (1*R*,2*S*)-(+)-*cis*-1-amino-2-indanol have been shown to catalyse TH reactions to afford chiral alcohols with enantiomeric excesses (ee) > 89%.[34] Separately, the conversion and enantioselectivities of base-assisted TH of ketones by osmium catalysts bearing amino acid ligands were found to be highly dependent on the nature of the substrate and the reaction temperature.[35]

We have reported a series of osmium complexes [Os(η^6-arene)(diamine)], analogous to the Noyori ruthenium complex. However, unlike their ruthenium counterparts, the active 16-electron Os(II) catalyst [Os(*p*-cymene)(TsDPEN-H_2)] **2** is highly stable in air, and can be isolated directly, while maintaining catalytic enantioselectivity and exceeding the turnover frequency of the analogous Ru(II) complex (Fig. 2).[16] Interestingly, both Ru and Os catalysts maintain catalytic activity in cancer cells, providing the first examples of in-cell TH catalysis (Ru) and in-cell ATH catalysis (Os) using synthetic catalysts, a new potential strategy for the treatment of resistant cancers using low dose metal therapy with a novel mechanism of action.[18,36]

The exploration of [Os(arene)(TsDPEN-H_2)] catalysts has so far been centred on reduction of ketones and subcellular targets for reduction.[17,19,37] The chemical

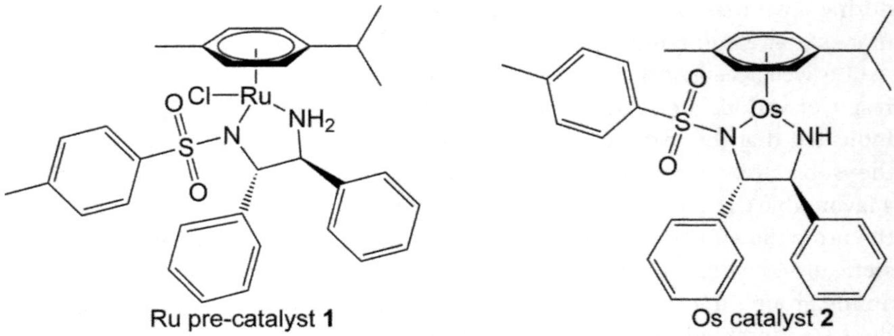

Fig. 2 Structures of the ruthenium pre-catalyst [Ru(*p*-cymene)(TsDPEN-H)Cl] 1 and active osmium catalyst [Os(*p*-cymene)(TsDPEN-H_2)] 2. Note that the pre-catalyst must eliminate HCl and thereby deprotonate the chelated ligand NH_2 to become the active $16e^-$ species.

mechanism of Os(II)-catalysed hydride transfer has generally been assumed to be similar to the accepted mechanism for Ru(II) analogues. The mechanism of the Os(II)-catalysed TH of pyruvic acid investigated using density functional theory (DFT) at the hybrid meta-GGA level M06 functional in conjugation with 6-31G(d) basis set for small atoms, and Stuttgart relativistic effective core potential basis set for Os (ECP60MDF), recently identified a proton-coupled hydride transfer mechanism, involving interaction between the catalyst η^6-arene and substrate carboxyl group, which influences the resulting enantioselectivity.[38]

In this work, we use full modelling of the Ru(II) and Os(II) [Ru/Os(*p*-cymene)(TsDPEN-H_2)] catalysts to investigate the mechanisms of their catalytic activity in ATH of the ketone acetophenone. Our calculations provide new insight into the transfer of hydride from formic acid to the metal, the recognition of acetophenone by the active hydride complex, and stereoselective hydride transfer along with a proton to the ketone. The roles of the four types of chiral centres present in the catalyst are discussed: axial chirality of the π-bonded *p*-cymene arene ligand, the two chiral C centres on the TsDPEN-H_2 backbone, chirality at the metal centre, and chirality of the 5-membered chelate ring. Such modelling provides unprecedented insights into the mechanism for catalysis of these metal-assisted transfer hydrogenation reactions, and in this case the subtle difference between Ru(II) and Os(II) catalysts.

Results

A possible catalytic cycle for the reduction of acetophenone by Ru(II) pre-catalyst **1** [Ru(*p*-cymene)(TsDPEN-H)Cl] and the Os(II) analogue of the active catalytic species [Os(*p*-cymene)(TsDPEN-H_2)] **2** in the presence of formic acid as a hydride donor was modelled using DFT calculations. First, we describe features of the chiral centres in the models we have constructed by DFT calculations.

Stereochemistry of the hydride intermediates

The two phenylated carbons of the diamine backbone of the chelated DPEN ligand are chiral, as is the 5-membered M–N–C–C–N′ chelate ring, which can adopt δ or λ configurations. In our modelling, this chelate ring conformation was found to be always λ for the *S*,*S*-chiral ligand; the opposite δ conformation appears to bring about more steric congestion due to closer approach of ligand phenyls and the N–SO_2–tolyl group.

In the case of Ru complex **1**, the pre-catalyst must first eliminate HCl to form the active (less stable) catalytic species. This step is not required for Os complex **2**, because it is isolable as a stable species and can be prepared directly by synthesis. For catalysis to proceed, a metal–hydride bond (Ru–H or Os–H) is formed. The hydride complex has a chiral metal centre (Fig. 3, Fig. S4, PDB1 and Fig. S5, PDB2).† Additionally, axial chirality arises from the orientation of the *p*-cymene ligand with respect to the *N*,*N*′-chelate (Fig. 3). The structure of the transition state for inversion of axial chirality of [M(*p*-cymene)(TsDPEN-H)H] is shown in the ESI (Fig. S1 and Fig. S6, PDB3),† the energy barrier for *p*-cymene rotation has been estimated to be 18 kJ mol^{-1} on going from S_a to R_a for the *R*-absolute configuration of the metal.

Metal chirality

Our modelling shows that the absolute configuration of the metal centre plays a crucial role in determining the electronic energy of the diastereoisomers of

(a) Metal = *R*, diamine = *S*,*S*, axial = S_a

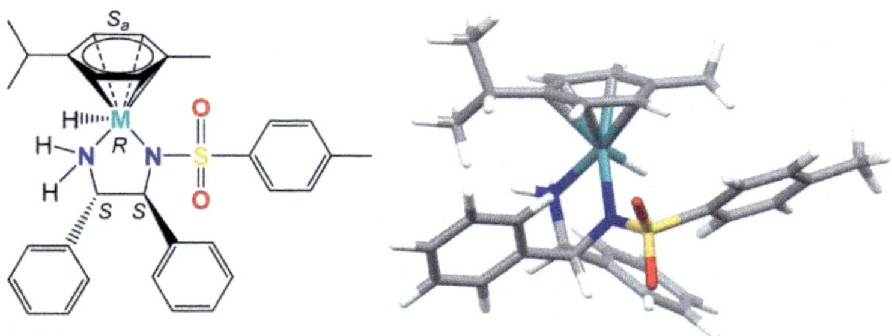

(b) Metal = *R*, diamine = *S*,*S*, axial = R_a

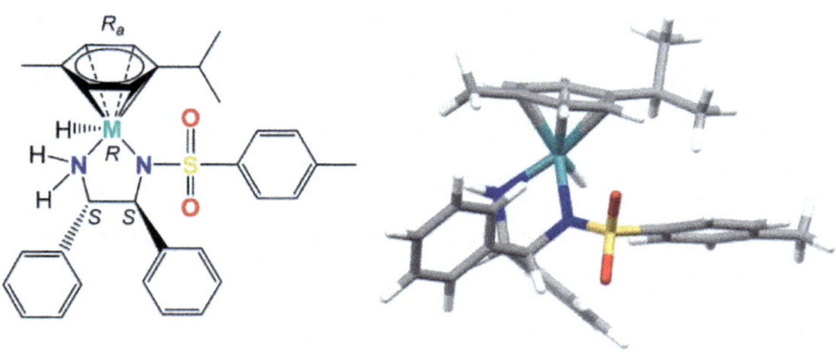

Fig. 3 Structures of the *R*-configured hydride complex [*R*-M(*p*-cymene)(*S*,*S*-TsDPEN-H)H], where M = Ru or Os showing the axial chirality involving rotation of the π-bonded *p*-cymene ligand: (a) S_a axial chirality (0 kJ mol^{-1}) and (b) R_a axial chirality (5 kJ mol^{-1}). Note that the chelated ligand contains a protonated NH group (as NH_2 in TsDPEN-H), which facilitates hydride transfer to the metal (Fig. S4 and S5, PDB1 and PDB2).†

complexes containing the (*S*,*S*)-ligand. The hydride diastereomers with *S* chirality (and consequent η^2 coordination) are of much higher energy for Ru and Os than for *R* chirality (η^6 coordination). The differences were slightly dependent on the axial chirality, being 111 and 112 kJ mol^{-1} for the R_a and S_a rotamers of Ru, respectively. The η^2 coordination for the Os complexes brings about even more significant energetic destabilisation, yielding values of 146 and 148 kJ mol^{-1}, for the R_a and S_a rotamers, respectively. This suggests that for complexes bearing the (*S*,*S*)-configured diamine, diastereomers bearing an *S*-configured metal centre,

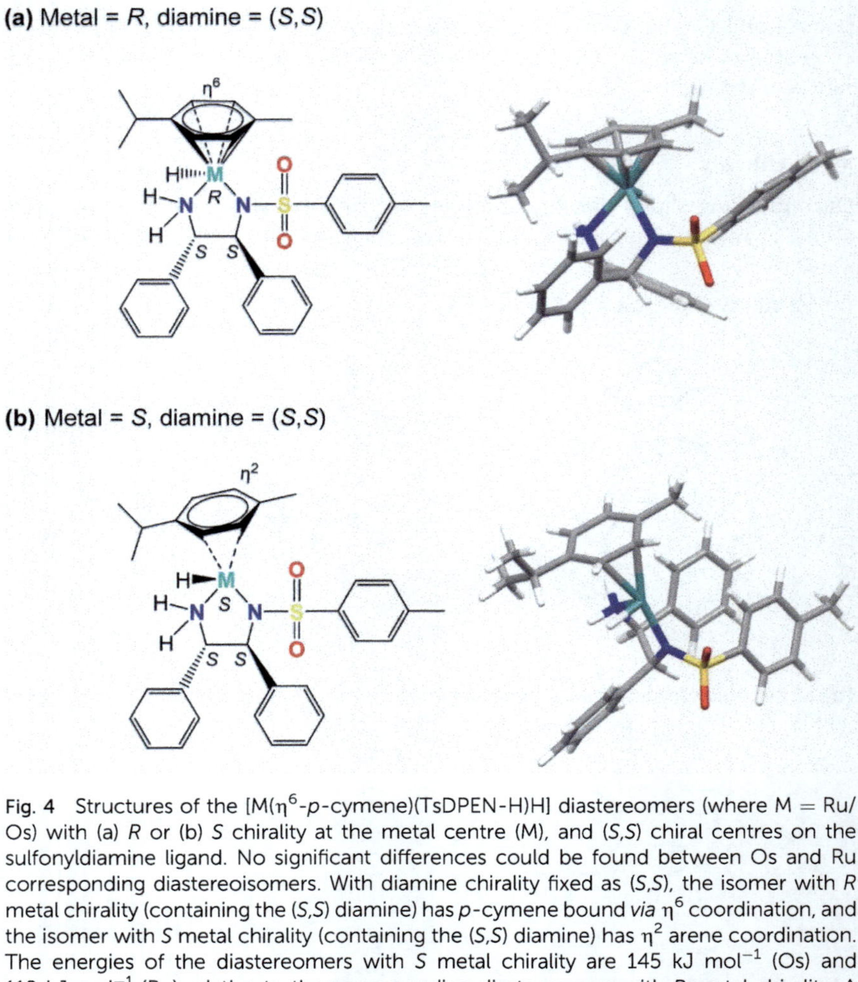

Fig. 4 Structures of the [M(η^6-*p*-cymene)(TsDPEN-H)H] diastereomers (where M = Ru/Os) with (a) *R* or (b) *S* chirality at the metal centre (M), and (*S*,*S*) chiral centres on the sulfonyldiamine ligand. No significant differences could be found between Os and Ru corresponding diastereoisomers. With diamine chirality fixed as (*S*,*S*), the isomer with *R* metal chirality (containing the (*S*,*S*) diamine) has *p*-cymene bound *via* η^6 coordination, and the isomer with *S* metal chirality (containing the (*S*,*S*) diamine) has η^2 arene coordination. The energies of the diastereomers with *S* metal chirality are 145 kJ mol^{-1} (Os) and 110 kJ mol^{-1} (Ru) relative to the corresponding diastereomers with *R* metal chirality. A possible reason for this dramatic change seems to be the higher steric clashes between *p*-cymene and the sulfonyldiamine ligand phenyl rings. In the case of the *R* metal chirality diastereoisomer (containing the (*S*,*S*) diamine), the *p*-cymene substituent is further from the phenyl group bound to the TsN-CH carbon than it is from the phenyl group bound to the NH_2-CH carbon. For *S* metal chirality, this congestion is reduced by *p*-cymene ring slippage (to η^2 coordination) with a concomitant shortening of M–N and M–hydride bonds (compare Fig. S7 and S8, PDB4 and PDB5).†

S(M), are less stable than their counterparts bearing *R*-configured metal centres, *R*(M). They are shown in Fig. 4 and the corresponding PDB files (Fig. S7, PDB4 and Fig. S8, PDB5) are in the ESI.†

Inspection of the calculated structures reveals a ring slippage on going from the *R*(M) to the *S*(M) diastereoisomer with a concomitant change of the cymene coordination mode from η^6 to η^2. Additionally, on changing from *R*(M) to the *S*(M) metal chirality, the Ru–H bond (for R_a axial chirality) shortens from 1.607 to 1.566 Å for Ru complex **1** and from 1.64 to 1.592 Å for Os complex **2.** On the other hand, the $M–N_{tosylate}$ bond shortens from 2.115 to 2.035 Å for the Ru complex and from 2.128 to 2.046 Å for the Os complex. Comparing *R*(M) and *S*(M) configurations, the Os–NH_2 bond shortens from 2.162 to 2.069 Å, while the Ru–NH_2 bond shortens from 2.136 to 2.052 Å, respectively. Apparently, the interchange of the hydride and arene ligands in the *S*-(*S*,*S*) diastereoisomer to give the *R*-(*S*,*S*)-diastereoisomer brings about very extensive steric clashes that are avoided by changing from pseudo-octahedral to strongly-distorted penta-coordination.

Axial chirality

Previous ^{1}H NMR studies revealed the formation of two Os–H singlets (−5.85 and −6.05 ppm) which changed from a 3 : 1 to 1.2 : 1 ratio over time,[16] suggesting that a diastereomeric mixture of hydrides was present. However, our calculations reveal $\Delta G > 100$ kJ mol^{-1} ($K < 3 \times 10^{-18}$) between these diastereomers, which suggests that the two ^{1}H NMR resonances cannot be attributed to the above mixture of diastereomers, and instead are related to further chirality in the molecule. This is most likely the axial chirality (R_a and S_a) with a preference for S_a (0 kJ mol^{-1}) *vs.* R_a (5 kJ mol^{-1}). This difference of 5 kJ mol^{-1} between these axial isomers would give a 1 : 4 ratio of ^{1}H NMR Os–H peaks, with 20% S_a and 80% R_a.

The transition state for the axial chirality and energy barrier for rotation of the *p*-cymene is *ca.* 18 kJ mol^{-1}, implying that rotation may be partially hindered. Hence, in the DFT modelling, both possible axial chiralities were considered (Fig. 3). In future work, the stereochemical relationship between the hydride and *p*-cymene substituents could be investigated by NOESY NMR.

The reaction profile of the reduction of acetophenone

The reaction profile for the reduction of acetophenone catalysed using the ruthenium pre-catalyst [Ru(*p*-cymene)(*S*,*S*-TsDPEN-H_2)Cl] was modelled with formic acid as the solvent. The species with minimal energy was found to be the hydride complex with the acetophenone bound *via* π–π stacking to the tosylate phenyl ring. The following species were identified (Fig. 5).

Species A [Ru(*p*-cymene)(*S*,*S*-TsDPEN-H_2)] + formic acid + acetophenone: The initial complex corresponds to the product formed after HCl abstraction from the pre-catalyst. The NH_2 group of the TsDPEN ligand is deprotonated (Ru–NH) and with a *ca.* 2.4 Å contact with H–C hydrogen of a HCO_2H molecule. Acetophenone is weakly bound to the complex, lying approximately parallel to formic acid and one of the phenyl rings of the nitrogen ligands (Fig. S10 and S11, PDB7 and PDB8† for both rotamers differing in axial chirality). These species lie 35–38 kJ mol^{-1} higher in energy than the hydride species, with R_a axial chirality being of slightly higher energy than S_a.

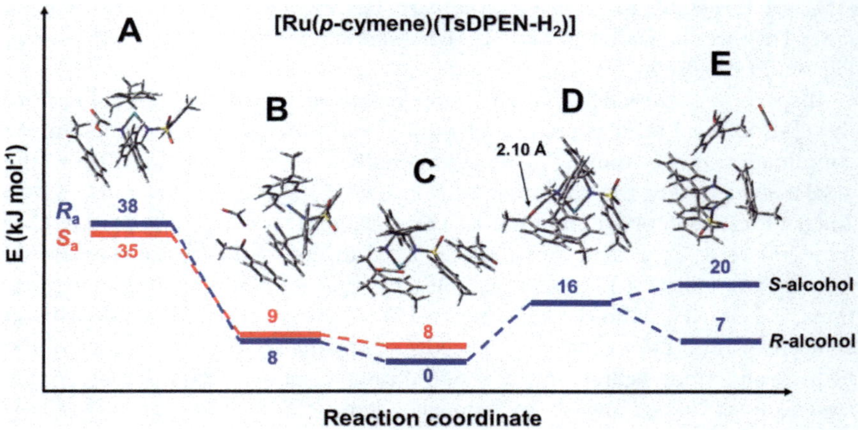

Fig. 5 Energy profile for the reduction of acetophenone catalysed by [Ru(*p*-cymene)(*S*,*S*-TsDPEN-H_2)], with formic acid as the hydride donor. Note the axial chirality was taken into account only for the first three steps. Species **A**: [Ru(*p*-cymene)(*S*,*S*-TsDPEN-H_2)] + formic acid + acetophenone (Fig. S10 and S11, PDB7 and PDB8†). Species **B**: [Ru(*p*-cymene)(*S*,*S*-TsDPEN-H)]$^+$ + formate + acetophenone (the white dot in the middle is showing the transfer of H^+), Fig. S12 and S13, PDB9 and PDB10.† Species **C**: [*R*-Ru(*p*-cymene)(*S*,*S*-TsDPEN-H)H] + CO_2 + acetophenone (Fig. S2 and S9: PDB6, Fig. S14 and S15: PDB11 and PDB12†). Species **D**: [*R*-Ru(*p*-cymene)(*S*,*S*-TsDPEN-H-acetophenone)H] + CO_2 (Fig. S16, PDB13†) Species **E**: [Ru(*p*-cymene)(*S*,*S*-TsDPEN-H_2)] + CO_2 + phenylethanol (Fig. S17 and S18, PDB14 and PDB15†). Note, bonds between Ru and *p*-cymene are omitted for clarity.

Species B [Ru(*p*-cymene)(*S*,*S*-TsDPEN-H)]$^+$ + formate + acetophenone: This species results from protonation of the NH group of the ligand (Fig. S12 and S13, PDB9 and PDB10†). The transferred proton has *ca.* 1.69 Å contact with a formate oxygen. The protonation leads to a significant lowering of the electronic energy, which is now only approximately 8 kJ mol^{-1} higher than that of the reference Ru–H hydride.

Remarkably, the next step (**B** to **C**) does not correspond to carboxylate *O*-coordination of formate to the metal, as previously proposed.[39] Our modelling revealed that imposing such a coordination in the starting structure leads to a movement of formate onto the rim of the catalyst (Ru–O distance > 3 Å) due to a stronger interaction between the formate oxygen and the diamine ligand NH_2, leading to a significant increase in energy of the species (due to the poor acidity of the amine). The same effect was established for the osmium analogue.

Species C [Ru(*p*-cymene)(*S*,*S*-TsDPEN-H)H] + CO_2 + acetophenone: The next structure in the catalytic cycle is the M–H hydride complex. As stated above, for the (*S*,*S*) chirality of the chelated ligand, only the η^6-*p*-cymene M(*R*)-enantiomers of the hydride complex are stable. Due to axial chirality, two diastereoisomers are possible, of which R_a chirality is the most stable species in the catalytic cycle, the S_a diastereoisomer being 8 kJ mol^{-1} higher in energy (note that with bound acetophenone this is contrary to the isolated hydride system discussed above). The PDB structures of both diastereoisomers differing in axial chirality are given in the ESI† (Fig. S14 and S15, PDB11 and PDB12) and are shown in Fig. 6. Two noteworthy features of these structures are that acetophenone is bound *via* π–π

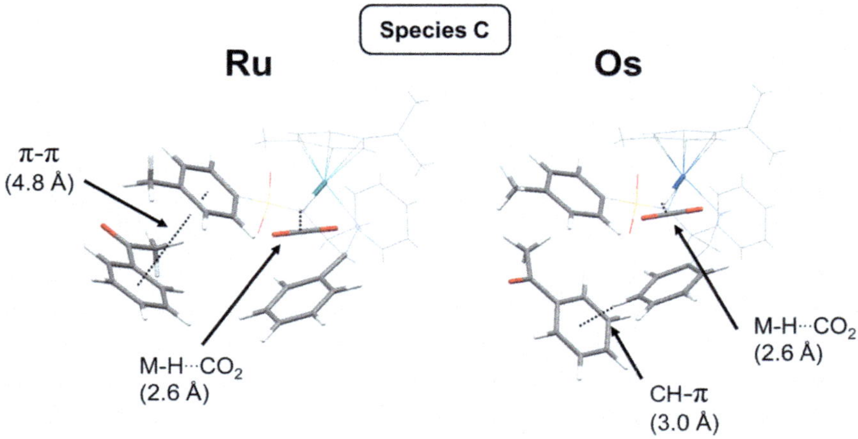

Fig. 6 Structures of species **C** (M = Ru or Os) showing the interaction of [M(η^6-*p*-cymene)](TsDPEN-H)H (with *R* metal chirality, (*S*,*S*) ligand chirality, and R_a chirality) with acetophenone (substrate). The relevant molecular fragments are highlighted for clarity. For the Ru system, π–π interaction between acetophenone and the tosylate ring is seen, whereas for the Os complex, both rings are shifted, and a pendant phenyl ring (from the diamine ligand) points towards the ketone phenyl ring (dashed line, CH···C(ring) distance of 3.04 Å). Both Ru and Os show interaction (2.6 Å) between the metal hydride and CO_2 (Ru: Fig. S15, PDB12; and Os: Fig. S24, PDB21†).

stacking to the tosylate phenyl, and there is a significant interaction of the formed CO_2 molecule with the M–H hydride, with a H–C distance of *ca.* 2.6 Å (Fig. 5 and 6).

Species D [Ru(*p*-cymene)(*S*,*S*-TsDPEN-H-acetophenone)H] + CO_2: The next structure in the catalytic cycle that we were able to identify by DFT modelling is the hydride complex with acetophenone anchored to the catalyst *via* a carbonyl-O···HNH hydrogen bond (2.127 Å). The structure of its axial R_a-diastereomer can be found in Fig. S16, PDB13.†

Other possibilities for substrate docking were explored, such as hydride–carbonyl contact, and the previously proposed simultaneous proximity of the substrate carbonyl group and the amine hydrogen/hydride pair (with two possible arrangements, *i.e.* hydride–carbon and hydride oxygen proximity).[16] All of these models led to movement of the ketone molecule onto the rim of the complex. It appears that these arrangements are energetically unfavourable due to both electronic and steric reasons.

Species E [Ru(*p*-cymene)(*S*,*S*-TsDPEN-H_2)] + CO_2 + phenylethanol: This final species corresponds to the reduced ketone (the alcohol) with *S*- or *R*-chirality and the initial catalyst described for species **A.** Interestingly, interactions between both molecules lead to a lower energy in the case of the [Ru(*p*-cymene)(*S*,*S*-TsDPEN-H_2)]/*R*-alcohol pair (8 kJ mol^{-1}) compared to that with the *S*-alcohol pair (20 kJ mol^{-1}). This is in contrast with the previously reported high enantioselectivity for formation of the *S*-alcohol with the (*S*,*S*)-ligand, and *R*-alcohol from the (*R*,*R*)-ligand. The most likely explanation is that the observed enantioselectivity arises from kinetic factors rather than the thermodynamic energies revealed by the DFT calculations (*vide infra*). Structures can be found in Fig. S17 and S18 (PDB14 and PDB15).†

The reaction profile for [Os(*p*-cymene)(*S*,*S*-TsDPEN-H_2)] is shown in Fig. 7. The energy profile for the osmium complex is similar to that of ruthenium. The first difference is that the energy of species **B**, corresponding to the second stage, *i.e.* protonation of the chelated sulfonyldiamine ligand by formic acid, is *ca.* 8 kJ mol^{-1} higher for the osmium complex. On the other hand, the energy of the structure with the N–H bound ketone (**D**) relative to the hydride structure **C** is 8 kJ mol^{-1} higher than for the ruthenium analogue. Apart from that, there are subtle differences in the way acetophenone interacts with the Ru and Os complexes. The corresponding structures are compared in Fig. 6. While for the Ru system, clear π–π interaction involving the acetophenone and tosylate ring is seen, for the Os complex, both rings are shifted and a chelated ligand phenyl ring points towards the ketone phenyl ring, revealing a C–H···aromatic ring interaction with C–H···C(ring) distance of 3.04 Å (dashed line in Fig. 6).

Mechanistic considerations

The structure identified for the ruthenium hydride complex with acetophenone bound to an NH_2 hydrogen of the ligand is a plausible starting point for the reduction. As shown in Fig. 8, the ketone and catalyst molecules are pre-organised so that the vector defined by the hydride and carbonyl carbon atom points towards one of the enantiotopic faces of the ketone. We have modelled several variants of the reduction starting from this assembly; the only one with

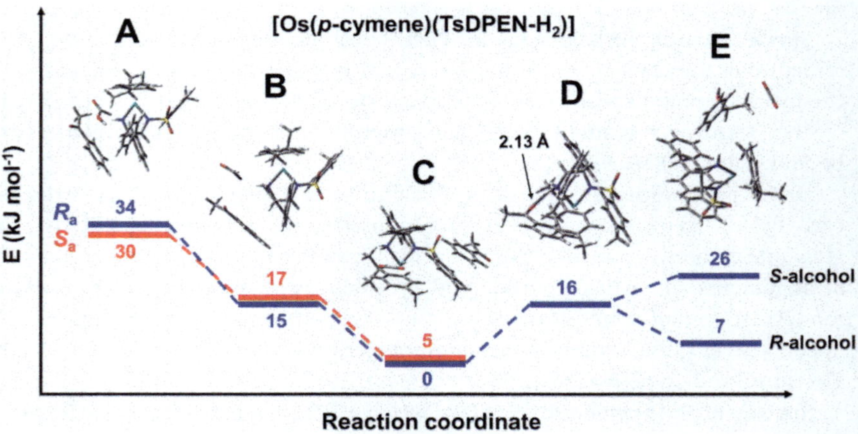

Fig. 7 Energy profile for the reduction of acetophenone catalysed by [Os(*p*-cymene)(*S*,*S*-TsDPEN-H_2)]. Note the axial chirality was taken into account only for the first three steps. Species **A**: [Os(*p*-cymene)(*S*,*S*-TsDPEN-H_2)] + formic acid + acetophenone (Fig. S19 and S20, PDB16 and PDB17†). Species **B**: [Os(*p*-cymene)(*S*,*S*-TsDPEN-H)]$^+$ + formate anion + acetophenone (Fig. S21 and S22, PDB18 and PDB19†). Species **C**: [Os(*p*-cymene)(*S*,*S*-TsDPEN-H)H] + CO_2 + acetophenone (Fig. S23 and S24, PDB20 and PDB21†). Note the different mode of interactions of acetophenone compared to the corresponding Ru–hydride. Species **D**: [Os(*p*-cymene)(*S*,*S*-TsDPEN-H-acetophenone)H] + CO_2 (Fig. S25, PDB22†). Species **E**: [Os(*p*-cymene)(*S*,*S*-TsDPEN-H_2)] + CO_2 + phenylethanol (Fig. S26 and S27, PDB23 and PDB24†).

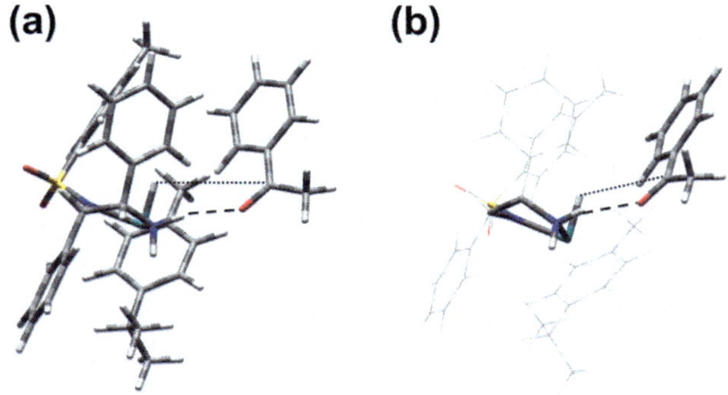

Fig. 8 (a) Pre-organisation of the [Ru(*p*-cymene)(*S*,*S*-TsDPEN-H)]/acetophenone assembly with CO···H_2N H-bonding (denoted by the dashed line), the initial stage of the catalytic cycle. Note that the vector defined by hydride H and carbonyl C atoms points towards one of the enantiotopic faces of the ketone. (b) Simplified scheme to highlight bonding interactions between catalyst diamine and substrate carbonyl (Fig. S16, PDB13†).

a reasonably low-energy structure involves a starting geometry in which only the hydride was set closer to the carbonyl carbon, while the amine proton is still bound to nitrogen, *i.e.* there is no concerted transfer of hydride and proton. Geometry optimisation led first to formation of the *S*-alkoxide bound to the NH_2 group of the ligand, and then a gradual transfer of the amino proton to form the alcohol. The final product corresponds to the recovered catalyst with *S*-phenylethanol (α-methylbenzyl alcohol) forming a contact to the NH group of the ligand *via* a hydrogen bond (OH···N distance of 2.41 Å) (see Fig. 8). The estimated transition state between this and the ketone bound to the hydride complex *via* NH···O hydrogen-bonding (structure **D** in Fig. 7) is shown in Fig. 9. It corresponds to the above-mentioned *S*-alkoxide bound to the NH_2 group of the ligand.

Within this model, the chirality of the reduced acetophenone is *S* (for the *S*,*S* absolute configuration of the chelated ligand) in line with experimental findings.[16] This agreement suggests that the poor acidity of the NH_2 group plays an important role in the mechanism, as was also seen above in the destabilisation of the Ru–O(formate) bond. It seems that this is even more pronounced for a hydride complex, while the dissociation of a proton from the NH_2 group would lead to an effective combined charge of −3 from the hydride and sulfonyldiamine N ligands. Thus, it is the initial transfer of hydride that then allows the transfer of the NH_2 proton to regenerate the active catalyst.

We checked the possibility of the exchange of the phenyl and methyl groups shown in Fig. 8 that might lead to the possibility of attack of hydride on the alternative enantiotopic face. We optimised the geometry of such a structure, as shown in Fig. 10. It is *ca.* 9 kJ mol^{-1} higher in energy than the molecule shown in Fig. 8. Significantly, it does not appear to be pre-organised in favour of a particular enantiomer of the reduction product.

This structure corresponds to another stereoisomer regarding the orientation of the ketone O···HN contact. The hydride ligand lies nearly perfectly in the plane of the ketone fragment of acetophenone, *i.e.* in the plane defining the enantiotopic faces. Considering the possibilities for different orientations of the *p*-

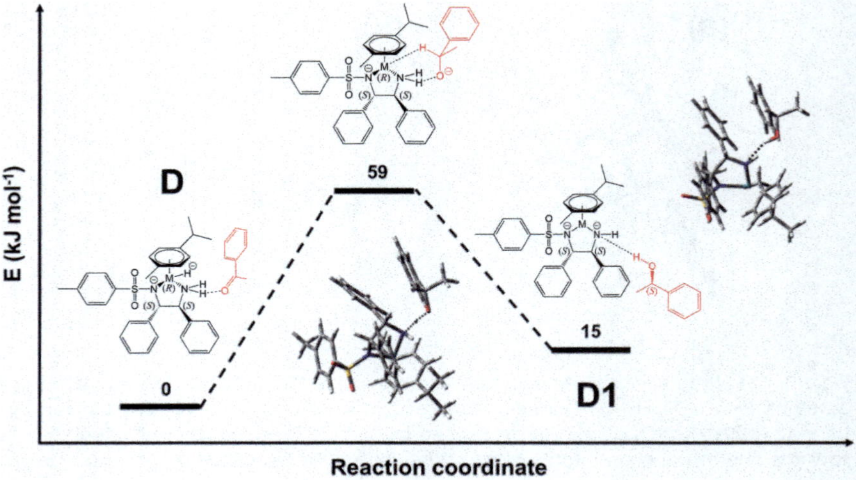

Fig. 9 QST-estimated structure for the transition state from species **D**, in which the acetophenone keto group is bound to the catalyst NH_2, to the recovered Ru-catalyst **D1**, in which the product alcohol is still bound *via* HN···HO bond. The transition state is estimated to lie 59 kJ mol^{-1} above the structure with minimal energy (**D**). The contacts indicated by dashed lines in the transition state are 1.32 Å for NH···OC (see Fig. S28, PDB25†) and 1.20 Å (O–H···N). For species **D**, the N–H···OC distance is 2.13 Å (Fig. S16, PDB13†), while O–H···N for species **D1** is 2.41 Å (Fig. S29, PDB26†).

cymene ring, leading to different axial chirality, at least four different stereoisomers of the complex with hydrogen-bonding from the catalyst amine to the bound ketone oxygen, would be expected. Each of them may reveal different energies and

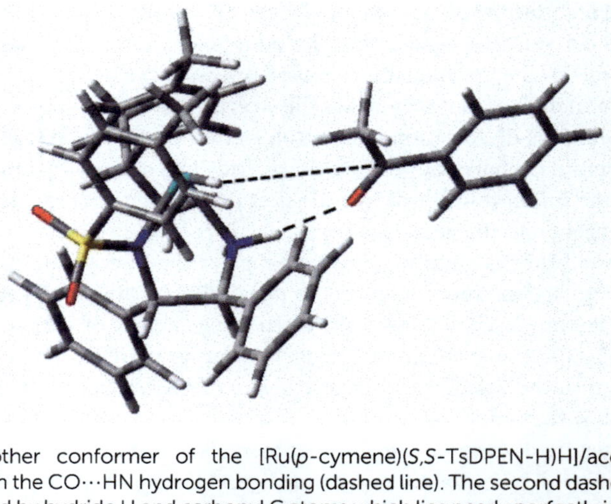

Fig. 10 Another conformer of the [Ru(*p*-cymene)(*S*,*S*-TsDPEN-H)H]/acetophenone assembly with the CO···HN hydrogen bonding (dashed line). The second dashed line is the vector defined by hydride H and carbonyl C atoms which lies nearly perfectly in the ketone plane, so there is no preference for either diastereotopic face, unlike the conformation shown in Fig. 8 (Fig. S28, PDB25†).

pre-organisation, leading to a given enantiomer of the product and corresponding to a different transition state.

Overall, our calculations suggest that the reduction of acetophenone using such catalysts proceeds first by docking of the ketone substrate onto an amine NH_2 proton of the chelated sulfonyldiamine ligand, followed by enantioselective hydride transfer onto a particular face of the ketone carbonyl carbon. Then, protonation occurs by transfer of a proton from the amine NH_2 to the ketone oxygen to release the enantiopure alcohol. Moreover, the hydride complex is highly dependent for stability on the presence of the amine NH_2 protons. The proposed sequence of steps and cycle for the reduction of acetophenone by the Ru(II) and Os(II) catalysts, and the stereoisomer preference with hydrogen-bonded ketone, corresponding to the molecule in Fig. 8, are shown in Fig. 11 and 12, respectively.

Discussion of experimental evidence

In the absence of experimental evidence, DFT calculations may lead to spurious possibilities. These DFT calculations serve to stimulate our thinking about the detailed mechanism of asymmetric transfer hydrogenation of ketones by half-sandwich Ru(II) and Os(II) arene complexes, especially for Os(II) catalysts, which have only recently been explored, and the design of experiments to provide evidence for the proposed pathways. In this section, we briefly discuss our calculations in the context of existing experimental data.

We have reported that reduction of acetophenone using the 16-electron Os(II) catalyst **2** yields the *S*-configured alcohol (*S*-phenylethanol) in the presence of the Os catalyst bearing an *S*,*S*-diamine ligand.[16] Similarly, the *R*-configured alcohol was obtained by reduction in the presence of the catalyst bearing an *R*,*R*-diamine ligand. Alcohol chirality was determined after workup and extraction, using chiral chromatography (GC-MS using a chrompac cyclodextrin-β-236M-19 column). DFT calculations for the *S*,*S*-catalyst initially indicated that the energy of the *S*-alcohol and the 16-electron catalyst is higher than the corresponding assembly with the *R*-

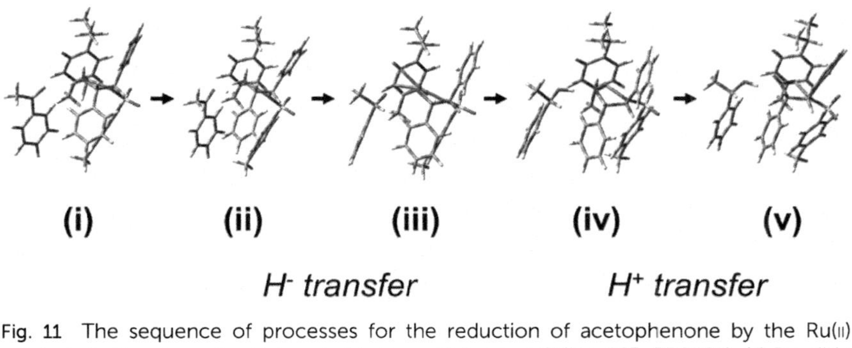

Fig. 11 The sequence of processes for the reduction of acetophenone by the Ru(II) catalyst starting from the H-bonded ketone. (i) Species **C** (Fig. S15, PDB12†). (ii) Possible transition state for hydride transfer towards the carbonyl group with preference for one enantiotopic face of the substrate ketone (*cf.* Fig. 8). (iii) The alkoxide is formed, with catalyst NH_2 group still retained (Fig. S29, PDB26†). (iv) Transition state for the amino proton transfer (Fig. S28, PDB25†). (v) Catalyst NH^- adopts trigonal planar geometry (Fig. S31, PDB28†).

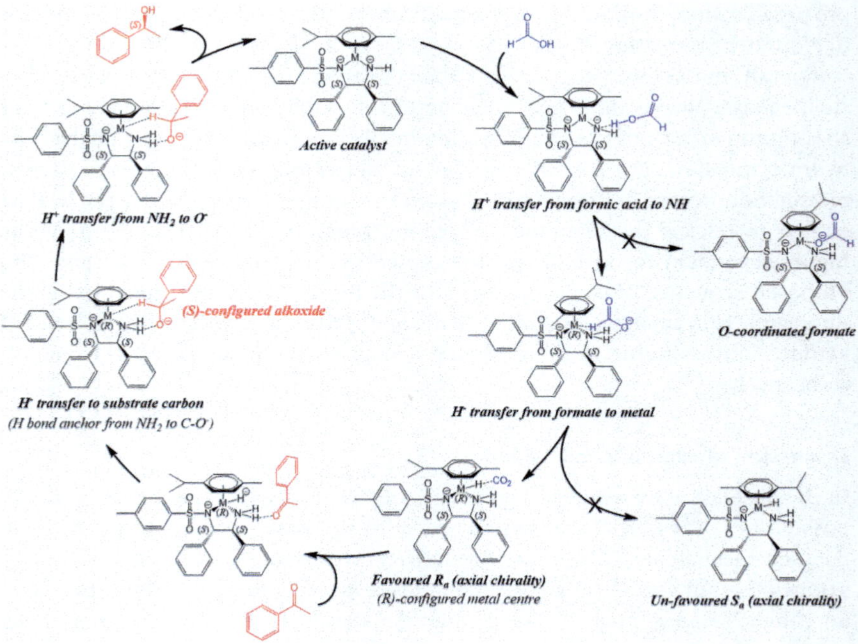

Fig. 12 The mechanism of asymmetric transfer hydrogenation catalysed by Ru(II) and Os(II) half-sandwich TsDPEN catalysts with formate as the hydride donor. The diamine ligand is always shown as (*S*,*S*)-configured. First, formic acid transfers H^+ to the catalyst M–NH, without prior formation of an *O*-coordinated intermediate. Hydride transfer then occurs from the formate anion to the metal, forming the hydride intermediate bearing a favoured (*R*)-configured metal centre and R_a axial chirality. The carbonyl of the substrate (acetophenone) forms a hydrogen bond with the catalyst NH_2 which serves to anchor the substrate molecule prior to the outer-sphere hydride transfer from M–H to the substrate carbon atom. After final H^+ transfer, the reduced substrate (*S*-phenylethanol) is released, in turn regenerating the active 16-electron catalyst.

alcohol, thus if the reaction outcome was purely governed by thermodynamics, the predicted enantioselectivity of the reduction would be opposite to experimental observations. It seems likely therefore that experimental enantioselectivity is driven by the kinetics of the hydride transfer.

Three possible orientations of substrate and catalyst were considered: (1) the structure with the minimum energy, in which the acetophenone ketone substrate is π–π stacked with the diamine pendant phenyl groups far away from the catalyst hydride and NH_2 (for Ru but not Os, Fig. 6), (2) O–HN bound ketone where the hydride is oriented towards one of the enantiotopic faces of the ketone (Fig. 8 for Ru), and (3) O–HN bound ketone with hydride oriented towards the plane of the ketone with no preference for either enantiotopic face of the ketone (Fig. 10 for Ru). The stereoisomers for orientations (2) and (3) are +16 and +25 kJ mol^{-1}, respectively, higher in energy than the π–π structure. In fact, the kinetically-favoured formation of the alcohol involves hydride transfer to the carbonyl with preference for one of the enantiotopic faces. In the case of orientation (2), this corresponds to the observed experimental enantioselectivity. Thus, the calculated stereochemical model for the TH of acetophenone agrees with the experimental data. Further DFT modelling will explore the transition states for all stereoisomers (*i.e.* two orientations of bound ketone, with two different axial chiralities, for both Ru and Os).

As a result of the present DFT calculations, we can now suggest that the two Os–H ^{1}H NMR resonances observed in spectra of the active hydride species (with chemical shifts of −5.89 and −6.04 ppm, and ^{187}Os satellites, $^{1}J(^{1}H, ^{187}Os) = 44$ Hz), can be assigned to the two possible axial conformers (R_a and S_a), and not to a mixture of Os–H diastereomers as originally proposed.[16] Calculated energies of the axial species suggest that they would be present in an 4 : 1 ratio, in agreement with experimental observations.

Future experimental studies of these Os(II) catalysts might involve the use of deuterium-labelled formate to probe kinetic isotope effects and the stereospecificity for H/D transfer. Such reported studies for Ru(II) have indicated that H transfer from formate to the ketone is rate-limiting, and (in agreement with the DFT calculations above) that H transfer is sequential rather than concerted.

Conclusions

Catalytic delivery of hydrogen to a substrate to carry out chemical reduction can be achieved by different systems (metal or organic) in a variety of ways. For metal-catalysed reactions, this could involve redox change at the metal centre, however in the case of transfer hydrogenation, the reduction occurs *via* an outer-sphere process, with no apparent direct interaction between the substrate and the catalyst metal centre. For Ru/TsDPEN systems, the TH mechanism is generally well understood: experiments have demonstrated transfer of deuterium (^{2}H) from formate to the substrate,[28] and many reports have explored the breadth of compatible substrates, including how particular substituents impact conversion and enantioselectivity. Various sources of hydride have been explored, including 2-propanol and formic acid in chemical systems,[5,25] as well as organic hydride donors, such as NADH, in biological systems.[36] More recently, computational studies have provided detailed insight into the subtler aspects of the ATH mechanism.[30,32]

We have modelled the catalytic reduction of acetophenone using the Noyori ruthenium TsDPEN catalyst, and its analogous osmium complex, in the presence of formic acid, revealing a novel interaction between the substrate and the chiral diamine ligand. We find the electronic energies of the diastereomers of the chiral diamine ligand to be dependent on the absolute configuration at the Ru/Os centre: The *S* diastereomer is less stable than the *R* diastereomer due to extensive steric clashes, resulting in ring slippage (from η^{6} to η^{2}) for the latter. Previous ^{1}H NMR studies revealed the formation of two Os–H singlets which changed from a 3 : 1 to 1.2 : 1 ratio over time, suggesting a mixture of these diastereomers. However, our calculations suggest that this not the case, and this observation arises instead as a consequence of axial chirality (*p*-cymene orientation).

Modelling of the reaction profile for the reduction of [Ru(*p*-cymene)(*S*,*S*-TsDPEN-H)Cl] and [Os(*p*-cymene)(*S*,*S*-TsDPEN-H_2)] with formic acid revealed acetophenone preferentially bound *via* π–π stacking to the tosylate phenyl ring, followed by protonation of the amino group of the ligand. Remarkably, our calculations suggest that the formation of Ru/Os–H does not occur *via* *O*-coordinated formate as proposed previously, instead involving a significant interaction between CO_2

product and the metal-hydride. In the final DFT structure in the catalytic cycle, acetophenone is anchored to the catalyst *via* a H-bond between the substrate carbonyl-O···HNH group. Further geometry optimisation revealed the formation of alkoxide bound to the NH_2 group, leading to gradual amino proton transfer, suggesting that hydride transfer itself allows the dissociation of the NH_2 to form the initial pre-catalyst. Moreover, our calculations suggest that substrate docking to the amine followed by hydride transfer occurs prior to substrate protonation. Thus, the presence of the amine proton is crucial for the stability of the hydride.

In future work, efforts will be made to identify the transition states for all possible (minimum four) diastereoisomers involving the H-bonded ketone, with calculation of the pathways along the identified internal reaction coordinate for both Ru and Os analogues, following the approach outlined previously.[29] It is envisaged that a detailed understanding of the reaction coordinates for both metals will provide mechanistic insights into why the osmium complexes can achieve higher rates of catalysis for ATH reduction compared to their well-established ruthenium analogues, even after accounting for dissociation of HCl from the pre-catalyst.

It will also be interesting to discuss and compare the mechanisms for transfer hydrogenation catalysed by the organometallic complexes described here, with those for purely organic catalysts, including natural enzymes. The effective (stereospecific) addition of H_2 to substrates may involve reductants ranging from H_2 itself, to sequential addition of $H^•$ (H atoms), electrons, and protons. For both types of catalyst, there are likely to be key roles for neighbouring groups in recognising and orienting the substrate, and transferring the reducing equivalents. The versatile ability of metal ions in transition metal coordination complexes to organise the structures and electronic properties of their ligands in the first and second coordination spheres, to generate 'H_2', as well as control the dynamics of substrate recognition and product release, through changes in coordination numbers and coordination geometries and oxidation states, is notable.

Materials and methods

DFT calculations

The calculations were performed with the Gaussian 16 packet.[40] The relevant structures were optimised using the TPSS functional,[41] and QZVP basis set.[42] Grimme's D3 dispersion correction was applied.[43] Gaussian's IEFPCM option was used to model the formic acid as solvent. Typically, the optimization of the geometry for each energy minima shown takes two to three months using the parallel computations with 24 Intel XEON SP 6126 (http://ark.intel.com/products/120483/) processors (*cf.* https://elwe.rhrk.uni-kl.de/elwetritsch/hardware.shtml). No imaginary frequencies were found for the optimized geometries. The data for the transition states of the cymene rotation and hydride transfer are estimated only as corresponding to not yet fully optimized geometries. All computational data are available from JAW on request.

Author contributions

The research was conceptualized by EMB, JPCC, JAW, VS and PJS. Formal analysis was carried out by JAW with methodology and investigation supported by EMB,

JPCC and PJS. Writing of the original draft, review and editing was carried out by EMB, JPCC, JAW and PJS.

Conflicts of interest

The authors have no conflicts of interest to declare.

Acknowledgements

We thank Diamond Light Source and Warwick Collaborative Postgraduate Research Scholarship (WCPRS, PhD award for E. M. B.), the Engineering and Physical Sciences Research Council (EPSRC grant no. EP/P030572/1), and Anglo American Platinum (for P. J. S.), for funding. V. S. and J. A. W. acknowledge support by the research initiative NANOKAT, as well as Allianz für Hochleistungsrechnen Rheinland–Pfalz (AHRP) for providing CPU-time within the project TUK-SPINPLUSVIB.

Notes and references

1 S. Ilic, A. Alherz, C. B. Musgrave and K. D. Glusac, *Chem. Soc. Rev.*, 2018, **47**, 2809–2836.
2 S. Banerjee and P. J. Sadler, *RSC Chem. Biol.*, 2021, **2**, 12–29.
3 T. Ohkuma, *Proc. Jpn. Acad., Ser. B*, 2010, **86**, 202–219.
4 C. A. Sandoval, T. Ohkuma, K. Muñiz and R. Noyori, *J. Am. Chem. Soc.*, 2003, **125**, 13490–13503.
5 R. Noyori and S. Hashiguchi, *Acc. Chem. Res.*, 1997, **30**, 97–102.
6 J. Václavík, P. Sot, B. Vilhanová, J. Pecháček, M. Kuzma and P. Kačer, *Molecules*, 2013, **18**, 6804–6828.
7 T. Ikariya and A. J. Blacker, *Acc. Chem. Res.*, 2007, **40**, 1300–1308.
8 T. Ikariya, K. Murata and R. Noyori, *Org. Biomol. Chem.*, 2006, **4**, 393–406.
9 M. Yamakawa, H. Ito and R. Noyori, *J. Am. Chem. Soc.*, 2000, **122**, 1466–1478.
10 M. Yamakawa, I. Yamada and R. Noyori, *Angew. Chem., Int. Ed.*, 2001, **40**, 2818–2821.
11 T. Ohkuma, N. Utsumi, K. Tsutsumi, K. Murata, C. Sandoval and R. Noyori, *J. Am. Chem. Soc.*, 2006, **128**, 8724–8725.
12 F. Chen, J. J. Soldevila-Barreda, I. Romero-Canelón, J. P. C. Coverdale, J.-I. Song, G. J. Clarkson, J. Kasparkova, A. Habtemariam, V. Brabec, J. A. Wolny, V. Schünemann and P. J. Sadler, *Dalton Trans.*, 2018, **47**, 7178–7189.
13 R. Soni, T. H. Hall, B. P. Mitchell, M. R. Owen and M. Wills, *J. Org. Chem.*, 2015, **80**, 6784–6793.
14 F. Chen, I. Romero-Canelón, J. J. Soldevila-Barreda, J. I. Song, J. P. C. Coverdale, G. J. Clarkson, J. Kasparkova, A. Habtemariam, M. Wills, V. Brabec and P. J. Sadler, *Organometallics*, 2018, **37**, 1555–1566.
15 Y. Fu, C. Sanchez-Cano, R. Soni, I. Romero-Canelon, J. M. Hearn, Z. Liu, M. Wills and P. J. Sadler, *Dalton Trans.*, 2016, **45**, 8367–8378.
16 J. P. C. Coverdale, C. Sanchez-Cano, G. J. Clarkson, R. Soni, M. Wills and P. J. Sadler, *Chem.–Eur. J.*, 2015, **21**, 8043–8046.

17 E. M. Bolitho, J. P. C. Coverdale, H. E. Bridgewater, G. J. Clarkson, P. D. Quinn, C. Sanchez-Cano and P. J. Sadler, *Angew. Chem., Int. Ed.*, 2021, **60**, 6462–6472.
18 J. P. C. Coverdale, I. Romero-Canelón, C. Sanchez-Cano, G. J. Clarkson, A. Habtemariam, M. Wills and P. J. Sadler, *Nat. Chem.*, 2018, **10**, 347–354.
19 E. M. Bolitho, N. G. Worby, J. P. C. Coverdale, J. A. Wolny, V. Schünemann and P. J. Sadler, *Organometallics*, 2021, **40**, 3012–3023.
20 S. Infante-Tadeo, V. Rodríguez-Fanjul, A. Habtemariam and A. M. Pizarro, *Chem. Sci.*, 2021, **12**, 9287–9297.
21 S. Betanzos-Lara, Z. Liu, A. Habtemariam, A. M. Pizarro, B. Qamar and P. J. Sadler, *Angew. Chem., Int. Ed.*, 2012, **51**, 3897–3900.
22 J. J. Soldevila-Barreda, A. Habtemariam, I. Romero-Canelón and P. J. Sadler, *J. Inorg. Biochem.*, 2015, **153**, 322–333.
23 M. J. Rogatzki, B. S. Ferguson, M. L. Goodwin and L. B. Gladden, *Front. Neurosci.*, 2015, **9**, 22.
24 L. Sellés Vidal, C. L. Kelly, P. M. Mordaka and J. T. Heap, *Biochim. Biophys. Acta, Proteins Proteomics*, 2018, **1866**, 327–347.
25 A. Fujii, S. Hashiguchi, N. Uematsu, T. Ikariya and R. Noyori, *J. Am. Chem. Soc.*, 1996, **118**, 2521–2522.
26 K.-J. Haack, S. Hashiguchi, A. Fujii, T. Ikariya and R. Noyori, *Angew. Chem., Int. Ed. Engl.*, 1997, **36**, 285–288.
27 R. Noyori, M. Yamakawa and S. Hashiguchi, *J. Org. Chem.*, 2001, **66**, 7931–7944.
28 X. Wu, J. Liu, D. Di Tommaso, J. A. Iggo, C. R. Catlow, J. Bacsa and J. Xiao, *Chem.–Eur. J.*, 2008, **14**, 7699–7715.
29 P. A. Dub and J. C. Gordon, *Dalton Trans.*, 2016, **45**, 6756–6781.
30 Y. Meftah, Y. Boumedjane, P. Fleurat-Lessard, F. Delbecq and C. Michel, *J. Organomet. Chem.*, 2021, **939**, 121765.
31 J.-W. Handgraaf and E. J. Meijer, *J. Am. Chem. Soc.*, 2007, **129**, 3099–3103.
32 P. Šot, M. Kuzma, J. Václavík, J. Pecháček, J. Přech, J. Januščák and P. Kačer, *Organometallics*, 2012, **31**, 6496–6499.
33 P. Crochet and V. Cadierno, *Inorganics*, 2021, **9**, 55.
34 J. W. Faller and A. R. Lavoie, *Org. Lett.*, 2001, **3**, 3703–3706.
35 R. A. Sarfraz, T. G. Kazi, S. Iqbal, H. I. Afridi, M. K. Jamali, N. Jalbani and M. B. Arain, *Appl. Organomet. Chem.*, 2008, **22**, 187–192.
36 J. J. Soldevila-Barreda, I. Romero-Canelón, A. Habtemariam and P. J. Sadler, *Nat. Commun.*, 2015, **6**, 6582.
37 J. P. C. Coverdale, H. E. Bridgewater, J.-I. Song, N. A. Smith, N. P. E. Barry, I. Bagley, P. J. Sadler and I. Romero-Canelón, *J. Med. Chem.*, 2018, **61**, 9246–9255.
38 W. Wang and X. Yang, *Chem. Commun.*, 2019, **55**, 9633–9636.
39 J. J. Soldevila-Barreda, P. C. A. Bruijnincx, A. Habtemariam, G. J. Clarkson, R. J. Deeth and P. J. Sadler, *Organometallics*, 2012, **31**, 5958–5967.
40 M. J. Frisch, G. W. Trucks, H. B. Schlegel, G. E. Scuseria, M. A. Robb, J. R. Cheeseman, G. Scalmani, V. Barone, G. A. Petersson, H. Nakatsuji, X. Li, M. Caricato, A. V. Marenich, J. Bloino, B. G. Janesko, R. Gomperts, B. Mennucci, H. P. Hratchian, J. V. Ortiz, A. F. Izmaylov, J. L. Sonnenberg, D. Williams-Young, F. Ding, F. Lipparini, F. Egidi, J. Goings, B. Peng, A. Petrone, T. Henderson, D. Ranasinghe, V. G. Zakrzewski, J. Gao, N. Rega, G. Zheng, W. Liang, M. Hada, M. Ehara, K. Toyota, R. Fukuda, J. Hasegawa,

M. Ishida, T. Nakajima, Y. Honda, O. Kitao, H. Nakai, T. Vreven, K. Throssell, J. J. A. Montgomery, J. E. Peralta, F. Ogliaro, M. J. Bearpark, J. J. Heyd, E. N. Brothers, K. N. Kudin, V. N. Staroverov, T. A. Keith, R. Kobayashi, J. Normand, K. Raghavachari, A. P. Rendell, J. C. Burant, S. S. Iyengar, J. Tomasi, M. Cossi, J. M. Millam, M. Klene, C. Adamo, R. Cammi, J. W. Ochterski, R. L. Martin, K. Morokuma, O. Farkas, B. Foresman and D. J. Fox, Gaussian, Inc., Wallingford CT, 2016.

41 J. Tao, J. P. Perdew, V. N. Staroverov and G. E. Scuseria, *Phys. Rev. Lett.*, 2003, **91**, 146401–146404.

42 F. Weigend, *Phys. Chem. Chem. Phys.*, 2006, **8**, 1057–1065.

43 S. Grimme, J. Antony, S. Ehrlich and H. Krieg, *J. Chem. Phys.*, 2010, **132**, 154104–154118.

Faraday Discussions

ROYAL SOCIETY OF CHEMISTRY

PAPER

A GPx-mimetic copper vanadate nanozyme mediates the release of nitric oxide from *S*-nitrosothiols†

Sourav Ghosh, Punarbasu Roy, Sanjay Prasad and Govindasamy Mugesh *

Received 4th October 2021, Accepted 6th December 2021

DOI: 10.1039/d1fd00067e

Although reactive oxygen and nitrogen species (ROS/RNS), such as hydrogen peroxide (H_2O_2), nitric oxide (NO), hydroxyl radicals ($OH^•$), superoxide (O_2^-) *etc.*, play crucial roles in redox biology and cellular signaling, higher concentrations of these species lead to oxidative and nitrosative stress, which are associated with various pathophysiological conditions like neurodegeneration, cardiovascular diseases and cancer. There is growing evidence that functional impairment of the endothelium is one of the first recognizable signs of the development of atherosclerotic cardiovascular disease. A decreased bioavailability of NO and increased generation of ROS are the two major molecular changes associated with endothelial dysfunction. Therefore, it is a viable strategy to increase the bioavailability of NO while reducing the amount of ROS to prevent the progression of cardiovascular diseases. In this paper, we discuss for the first time that copper vanadate (CuV_2O_6) can not only release NO from *S*-nitrosothiols but can also control the ROS levels by functionally mimicking the antioxidant enzyme glutathione peroxidase (GPx) at physiological pH. We used several imaging techniques and spectroscopic measurements to understand the catalysis on the surface of the material during the reactions. The denitrosylation, as well as GPx-like activity, by CuV_2O_6 can be carried out multiple times without affecting the catalytic activity.

Introduction

Reactive oxygen species (ROS), such as hydrogen peroxide (H_2O_2), hydroxyl radicals ($^•OH$) and superoxide (O_2^-), are reactive intermediates of molecular oxygen that act as important secondary messengers within cells at physiological concentrations.[1–5] However, excessive generation of ROS or an imbalance between the generation of reactive ROS and the antioxidant defense systems cause oxidative stress, resulting in damage to biomolecules such as DNA, proteins, and

Department of Inorganic & Physical Chemistry, Indian Institute of Science, Bangalore, 560012, India. E-mail: mugesh@iisc.ac.in

† Electronic supplementary information (ESI) available. See DOI: 10.1039/d1fd00067e

lipids.[6–9] In the long term, these effects can lead to various pathophysiological conditions like neurodegeneration, HIV activation, diabetes, cardiovascular diseases (CVDs), chronic kidney diseases, cancer, and aging.[6–16] There is growing evidence that the functional impairment of the endothelium is one of the first recognizable signs of the development of atherosclerosis and is present long before the occurrence of atherosclerotic cardiovascular disease, one of the leading causes of morbidity and mortality worldwide.[17–21] The endothelium, a single layer of cells, covers the lumen of blood vessels to separate the vascular wall from the circulating blood.[22–24] Damage to the endothelium promotes endothelial dysfunction, characterized by increased vascular reactivity, lipid deposition, thrombus formation, platelet aggregation and vasoconstriction *etc.*

There are several potential pathways that contribute to endothelial dysfunction in patients with cardiovascular diseases. Many reports suggest that a balance between the production of nitric oxide (NO) and ROS is important for the proper functioning of the endothelial cells. NO, a potent endothelium-derived relaxing factor (EDRF), is synthesized by endothelial NOS (nitric oxide synthase) through a two-step process from L-arginine to *N*-hydroxyl-L-arginine.[25] It performs signaling through numerous downstream pathways *via* guanylate cyclase to generate cGMP, and by the direct *S*-nitrosylation of cysteine residues in proteins.[26–32] The modification of proteins further activates signaling molecules, such as the transcription factors NF-κB and AP-1, and inhibits Ca^{2+}-mediated vasoconstriction. Reduced NO production and/or increased NO degradation by reactive oxygen species (ROS) mark the onset of endothelial dysfunction.[17–21,33] Enzymes like nicotinamide adenine dinucleotide phosphate (NADPH) oxidases and xanthine oxidase, as well as alterations in mitochondrial respiratory chain complexes, are the sources of several oxidative stress-related chemical species. Additionally, oxidative stress uncouples eNOS and generates ROS, specifically the superoxide anion (O_2^-), which either gets converted to H_2O_2 by SOD or reacts with NO to the form of peroxynitrite $ONOO^-$.[17,34,35] Peroxynitrite promotes the nitration of the protein tyrosine residues and contributes to dysfunction and the death of endothelial cells. Furthermore, the cross talk between ROS generating processes and the imbalance of NO causes severe oxidative stress, which accelerates the progression of atherosclerosis and cardiovascular diseases. Therefore, increasing the bioavailability of NO and the protection of the cells against excess reactive oxygen and nitrogen species are two necessary requirements for the prevention of endothelial dysfunction related diseases. In this paper, we demonstrate for the first time that the CuV_2O_6 (CuV) nanozyme exhibits GPx-like antioxidant activity and mediates the release of NO from *S*-nitrosothiols at physiological pH.

Experimental section

Glutathione peroxidase (GPx)-like activity of nanomaterials

The standard UV–vis spectroscopy based GPx activity assay was used to monitor the GPx-like activity of the nanomaterials.[36] The decrease in the absorbance of reduced nicotinamide adenine dinucleotide phosphate (NADPH) at 340 nm was monitored to measure the rate of the reaction catalyzed by the nanomaterials. The reaction mixture, with a total volume of 1 mL, was composed of the catalyst (20 μg mL^{-1}), GSH (2 mM), NADPH (0.2 mM), GR (~1.7 U) and H_2O_2 (200 μM) in 100 mM, pH 7.4 phosphate buffer. All the assays, along with the respective controls, were carried out at 25 °C. The kinetics were studied by varying the concentration of both the substrates, H_2O_2 (0–400 μM) and GSH (0–7 mM), consecutively at fixed concentrations of GR and NADPH in phosphate buffer (100 mM, pH 7.4). Sigma plots were obtained by plotting V *vs.* $[S]$ (eqn (1)), which follows Michaelis–Menten kinetics. Michaelis–Menten parameters, such as the substrate binding affinity (K_M) and maximum reaction velocity (V_{max}), were obtained from the Lineweaver–Burk plots (eqn (2)), where V is the rate of the reaction and [S] is the concentration of substrate (in our case, H_2O_2 and GSH).

$$V = \frac{V_{\max}[S]}{K_M + [S]} \tag{1}$$

$$\frac{1}{V} = \frac{K_M}{V_{\max}}\frac{1}{[S]} + \frac{1}{V_{\max}} \tag{2}$$

Measurement of the denitrosylation of *S*-nitrosothiols (RSNO) by nanomaterials

The denitrosylation of *S*-nitroso-*N*-acetyl penicillamine (SNAP) and *S*-nitrosoglutathione (GSNO) was followed by monitoring the decrease in the absorbance at 335 nm over time using a UV–vis spectrophotometer.[37–39] The reactions were carried out in a 1 mL quartz cuvette. For the denitrosylation of SNAP, 1 mL of the reaction mixture contained the catalyst (4 ng μL^{-1}) and SNAP (0.5 mM). For the denitrosylation of GSNO, 1 mL of the reaction mixture contained the catalyst (20 ng μL^{-1}) and GSNO (1.0 mM). In the case of GSNO, the reactions were carried out in the presence of GSH as well. The standard concentration of GSH used was 0.5 mM in the reaction mixture. The rate of denitrosylation was calculated from the linear region of the absorbance *vs.* time plot by the following equation. The molar extinction coefficients are given in Table S1.†

$$\text{Rate of denitrosylation} = \frac{\text{rate of decrease in the absorbance }(\text{s}^{-1})}{\text{molar extinction coefficient }(\text{M}^{-1}\ \text{cm}^{-1})} \times 10^9\ \text{nM m}^{-1}$$

Detection of nitric oxide using fluorescence assays

A smaller tube containing CuV (50 ng μL^{-1}) and GSH (5 mM) in 0.1 M phosphate buffer with pH 7.4 was carefully placed inside a bigger vial containing DAN (2,3-diaminonaphthalene) (1.0 mM in DMSO) solution so that two solutions remained separated, and the bigger tube was sealed with septum and paraffin. A solution of GSNO (20 mM) was injected into the inner tube. The vials were shaken occasionally for 45 min to allow for the generation of nitric oxide. The *in situ* generated nitric oxide diffuses into the bigger vial containing DAN in DMSO and reacts with the DAN to form 2,3-naphthotriazole (NATH), which is weakly fluorescent. When NATH reacts with the NaOH solution, it forms the 2,3-naphthotriazole anion (NAT), which is strongly fluorescent. In our experiments, after the incubation, 15 μL of the solution from the outer vial was taken and diluted in 3 mL of an alkaline solution of a DMSO + H_2O mixture. The final concentration of the NAT in the cuvette was 5 μM. As controls, reactions were carried out in the absence of GSH or both CuV and GSH.[40]

Detection of nitric oxide by EPR spectroscopy

The release of nitric oxide was probed using bis(*N*,*N*-diethyldithiocarbamato) iron(II) ($(dtc)_2Fe^{II}$) by EPR spectroscopy. A smaller Eppendorf tube containing GSNO (20 mM in 0.1 M phosphate buffer pH 7.4) and CuV nanomaterials (50 μg mL^{-1}) was placed carefully inside a bigger Eppendorf containing $(dtc)_2Fe^{II}$ (in DCM) so that the two solutions remained separated. The larger Eppendorf tube was kept sealed, and the setup was kept undisturbed for 45 min to allow for the *in situ* generated NO to diffuse into the DCM solution, react with $(dtc)_2Fe^{II}$ and form $(dtc)_2FeNO$. After 45 min, 200 μL of the DCM solution from the bigger Eppendorf tube was transferred to a quartz EPR tube and analyzed by X-band EPR spectroscopy. The room temperature EPR spectrum exhibited three lines, which indicated the formation of the $(dtc)_2FeNO$ complex and, therefore, confirmed the release of NO during CuV mediated denitrosylation.[41]

The conditions used during EPR-spectroscopy measurement were as follows: central field = 336.0 mT, mod. width = 1.0 × 1 mT, sweep width = 5 × 10 mT, time constant = 0.03 s. The g-values and frequencies used for EPR are summarized in Table S2.†

FT-Raman and FT-IR spectroscopy after catalysis

The catalysis was performed in 0.1 M phosphate buffer at 25 °C. After each time point, the catalyst was recovered from the reaction mixture by centrifugation and washed with acetone several times. The recovered catalyst was used for the spectroscopic studies. FT-Raman spectroscopy was performed using a Renishaw inVia™ Raman microscope (Renishaw Inc, UK). The FT-IR spectra were recorded on PerkinElmer instrument using a KBr pellet made with the nanomaterials recovered from the reaction mixture.

Isothermal titration calorimetry (ITC)

The stock solution of the catalyst (2 mg mL^{-1}) was prepared by dispersing the catalyst in isopropyl alcohol (IPA) and the stock solution of GSH was prepared in 0.1 M sodium phosphate buffer with pH 7.4. ITC experiments were performed using a Microcal VP-ITC at 25 °C. The catalyst (20 ng μL^{-1}) was dispersed in ultrapure water and loaded onto the sample cell. The ligand solution (GSH, 1.0 mM), loaded in the injection syringe, was titrated into the sample cell as a series of injections (aliquots of 10 μL) separated by 4 min interval with constant stirring at 307 RPM. Control titrations with the ligand into water in the sample cell and water titrated with the catalyst in the sample cell were carried out to confirm that the heat of dilution was negligible. The data were fitted with Microcal Origin 5 software. After subtraction of the respective controls from the main titration profile, the binding constant (K_a) and enthalpy (ΔH) were obtained by fitting the titration curves to a one-site binding model.

Results and discussion

Synthesis and characterization of the nanomaterials

Three metal vanadates, CuV_2O_6 (CuV), MnV_2O_6 (MnV) and ZnV_2O_6 (ZnV), and two metal oxides, CuO and Cu_2O, were synthesized by hydrothermal and precipitation methods, respectively (see the ESI†). The synthesized materials were

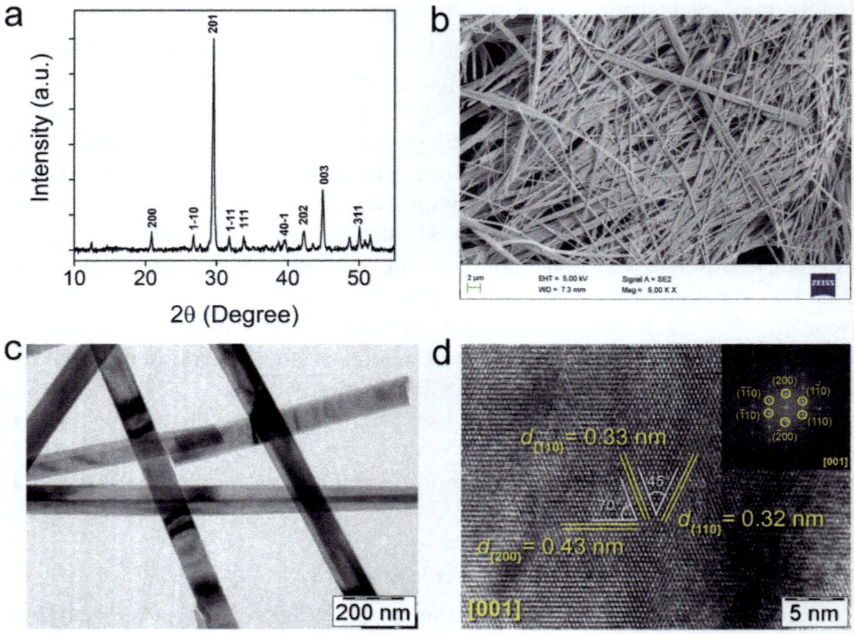

Fig. 1 (a) PXRD of the CuV_2O_6 nanowires. The peaks matched well with standard JCPDS pattern (45-1045) of triclinic CuV_2O_6. (b) and (c) Scanning electron microscopy (SEM) and transmission electron microscopy images of CuV, respectively. The images show that the synthesized material has a wire-like morphology with an average length and diameter of 1–2 μm and 150–250 nm, respectively. (d) High resolution TEM (HRTEM) and its corresponding FFT pattern (inset). The lattice fringes observed in HRTEM are the (1–10), (110) and (200) planes with *d*-spacings of 0.33 nm, 0.32 nm and 0.43 nm, respectively. The interfacial angles of (1–10) and (110) are 45.0°, and similarly the interfacial angle between (200) and (1–10) is 70.0°. These three planes in the HRTEM fall in the common zone axis [001].

characterized using various imaging and spectroscopic techniques. In the powder X-ray diffraction (PXRD) patterns, sharp peaks were observed for all the materials, which indicated the crystalline nature of the synthesized nanomaterials (Fig. 1a and S2†). For CuV, the PXRD pattern was indexed using the standard JCPDS pattern of triclinic CuV_2O_6, and for MnV, ZnV and CuO, the phases were observed to be monoclinic.[42–45] Scanning electron microscopy (SEM) and transmission electron microscopy (TEM) images were taken to visualize the morphologies of the materials and it was observed that all the materials except Cu_2O formed wire-like morphologies (Fig. 1b, c, S1 and S3†). The elemental compositions of all the vanadates were investigated by the energy dispersive spectroscopy (EDS) spectra, which show characteristic peaks arising from the inner K shells of the respective elements (Fig. S4†). Further X-ray mapping images confirmed the elemental purity, as well as the homogenous distribution of atoms in the materials (Fig. S5–S7†). The well resolved lattice fringes observed in the HRTEM images of all the nanomaterials confirmed the highly crystalline nature of the materials, which is in accordance with the result of the PXRD (Fig. 1a and S2†).

To understand the nature of the bonding between the metal and oxygen atoms, FT-Raman and FT-IR spectra were recorded for all the materials. The peaks in the ranges of 700 to 950 cm^{-1} (Raman spectra) and 800 to 1000 cm^{-1} (FT-IR spectra) are observed due to the stretching vibrations of the V–O bonds of M–vanadates. In the FT-IR spectra, the peak below 600 cm^{-1} corresponds to the M–O stretching vibrations of MV_2O_6. The detailed analyses of the peaks of the FT-Raman and FT-IR spectra are given in Fig. S10 and S11, respectively.†[46] To investigate the detailed composition and oxidation states of the constituent elements of CuV, XPS was carried out (Fig. S12 and S13†). No impurity peaks except carbon and nitrogen were identified in the wide range XPS spectrum of CuV, which confirmed that the material was pure and contained copper, vanadium and oxygen (Fig. S12†). The binding energies (BE) and full width at half maxima (FWHM) for the V $2p_{3/2}$ and V $2p_{1/2}$ peaks corresponded to the +5 state of V in CuV. The characteristic doublet (at 934.16 eV for Cu $2p_{3/2}$ and 954.01 eV for Cu $2p_{1/2}$) with a shake-up satellite peak (at 944 eV) observed in the narrow range XPS of copper is the signature of the +2 oxidation state of Cu. The deconvolution of the O 1s resulted in two asymmetric peaks; the intense one centered at 530.33 eV was due to the metal–oxygen (M–O) bond and the other at 531.69 eV was assigned to the metal bound hydroxyl group (M–OH). The binding energies and full width at half maxima (FWHM) values from the detailed analysis of the XPS spectra are given in Fig. S13 and Table S3, respectively.†[36,47–51]

Glutathione peroxidase (GPx)-like activity of CuV

After the comprehensive characterization of the materials, standard glutathione reductase (GR)-coupled GPx assays (Fig. 2a) were performed to examine the enzyme-like activity of the nanomaterials using UV–vis spectrophotometry.[36] The decrease in the absorbance of NADPH (at 340 nm) was monitored to measure the rate of reaction. V_2O_5 nanowires (VNw) were used as a positive control during the course of reaction and it was observed that among the three vanadates (MnV, ZnV and CuV), the copper-based material showed the highest GPx-like activity, although the activity was found to be lower than that of VNw (Fig. 2a and b). To rule out the possibility of non-specific reactions causing the decrease in the absorbance at 340 nm, various control experiments were performed. To further understand the specific GPx-mimetic nature of the nanomaterial CuV, it was important to examine whether the material could functionally mimic other antioxidant enzymes such as superoxide dismutase (SOD) or catalase.

Briefly, the SOD mimetic activity of CuV was checked using a dye, WST-1 (2-(4-iodophenyl)-3-(4-nitrophenyl)-5-(2,4-disulfophenyl)-2*H*-tetrazolium), coupled with xanthine/xanthine oxidase (XO), where superoxide (O_2^-), generated by the reaction of xanthine/xanthine oxidase, can reduce WST-1 to spectrophotometrically detectable formazan at 440 nm.[52] The percentage of formazan production did not change with an increasing concentration of CuV (0–40 ng μL^{-1}), indicating that CuV did not show SOD-like activity (Fig. S14†). Similarly, to investigate the catalase-like activity of the materials, the decrease of the absorbance of H_2O_2 at 240 nm was followed using a UV–vis spectrophotometer (Fig. S14c†). It was observed that except for MnO_2, which is a known catalase mimetic, none of the materials showed catalase-like activity. These results again confirm that CuV

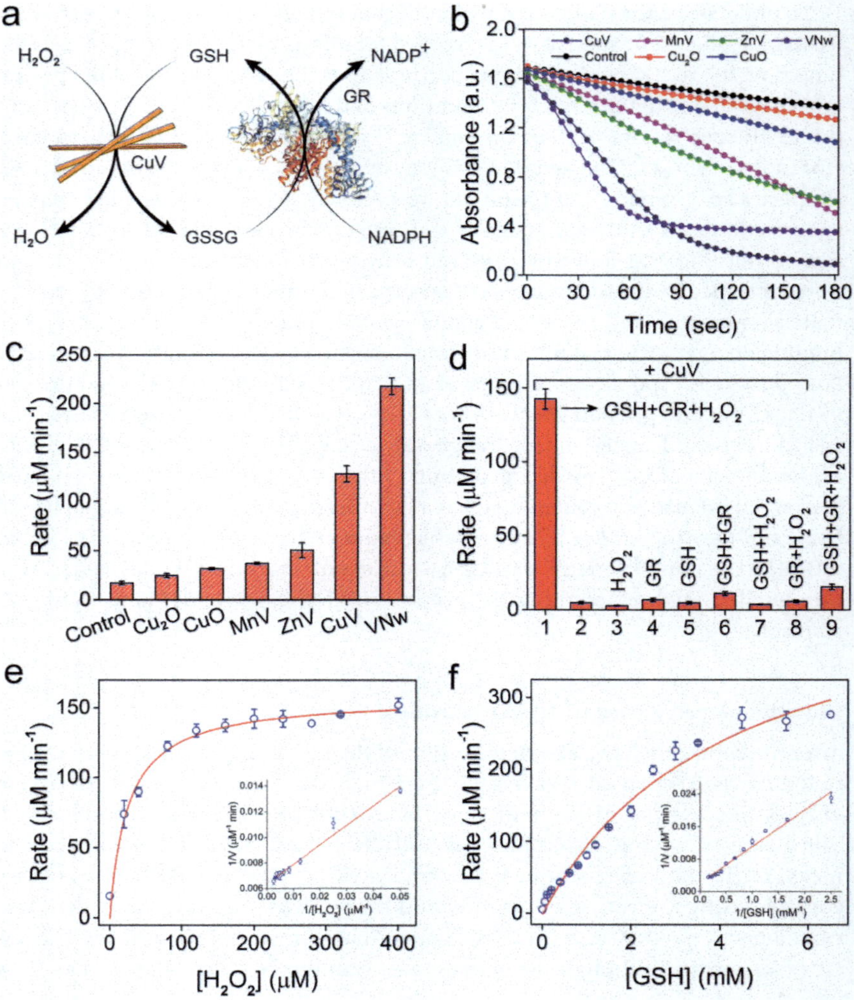

Fig. 2 (a) Schematic of the glutathione reductase (GR) coupled GPx assay used to measure the activity of the nanozyme. (b) Monitoring the decrease of the absorbance of NADPH (at 340 nm) to examine the GPx-like activity of different metal vanadate and metal oxide nanomaterials. (c) Bar diagram shows a comparison of the initial rates of reaction catalyzed by the different nanomaterials. The reaction mixture had a total volume of 1 mL, composed of the catalyst (20 ng μL^{-1}), GSH (2 mM), NADPH (0.3 mM), GR (~1.7 U) and H_2O_2 (200 μM) in phosphate buffer (100 mM, pH 7.4), and the catalysis was carried out at 25 °C. (d) Comparison of the initial rates for the GPx-like activity of CuV, along with the respective controls. The reaction mixture had a total volume of 1 mL, composed of the catalyst (20 ng μL^{-1}), GSH (2 mM), NADPH (0.2 mM) and GR (~1.7 U), H_2O_2 (200 μM) in phosphate buffer (100 mM, pH 7.4). All the assays, along with the respective controls, were carried out at 25 °C. (e) Michaelis–Menten plot and Lineweaver–Burk plot (inset) obtained from varying the concentration of H_2O_2 (0–400 μM) in the presence of CuV (20 ng μL^{-1}), GSH (2 mM), NADPH (0.2 mM) and GR (~1.7 units) in phosphate buffer (100 mM, pH 7.4) at 25 °C. (f) Michaelis–Menten plot and Lineweaver–Burk plot (inset) obtained by varying the concentration of GSH (0–7 mM) in the presence of CuV (20 ng μL^{-1}), H_2O_2 (200 μM), NADPH (0.2 mM) and GR (~1.7 units) in phosphate buffer (100 mM, pH 7.4) at 25 °C.

functionally mimics the GPx enzyme and can reduce H_2O_2 to H_2O only in the presence of GSH (Fig. 2d). GPx-like activity assays were also performed by varying the concentration of CuV (0 to 30 ng μL^{-1}) and a proportional dependence of the initial rate was observed with first-order reaction kinetics (Fig. S15†). GPx requires both GSH and H_2O_2 as substrates, simultaneously, to metabolize H_2O_2 by converting two molecules of GSH to its oxidized form, GSSG. To understand the substrate binding at the surface of CuV, we studied the effect of H_2O_2 and GSH on the reaction rates and determined the kinetic parameters, such as the Michaelis constant (K_M) and maximum velocity (V_{max}), using various concentrations of H_2O_2 (0–400 μM) and GSH (0–7 mM); typical Michaelis–Menten kinetics was observed for both the substrates (Fig. 2e and f). The kinetic parameters were obtained from the corresponding Lineweaver–Burk plots. For H_2O_2, the K_M (μM) and V_{max} (μM min^{-1}) values obtained for CuV (23.83 and 156.25, respectively), were lower than those of VNw (44.4 and 192.3, respectively) reported from our group.[53] The results indicate that the CuV surface has a higher affinity towards H_2O_2 compared to that of VNw. Furthermore , the K_M value reported for the GPx enzyme was 25 μM in the case of H_2O_2 as a substrate, which was comparable to CuV's K_M value for H_2O_2, which was 23.83 μM.[67–69] These values are low and they indicate that the affinities of CuV and the GPx enzyme toward H_2O_2 are high, although there is a huge difference between the native enzyme and the nanozyme in terms of the structure and mechanism. The higher GPx-like activity ($\sim$140 μM min^{-1}) of CuV is also attributed to its higher affinity towards H_2O_2, leading to the broad coverage of the surface of CuV with H_2O_2 molecules, transforming GSH to GSSH and exhibiting steady state kinetics.

Natural biological pathways include the enzymes working in a cascade. Although nanozymes are known to be incompatible for cascade reactions with some other enzymes,[54–56] previously, we reported that different morphologies of V_2O_5 nanomaterials have the compatibility to work with glucose oxidase (GO) and GR in a series.[53] To check the compatibility of CuV with GO, we performed similar experiments and, interestingly, it was observed that CuV can also use H_2O_2 which was *in situ* generated by glucose and GO (Fig. S16†). As CuV exhibits efficient catalysis when coupled with GO and GR, the material may be useful for the sensing and quantification of glucose level in the food industry, medical monitoring and treatment of patients.

Denitrosylation activity of CuV and confirmation of NO release at physiological pH

The time-dependent denitrosylation of *S*-nitrosothiols (RSNO) by different nanomaterials was monitored spectrophotometrically by following the decrease in the absorbance of RSNO at 335 nm.[37–39] *S*-nitroso-*N*-acetyl penicillamine (SNAP), *S*-nitrosoglutathione (GSNO) and *S*-nitrosocysteine (CYSNO) were used as substrates for a comparison of the denitrosylation by different materials. As shown in Fig. 3a and b, only CuV and CuO were able to mediate the denitrosylation of SNAP. The rate of denitrosylation observed for CuV was found to be comparable to that of CuO, indicating that both materials may follow similar mechanisms for their activity. The denitrosylation activity of CuV was again validated by using other RSNO *i.e.* GSNO and CYSNO (Fig. S17†). As CuV uses GSH as one of the substrates to perform its GPx-like activity, we studied the

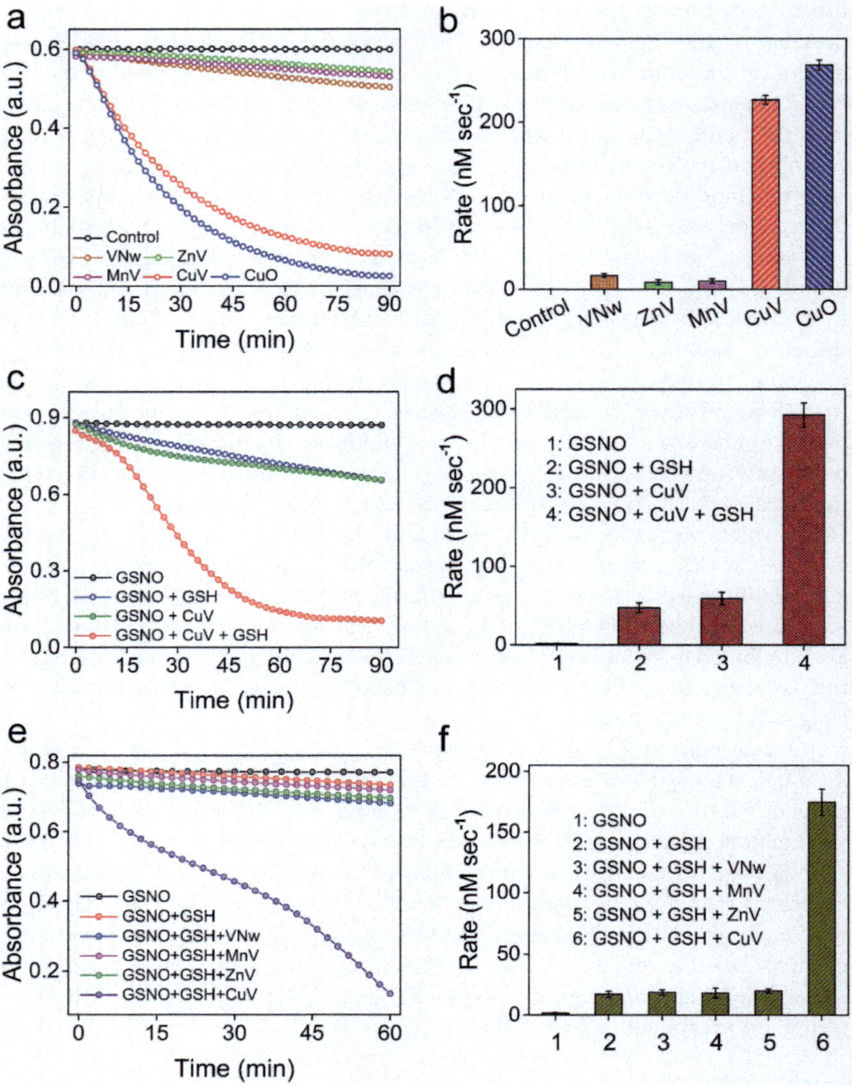

Fig. 3 Representative graph of the denitrosylation of SNAP by different M-oxide nanomaterials. (a) The decrease in the absorbance of SNAP was monitored at 335 nm by a UV–vis spectrophotometer. Assay conditions: SNAP (500 μM), catalyst (4 ng μL^{-1}), 0.1 M sodium phosphate buffer (pH 7.4), 25 °C. (b) The bar diagram shows a comparison of the rates of the denitrosylation of SNAP by the nanomaterials. (c) Representative graph of the denitrosylation of GSNO in the presence of different constituents. Assay conditions: GSNO (1.0 mM), catalyst (20 ng μL^{-1}), GSH (0.5 mM), 0.1 M sodium phosphate buffer (pH 7.4), 25 °C. In the absence of GSH and/or CuV, the rate of denitrosylation was observed to be very slow. (d) The bar diagram shows a comparison of the rate of the denitrosylation of GSNO in different conditions. (e) Representative graph of the denitrosylation of GSNO by different M-oxides in the presence of fixed concentrations of GSNO (1.0 mM), GSH (0.5 mM) and the catalyst (20 ng μL^{-1}). None of the M-oxides except CuV showed considerable denitrosylation, indicating the reaction between GSH and the surface of CuV. (f) Comparison of the rate of denitrosylation of GSNO (1.0 mM) by different M-oxides in the presence of GSH (0.5 mM) and the catalyst (20 ng μL^{-1}).

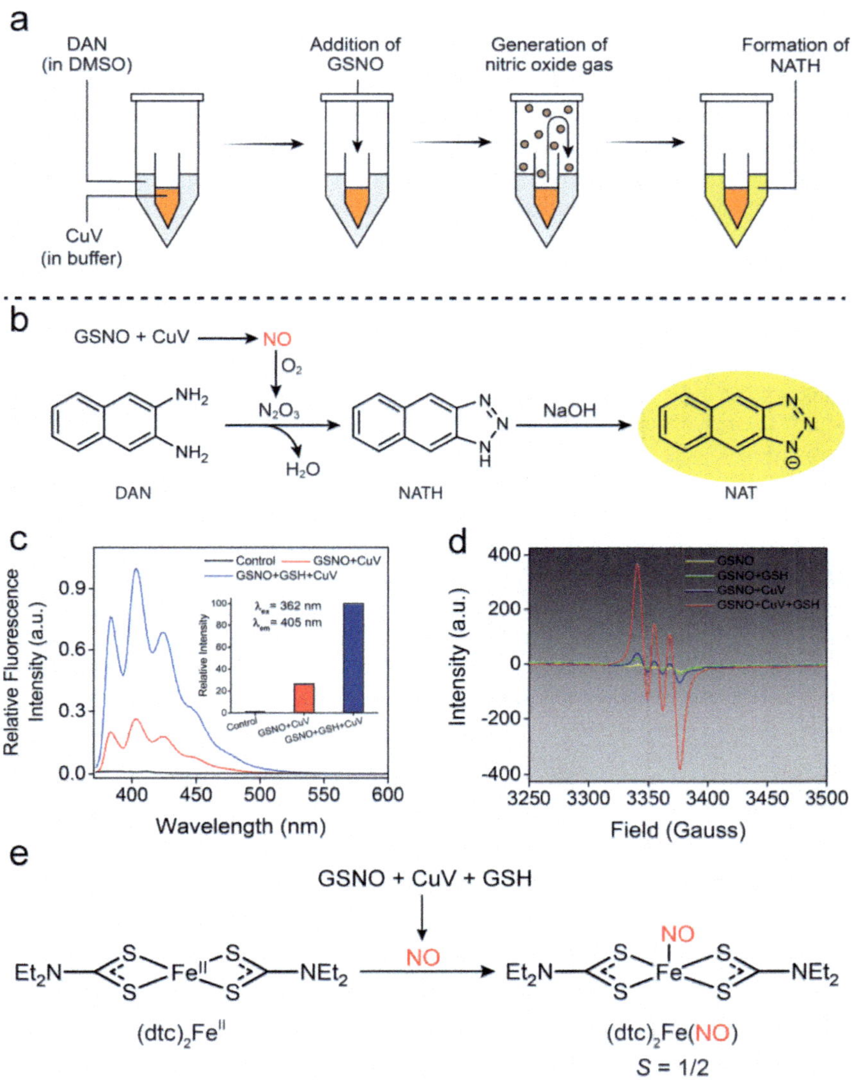

Fig. 4 (a) Schematic representation of the experimental setup to detect NO and the steps involved in the assay. (b) Reaction scheme for the formation of the 2,3-naphthotriazole anion (NAT) from 2,3-diaminonaphthalene (DAN). (c) Fluorescence spectra of DAN (5 μM) in DMSO solutions, using GSNO after treating with NaOH solution (black: GSH and CuV both absent in the buffer; red: CuV present but GSH absent in the buffer; blue: GSNO and CuV both present in the buffer). Strong fluorescence was observed only when both GSH and CuV were present in the buffer, which indicates more production of NO from GSNO, which in turn results in the formation of NATH (inset: relative fluorescence intensities). (d) Room temperature EPR spectra of the [$(dtc)_2FeNO$] complex. A strong EPR signal was detected from GSNO when both CuV and GSH were present, indicating that GSH has an important role in the enhanced denitrosylation. (e) Schematic representation of the formation of the $(dtc)_2FeNO$ complex upon reacting with NO and $(dtc)_2Fe$.

denitrosylation of GSNO in the presence of GSH. It was observed that the rate of denitrosylation of GSNO by CuV was enhanced in the presence of GSH (Fig. 3c and d). In the case of the control reaction where GSH was not used (GSNO + CuV), the denitrosylation was found to be very slow, indicating that the activation of the CuV surface by GSH is important for the activity. To understand the activation of the surface by GSH, one more experiment was performed for the denitrosylation of GSNO using different M-oxides in the presence of GSH (Fig. 3e and f). Interestingly, it was found that only CuV exhibited significant denitrosylation among them, implying that the reduction of surface Cu(II) ions by GSH is important for the enhanced activity.

Earlier reports have suggested that the denitrosylation of RSNO produces nitric oxide (NO).[57–62] We reported that the Cu_2O polymorph-mediated denitrosylation of RSNO indeed generates NO as one of the major products.[63] We demonstrated an experimental set-up for the detection of NO produced from the denitrosylation reactions. Using a similar set-up, we employed a 2,3-diaminonaphthalene (DAN)-based fluorescence method to detect the formation of NO during the denitrosylation of GSNO by CuV (Fig. 4a).[40]

Briefly, a smaller vial containing CuV (50 μg mL^{-1} in 0.1 M phosphate buffer pH 7.4) and GSH (5 mM) was placed inside a larger vial containing DAN (1.0 mM in DMSO) solution so that the two solutions did not mix with each other. The larger tube was then sealed and a solution of GSNO (20 mM) was injected into the smaller vial. The setup was left undisturbed for 45 min for the denitrosylation of GSNO. The *in situ* generated NO diffuses into the larger vial and reacts with DAN to produce 2,3-naphthotriazole (NATH), which is weakly fluorescent. In basic medium, NATH is converted to the 2,3-naphthotriazole anion (NAT), which is highly fluorescent (Fig. 4b).[40] We observed a strong fluorescence intensity of the outer solution when the inner vial contained GSNO, GSH and CuV. A relatively lower fluorescence intensity was observed in the case where CuV was added without GSH, and an almost negligible fluorescence intensity was observed when GSH and CuV were both absent (Fig. 4c). These results indicated that the CuV-mediated denitrosylation of GSNO produces NO and the rate of denitrosylation is increased in the presence of GSH.

The release of NO was again confirmed by electron paramagnetic resonance (EPR) spectroscopy using a similar experimental set-up. For this purpose, the $(dtc)_2Fe^{II}$ complex was used for trapping the NO released from GSNO to produce the EPR active $(dtc)_2FeNO$ (Fig. 4e). The room temperature EPR spectrum shows three lines ($S = 1/2$) at $g = 2.01$ with A $(^{14}N) = 13.0$ G, which is characteristic of the $(dtc)_2FeNO$ complex.[41] The weak EPR signal observed for GSNO in the absence of CuV and GSH indicated that the release of NO from GSNO alone is very slow and it leads to the formation of only a minor amount of $(dtc)_2FeNO$ complex. A very strong EPR signal was detected when GSNO was treated with both GSH and CuV as compared to those where GSNO was treated with either GSH or CuV under identical experimental conditions (Fig. 4d). These results indicated more production of NO and thereby the $(dtc)_2FeNO$ complex from the solution where both GSH and CuV were present in comparison to the controls where one of the constituents was absent. Both the fluorescent and EPR experiments suggest that GSH has an important role in the CuV-mediated enhanced denitrosylation of *S*-nitrosothiols.

Understanding the denitrosylation of GSNO by CuV in the presence of GSH

After confirming the release of NO during denitrosylation, it was thought to be worthwhile to understand the role of GSH further in the process. The denitrosylation of GSNO was monitored at varying concentrations of GSH by keeping the concentrations of GSNO and CuV fixed (Fig. 5a). Interestingly, it was observed that the rate of denitrosylation increased with an increase in the concentration of GSH, indicating the reduction of a greater amount of surface Cu atoms (Cu^{2+} to Cu^{1+}).[57,58] Further experiments indicated that the denitrosylation activity directly depends on the amount of CuV used for the experiments, indicating a direct correlation between the number of copper atoms and the rate of the denitrosylation reaction (Fig. S18b†). To check the effect of the substrate (GSNO) on the denitrosylation, we varied the concentration of GSNO at fixed concentrations of CuV and GSH (Fig. S18a†). Surprisingly, complete denitrosylation was not observed for a lower concentration of GSNO, probably due to the side reaction known as *trans*-nitrosation which has been previously reported in the literature.[38,59,60] Complete denitrosylation was observed when the ratio between GSNO

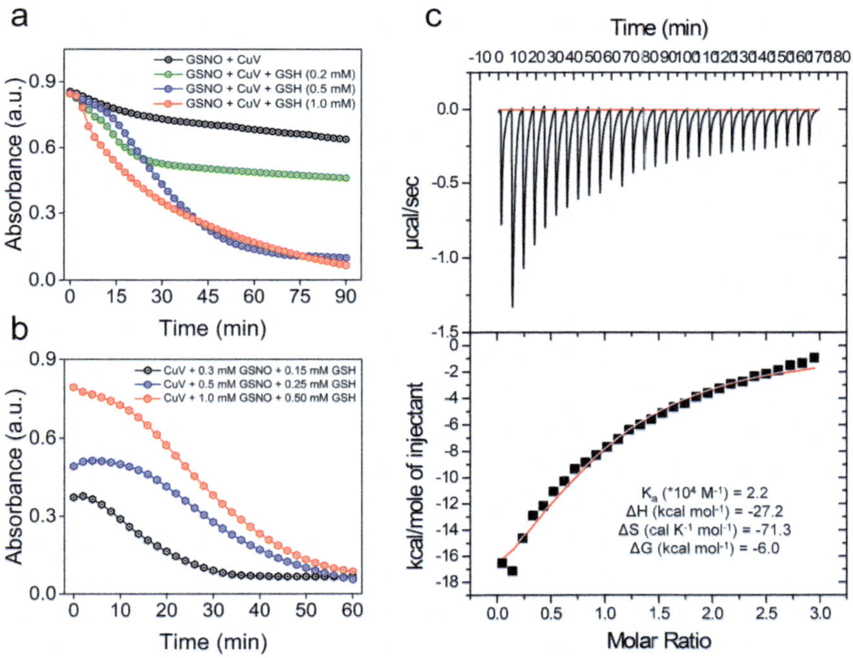

Fig. 5 (a) Denitrosylation of GSNO with varying concentrations of GSH. Assay conditions: GSNO (1.0 mM), catalyst (20 ng μL^{-1}), GSH (0–1.0 mM) and 0.1 M sodium phosphate buffer (pH 7.4) at 25 °C. (b) De-nitrosylation reaction by CuV with varying concentrations of GSNO and GSH but at a fixed ratio of GSNO and GSH. Assay conditions: GSNO (0.3–1.0 mM), catalyst (20 ng μL^{-1}), GSH (0.15–0.5 mM) and 0.1 M sodium phosphate buffer (pH 7.4) at 25 °C. (c) Isothermal titration calorimetry (ITC) profile for the binding of GSH (1.0 mM) at 298 K by CuV. The original titration profile (top) and the integrated heat (bottom) of each reaction are shown. Thermodynamic parameters (inset) were obtained using a one-site binding model and the concentration of the catalyst was 20 ng μL^{-1} or 0.075 mM, considering the molar mass of CuV as 261.43 g mol^{-1}.

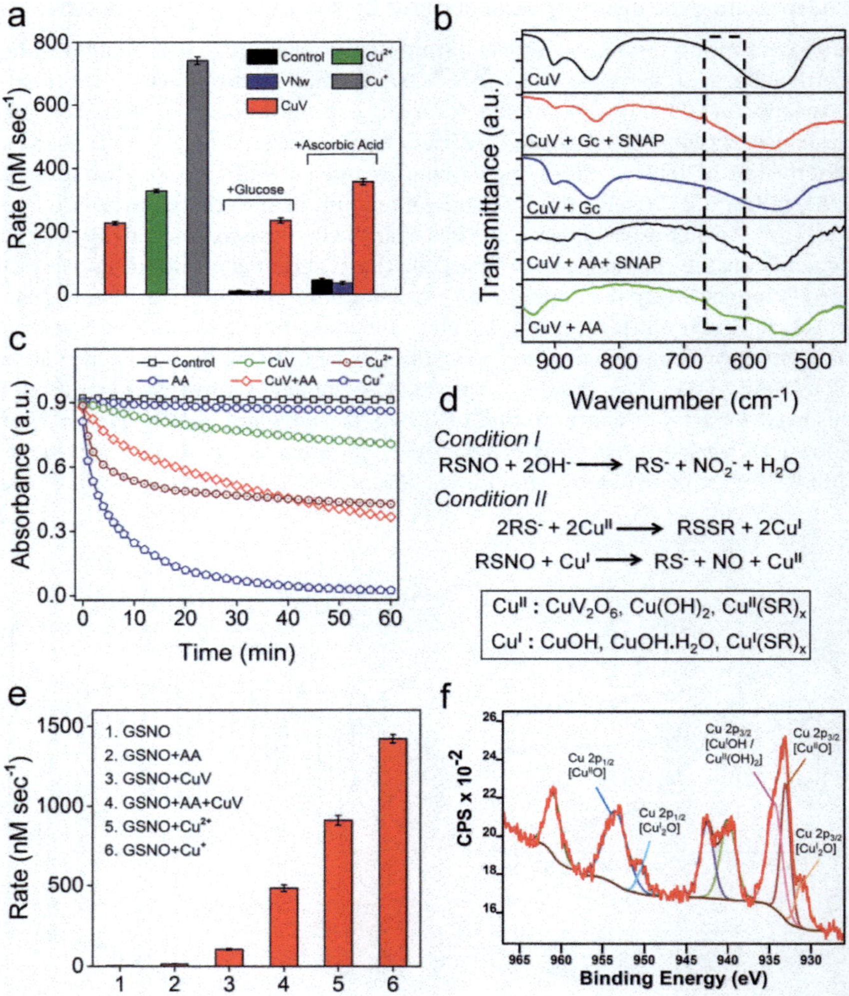

Fig. 6 (a) Denitrosylation of SNAP by CuV in the presence of reducing agents *i.e.* glucose (Gc) and ascorbic acid (AA). Bar diagram showing the comparisons of the rate of denitrosylation of SNAP under different conditions. Assay conditions: SNAP (0.5 mM), catalyst (4 ng μL^{-1}), glucose (0.5 mM) or ascorbic acid (0.005 mM), 0.1 M sodium phosphate buffer (pH 7.4), 25 °C. The denitrosylation was checked without CuV and in presence of 20 μM Cu^{2+} ($CuCl_2 \cdot 2H_2O$) or Cu^+ (CuCl) ions under the same experimental conditions, where Cu^+ showed faster denitrosylation than Cu^{2+}. Since glucose is a mild reducing agent, the presence of glucose did not cause any significant difference in the rate of denitrosylation. The presence of ascorbic acid must have reduced Cu^{2+} to Cu^{1+} species on the surface of CuV, which enhanced the rate of denitrosylation. (b) FT-IR spectra of CuV in the presence of SNAP and AA/Gc. Since human blood contains AA and Gc in the micromolar and millimolar range, respectively, 10 μM AA and 5.0 mM Gc were used for the study. The small broad peak observed (in CuV + AA) at 633 cm^{-1} is characteristic of the Cu^{1+} state. (c) Denitrosylation of GSNO by CuV in the presence of ascorbic acid (assay conditions: GSNO (1.0 mM), catalyst (20 ng μL^{-1}), ascorbic acid (0.005 mM), 0.1 M sodium phosphate buffer, pH 7.4, 25 °C). (d) Mechanism of Cu-ion mediated NO release from RSNO, involving a redox cycle between Cu(I) and Cu(II) species. The conversion of Cu(II) to Cu(I) requires reducing agents. (e) Comparisons of the rate of denitrosylation of GSNO under different

and GSH was kept constant in the denitrosylation process, confirming that an excess amount of GSH was responsible for the side reaction *i.e. trans*-nitrosation and thereby incomplete denitrosylation (Fig. 5b). The above experiments suggested that there is a favorable interaction of GSH with the surface of CuV.

To further investigate the interaction and specificity of GSH binding with CuV, isothermal titration calorimetry (ITC) experiments were performed with both CuV and VNw (Fig. 5c and S19b†). It can be observed from Fig. 5c that GSH interacted with the surface of CuV and, from the thermodynamic parameters, it is evident that the binding of GSH with CuV was energetically favorable, but no such interactions were observed with VNw. This observation indicated that the Cu atoms are responsible for the binding of GSH on the surface of CuV. These results were in accordance with the earlier observations (Fig. 4e and f), where the increased rate of the denitrosylation of GSNO was observed for CuV in the presence of GSH, unlike other metal oxides or metal vanadates. Another ITC experiment was performed at a higher concentration of GSH (2.0 mM), keeping the concentration of CuV constant. Interestingly, it was found that at a higher concentration of GSH, the saturation comes faster, and the K_a is ~5 times more than the K_a observed at a lower concentration (1.0 mM) of GSH. This result indicates that the binding of one molecule of GSH on the surface of CuV favors the binding of other molecules of GSH to make the reduction (Cu^{2+} to Cu^{1+}) of the surface faster and thereby there is a faster rate of denitrosylation, which correlates well with the earlier observation presented in Fig. 5a.

Mechanistic investigations for both denitrosylation and GPx-like activity

Earlier reports have suggested that Cu ion-mediated denitrosylation involves the redox change of Cu^{1+} to Cu^{2+} species and Cu^{1+} has been identified as the active intermediate to perform the catalysis.[57,60,61] Furthermore, it is known in the literature[57–62] that the rate of denitrosylation by Cu^{1+} is much faster than that of Cu^{2+}, which was also observed in the denitrosylation reactions with SNAP and GSNO (Fig. 6a and e). Additionally, we observed that the rate of denitrosylation by CuV is faster in the presence of GSH (Fig. 3c and d), suggesting that GSH may be acting as a reducing agent to perform the redox change (Cu^{2+} to Cu^{1+}) on the surface of CuV. To further understand the effect of reducing agents, SNAP was used as a model substrate for denitrosylation in the presence of two biologically relevant reducing agents *i.e.* glucose (Gc) and ascorbic acid (AA). As shown in Fig. 6a, a faster rate of denitrosylation was observed in the presence of ascorbic acid as compared to that of glucose (Fig. S20†). In both cases, VNw did not show any noticeable denitrosylation reaction, which is in agreement with the hypothesis that redox state change of Cu ions on the surface of CuV is responsible for the denitrosylation (Fig. 6a).

To probe the nature of the intermediates formed in the presence of reducing agents, several FT-Raman and FT-IR studies were performed with different concentrations of ascorbic acid (AA) and glucose (Gc) (Fig. S21†). When CuV was

conditions. (f) Short range XPS of Cu 2p recorded on CuV which was recovered after reacting with AA (5 μM). The peaks were deconvoluted and assigned according to the binding energies and FWHM. The binding energy of C1s (284.6 eV) was used as a standard to calibrate the spectra.

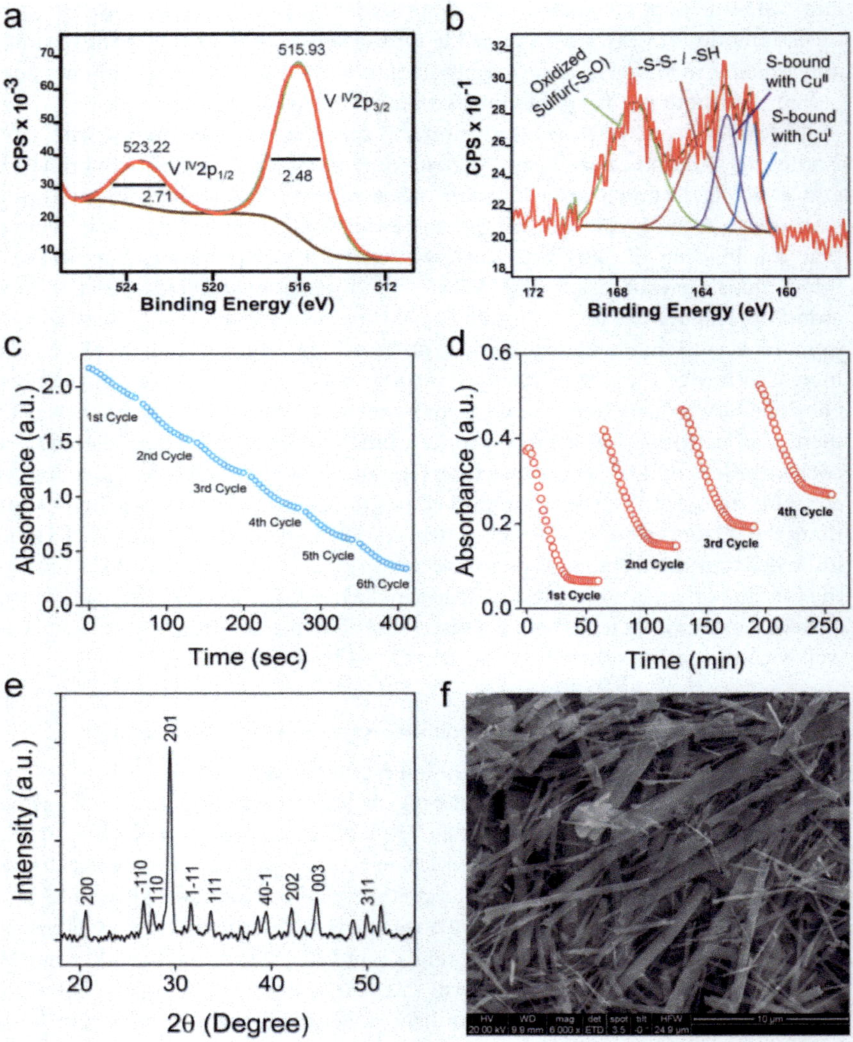

Fig. 7 (a and b) Short-range XPS spectra of V 2p and S 2p, respectively, recorded after treating CuV with GSH (10 mM). The peaks were deconvoluted and assigned according to their binding energies and FWHM. Vanadium was completely reduced from V^{5+} to V^{4+} species and GSH was oxidized, forming various kind of oxidized species of sulfur and metal bound sulfur, especially Cu bound sulfur, mentioned in (b). (c) The recyclability of CuV was monitored time-dependently by following the decrease in the absorbance of NADPH in multiple cycles. Fresh H_2O_2 was added to the reaction mixture at the end of one cycle, up to six cycles. Assay conditions: sodium phosphate buffer (100 mM, pH 7.4), GSH (2 mM), NADPH (0.4 mM), catalyst (20 ng μL^{-1}), GR (~1.7 U) and H_2O_2 (25 μM) at 25 °C. (d) Recyclability of CuV (20 ng μL^{-1}) after up to four cycles of denitrosylation, using 0.5 mM GSNO in each cycle. Reactions were carried out in the presence of 0.25 mM GSH in 0.1 M phosphate buffer (pH 7.4) at 25 °C. Increase in the initial absorbance for each cycle is due to the residual GSNO after the completion of one cycle. (e and f) PXRD and SEM images of CuV, respectively, after GPx-like enzymatic catalysis, showing that the crystallinity, as well as the morphology of the material, was still intact after catalysis.

treated with a higher concentration (10 μM) of AA, a small peak at ~550 cm^{-1} was observed, characteristic of a +1 oxidation state of surface Cu atoms (Fig. S21a†).[47–50] Similarly, in the FT-IR spectra, a small broad peak observed at 633 cm^{-1} when CuV was reacted with ascorbic acid (AA) could be ascribed to the Cu(I)–O vibration (Fig. 6b). This peak disappeared after the complete denitrosylation of SNAP, indicating the formation of Cu(II) from Cu(I).[48–50] However, because of the mild reducing ability of glucose, no additional peaks for Cu(I) species were observed in either of the experiments (Fig. 6b and S21b†). This may explain the reason why glucose did not influence the rate of denitrosylation of SNAP (Fig. S20†). Similar results were obtained when GSNO was used as a substrate for denitrosylation by CuV in the presence of AA (Fig. 6c and e). The chemical species were identified from the XPS data of Cu 2p, which confirmed the presence of Cu(I) intermediates along with Cu(II) on the surface of CuV (Fig. 6f). Additionally, the corresponding O 1s spectrum obtained for the material indicated the presence of metal-bound hydroxo species (Fig. S23†). In another experiment, CuV was treated with GSH and then used for the denitrosylation of GSNO. The catalyst was recovered after each step and FT-Raman spectra were recorded. The peak at 481 cm^{-1}, corresponding to $Cu(OH)_2$, appeared immediately after the treatment with GSH, but the peak disappeared shortly afterwards with a new peak appearing at ~550 cm^{-1}, probably due to the formation of Cu^I species upon the reduction of Cu^{II} by GSH (Fig. S22a†).[50,63] Similar results were obtained from the FT-IR experiments as well. The small broad peak observed at 635 cm^{-1} can be ascribed to the existence of mixed oxidation states (+I and +II) of Cu ions (Fig. S22b†). These results confirm that the redox change of Cu^{2+} to Cu^{1+} was the responsible for the denitrosylation reactions and Cu^{1+} is the active intermediate for the catalysis. This correlates well with the proposed mechanism of NO release in the presence of free thiols (GSH, NAP *etc.*), involving the initial reduction of Cu(II) to Cu(I) by the thiolates (RS^-) produced in the first step of the denitrosylation reaction, which can be oxidized back to Cu(II) by RSNO (Fig. 6d).[60–63] In the absence of any reducing agents, the denitrosylation reaction is very slow because Cu^{2+} needs to be reduced first to Cu^{1+} by RSNO, as shown in Fig. 6d, condition I.[61] The facile redox changes between the Cu(II) and Cu(I) on the surface of CuV were further confirmed by carrying out the denitrosylation reactions for multiple cycles without any loss of catalytic activity (Fig. 7d and S25b†).

To understand the mechanism of the GPx-like activity of CuV, time-dependent FT-Raman and FT-IR spectra of CuV were recorded upon the reaction with H_2O_2 (Fig. S27†). The formation of V-peroxido species was not observed in both the spectra, indicating that CuV follows a different mechanism for GPx-like activity than V_2O_5 nanozymes.[53] CuV showed recyclability for multiple cycles (Fig. 7c and S25a†) when fresh H_2O_2 was added externally after the completion of one cycle, and this indicates the stability and robustness of the material during catalysis, which was also reaffirmed by the PXRD patterns and SEM and TEM images of CuV after catalysis (Fig. 7e, f, S25c and d†). As Cu is known to generate hydroxyl radicals upon reacting with H_2O_2, it was important to check whether CuV generates hydroxyl radicals during its GPx-like activity. For this, a TPA-based fluorescence method was utilized,[64] which confirmed that CuV does not generate hydroxyl radicals during its GPx-like activity (Fig. S28†). To further investigate the nature of the intermediate formed during catalysis, CuV was reacted with GSH and narrow range XPS spectra of V 2p, Cu 2p and S 2p were recorded (Fig. 7a, b, and S24†, respectively). The spectra indicated the reduction of V(V) to V(IV) which could be oxidized back to V(V) in the presence of excess H_2O_2 (Fig. S26†).[42,48]

During the reduction of vanadium, Cu(II) ions also get reduced to Cu(I) and the sulfur of the thiol gets oxidized. From the deconvoluted spectra of S 2p, we obtained peaks for the metal-bound sulfur, mostly Cu-bound sulfide (Fig. 7b and Table S6†).[41,65,66] These results indicate that the GPx-like activity of CuV is due to the change of the redox state of the vanadium center from +V to + IV and *vice versa*.

Conclusions

In conclusion, we report the development of a novel nanozyme copper vanadate (CuV_2O_6) that not only can reduce H_2O_2 to H_2O by following the classical GPx-like mechanism, but also can mediate the release of NO from *S*-nitrosothiols at physiological pH. The nanozyme catalyzes the denitrosylation, as well as exhibiting GPx-like activity multiple times without affecting its catalytic activity. The increase in the rate of denitrosylation of GSNO in the presence of GSH confirms that the interaction between GSH and the surface of the nanomaterials is important for the activity. ITC experiments confirmed that the nanomaterial–GSH interactions occur at the surface Cu sites. Various imaging techniques and spectroscopic measurements were used to understand the catalysis at the nanomaterial surface. These experiments reveal that the changes in the redox states of Cu atoms from the + I to +II oxidation states and V atoms from +IV to +V are important for the denitrosylation reaction and GPx-like activity, respectively. The decreased bioavailability of nitric oxide and increased generation of reactive oxygen species (ROS) are the major reasons for the impairment of endothelial cells, leading to several vascular diseases. Materials that can improve the levels of NO, as well as combating ROS in these cells, are likely to be potential candidates for the treatment of vascular disorders.

Conflicts of interest

The authors declare no conflict of interest.

Acknowledgements

This study was supported by the Science and Engineering Research Board (SERB, EMR/IISc-01/2016) and Nano Mission (SR/NM/NS-1380/2014), Department of Science and Technology (DST), New Delhi. S. G. acknowledges the Indian Institute of Science, Bangalore, for a research fellowship. S. G. thanks Estak and Arnab for helping with the EPR spectroscopy with the X-band JEOL (JES FA200) instrument. The authors thank Prof. S. Umapathy and Prof. P. Thilagar for providing the Raman spectroscopy and UV–vis spectroscopy facilities, respectively. We also thank the AFMM Facility and CeNSE, IISc for the spectroscopic and microscopic facilities. S. G. thanks Aparna for helping with the ITC experiments. G. M. acknowledges the SERB/DST for the award of the J. C. Bose National Fellowship (SB/S2/JCB-067/2015).

References

1 V. Adler, Z. Yin, K. D. Tew and Z. Ronai, *Oncogene*, 1999, **18**, 6104.
2 B. D'Autréaux and M. B. Toledano, *Nat. Rev. Mol. Cell Biol.*, 2007, **8**, 813.
3 K. Bedard and K. H. Krause, *Physiol. Rev.*, 2007, **87**, 245.
4 H. S. Marinho, C. Real, L. Cyrne, H. Soares and F. Antunes, *Redox Biol.*, 2014, **2**, 535.
5 H. Sies, *Redox Biol.*, 2017, **11**, 613.
6 H. Sies, *Angew. Chem., Int. Ed.*, 1986, **25**, 1058.
7 K. J. Davies, *J. Biol. Chem.*, 1987, **262**, 9895.
8 A. M. Adly, *J. Immunol.*, 2010, **3**, 129.
9 R. Mittler, *Trends Plant Sci.*, 2017, **22**, 11.
10 K. J. Barnham, C. L. Masters and A. I. Bush, *Nat. Rev. Drug Discovery*, 2004, **3**, 205.
11 K. Apel and H. Hirt, *Annu. Rev. Plant Biol.*, 2004, **55**, 373.
12 T. Finkel and N. J. Holbrook, *Nature*, 2000, **408**, 239.
13 H. Selye, *Science*, 1955, **122**, 625.
14 H. Sies, *Redox Biol.*, 2015, **4**, 180.
15 O. I. Aruoma, *J. Am. Oil Chem. Soc.*, 1998, **75**, 199.
16 M. Morita, Y. Naito, T. Yoshikawa and E. Niki, *Redox Biol.*, 2016, **8**, 127.
17 T. Heitzer, T. Schlinzig, K. Krohn, T. Meinertz and T. Munzel, *Circulation*, 2001, **104**, 2673.
18 H. Cai and D. G. Harrison, *Circ. Res.*, 2000, **87**, 840.
19 J. Davignon and P. Ganz, *Circulation*, 2004, **109**, 27.
20 S. Sitia, L. Tomasoni, F. Atzeni, G. Ambrosio, C. Cordiano, A. Catapano, S. Tramontana, F. Perticone, P. Naccarato, P. Camici, E. Picano, L. Cortigiani, M. Bevilacqua, L. Milazzo, D. Cusi, C. Barlassina, P. Sarzi-Puttini and M. Turiel, *Autoimmun. Rev.*, 2010, **9**, 830.
21 M. A. Gimbrone and G. G. Cardeña, *Circ. Res.*, 2016, **118**, 620.
22 T. F. Lüscher and M. Barton, *Clin. Cardiol.*, 1997, **20**, 3–10.
23 T. F. Lüscher, *Am. J. Hypertens.*, 1990, **3**, 317.
24 J. R. Vane, E. E. Anggard and R. M. Botting, *N. Engl. J. Med.*, 1990, **323**, 27.
25 W. K. Alderton, C. E. Cooper and R. G. Knowles, *Biochem. J.*, 2001, **357**, 593.
26 S. A. Lipton, Y. B. Choi, Z. H. Pan, S. Z. Lei, H. S. V. Chen, N. J. Sucher, J. Loscalzo, D. J. Singel and J. S. Stamler, *Nature*, 1993, **364**, 626.
27 G. Melino, F. Bernassola, R. A. Knight, M. T. Corasaniti, G. Nistic and A. Finazzi-Agro, *Nature*, 1997, **388**, 432.
28 C. Bogdan, *Nat. Immunol.*, 2001, **2**, 907.
29 P. Vallance and J. Leiper, *Nat. Rev. Drug Discovery*, 2002, **1**, 939.
30 A. R. Butler and R. Nicholson, *Life, Death and Nitric Oxide*, Royal Society of Chemistry, Cambridge, 1st edn, 2003, vol. 154.
31 B. Limaa, M. T. Forrester, D. T. Hess and J. S. Stamler, *Circ. Res.*, 2010, **106**, 633.
32 M. Benhar, M. T. Forrester and J. S. Stamler, *Nat. Rev. Mol. Cell Biol.*, 2009, **10**, 721.
33 G. N. Levine, B. Frei, S. N. Koulouris, M. D. Gerhard, J. F. Keaney and J. A. Vita, *Circulation*, 1996, **93**, 1107.
34 K. Ravi, L. A. Brennan, S. Levic, P. A. Ross and S. M. Black, *Proc. Natl. Acad. Sci. U. S. A.*, 2004, **101**, 2619.

35 C. A. Chen, T. Y. Wang, S. Varadharaj, L. A. Reyes, C. Hemann, M. A. Talukder, Y. R. Chen, L. J. Druhan and J. L. Zweier, *Nature*, 2010, **468**, 1115.
36 A. A. Vernekar, D. Sinha, S. Srivastava, P. U. Paramasivam, P. D'Silva and G. Mugesh, *Nat. Commun.*, 2014, **5**, 5301.
37 R. J. Singh, N. Hogg, J. Joseph and B. Kalyanaraman, *J. Biol. Chem.*, 1996, **271**, 18596.
38 D. L. H. Williams, *Acc. Chem. Res.*, 1999, **32**, 869.
39 G. Stubauer, A. Giuffrè and P. Sarti, *J. Biol. Chem.*, 1999, **274**, 28128.
40 R. Esquembre, I. Pastor, R. Mallavia and C. R. Mateo, *J. Photochem. Photobiol., A*, 2005, **173**, 384.
41 S. Kundu, W. Y. Kim, J. A. Bertke and T. H. Warren, *J. Am. Chem. Soc.*, 2017, **139**, 1045.
42 H. Ma, S. Zhang, W. Ji, Z. Tao and J. Chen, *J. Am. Chem. Soc.*, 2008, **130**, 5361.
43 Y. Sun, C. Li, L. Wang, Y. Wang, X. Ma, P. Maa and M. Song, *RSC Adv.*, 2012, **2**, 8110.
44 Q. Zhang, K. Zhang, D. Xu, G. Yang, H. Huang, F. Nie, C. Liu and S. Yang, *Prog. Mater. Sci.*, 2014, **60**, 208.
45 M. Sabbaghann, J. Beheshtian and R. N. Liarjdame, *Mater. Lett.*, 2015, **153**, 1.
46 Y. Wei, C. W. Ryu, G. Chen and K. B. Kim, *Electrochem. Solid-State Lett.*, 2006, **9**, A487.
47 S. Poulston, P. M. Parlett, P. Stone and M. Bowker, *Surf. Interface Anal.*, 1996, **24**, 811.
48 M. C. Biesinger, L. W. M. Laua, A. R. Gerson and R. C. Smart, *Appl. Surf. Sci.*, 2010, **257**, 887.
49 A. Sahai, N. Goswami, S. D. Kaushik and S. Tripathi, *Appl. Surf. Sci.*, 2016, **390**, 974.
50 Y. Deng, A. D. Handoko, Y. Du, S. Xi and B. S. Yeo, *ACS Catal.*, 2016, **6**, 2473.
51 W. Hou, Q. Wang, Z. Guo, J. Li, Y. Zhou and J. Wang, *Catal. Sci. Technol.*, 2017, **7**, 1006.
52 A. V. Peskin and C. C. Winterbourn, *Clin. Chim. Acta*, 2000, **293**, 157.
53 S. Ghosh, P. Roy, N. Karmodak, E. D. Jemmis and G. Mugesh, *Angew. Chem., Int. Ed.*, 2018, **130**, 4600.
54 H. Wei and E. Wang, *Chem. Soc. Rev.*, 2013, **42**, 6060.
55 J. Wu, X. Wang, Q. Wang, Z. Lou, S. Li, Y. Zhu, L. Qina and H. Wei, *Chem. Soc. Rev.*, 2019, **48**, 1004.
56 Y. Huang, J. Ren and X. Qu, *Chem. Rev.*, 2019, **119**, 4357.
57 D. R. Noble and D. L. H. Williams, *J. Chem. Soc., Perkin Trans. 2*, 2002, 1834.
58 M. M. Melzer, E. Li and T. H. Warren, *Chem. Commun.*, 2009, 5847.
59 L. L. Perissinotti, A. G. Turjanski, D. A. Estrin and F. Doctorovich, *J. Am. Chem. Soc.*, 2005, **127**, 486.
60 K. A. Broniowska and N. Hogg, *Antioxid. Redox Signaling*, 2012, **17**, 969.
61 A. P. Dicks, H. R. Swift, D. L. H. Williams, A. R. Butler, H. H. Al-Sa'doni and B. G. Cox, *J. Chem. Soc., Perkin Trans. 2*, 1996, 481.
62 C. Toubin, D. Y. H. Yeung, A. M. English and G. H. Peslherbe, *J. Am. Chem. Soc.*, 2002, **124**, 14816.
63 S. Ghosh, P. Roy, S. Prasad and G. Mugesh, *Chem. Sci.*, 2019, **10**, 5308.
64 C. Hao, A. Qu, L. Xu, M. Sun, H. Zhang, C. Xu and H. Kuang, *J. Am. Chem. Soc.*, 2019, **141**, 1091.

65 S. Zhang, N. Çelebi-Ölçüm, M. M. Melzer, K. N. Houk and T. H. Warren, *J. Am. Chem. Soc.*, 2013, **135**, 16746.
66 Y. Ju-Nam, Y. S. Chen, J. J. Ojeda, D. W. Allen, N. A. Cross, P. H. E. Gardiner and N. Bricklebank, *RSC Adv.*, 2012, **2**, 10345.
67 D. E. Paglia and W. N. Valentine, *J. Lab. Clin. Med.*, 1967, **70**, 158.
68 L. Flohé and I. Brand, *Biochim. Biophys. Acta*, 1969, **191**, 541.
69 X. Mu, J. Wang, H. He, Q. Li, B. Yang, J. Wang, H. Liu, Y. Gao, L. Ouyang, S. Sun, Q. Ren, X. Shi, W. Hao, Q. Fei, J. Yang, L. Li, R. Vest, T. Wyss-Coray, J. Luo and X. D. Zhang, *Sci. Adv.*, 2021, **7**, eabk1210.

Faraday Discussions

PAPER

Exploring hitherto uninvestigated reactions of the fatty acid peroxygenase CYP152A1: catalase reaction and Compound I formation†

Hiroki Onoda, [ab] Shota Tanaka,[a] Yoshihito Watanabe [a] and Osami Shoji *[a]

Received 1st October 2021, Accepted 27th October 2021
DOI: 10.1039/d1fd00065a

CYP152A1 (cytochrome $P450_{BS\beta}$) is a fatty acid peroxygenase, which specifically catalyses the oxidation of long-chain fatty acids using hydrogen peroxide as an oxidant. We have found that CYP152A1 possesses catalase activity, which competes with the hydroxylation of long-chain fatty acids, the oxidation of non-native substrates, and haem degradation. Using hydrogen peroxide, Compound I of CYP152A1 could not be observed, due to its swift decomposition *via* catalase activity, where Compound I reacts with another molecule of hydrogen peroxide to form O_2. In contrast, a clear spectral change indicative of Compound I formation was observed when *m*CPBA was employed as the oxidant. This work presents valuable insights into an important role for the catalase activity of CYP152A1 in avoiding enzyme deactivation when no substrate is available for oxidation.

Introduction

Hydrogen peroxide is regarded as an ideal renewable oxidant, owing to its low environmental burden. The fatty-acid peroxygenase cytochrome $P450_{BS\beta}$ (CYP152A1, EC. 1.11.2.4), which was first discovered in 1999 by Matsunaga *et al.*,[1] catalyses the oxidation of long-chain fatty acids by utilising hydrogen peroxide as an oxidant.[1] CYP152A1 has a high affinity for hydrogen peroxide when compared with other peroxides, such as t-butylhydroperoxide[1] and cumene hydroperoxide.[2] The crystal structure of the palmitic acid-bound form of CYP152A1 (ref. 3) suggests a unique catalytic mechanism: (1) catalysis commences upon the interaction of the terminal carboxylate of the long-chain fatty acid with Arg242

[a]*Department of Chemistry, Graduate School of Science, Nagoya University, Furo-cho, Chikusa-ku, Nagoya 464-0802, Japan. E-mail: shoji.osami@a.mbox.nagoya-u.ac.jp*
[b]*Department of Medical Life Science, Graduate School of Medical Life Science, Yokohama City University, Suehiro-cho, Tsurumi-ku, Yokohama 230-0045, Japan*

† Electronic supplementary information (ESI) available. See DOI: 10.1039/d1fd00065a

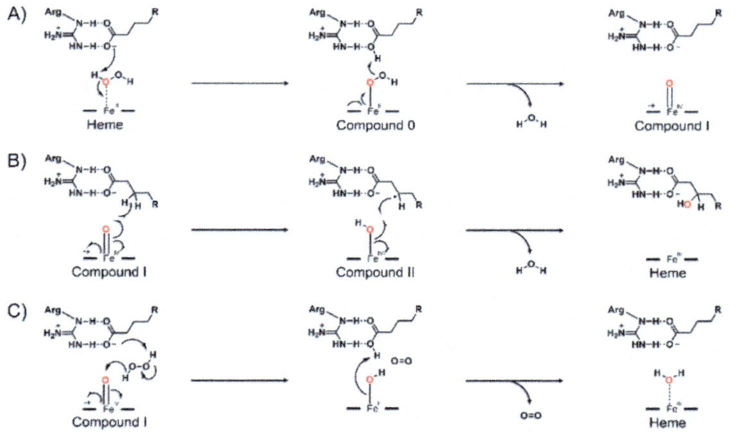

Scheme 1 Proposed reaction mechanisms of the CYP152 family of enzymes: Compound I formation (A), long-chain fatty acid hydroxylation (B) and the catalase reaction (C).

situated above the haem; (2) the general acid–base function of the fatty acid–Arg242 salt bridge promotes the generation of the oxyferryl porphyrin π-cation radical active species, Compound I, from hydrogen peroxide (Scheme 1A);[4–6] (3) the long-chain fatty acid is then oxidised by Compound I (Scheme 1B).[6] As CYP152A1 remains inactive unless a suitable fatty acid binds to the active site in both mechanisms, CYP152A1 displays a high substrate specificity for long-alkyl-chain fatty acids. The amino acids involved in salt bridge formation and the determination of substrate specificity are widely conserved throughout the CYP152 family of enzymes, such as CYP152B1,[7] CYP152K6,[8] CYP152L1,[9] and CYP152N1.[10]

We have previously reported that CYP152A1 can catalyse the oxidation of non-native substrates such as guaiacol,[11] thioanisole,[12] styrene,[11] ethylbenzene,[11] and 1-methoxy-naphthalene (1-MeO-Nap)[13,14] in the presence of short-alkyl-chain fatty acids, such as heptanoic acid, as "decoy molecules". This substrate mis-recognition system using decoy molecules has also been successfully expanded to P450BM3 monooxygenase in benzene hydroxylation,[15] high-pressure gas hydroxylation,[16] intracellular reactions,[17] and carbene insertion reactions.[18] The crystal structure of CYP152A1 binding to heptanoic acid (PDB: 2ZQX) reveals that the carboxylate of heptanoic acid interacts with Arg242, permitting the generation of Compound I.[19] Owing to its short length, the oxidation of heptanoic acid does not proceed, making Compound I available for the oxidation of non-native substrates.[20] In our research on the oxidation of non-native substrates using decoy molecules, we could not observe any spectral changes assignable to Compound I of CYP152A1 by stopped-flow spectroscopy. As Compound I is an extremely reactive and short-lived intermediate (2.2–42 ps),[21] we speculated that Compound I of CYP152A1 may have reacted further with hydrogen peroxide to yield molecular oxygen (Scheme 1C). This catalase activity may play an important role in protecting haem from degradation. Herein, we report the catalase activity of CYP152A1 in conjunction with the decomposition of haem. We also report the detection of CYP152A1 Compound I when *m*CPBA was used as the oxidant.

Results and discussion

Guaiacol oxidation using several peroxides

The peroxidase activity of CYP152A1 in the presence of several different peroxides was examined by spectroscopically monitoring guaiacol oxidation. In the case of hydrogen peroxide, peroxidase activity was only observed in the presence of a short-alkyl-chain carboxylic acid, such as heptanoic acid (Fig. 1A). The addition of hypochlorous acid (Fig. 1C), peroxylauric acid (Fig. 1D) and *m*CPBA (Fig. 1E) elicited peroxidase activity, whilst cumene hydroperoxide (Fig. 1B) did not. Regardless of the presence of heptanoic acid, the oxidation of guaiacol was catalysed by the addition of peroxylauric acid or *m*CPBA, indicating that these peroxides alone can generate the active species. No guaiacol oxidation was detected by 100 μM of these peroxides without the addition of CYP152A1. CYP152A1 has the potential to produce haem active species utilising peroxy acid as a substitute for hydrogen peroxide and fatty acids.

Compound I formation

The active species generated upon the addition of *m*CPBA were detected *via* stopped-flow spectroscopy (Fig. 2A and B). Within the first 50 ms, the intensity of the peak at 368 nm increased, whereafter it gradually decreased (Fig. 2C and D). The absorption spectrum of CYP152A1 Compound I was calculated using the singular value decomposition (SVD) method.[4,22,23] The deconvoluted spectrum displayed characteristic absorption peaks corresponding to Compound I at 368 nm and at around 690 nm, which are essentially identical to those observed for CYP119.[4] A similar spectrum has been reported for CYP152L1 Compound I, following the addition of hydrogen peroxide and perdeuterated fatty acids.[5,6,24]

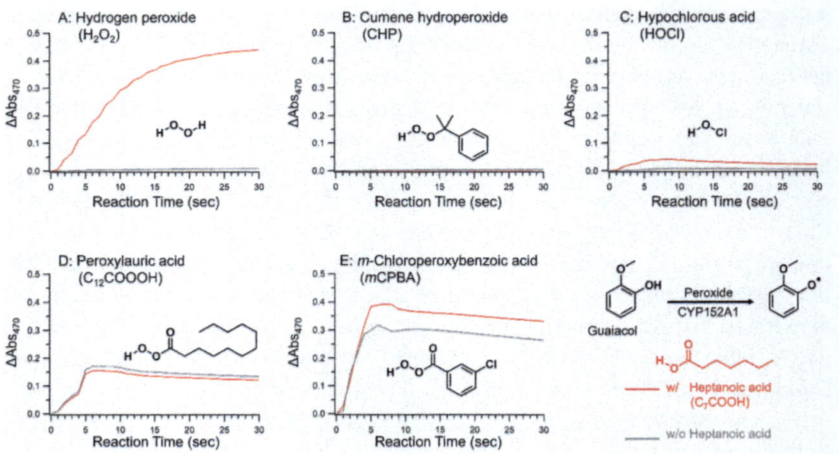

Fig. 1 Guaiacol oxidation catalysed by CYP152A1 in the presence (red line) and absence (grey line) of heptanoic acid as the decoy molecule, monitored by the formation of the oxidised product with absorption at 470 nm. Hydrogen peroxide (A), cumene hydroperoxide (B), hypochlorous acid (C), peroxylauric acid (D) and *m*-choloroperoxybenzoic acid (E) were used as oxidants.

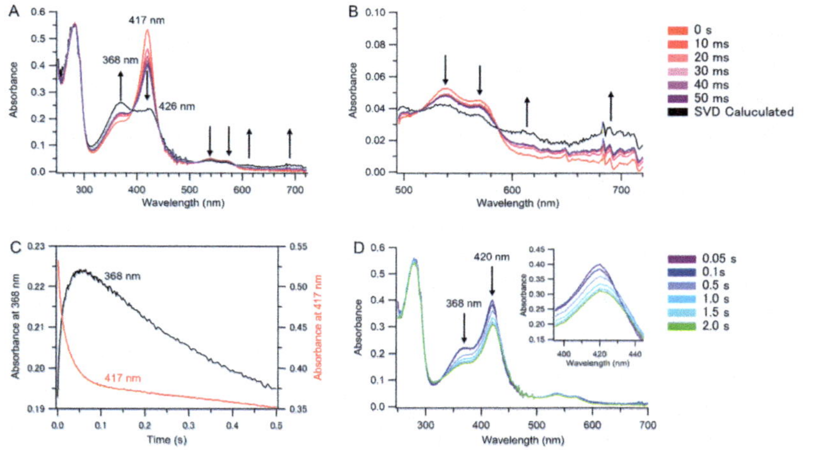

Fig. 2 Time-dependent UV–vis spectral changes of CYP152A1 upon the addition of *m*CPBA measured by stopped-flow technique, (A and B) 0 ms to 50 ms and (D) 50 ms to 2 s timeframes. The black lines in (A) and (B) are the spectra calculated by the SVD method from the absorption spectra at 0, 10, 20, 30, 40 and 50 ms. Time-dependent absorbance changes at 368 nm (Compound I) and 417 nm (ferric haem thiolate Soret band) are shown in (C).

The spectrum obtained by SVD reveals an additional peak at 426 nm, which suggests that Compound II (Fe^{4+}–OH) was generated from Compound I in the same way as for APO[25] and CYP152L1.[6]

Catalase reaction

We speculated that because Compound I could not be detected when hydrogen peroxide was used as the oxidant, Compound I of CYP152A1 may have reacted further with hydrogen peroxide to yield molecular oxygen (catalase activity). Horseradish peroxidase (HRP, EC 1.11.1.7), which catalyses the one-electron oxidation of ferulic acid in lignin biosynthesis, exhibits weak catalase activity in the absence of a suitable substrate.[24] The catalase activity of HRP is considered to be a protective mechanism against enzyme inactivation, such as haem degradation caused by the decomposition of the active oxygen species. The catalase activity of CYP152A1 in the presence and absence of a series of fatty acids with different alkyl chain lengths was evaluated by monitoring oxygen evolution on a polarographic oxygen sensor. In the absence of a fatty acid, some catalase activity was observed (Table 1 and Fig. S1†). Interestingly, CYP152A1 exhibited catalase activity even in the presence of myristic acid, indicating that both reactions proceed in competition. Catalase activity was further increased by the addition of short-alkyl-chain fatty acids, with heptanoic acid being the best stimulator of catalase activity. Short-alkyl-chain fatty acids also function as decoy molecules, *i.e.* as stimulators of non-native catalytic activity. Catalase activity of CYP152A1 was found to be proportional to the propensity of decoy molecules to promote non-native substrate oxidation, with the best decoy molecules also eliciting the highest catalase activity (Table 1 and Fig. S1†). The activation tendency caused by the alkyl-chain length implies that the catalase reaction

Table 1 Turnover frequency (TOF) (min^{-1}) of the catalase reaction and various oxidation reactions catalysed by CYP152A1 in the presence of a series of fatty acids

Fatty acid	Catalase reaction	Peroxidation reaction[11]	Sulphoxidation[12]	Epoxidation[11]	Benzylic hydroxylation[11]	Aromatic hydroxylation[14]
	H_2O_2	Guaiacol	Thioanisole	Styrene	Ethylbenzene	1-MeO-Nap
None	12.5 ± 0.6	N.D.	24 ± 2	N.D.	N.D.	1.7 ± 0.2
Acetic acid (C_2)	16.3 ± 0.6	N.D.	—	—	—	8.0 ± 0.3
Propionic acid (C_3)	18.1 ± 0.6	26	270 ± 20	—	—	11.6 ± 0.5
Butanoic acid (C_4)	37.8 ± 1.2	230	260 ± 80	140 ± 30	11 ± 1	10.8 ± 0.3
Pentatonic acid (C_5)	42.2 ± 0.9	1900	270 ± 50	317 ± 14	20 ± 2	34 ± 2
Hexanoic acid (C_6)	59.7 ± 1.2	2420	390 ± 50	**330 ± 60**	24 ± 2	61 ± 3
Heptanoic acid (C_7)	**75.3 ± 1.2**	**3750**	**510 ± 70**	290 ± 40	**28 ± 4**	**103 ± 2**
Octanoic acid (C_8)	62.8 ± 0.6	2490	250 ± 20	150 ± 30	10 ± 3	76 ± 3
Nonanoic acid (C_9)	50.0 ± 0.9	2380	—	—	—	—
Decanoic acid (C_{10})	43.5 ± 0.9	2360	—	—	—	—
Myristic acid (C_{14})	37.8 ± 0.6	14	—	—	—	—

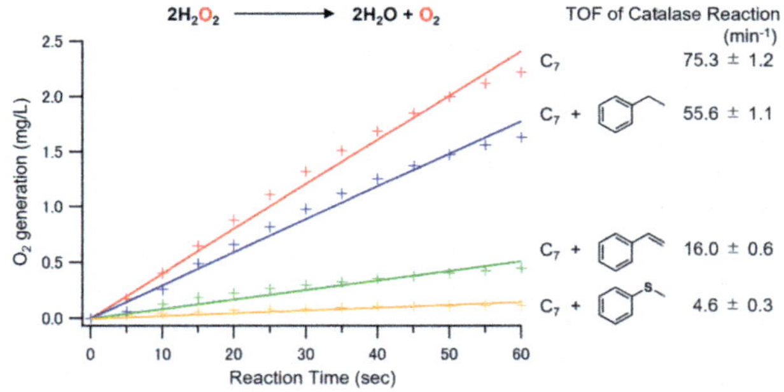

Fig. 3 Time course of O_2 generation due to the catalase activity of CYP152A1 in the presence of heptanoic acid (C_7) (red). The effect of the co-addition of non-natural substrates, ethylbenzene (blue), styrene (green) and thioanisole (yellow), was also examined. The turnover frequencies (TOFs) of the catalase reactions are shown on the right.

utilises the same haem active species (Compound I) as the oxidation reactions of the non-native substrate. It is noteworthy to mention here that even in the presence of non-native substrates, catalase activity could still be observed with heptanoic acid (Fig. 3). The generation of O_2 was throttled by the addition of a non-native substrate (Fig. 3). The catalase reaction rate decreased as the peroxygenation rate of non-native substrates increased (ethylbenzene: 28 min^{-1}, styrene: 290 min^{-1}, thioanisole 510 min^{-1}). These results highlight that the catalase reaction competes with short-chain fatty acid triggered peroxygenation reactions. A Michaelis–Menten kinetic study of catalase activity in the presence and absence of fatty acids (Table 2 and Fig. S2†) revealed that the K_m for H_2O_2 was unchanged in the presence (4.3 mM) and absence (4.4 mM) of myristic acid. With heptanoic acid, however, the K_m was markedly reduced (1.7 mM), indicating a higher affinity for H_2O_2. The maximum turnover number (k_{cat}) of CYP152A1 with heptanoic acid is comparable with that of HRP[26] and 2–4 orders of magnitude slower than that of catalase from beef liver[27] and catalase peroxidase (katG).[28]

Haem degradation

Upon the addition of H_2O_2 to CYP152A1, the intensity of the Soret band started to decrease (Fig. 4). This showed that haem degradation and the catalase reaction proceeded together. The degradation of haem could be decelerated by the addition of myristic acid (Fig. 4B), whilst the addition of heptanoic acid accelerated

Table 2 Michaelis–Menten kinetics parameters of the catalase reaction

Enzyme	k_{cat} (s^{-1})	K_m (mM)	k_{cat}/K_m (mM^{-1} s^{-1})
CYP152A1 w/o fatty acid	0.44	4.4	0.10
CYP152A1 w/heptanoic acid	1.87	1.7	1.07
CYP152A1 w/myristic acid	1.31	4.3	0.31
Catalase from beef liver[27]	2.12×10^5	93	2.3×10^4
KatG[28]	3.50×10^3	4.1	8.5×10^2
HRP-C[26]	1.78	4.0	0.45

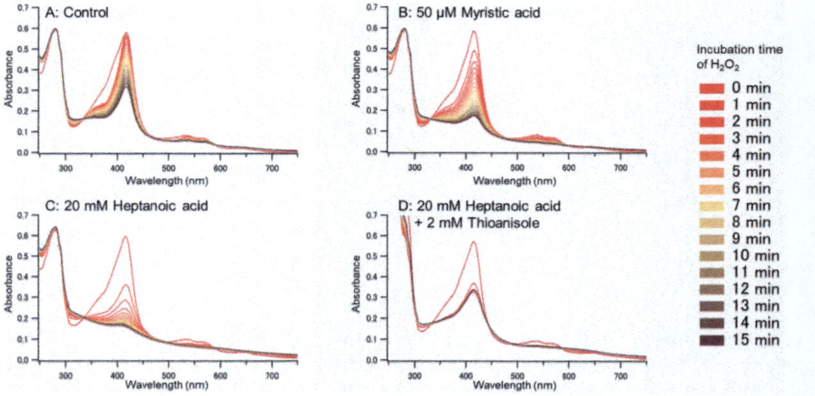

Fig. 4 Haem degradation by CYP152A1 in the presence of 4 mM H_2O_2. UV–vis spectral changes of CYP152A1 without substrate (A), with 50 μM myristic acid (B), with 20 mM heptanoic acid (C), and with 20 mM heptanoic acid and 2 mM thioanisole (D).

haem degradation (Fig. 4C). In the presence of heptanoic acid, half of the haem was decomposed within one minute. Degradation could also be inhibited by the addition of thioanisole (Fig. 4D). These results showed that haem degradation competes with fatty acid oxidation and non-native substrate oxidation. In the case that a viable substrate is not available for oxidation, the active species leads to haem degradation. Thioanisole can be regarded as an efficient scavenger to prevent haem degradation (Fig. 4D).

Conclusions

In conclusion, we have established that the fatty-acid-specific peroxygenase CYP152A1 possesses catalase activity and that this competes with the hydroxylation of long-chain fatty acids, the oxidation of non-native substrates, and haem degradation. In the absence of a substrate, the remaining haem active species reacts with H_2O_2 to prevent undesirable haem degradation. The catalase activity of CYP152A1 may play an important role in avoiding enzyme deactivation in the absence of a viable substrate. In addition, we have demonstrated that *m*CPBA can induce the formation of Compound I by itself and does not require the addition of a substrate and H_2O_2.

Experimental section

Materials

All chemical reagents were purchased from commercial sources and, except for styrene, used without further purification. The initial concentration of hydrogen peroxide (H_2O_2) was determined by iodometric titration. Potassium dihydrogenphosphate, dipotassium phosphate, potassium chloride, propionic acid (C_3), hexanoic acid (C_6), octanoic acid (C_8), lauric acid (C_{12}), cumene hydroperoxide (CHP), *m*-chloroperoxybenzoic (*m*CPBA), acid ethylbenzene and glycerol were obtained from Nacalai Tesque. Acetic acid (C_2), heptanoic acid (C_7), hydrogen peroxide, sodium hypochlorite, guaiacol and IPTG were obtained from

Wako. Styrene, butanoic acid (C_4), pentanoic acid (C_5), nonanoic acid (C_9) and decanoic acid (C_{10}) were obtained from TCI. Myristic acid (C_{14}), thioanisole, and thrombin were obtained from Aldrich. Peroxylauric acid was prepared according to a reported procedure[29] and the purity (98%) was checked by the iodometric method.[30]

Purification of CYP152A1

The recombinant CYP152A1 with a 6x histidine tag at its N-terminal was expressed in *Escherichia coli* M15 [pQE30t-CYP152A1] and purified by nickel chelate affinity chromatography. The histidine tag was cleaved by treatment with thrombin. These P450s were further purified by gel filtration using BioAssist eZ according to a reported procedure,[2] and stored in 100 mM potassium phosphate buffer with 20% glycerol at −80 °C. The concentrations of these P450s were determined by the CO-different spectra or pyridine ferro-haemochrome method.[31]

Guaiacol oxidation

The reaction mixture contained 0.1 M potassium phosphate (pH 7.0, 25 °C), 1 μM of CYP152A1, 20 or 0 mM of hexanoic acid, 4 mM of guaiacol and 100 μM of peroxides. Guaiacol oxidation was evaluated by the absorbance change at 470 nm on a UV-2600 (Shimadzu, Japan).

Compound I detection

The detection of CYP152A1's Compound I was performed on an RSP-1000 stopped-flow spectrophotometer (Unisok, Japan). The UV–vis spectra were evaluated every 1 ms for 10 s, immediately after 30 μM of *m*CPBA was mixed into the CYP152A1 solution.

Catalase activity

A reaction mixture of 0.1 M potassium phosphate (pH 7.0, 25 °C) containing 1 μM CYP152A1 was prepared and 1485 μL of this was transferred to the trace dissolved oxygen analyser (TD-51). The reaction was initiated by the addition of 15 μL of H_2O_2 (final concentration 0.1–4.0 mM) after the incubation for stabilisation of O_2 concentration. The final concentrations of fatty acid were 20 mM for short-chain fatty acids (acetic acid, propionic acid, butanoic acid, pentanoic acid, hexanoic acid, and hexanoic acid), 10 mM for medium-chain fatty acids (octanoic acid and nonanoic acid), 5 mM for decanoic acid, and 50 μM for myristic acid. k_{cat} and K_m for hydrogen peroxide were calculated *via* the Michaelis–Menten equation in the same way as for other catalases.[27,28]

H_2O_2 consumption

A 150 μL reaction mixture containing 0.1 M potassium phosphate (pH 7.0, 25 °C), 1 μM of CYP152A1, 20 mM of hexanoic acid or 50 μM of myristic acid, and 4 mM H_2O_2 was incubated for 5, 15, 30, 60, or 180 seconds, whereafter it was quenched by the addition of 30 μL of 2 M HCl. 5 μL of the resulting solution was added to 995 μL of the iodometry solution (10 μg mL^{-1} of HRP in 50 mM sodium acetate buffer, pH 5.0). The concentration of H_2O_2 was estimated from the concentration of I^{3-} in 50 mM acetate buffer (ε_{353} = 26 400 M^{-1} cm^{-1}).[32]

Haem degradation

The reaction mixture contained 0.1 M potassium phosphate (pH 7.0, 25 °C), 7 μM of CYP152A1, 20 mM of hexanoic acid or 50 μM of myristic acid, and 4 mM H_2O_2. The UV–vis spectra were recorded on a UV-2600 (Shimadzu, Japan).

Abbreviations

CYP or P450	Cytochrome P450
CYP152s	Cytochrome P450 152 family enzymes (fatty acid peroxygenases, EC 1. 11. 2. 4)
CYP152A1	Cytochrome P450 isolated from *Bacillus subtilis* IFO14144 ($P450_{BS\beta}$)
CYP152B1	Cytochrome P450 isolated from *Clostridium acetobutylicm* ATCC824 ($P450_{SP\alpha}$)
CYP152K6	Cytochrome P450 isolated from *Bacillus methanolicus* MGA3
CYP152L1	Cytochrome P450 isolated from *Jeotgalicoccus* sp. 8456 ($OleT_{JE}$)
CYP152N1	Cytochrome P450 isolated from *Exiguobacterium* sp. AT1b ($P450_{Ex\alpha}$)

Conflicts of interest

There are no conflicts to declare.

Acknowledgements

This work was supported, in part, by a Grant-in-Aid-for Scientific Research on Innovative Areas "Precise Formation of Catalyst Having a Specified Field for Use in Extremely Difficult Substrate Conversion Reactions" to O. S. (15H05806), a Grant-in-Aid for Scientific Research (B) to Y. W. (17H03087) from the Ministry of Education, Culture, Sports, Science and Technology (Japan), and a Grant-in-Aid for JSPS Research Fellow to H. O. (16J02846) from the Japan Society for the Promotion of Science (Japan).

Notes and references

1 I. Matsunaga, A. Ueda, N. Fujiwara, T. Sumimoto and K. Ichihara, Characterization of the ybdT gene product of *Bacillus subtilis*: novel fatty acid beta-hydroxylating cytochrome P450, *Lipids*, 1999, **34**(8), 841–846.
2 I. Matsunaga, A. Ueda, T. Sumimoto, K. Ichihara, M. Ayata and H. Ogura, Site-directed mutagenesis of the putative distal helix of peroxygenase cytochrome P450, *Arch. Biochem. Biophys.*, 2001, **394**(1), 45–53.
3 D. S. Lee, A. Yamada, H. Sugimoto, I. Matsunaga, H. Ogura, K. Ichihara, S. Adachi, S. Y. Park and Y. Shiro, Substrate recognition and molecular mechanism of fatty acid hydroxylation by cytochrome P450 from *Bacillus subtilis*. Crystallographic, spectroscopic, and mutational studies, *J. Biol. Chem.*, 2003, **278**(11), 9761–9767.
4 J. Rittle and M. T. Green, Cytochrome P450 Compound I: capture, characterization, and C-H bond activation kinetics, *Science*, 2010, **330**(6006), 933–937.

5 J. L. Grant, C. H. Hsieh and T. M. Makris, Decarboxylation of fatty acids to terminal alkenes by cytochrome P450 Compound I, *J. Am. Chem. Soc.*, 2015, **137**(15), 4940–4943.

6 J. L. Grant, M. E. Mitchell and T. M. Makris, Catalytic strategy for carbon–carbon bond scission by the cytochrome P450 OleT, *Proc. Natl. Acad. Sci. U. S. A.*, 2016, **113**(36), 10049–10054.

7 T. Fujishiro, O. Shoji, S. Nagano, H. Sugimoto, Y. Shiro and Y. Watanabe, Crystal structure of H_2O_2-dependent cytochrome P450SPalpha with its bound fatty acid substrate: insight into the regioselective hydroxylation of fatty acids at the alpha position, *J. Biol. Chem.*, 2011, **286**(34), 29941–29950.

8 H. M. Girvan, H. Poddar, K. J. McLean, D. R. Nelson, K. A. Hollywood, C. W. Levy, D. Leys and A. W. Munro, Structural and catalytic properties of the peroxygenase P450 enzyme CYP152K6 from *Bacillus methanolicus*, *J. Inorg. Biochem.*, 2018, **188**, 18–28.

9 S. Matthews, J. D. Belcher, K. L. Tee, H. M. Girvan, K. J. McLean, S. E. Rigby, C. W. Levy, D. Leys, D. A. Parker, R. T. Blankley and A. W. Munro, Catalytic Determinants of Alkene Production by the Cytochrome P450 Peroxygenase OleTJE, *J. Biol. Chem.*, 2017, 762336.

10 H. Onoda, O. Shoji, K. Suzuki, H. Sugimoto, Y. Shiro and Y. Watanabe, α-Oxidative decarboxylation of fatty acids catalysed by cytochrome P450 peroxygenases yielding shorter-alkyl-chain fatty acids, *Catal. Sci. Technol.*, 2018, **8**(2), 434–442.

11 O. Shoji, T. Fujishiro, H. Nakajima, M. Kim, S. Nagano, Y. Shiro and Y. Watanabe, Hydrogen peroxide dependent monooxygenations by tricking the substrate recognition of cytochrome P450BSbeta, *Angew. Chem., Int. Ed.*, 2007, **46**(20), 3656–3659.

12 T. Fujishiro, O. Shoji and Y. Watanabe, Non-covalent modification of the active site of cytochrome P450 for inverting the stereoselectivity of monooxygenation, *Tetrahedron Lett.*, 2011, **52**(3), 395–397.

13 O. Shoji, C. Wiese, T. Fujishiro, C. Shirataki, B. Wunsch and Y. Watanabe, Aromatic C-H bond hydroxylation by P450 peroxygenases: a facile colorimetric assay for monooxygenation activities of enzymes based on Russig's blue formation, *JBIC, J. Biol. Inorg. Chem.*, 2010, **15**(7), 1109–1115.

14 H. Onoda, O. Shoji and Y. Watanabe, Acetate anion-triggered peroxygenation of non-native substrates by wild-type cytochrome P450s, *Dalton Trans.*, 2015, **44**(34), 15316–15323.

15 K. Yonemura, S. Ariyasu, J. K. Stanfield, K. Suzuki, H. Onoda, C. Kasai, H. Sugimoto, Y. Aiba, Y. Watanabe and O. Shoji, Systematic Evolution of Decoy Molecules for the Highly Efficient Hydroxylation of Benzene and Small Alkanes Catalyzed by Wild-Type Cytochrome P450BM3, *ACS Catal.*, 2020, **10**(16), 9136–9144.

16 S. Ariyasu, Y. Kodama, C. Kasai, Z. Cong, J. K. Stanfield, Y. Aiba, Y. Watanabe and O. Shoji, Development of a High-Pressure Reactor Based on Liquid-Flow Pressurisation to Facilitate Enzymatic Hydroxylation of Gaseous Alkanes, *ChemCatChem*, 2019, **11**(19), 4709–4714.

17 M. Karasawa, J. K. Stanfield, S. Yanagisawa, O. Shoji and Y. Watanabe, Whole-Cell Biotransformation of Benzene to Phenol Catalysed by Intracellular Cytochrome P450BM3 Activated by External Additives, *Angew. Chem., Int. Ed.*, 2018, **57**(38), 12264–12269.

18 K. Suzuki, Y. Shisaka, J. K. Stanfield, Y. Watanabe and O. Shoji, Enhanced cis- and enantioselective cyclopropanation of styrene catalysed by cytochrome P450BM3 using decoy molecules, *Chem. Commun.*, 2020, **56**(75), 11026–11029.
19 O. Shoji, T. Fujishiro, S. Nagano, S. Tanaka, T. Hirose, Y. Shiro and Y. Watanabe, Understanding substrate misrecognition of hydrogen peroxide dependent cytochrome P450 from *Bacillus subtilis*, *JBIC, J. Biol. Inorg. Chem.*, 2010, **15**(8), 1331–1339.
20 O. Shoji and Y. Watanabe, Peroxygenase reactions catalyzed by cytochromes P450, *JBIC, J. Biol. Inorg. Chem.*, 2014, **19**(4–5), 529–539.
21 C. H. Hsieh, X. Huang, J. A. Amaya, C. D. Rutland, C. L. Keys, J. T. Groves, R. N. Austin and T. M. Makris, The Enigmatic P450 Decarboxylase OleT Is Capable of, but Evolved To Frustrate, Oxygen Rebound Chemistry, *Biochemistry*, 2017, **56**(26), 3347–3357.
22 G. H. Golub and C. Reinsch, Singular value decomposition and least squares solutions, *Numer. Math.*, 1970, **14**(5), 403–420.
23 J. Rittle, J. M. Younker and M. T. Green, Cytochrome P450: the active oxidant and its spectrum, *Inorg. Chem.*, 2010, **49**(8), 3610–3617.
24 C. M. Casadei, A. Gumiero, C. L. Metcalfe, E. J. Murphy, J. Basran, M. G. Concilio, S. C. M. Teixeira, T. E. Schrader, A. J. Fielding, A. Ostermann, M. P. Blakeley, E. L. Raven and P. C. E. Moody, Neutron cryo-crystallography captures the protonation state of ferryl heme in a peroxidase, *Science*, 2014, **345**(6193), 193–197.
25 X. Wang, R. Ullrich, M. Hofrichter and J. T. Groves, Heme-thiolate ferryl of aromatic peroxygenase is basic and reactive, *Proc. Natl. Acad. Sci. U. S. A.*, 2015, **112**(12), 3686–3691.
26 J. Hernandez-Ruiz, M. B. Arnao, A. N. Hiner, F. Garcia-Canovas and M. Acosta, Catalase-like activity of horseradish peroxidase: relationship to enzyme inactivation by H2O2, *Biochem. J.*, 2001, **354**(1), 107–114.
27 J. Switala and P. C. Loewen, Diversity of properties among catalases, *Arch. Biochem. Biophys.*, 2002, **401**(2), 145–154.
28 C. Jakopitsch, M. Auer, G. Regelsberger, W. Jantschko, P. G. Furtmuller, F. Ruker and C. Obinger, Distal site aspartate is essential in the catalase activity of catalase-peroxidases, *Biochemistry*, 2003, **42**(18), 5292–5300.
29 F. P. Greenspan, R. J. Gall and D. G. MacKellar, Preparation and Properties of Perlauric Acid, *J. Org. Chem.*, 1955, **20**(2), 215–217.
30 R. W. Ramette and R. W. Sandford, Thermodynamics of Iodine Solubility and Triiodide Ion Formation in Water and in Deuterium Oxide, *J. Am. Chem. Soc.*, 1965, **87**(22), 5001–5005.
31 E. A. Berry and B. L. Trumpower, Simultaneous Determination of Hemes-a, Hemes-B, and Hemes-C from Pyridine Hemochrome Spectra, *Anal. Biochem.*, 1987, **161**(1), 1–15.
32 E. J. Hart and A. Henglein, Free-Radical and Free Atom Reactions in the Sonolysis of Aqueous Iodide and Formate Solutions, *J. Phys. Chem.*, 1985, **89**(20), 4342–4347.

Faraday Discussions

ROYAL SOCIETY OF CHEMISTRY

PAPER

Computationally driven design of an artificial metalloenzyme using supramolecular anchoring strategies of iridium complexes to alcohol dehydrogenase†

Floriane L. Martins, Anca Pordea * and Christof M. Jäger *

Received 15th October 2021, Accepted 29th October 2021
DOI: 10.1039/d1fd00070e

Artificial metalloenzymes (ArMs) confer non-biological reactivities to biomolecules, whilst taking advantage of the biomolecular architecture in terms of their selectivity and renewable origin. In particular, the design of ArMs by the supramolecular anchoring of metal catalysts to protein hosts provides flexible and easy to optimise systems. The use of cofactor dependent enzymes as hosts gives the advantage of both a (hydrophobic) binding site for the substrate and a cofactor pocket to accommodate the catalyst. Here, we present a computationally driven design approach of ArMs for the transfer hydrogenation reaction of cyclic imines, starting from the $NADP^+$-dependent alcohol dehydrogenase from *Thermoanaerobacter brockii* (TbADH). We tested and developed a molecular docking workflow to define and optimize iridium catalysts with high affinity for the cofactor binding site of TbADH. The workflow uses high throughput docking of compound libraries to identify key structural motifs for high affinity, followed by higher accuracy docking methods on smaller, focused ligand and catalyst libraries. Iridium sulfonamide catalysts were selected and synthesised, containing either a triol, a furane, or a carboxylic acid to provide the interaction with the cofactor binding pocket. IC_{50} values of the resulting complexes during TbADH-catalysed alcohol oxidation were determined by competition experiments and were between 4.410 mM and 0.052 mM, demonstrating the affinity of the iridium complexes for either the substrate or the cofactor binding pocket of TbADH. The catalytic activity of the free iridium complexes in solution showed a maximal turnover number (TON) of 90 for the reduction of salsolidine by the triol-functionalised iridium catalyst, whilst in the presence of TbADH, only the iridium catalyst with the triol anchoring functionality showed activity for the same reaction (TON of 36 after 24 h). The observation that the artificial metalloenzymes developed here lacked stereoselectivity demonstrates the need for the further investigation and optimisation of

Sustainable Process Technologies, Faculty of Engineering, University of Nottingham, Nottingham, UK. E-mail: christof.jaeger@nottingham.ac.uk; anca.pordea@nottingham.ac.uk

† Electronic supplementary information (ESI) available. See DOI: 10.1039/d1fd00070e

the ArM. Our results serve as a starting point for the design of robust artificial metalloenzymes, exploiting supramolecular anchoring to natural NAD(P)H binding pockets.

1. Introduction

Enzyme catalysis can provide sustainable synthetic alternatives to high-value chemicals.[1] The application of biocatalysis can be expanded by engineering enzymes to incorporate non-natural activities. To achieve this, a promising strategy is the development of artificial metalloenzymes (ArMs).[2] These advanced chemo-enzymatic hybrids are created by the incorporation of a non-native metal catalytic functionality into a protein scaffold imparting reaction selectivity (substrate, regio-, or stereo-selectivity). ArMs have the potential to address key challenges in green chemistry: tunable reaction control, renewability, enantioselectivity and environmentally friendly process conditions.[3] The incorporation of the metal catalyst inside a protein can be achieved using dative, covalent, or supramolecular anchoring approaches.[2] Though dative anchoring strategies of a metal ion have been widely used in ArM design,[4] they are a challenging option with the need to precisely design an appropriate binding site inside the enzyme. The supramolecular anchoring of a metal complex of an organic ligand, designed to have affinity to the protein, is the most successful strategy to date, exemplified by the versatile biotin–(strep)avidin[5] and the LmrR multidrug resistance regulator systems.[6]

The efficiency of ArMs has constantly improved over the past decades, due to advancements in both experimental protein engineering technologies and in computational aided approaches.[7–9] Amongst the latter, computational tools for the *de novo* design and redesign of binding sites for metal cofactors have greatly contributed to the rapid design of ArMs with novel or improved functionality.[9] Such tools are focused either on the design of non-native metal binding or on the design of catalytic functionality within the newly created metalloproteins, using software such as the Rosetta suit of programmes,[10,11] which combines computational tools applicable for methods like molecular docking, structure prediction, and *de novo* protein design.

Computationally driven *de novo* (re-)design strategies of metal containing active sites were demonstrated successfully for a number of examples (see also ref. 9). Rosetta Match[12] was applied to optimise the design of a novel abiological hetero-Diels–Alderase, based on density functional theory (DFT) informed transition state (TS) model structures;[13] a novel organophosphate hydrolase was also designed *de novo* by first defining a suitable TS and subsequently redesigning the active site of a natural deaminase.[14]

With a focus on supramolecular catalyst anchoring, software such as RosettaDesign[15] has been used to fine tune the binding of the ligand, as demonstrated in the tailoring of the amino acid sequence of human carbonic anhydrase II to bind an iridium metal complex catalyst with high affinity, thus improving the catalytic performance of the resulting ArM.[16] Molecular docking is another useful method of choice for evaluating and designing protein–ligand interactions between the guest catalyst and the protein environment, particularly when combined with methods with higher accuracy (*e.g.* QM/MM, MD). Docking has commonly been employed as a second step in the design to gain insight into the experimental results from first-generation ArMs and to define key residues influencing optimal and directed substrate binding and thus enzymatic activity and substrate selectivity.[13,17–19] Computational analysis of the defined complexes can also guide the selection of bioconjugation sites in the case of the covalent binding of the metal complex to the protein.[20,21] In other approaches, molecular docking in combination with QM and MD simulations was used to control the activity and selectivity of ArMs by understanding substrate binding affinity, flexibility, and orientation.[22,23]

Despite the above examples and the similarity of the approach to computational drug design, where chemical structures are non-covalently incorporated within proteins as strongly binding inhibitors,[24] docking is not routinely used in ArM design. This is because binding affinity alone is not sufficient to generate a working ArM. The correct orientations of the incorporated metal catalyst and of the binding substrate(s) to form the Michaelis complex are equally important and thus must be evaluated by computational approaches. Since docking is only an approximate method, more accurate (but more costly) molecular dynamics coupled free energy perturbation (FEP) simulations and QM and QM/MM evaluations should also be considered.

Whilst most successful, the design of ArMs based on supramolecular anchoring relies on non-enzymatic protein structures with a large hydrophobic pocket which is able to accommodate both the metal catalyst and the substrate. An alternative is the use of cofactor-dependent enzymes, where the native cofactor is replaced with a non-native catalytic moiety.[25] For example, the heme-binding pocket in P450 was used to accommodate Ir-porphyrins to catalyse C–C and C–N bond formations.[26] In our group, we developed ArMs for the reduction of nicotinamide cofactors, starting from the wild type alcohol dehydrogenase from *Thermoanaerobacter brockii* (TbADH WT), by covalently binding rhodium piano stool complexes at the place of the catalytic zinc site.[21,27] In this work, we aim to simplify the ArMs by designing catalysts for supramolecular binding to avoid the need for the covalent modification of the protein. So far, ArMs based on supramolecular anchoring rely on systems where a natural affinity between the protein and ligand is a pre-requisite, thereby limiting the protein and/or ligand choice;

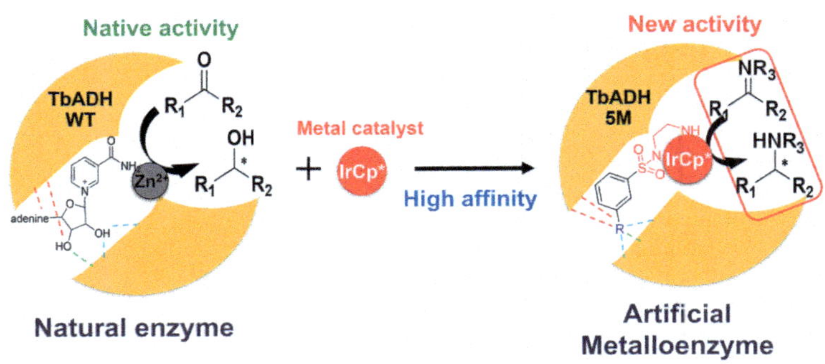

Fig. 1 Schematic representation of the design of ArMs: left, the natural catalysis in TbADH WT involving $NADP^+$ and a catalytic Zn^{2+}; right, the supramolecular insertion of an iridium piano stool complex (red) inside the $NADP^+$ cofactor pocket and Zn^{2+} binding region of TbADH 5M with high affinity induced by the anchor substructure (R).

whereas systematic rational strategies to specifically design the metal catalysts for supramolecular anchoring to a selected protein scaffold are yet to be developed.

To address this, we present a critically evaluated and refined docking strategy for the design of anchoring functionalities to provide strong interactions between d^6 piano stool metal catalysts and TbADH. In this design, the $NADP^+$ cofactor binding site was used to accommodate the catalyst, as schematically shown in Fig. 1. This site has a relatively high binding affinity for the native cofactor (*e.g.* $K_d^{NADP^+} = 9\ \mu M$)[28] due to a network of hydrogen bond interactions, in addition to being in proximity to the hydrophobic substrate binding pocket. We use the mutant TbADH 5M (H59A–D150A–C203S–C283A–C295A) as a starting point, where catalytic Zn^{2+} was removed *via* mutations of two coordinating residues to alanine.[27] We selected sulfonamide-based iridium d^6 piano stool catalysts for the study, because they are highly active in imine reduction and were previously successfully used in the transfer hydrogenation of imine by ArMs (Fig. 1).[29–31] Following computational design, the most promising catalyst structure was synthesised and evaluated for the formation of a functional ArM.

2. Results and discussion

2.1 Design and evaluation of the docking protocol

Before refining a docking protocol applicable to the design of suitable anchor structures for catalysts to ADH-based ArMs, it was necessary to evaluate whether an approximate method, such as docking, could result in sufficiently accurate predictions of anchor structure binding affinities and orientations. Therefore, we evaluated different docking protocols for their ability to predict the crystallographically known $NADP^+$ cofactor binding to TbADH. This evaluation was additionally used to analyse key binding interactions informing the subsequent anchor design.

Different docking accuracy levels were analysed. The standard precision (SP) docking, as implemented in Glide[32,33], was compared against the more costly extra precision (XP) docking protocols and the most sophisticated MM-GBSA (molecular mechanics generalized Born and surface area) docking refinement to estimate the relative binding free energies. The crystal structure of TbADH WT co-crystallised with $NADP^+$ was selected from the Protein Data Bank (PDB) for docking studies (PDB 1YKF[34]). For the subsequent docking approaches in the search of high affinity anchors for transition metal catalysts, TbADH 5M was additionally prepared without the catalytic zinc ion, to make space in the active site for the inserted transition metal, by the creation of five mutations to remove Zn^{2+}.

Table 1 Natural cofactor re-docking scores calculated by Glide XP (kcal mol^{-1}) and MM-GBSA (ΔG in kcal mol^{-1}). RMSD (Å) values compared to the co-crystallised $NADP^+$ cofactor in the X-ray structure of TbADH WT

Enzyme	Glide XP	MM-GBSA	RMSD
TbADH WT	−10.1	−76.4	1.7
TbADH 5M	−10.7	−59.5	1.8

The co-crystallised natural cofactor $NADP^+$ was re-docked into the cofactor binding site of both TbADH variants. The calculated binding affinities at the different docking accuracy levels are presented in Table 1. The results demonstrate high affinity binding in all cases. The scores were similar for both enzymes, with the MM-GBSA evaluation indicating slightly better binding to TbADH WT. This approach led to very good agreement with the two structures of TbADH WT and TbADH 5M, with an all-atom RMSD of < 2 Å for the bound cofactor (see Fig. S1 in the ESI†). Between the two enzymes, TbADH 5M docking positioned the nicotinamide cofactor slightly closer to the substrate binding site than that for TbADH WT.

Following validation of the docking procedure, an iterative docking workflow was developed to design iridium complexes that could be accommodated *via* supramolecular interactions within the cofactor binding site of TbADH. Initially, an efficient and computationally affordable screening method of a diverse library of chemical structures was implemented to find anchor structures providing high affinity binding to TbADH. The highest affinity anchor structures were

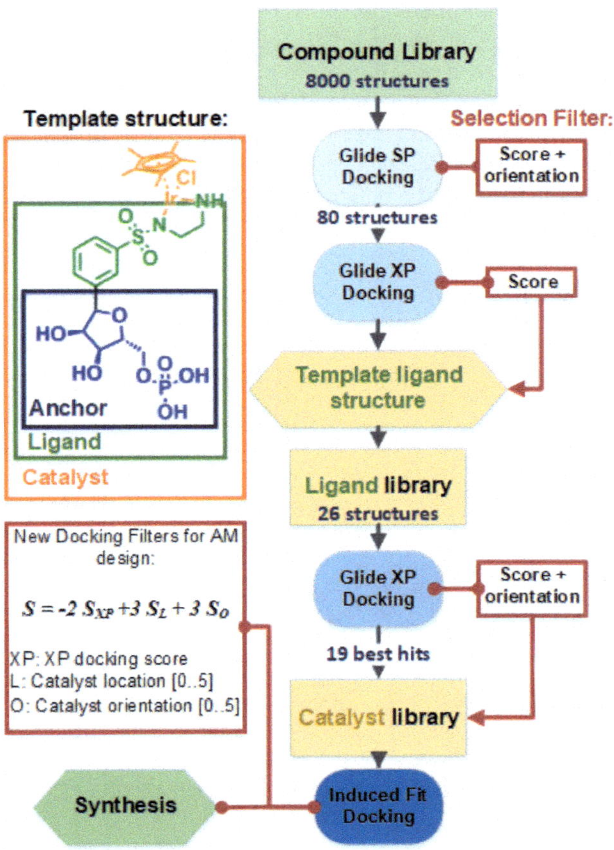

Fig. 2 Overall docking and library design protocol used to filter and optimise the anchor, ligand, and catalyst structures. Template structures are shown with definitions of the anchor, ligand, and catalyst structure, as presented in the text.

subsequently incorporated within whole catalyst structures with a piano stool Ir(Cp*)–sulfonamide functionality, which were computationally evaluated both for their binding affinity and catalytically relevant positioning.

The general docking protocol followed three major steps, as shown in Fig. 2: (1) a large compound library was screened to identify patterns of functional groups responsible for high affinity, leading to a template ligand structure; (2) anchor structures identified by the initial screening were appended to the sulfonamide functionality to create a ligand library of 26 compounds, which were evaluated with higher precision docking methods; (3) 19 of the highest affinity ligands were selected and analysed for their ease of synthesis and a final library of metal catalyst structures was then generated and docked using induced fit docking. Additional filters were applied to evaluate not only the binding affinity, but also the optimal positioning and orientation for catalysis. From this final docking step, six structures were identified as potentially suitable, and one was finally selected for chemical synthesis and testing.

2.1.1 Virtual high-throughput anchor screening and selection. In the first step of the protocol in Fig. 2, libraries of potential anchor structures were screened by applying a first raw filter based on the Glide SP protocol. For this high throughput approach, around 8000 structures of available chemical compounds and their conformers from the Schrödinger and ZINC library[35] (ZINC12; glide drug-like ligand decoy sets) were chosen, representing the breath of synthetically available organic chemistry in the field of drug discovery. The 80 best ranked and correctly orientated compounds (1% of the library) were subsequently selected and re-evaluated through Glide XP docking. To systematically analyse the results of the large library docking, the outcomes were compared and evaluated against docking of fragments of the natural cofactors NAD(P)H (Fig. S2 and Table S1 in the ESI†).

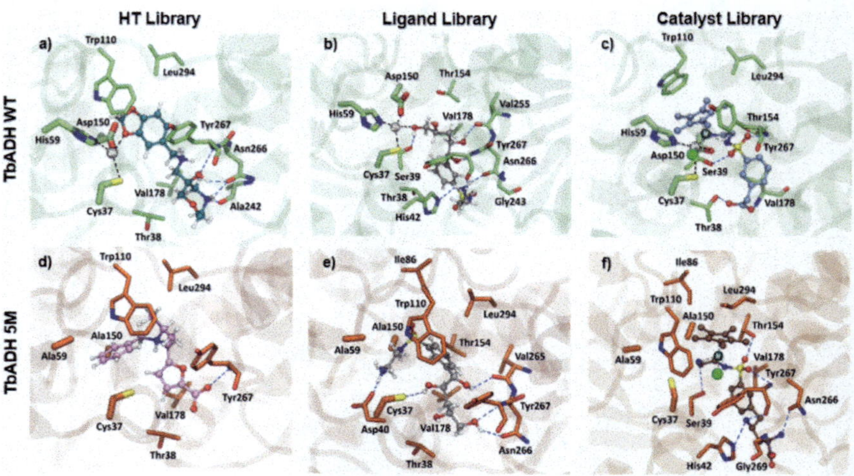

Fig. 3 Pictures of the best ranked compounds from ligand libraries screened through XP and IFD docking in TbADH WT and TbADH 5M, respectively: (a) and (d) best ranked ZINC library compounds; (b) and (e) best ligands docked through XP docking ((b) **1**, e: **1**) best metal complexes ranked by IFD score ((c) **Ir2**, (f) **Ir15**).

Both strategies led to similar observations. The best ranking structures from the high throughput docking and the fragment-based approach both demonstrated the importance of hydrophilic functional group anchoring being present in the ribose and phosphate binding regions of $NADP^+$ (see Fig. 3 for the best ranked docking structures). The best ranked structures all shared a hydrophilic centre composed of amino, carboxyl or hydroxyl groups, forming several hydrogen bonding interactions with residues such as Val178, Ser39, Asn266 or Tyr267, which correspond to the $NADP^+$ interactions found in the TbADH crystal structure[34] (Fig. S3 in the ESI†).

On the other hand, the fragment-based docking indicated little affinity of the adenosine for the respective binding locations, suggesting a low relevance of high affinity binding. Also, the nicotinamide ring showed no interaction in all the different fragment-based docking experiments. This strongly aligned with the library docking, which ranked hydrophobic structures incorporating aromatic rings lowest. Additionally, larger hydrophobic entities tended to be positioned in the hydrophobic substrate pocket, particularly in TbADH 5M. This would likely also prevent substrate binding and needed to be addressed in the subsequent steps.

2.1.2 Creation of a ligand template structure and small library docking evaluation. Based on the screening results, a general template structure for the design of a small organic ligand library was identified (Fig. S4 in the ESI†). The anchoring part of this template structure consisted of principal mimics of the pyrophosphate and ribose part of $NADP^+$ to create a high affinity with polar or hydrogen bonding interactions inside the enzymes (see Fig. 2) through hydroxyl and carboxyl groups. Modifications were then applied to the length, position, number and nature of these hydrophilic functional groups to generate a library of anchors. For validation of the protocol, hydrophobic structures were also designed, but were anticipated to have lower affinity.

As the pyridinium ring demonstrated little effect on the binding affinity, it was adapted to a phenyl ring, with an eye on the subsequent simpler chemical synthesis. As the last feature of the lead structures, the sulfonamide group was attached to the phenyl ring, which served to bind the transition metal. An initial screening of the library additionally revealed that the *meta* position of substituents on the phenyl delivered the best affinity scores and this was thus chosen for the subsequent library design.

The designed library of 26 structures is presented in the ESI (Fig. S5 in the ESI†). For all of the structures, all possible conformations and configurations were generated and docked inside both TbADH WT and TbADH 5M, using a combined SP and XP workflow (Table S2 in ESI†). The best binding structures all had hydroxyl and carboxyl groups in their anchoring part. This matched the aim to substitute ribose (by hydroxyl) and phosphate (by carboxylic acid) for easier chemical synthesis. Conversely, compounds with only one hydrophilic group and hydrophobic groups (*e.g.* **12** and **23**, Fig. S5 in the ESI†) were ranked lowest.

We observed that most configurations of the same anchored structures delivered similar XP scores and only the strongest binding configurations were selected for the next iteration of the docking protocol (see the example for ligand **4** in Table S3 in the ESI†). A comparison between TbADH WT and TbADH 5M showed overall slightly lower XP scores for the 5M mutant, indicating a higher binding affinity. Despite this difference, many of the first ranked structures were common for both enzymes, with only one different structure within the first five highest ranked structures. Analysis of the positions remarkably showed inverted orientations and unfavourable positioning of the majority of ligand structures in the case of TbADH WT, whilst all TbADH 5M bound ligands displayed an orientation favourable for catalysis, similar to the natural cofactor $NADP^+$.

2.1.3 Iridium complex docking and selection. The 19 best ranked and positioned ligands from previous SP/XP screening were then metalated with the d^6 piano stool iridium [IrClCp*] moiety and were subsequently docked inside the enzymes using induced fit docking (IFD) to increase the accuracy by accounting for the flexibility of the amino acid sidechains in the binding pocket (Fig. S6 in the ESI†). In order to systematically account for the correct positioning and orientation for catalysis, we introduced an additional score to be applied to the final docking pose and energy, which included the likeliness to bind in catalytic availability (Fig. 2). The final score (designated IFD(S)) incorporated the binding affinity represented by the Glide XP score (S_{XP}) from the IFD docking, the location of the ligand in the pocket (S_L) and the orientation of the ligand in the pocket (S_O). Since we reasoned that the correct positioning of the ligand for catalysis outweighed small differences in the binding affinities, the influence of positional parameters was privileged over the affinity approximation (see details in Section 1.4.1 of the ESI†). The resulting final score for each structure allowed a ranking and selection of several potential high affinity catalysts (TbADH 5M in Table 2 and TbADH WT in Table S4 in the ESI†).

From the 19 docked metal complexes fed through the IFD docking, 7 catalysts demonstrated significantly repulsive interactions induced by incorporating the metalation and were subsequently removed from the analysis. The ranking of the remaining 12 catalysts was highly influenced by the nature of the anchoring substituent, which directed the metal complex [IrClCp*] into a catalytic position. This was also demonstrated by a distinctly different ranking order when taking only the XP score from the IFD docking compared to the additionally applied new scoring scheme (Table 2). Thus, our best ranking catalyst **Ir15** only scored fifth in pure IFD docking (IFD(XP)) while the best IFD docking score (**Ir2**) was ranked

Table 2 Ranking of iridium catalysts from IFD docking inside TbADH 5M by applying the IFD(S) score defined in Fig. 2 compared to the IFD XP score (IFD(XP), kcal mol^{-1})

Iridium catalyst	IFD(S) score	Iridium catalyst	IFD(XP) score
Ir15	54.0	**Ir2**	−12.8
Ir3	53.6	**Ir1**	−12.2
Ir1	53.4	**Ir22**	−11.2
Ir4	52.6	**Ir13**	−11.6
Ir14	50.4	**Ir15**	−9.5
Ir9	46.6	**Ir6**	−9.5
Ir10	43.8	**Ir3**	−9.3
Ir13	43.6	**Ir14**	−9.2
Ir8	42.8	**Ir4**	−8.8
Ir2	42.6	**Ir9**	−8.8
Ir6	41.0	**Ir8**	−8.4
Ir22	39.4	**Ir10**	−7.4

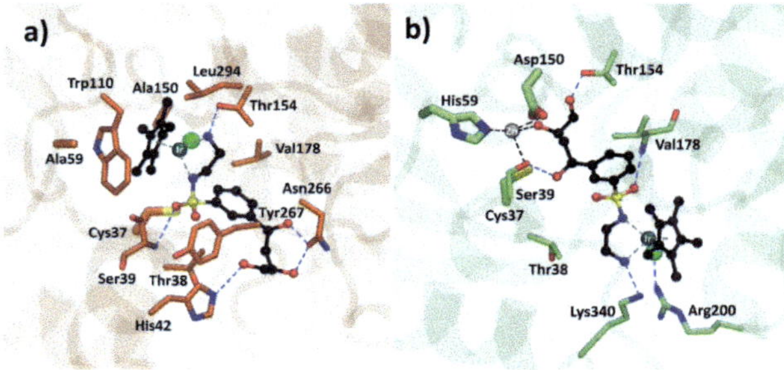

Fig. 4 Metal complex **Ir4** docked into (a) TbADH 5M with a catalytic orientation and (b) into TbADH WT with an inverted position.

down by applying the additional score correction due to the unfavourable orientation in the binding site, most likely not leading to catalytic activity.

Similar to the previous ligand docking, TbADH 5M presented higher affinity XP and IFD scores than those of TbADH WT. The low scoring results from the TbADH WT results were obtained because of the mainly reversed position of the metal complexes inside the cofactor binding pocket. Most likely this was due to the Zn coordination site occupying the space needed to fit the catalyst and preventing ideal binding.

Analysis of the structure poses in TbADH 5M demonstrated a general orientation of the metal toward the substrate pocket, except for few inverted poses. Fig. 4 shows an example of **Ir4** displaying a high IFD(S) score with its crucial

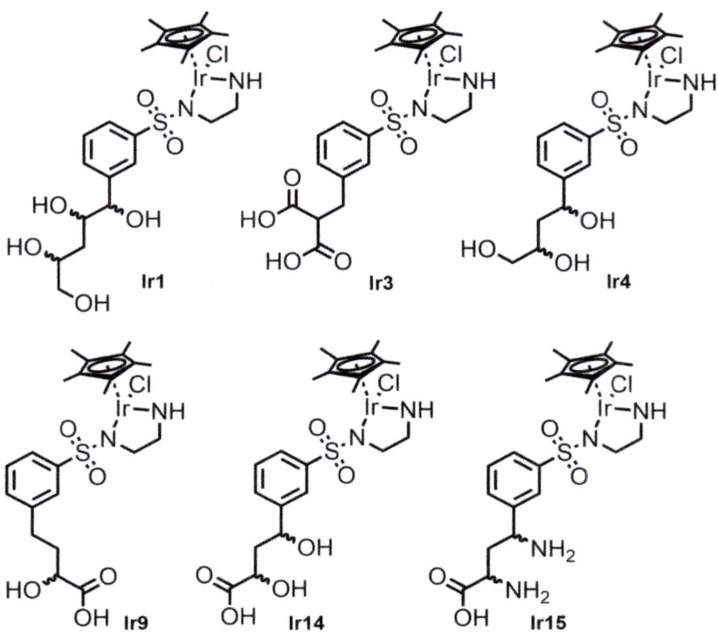

Fig. 5 Selected metal complex hit structures from the docking study, with expected high affinities for TbADH 5M.

binding interactions in the active site of TbADH 5M, whilst there is an inverted orientation in TbADH WT. All of the high affinity complexes demonstrated the key interactions with residues like Thr38, Ser39, Asn266 or Tyr247.

Fig. 5 shows the structures of the six best metal complexes based on the IFD(S) results for TbADH 5M. These structures were expected to bind with high affinity to TbADH 5M and to additionally accommodate a hydrophobic substrate in the catalytic position, therefore representing attractive options for constructing an ADH-based ArM using the supramolecular binding approach. From these structures, we proceeded to select and synthesise one example for experimental validation, based on the ease of synthesis.

2.2 Synthesis of iridium complexes

From the preceding docking analysis, the iridium complex **Ir14**, bearing a 2,4-dihydroxy carboxylic acid anchor, was initially selected for experimental validation, given its high IFD(S) score and its short retro-synthesis path. The synthesis of **Ir14** was attempted from either the Boc or the Cbz-protected *N*-sulfonyl-ethylenediamine-2,4-diketoester structure (compounds **28** and **32** in Fig. 6). This diketoester was obtained as a *E*/*Z* stereoisomer mixture of the mono-enol form *via* the Claisen condensation of the acetophenone derivatives **27**/**31** and diethyl oxalate under basic conditions (Fig. 6). However, given the lack of the stability of the 2,4-diketoester functionality under both acidic Boc deprotection and palladium catalysed Cbz deprotection conditions (leading to self-Claisen, retro-Claisen or hydrogenolysis by-products), a refinement of the target ligand structure was necessary. The tri-hydroxy ligand **Ir4**, which was also found to bind with high scores in the preceding docking approach, was synthesised by Pd-

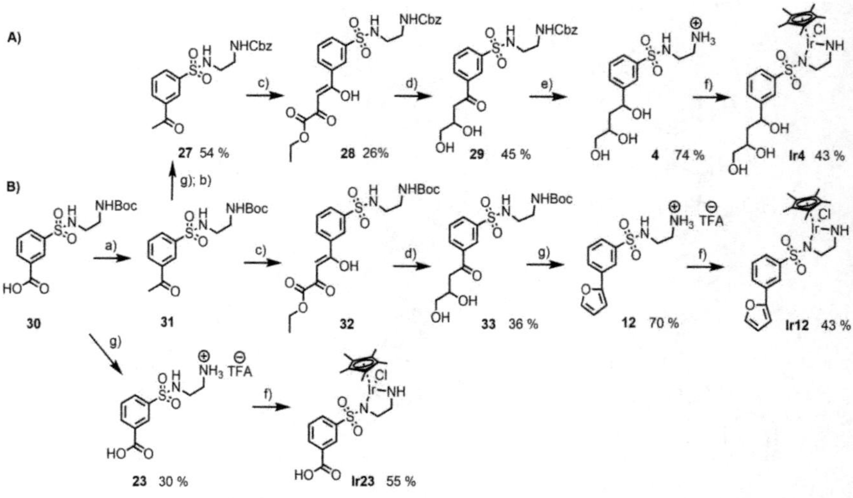

Fig. 6 Synthesis of the ligands and iridium complexes used in this study: (A) compounds **4** and **Ir4**, starting from the Cbz-protected *N*-sulfonyl-ethylenediamine precursor; (B) compounds **12**, **23**, **Ir12** and **Ir23**, starting from the Boc-protected *N*-sulfonyl-ethylenediamine precursor (a) (i) $SOCl_2$, CH_3NHOCH_3, Et(i-NPr)$_2$, CH_2Cl_2; (ii) CH_3MgBr, THF; (b) Na_2CO_3, Cbz-Cl, H_2O/Et_2O; (c) *t*-BuOLi, $(CO_2CH_2CH_3)_2$, THF; (d) $LiAlH_4$, THF; (e) Pd/C 10%, H_2, MeOH; (f) $[IrCl_2Cp^*]_2$, Et_3N, MeOH; (g) TFA, CH_2Cl_2.

catalysed reduction and the concomitant deprotection of the Cbz-protected 2,4-diketoester **28** with H_2 (Fig. 6A). Interestingly, when the same synthesis was attempted from the Boc-protected precursor **32**, the deprotection of intermediate **33** with trifluoroacetic acid led to cyclisation of the anchor with the formation of a furan ring to give compound **12** (Fig. 6B). This behaviour was similar to the Paal–Knorr reaction, involving the acid catalysed cyclisation of a 1,4-diketone in its mono-enol form, followed by dehydration to form a furan.[36] The synthesis of furan derivatives from a 1,4-keto-alcohol intermediate was previously demonstrated in acidic conditions,[37] as was the TFA-catalysed deprotection followed by the cyclisation of *tert*-butyl acetoacetate derivatives.[38] Compound **12** and its corresponding iridium complex **Ir12** were also prepared to be further tested in artificial metalloenzyme formation, as they had a conveniently available hydrophobic and aromatic anchor substructure to be compared with the hydrophilic triol anchor of **4** for its binding and catalytic potential. Given the similarity of the XP scores for the different configurations of ligand **4** we did not consider stereoconfiguration in the synthesis. Furthermore, intermediate **23** and its corresponding iridium complex **Ir23** were prepared, bearing a single hydrophilic (carboxylate) functionality as an anchor. As expected, docking analysis showed a very low ranking in all screenings and a reverted position for the metal complexes (see Table S5 and Fig. S7 in the ESI†).

2.3 Artificial metalloenzymes by the supramolecular anchoring of iridium complexes to TbADH

2.3.1 Evidence of Ir complex binding to TbADH. To assess the ability of the designed sulfonamide ligands (**4, 12, 23**) and iridium complexes (**Ir4, Ir12, Ir23**) to form ArMs, their supramolecular bioconjugation to TbADH was investigated. Initial attempts to measure the binding affinity by isothermal calorimetry (ITC) led to inconclusive results due to the high concentration of protein necessary to show ITC signals, even in the presence of the natural cofactor (100 μM protein and 2.1 mM $NADP^+$ or compound **32**). Thus, the binding affinity was investigated by indirect means using inhibition assays. Considering the structures as competitive inhibitors of $NADP^+$, the IC_{50} for each sulfonamide ligand and iridium complex was evaluated in combination with TbADH WT. For this assessment, the functional enzyme holding the catalytic zinc had to be chosen to facilitate butan-2-ol

Table 3 Inhibition of the catalytic oxidative activity of TbADH WT by the designed ligands and iridium complexes, reported as IC_{50} (mM). The IC_{50} corresponds to the concentration of inhibitor required to inhibit half of the enzyme activity, in competition with $NADP^+$. Assay conditions: $NADP^+$ (0.5 mM), *rac*-butan-2-ol (150 mM), inhibitor (0.001–40 mM), TbADH WT (1 μg) in 100 mM Tris–HCl buffer at pH 8, 40 °C. The results are presented including the standard error of the mean (SEM) of triplicate measurements

Ligand	IC_{50} (mM)
4	>10
Ir4	0.248 ± 0.003
12	16.050 ± 0.390
Ir12	0.052 ± 0.012
23	5.459 ± 0.430
Ir23	4.410 ± 0.035

oxidation by $NADP^+$, taken as reference reaction. The inhibitory potential of the ligands and complexes provided an insight into their binding capacity, using low concentrations of enzymes and small molecules (Table 3).

All of the sulfonamide ligands showed IC_{50} values higher than 5 mM, indicating lower affinities for TbADH in comparison with the native $NADP^+$ cofactor, as expected from the docking prediction. However, in contrast to docking, the inhibition assays suggested a lower affinity of TbADH for the tri-hydroxy ligand **4** compared to that of ligand **23**, which bears a mono-carboxyl anchor. One potential explanation for these results is that the assumption of competitive inhibition of the cofactor binding site was not appropriate in this case. In addition, ligand **4** also presented a mixture of stereoisomers of which some might bind with different binding affinities or in other locations, not adding to competitive inhibition, although docking suggests a similar affinity of the isomers for the cofactor binding pocket. The binding of smaller or hydrophobic anchors at positions where they hinder substrate approach and/or binding, but outside the cofactor binding pocket, may also explain the higher inhibition potential of these compounds.

All of the iridium complexes showed better inhibitory potential than their corresponding sulfonamide ligands alone, with IC_{50} values between 0.052 (**Ir12**) and 4.410 mM (**Ir23**). The iridium functionality was not expected to significantly add to the binding affinity, as its bulky nature is more of a challenge to accommodate in the pocket. This was also demonstrated by the observation from docking that in the presence of the Zn^{2+} ion in TbADH WT, the structures were inverted, positioning the metal toward the pocket entrance, in order to fit the pocket. Therefore, it can be speculated that the iridium piano stool functionality has a higher impact on enzyme inhibition than the binding conferred by the anchor. One explanation of this could be that the coordination of the iridium piano stool moiety to the surface exposed coordinating (cysteine or histidine) residues, leading to enzyme deactivation, as previously shown with rhodium complexes.[39] Furthermore, the slightly better inhibition demonstrated by **Ir12** over that of the tri-hydroxy functionalised **Ir4** could be attributed again to alternative binding modes with different orientations not predicted by docking that could add to the inhibition.

A detailed comparison of the binding within TbADH WT and the 5M mutant is difficult. This is because only binding to TbADH WT, which contains the Zn-binding site, could be assessed by inhibition studies. Moreover, inversion of binding was observed when docking the iridium structures to TbADH WT, compared to the mutant 5M. In conclusion, whilst inhibition by the designed iridium complexes was observed and indicated an interaction between the designed complexes and TbADH WT, certainty on the exact location and orientation of the bound structures and on the affinity of TbADH 5M for the iridium complexes could not be defined because of the indirect analysis.

2.3.2 Imine reduction activity of Ir complexes conjugated to TbADH. The catalytic performances of the synthesised iridium complexes **Ir4**, **Ir12** and **Ir23**, either alone or combined with TbADH WT and TbADH 5M, were investigated for the transfer hydrogenation of the cyclic imine salsolidine (6,7-dimethoxy-1-methyl-3,4-dihydroisoquinoline) in buffer at pH 7, in accordance to previously published conditions.[40] The most active of the free iridium catalysts was **Ir4** (Table 4, entries 1–3).

Table 4 Transfer hydrogenation of salsolidine. Ir catalyst = **Ir4**, **Ir12**, or **Ir23**, either free or in combination with TbADHs. Reaction conditions: 50 μM iridium, 50 μM TbADHs, 4 mM salsolidine, 300 mM HCOONa, 100 mM PBS buffer pH 7, 38 °C. The experiments were run in triplicate for 24 h and the results presented are the turnover frequency (TOF) and turnover number (TON) ± SEM. SEM = standard error of the mean, calculated from triplicate measurements

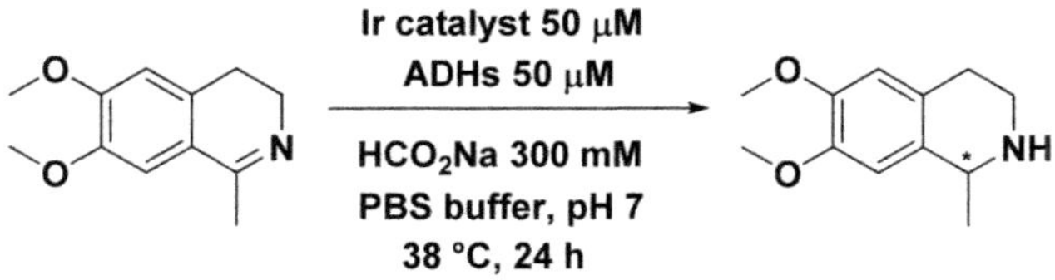

Entry	TbADH variant	Ir catalyst	TON	TOF (min^{-1})
1	—	**Ir4**	90 ± 1.04	0.062 ± 0.001
2	—	**Ir12**	66 ± 1.45	0.046 ± 0.001
3	—	**Ir23**	59 ± 8.39	0.041 ± 0.006
4	5M	**Ir4**	36 ± 3.21	0.025 ± 0.002
5	WT	**Ir4**	1.6 ± 0.59	0.001 ± 0.000
6	5M	**Ir12**	0	—
7	WT	**Ir12**	1.7 ± 0.55	0.001 ± 0.000
8	5M	**Ir23**	0	—
9	WT	**Ir23**	0	—
10	5M (100 μM)	**Ir4**	34 ± 1.34	0.023 ± 0.001
11	5M	**Ir4** (200 μM)	79 ± 3.18	0.055 ± 0.002
12	WT	**Ir4** (200 μM)	81 ± 0.80	0.056 ± 0.001

Only complex **Ir4** showed activity in the presence of TbADH 5M and no stereoselectivity was observed (Table 4, entries 4, 6 and 8). Furthermore, when the zinc-containing enzyme (TbADH WT) was used, the activity of **Ir4** was drastically reduced and this was the case with the other catalysts as well (Table 4, entries 5, 7 and 9).

Two different scenarios could explain the lack of activity in the presence of the WT and the mutual inactivation of the enzyme and catalyst: binding to the surface exposed coordinating residues and binding in the cofactor site, but without space for the imine substrate, or in the wrong orientation, as suggested by docking. Both of these scenarios would lead to the iridium complexes not being available as active free catalysts in solution, implying that any activity in the presence of the protein was likely to come from a TbADH-bioconjugated catalyst.

In addition, there was no difference in the surface exposed coordinating residues between the TbADH WT and 5M variants, which means that the difference in activity observed with the two variants cannot come from bioconjugation to surface exposed residues. These observations, alongside the docking results, suggest that the iridium catalyst **Ir4** binds within the cofactor site in the orientation suggested by docking, with the catalytic part towards the hydrophobic substrate pocket of the enzyme. When bound to 5M, there was enough space for the imine substrate to react, whilst within the zinc-occupied WT enzyme, the imine substrate could not access the iridium catalyst. Iridium complexes **Ir12** and **Ir23** may bind to the cofactor binding site, but not in active poses or orientations, which would explain their inhibitory potential for TbADH WT and their lack of activity within TbADH

5M. This theory was supported by the previous docking studies, which predicted the insertion of hydrophobic structures inside the substrate pocket (Fig. 3d).

To further confirm the theory of the supramolecular binding of the iridium catalyst to TbADH, the concentrations of both TbADH 5M and of the catalyst **Ir4** were varied, in order to perturb the equilibrium towards the bound state of the catalyst. When the concentration of TbADH 5M was doubled from the initial stoichiometric ratio, the turnover number remained similar (Table 4, entry 10 compared to 5), suggesting that most of the available binding sites were occupied with the iridium catalyst. This could mean either that the active catalyst binds to TbADH 5M with very high affinity, or that not all binding sites were occupied, therefore providing an excess of binding sites even at stoichiometric monomer : Ir conditions. On the other hand, when the concentration of the metal catalyst was increased four times, an increased TON was observed with both TbADH WT and 5M (Table 4, entries 11 and 12), with the increase in activity likely coming from the free catalyst in solution. Taken together, these results further suggest the formation of a functional artificial metalloenzyme for imine reduction by the supramolecular incorporation of the iridium complex **Ir4** within the $NADP^+$ binding site of the zinc devoid TbADH variant 5M.

3. Conclusions

This work demonstrates how docking methodologies can be applied and optimised to design ArMs based on the supramolecular anchoring of imine reduction catalysts into the $NADP^+$-binding site of alcohol dehydrogenase scaffolds. The computational workflow started from high throughput virtual screening and was refined to the docking of a selected catalyst library. The refined protocol focused on more accurate calculations to predict a transition metal catalyst with catalytic positioning.

The creation of an active ArM was demonstrated for the transfer hydrogenation of salsolidine, however the lack of stereoselectivity represents a drawback. Nonetheless, from the combination of competition experiments (indicating binding affinity) and catalytic activity, there is clear evidence of a functional ArM, strongly suggesting that the catalyst is bound to a defined site that can also accommodate the substrate. Although docking suggested that the iridium piano stool moiety would be oriented towards the substrate binding pocket, it cannot be ruled out that catalyst **4** binds towards the surface of the enzyme or is involved in multiple binding modes involving all its different stereoisomers.

The results presented here show that the first step to create this type of ADH-based ArMs needs further experimental and computational assessment and optimisation. While this study focused on improving the potential and workflow for docking applications in connection to ArM design, future computational work needs to target more advanced and accurate computational methods to test and challenge the original predictions.

Molecular dynamics (MD) simulations will give insights into the flexibility and integrity of the protein catalyst complexes. Catalyst and substrate binding and unbinding simulations will be able to give further into the binding strength, flexibility and alternative binding modes. Finally, the definition of transition state structures for the simulated binding arrangements will make it possible to optimise and re-design the surrounding binding pocket for better TS stabilisation and introduction of stereo selectivity.

4. Experimental

4.1 Docking calculation procedures

The Schrödinger suite[41] of programs and the Maestro graphical interphase were used for the design of the molecular docking protocols with the default parameters defined in the Schrödinger program. All compounds included in this work were prepared using Ligprep from the Schrödinger suite,[42] with the OPLS3 force field.[43] The generation of all possible protonation and ionisation state combinations was performed for each ligand using Epik23 in aqueous solution at pH values of 7.0 ± 2.0. Whenever the ligand contained a ribose, the same stereochemistry as that of the natural cofactor was kept. Structures containing iridium metal were created using the Maestro interface to build a pyramidal metal centre, substituted with chloride and a pentamethylcyclopentadienyl (Cp*) moiety, applying fixed bonds of zeroth bond order. The complex structures were then minimized within the Maestro graphical interphase.

The crystal structure of TbADH was obtained from the Protein Data Bank (PDB 1YKF2 (ref. 34)). The monomer of the protein was additionally prepared without the natural zinc ion in the active site. From the crystal structure of TbADH, several mutations were created, Cys283Ala, His59Ala, Asp150Ala, Cys295Ala and Cys203Ser, to yield the structure TbADH 5M. Protein structures were prepared using the Maestro Protein Preparation Wizard in the Schrödinger suite.[44] Missing hydrogen atoms and residue side chains were added to the structures by Prime refinement.[41] During the refinement, water molecules with less than three hydrogen bonds to other atoms were removed, which resulted in no more water in the binding site. The selection of the position of hydroxyl and thiol hydrogens, the protonation/tautomer states and the "flip" assignment of aspartic acid, glutamic acid, arginine, lysine and histidine were adjusted at pH = 7.0 using PROPKA.[45] Finally, the structures were minimized using the OPLS3 force field with a RMSD = 0.3 Å maximal displacement of non-hydrogen atoms as the convergence criterion.

For non-covalent docking using Glide software, the grid box was defined from the optimized protein structure at the centroid of the active site (20 Å radius around the co-crystallized ligand). In the section involving transition metal complex docking, the enclosure box was defined from the optimized protein structure without zinc by a selection of amino acid residues (TbADH 37, 59, 60, 42, 151, 178, 295, 266 and 150). No constraints were added. The standard settings of a van der Waals scaling factor of 1.0 for nonpolar atoms was kept. Nonpolar atoms were defined with an absolute value of partial atomic charges of $\leq$0.25.

The prepared ligand conformers without the metal were flexibly docked into the receptor grid using Glide SP and Glide XP procedures, with the receptor kept rigid.[32,33,46] The default parameters as defined in the Schrodinger program were kept. OPLS3 (ref. 43) was used as the force field and the variation of conformation was applied. The estimations of the free energy binding of each ligand for the enzymes were ranked. The structures with the best Glide XP score were used in the MM-GBSA calculations. The VSGB solvation model (surface generalized Born model and variable dielectric) was selected to approximate the solvation free energy and correct the interactions (H-bond, hydrophobic, π–π).[47] Flexibility was added to residues at 5.0 Å from the ligand position in the binding pocket. Prime minimisation was used on the residues selected in the flexible region, without any constraints. For the IFD docking of different metal complexes, the protein residues within 5.0 Å of the catalyst were kept flexible using Prime. The best scores were selected for analysis.

4.2 Synthesis of the ligands and metal complexes

For the detailed and quantified synthesis protocol, see Section 3.1 of the ESI.†

4.2.1 Benzyl (2-((3-acetylphenyl)sulfonamido)ethyl)carbamate (27). To a cooled solution of ketone **31** (1 eq., 2.87 mmol, 987 mg) in dichloromethane (47 mL), trifluoroacetic acid was added dropwise under stirring (12.7 eq., 36.45 mmol, 2.8 mL). The solution was then stirred overnight at room temperature and monitored by TLC (CH_2Cl_2 95/5 CH_3OH + traces of acetic acid). The solution was evaporated under vacuum, after which TFA traces were removed by the repeated addition and evaporation of dichloromethane under vacuum.[48] The obtained deprotected ketone (1 eq., 0.744 mmol, 253 mg) was dissolved in a bi-phasic diethyl ether (3 mL) and water (3.72 mL) mixture, to which Na_2CO_3 was added (3 eq., 2.23 mmol, 237 mg). The solution was cooled on ice before the dropwise addition of benzylchloroformate (1 eq., 0.744 mmol, 0.106 mL). The solution was stirred at room temperature for 2.5 h and monitored by TLC (ethyl acetate 80/20 hexane). The two phases were then separated, and the organic layer was washed with brine and water, dried over $MgSO_4$, and filtered. The solvent was removed under vacuum and the resulting residue was purified by flash column chromatography on silica gel (ethyl acetate 80/20 hexane). The desired compound **27** was collected as a yellow oil.[49]

4.2.2 *tert*-Butyl (2-((3-acetylphenyl)sulfonamido)ethyl)carbamate (31). This compound was synthesised by adapting a previously described protocol.[50,51] To a cooled solution of the compound 30 (1 eq., 0.145 mmol, 50 mg) in anhydrous dichloromethane (3 mL), thionyl chloride was added dropwise under stirring (7 eq., 1.016 mmol, 73 μL). The solution was then stirred at room temperature for 3 h. The reaction was cooled at 0 °C before adding *N*,*N*-diisopropylethylamine dropwise (4 eq., 0.58 mmol, 0.101 mL), followed by *N*,*O*-dimethylhydroxylamine (2 eq., 0.290 mmol, 28 mg). The reaction mixture was then stirred at room temperature overnight. The reaction was treated with saturated aqueous $NaHCO_3$ (pH = 7), then washed twice with water, dried over Na_2SO_4, filtered, and evaporated under vacuum.

The crude extract was purified by flash column chromatography on silica gel (ethyl acetate 60/40 hexane) and directly used in the next step. To a cooled solution of the previously purified compound (1 eq., 0.098 mmol, 38 mg) in anhydrous THF (2 mL), a solution of methylmagnesium bromide was added dropwise under nitrogen (3 eq., 0.294 mmol, 0.15 mL). The solution was then stirred at room temperature for 3 h. The reaction was monitored by TLC (ethyl acetate 60/40 hexane). The solution was treated with saturated aqueous $NaHCO_3$ (pH = 7) and extracted twice with ethyl acetate. The organic layers were washed with brine and water, then dried over Na_2SO_4, filtered, and evaporated under vacuum. The crude extract was purified by flash column chromatography on silica gel (ethyl acetate 60/40 hexane) to afford product **31** as a yellow oil.

4.2.3 General procedure for the synthesis of compounds 28 and 32. These compounds were synthesised by adapting a previously described protocol.[52] To a cooled solution of *t*-BuOLi 1 M (2.5 eq., 1.65 mmol, 1.65 mL) in anhydrous THF (5 mL), diethyl oxalate was added dropwise under nitrogen (2 eq., 1.36 mmol, 0.2

mL). The solution was stirred at 0 °C for 10 minutes before ketone **27** (1 eq., 0.66 mmol, 226 mg) in anhydrous THF (4 mL) was added dropwise. The mixture was stirred for 1.5 h at room temperature. The reaction was monitored by TLC (ethyl acetate 50/50 hexane) and treated with saturated aqueous $NaHCO_3$ (pH = 6). The aqueous phase was extracted twice with dichloromethane. The organic layers were washed with brine and water, dried over Na_2SO_4 and filtered. The solvent was removed under vacuum and the resulting residue was purified by flash column chromatography on silica gel (DCM 60/40 MeOH).

4.2.4 General procedure for the synthesis of compounds 29 and 33. These compounds were synthesised by adapting a previously described protocol.[53] To a cooled stirred mixture of the *N*-Boc-2,4-diketoester **32** (1 eq., 1.36 mmol, 601 mg) in anhydrous THF (44 mL) under nitrogen, lithium aluminium hydride was slowly added (4 eq., 5.44 mmol, 206 mg). The reaction mixture was stirred for 3 h on ice and monitored by TLC (ethyl acetate 80/20 hexane). The reaction was treated with ethyl acetate and water at 0 °C. The aqueous phase was extracted twice with ethyl acetate. The organic layers were washed with brine and water, dried over $MgSO_4$ and filtered. The solvent was removed in vacuum and the resulting residue was purified by flash column chromatography on silica gel (ethyl acetate 50/50 hexane).

4.2.5 *N*-(2-Aminoethyl)-3-(3,4-dihydroxybutanoyl)benzenesulfonamide (4). This compound was synthesised by adapting a previously described protocol.[54] To a Pd/C (1.3 eq., 46 mg, 0.43 mmol) powder purged under N_2, the Cbz-protected ketodiol **29** was added (1 eq., 0.39 mmol, 172 mg) in methanol (6 mL). The solution was then flushed with H_2 gas and stirred under H_2 atmosphere (balloon) at room temperature for 24 h. The suspension was filtered through a Celite pad and washed with MeOH (2 × 5 mL). The filtrate was removed under vacuum to afford the desired compound **4** as a yellow oil.

4.2.6 General procedure for synthesis of compounds 12 and 23. To a cooled solution of the ketodiol **33** (1 eq., 0.54 mmol, 220 mg) in dichloromethane (0.72 mL), trifluoroacetic acid was added dropwise under stirring (11 eq., 5.99 mmol, 0.46 mL). The solution was then stirred overnight at room temperature and monitored by TLC (ethyl acetate 60/40 hexane). The solution was evaporated under vacuum, after which TFA traces were removed by the repeated addition and evaporation of dichloromethane under vacuum.[48]

4.2.7 General procedure for the synthesis of compounds Ir4, Ir12 and Ir23. The iridium complexes were synthesised by adapting a previously described protocol.[55] To a suspension of pentamethylcyclopentadienyl iridium(III) chloride dimer (1 eq., 0.063 mmol, 50 mg) in MeOH (2.3 mL), the deprotected triol **4** (3 eq., 0.19 mmol, 46 mg) in MeOH (1 mL) solution was added . Triethylamine (4 eq.) was then added dropwise until the solution turned a clear yellow. The solution was then stirred for 1 h and the solvent was removed under vacuum. The resulting residue was purified by flash column chromatography on silica gel (DCM 98/2 MeOH to 100% MeOH).

4.3 Artificial metalloenzyme characterisation and catalytic assays

4.3.1 Enzyme expression. Recombinant overexpression of all enzymes was carried out in *E. coli* BL21(DE3). Wild-type TbADH and mutant TbADH 5M (H59A–D150A–C203S–C283A–C295A) were purified by affinity chromatography and routinely characterized by SDS-PAGE and enzymatic activity. Protein concentrations was determined by Bradford assays. These procedures were performed as previously described.[27]

4.3.2 Inhibition assays. The assays were performed in the oxidative direction.[56] The following reagents were added to a 1 mL cuvette: enzyme (TbADH WT 1.0 μg mL^{-1}, final conc.), cofactor ($NADP^+$ at 0.5 mM, final conc.) and substrate (butan-2-ol 150 mM, final conc.) in 100 mM Tris–HCl buffer pH 8. Inhibitors **4, 12, 23, Ir4, Ir12** and **Ir23** were prepared as concentrated stock solutions in buffer and added at final concentrations from 0.001 mM to 40 mM. The reaction was initiated by the addition of the enzyme. The increase of absorbance was monitored at 340 nm and at 40 °C for TbADH WT. Each experiment was run in triplicate and corrected for background activity without the enzyme. The specific activity and the subsequent IC_{50} of each inhibitor were calculated using the extinction coefficient for NADPH 6220 M^{-1} cm^{-1}.

4.3.3 Catalytic assays. The following assays were adapted from a previous literature procedure.[40] The commercially available imine substrate used was 6,7-dimethoxy-1-methyl-3,4-dihydroisoqunoline and the corresponding amine product was 6,7-dimethoxy-1-methyl-1,2,3,4-tetrahydroisoquinoline. Iridium complexes **Ir4, Ir12** and **Ir23** were used as catalysts alone or in mixtures with either of the two enzymes TbADH WT or TbADH 5M. The commercially available imine substrate and amine product were prepared in 100 mM phosphate-buffered saline (PBS) pH 7 at 10 mM. The iridium complexes and the enzymes were prepared in PBS buffer at pH 7 in 2 mM concentration and diluted to 250 μM concentration. Sodium formate was prepared at a concentration of 1 M in 100 mM PBS pH 7 buffer. The following reagents were added to a vial: sodium formate (300 mM final conc.), the imine substrate (4 mM, final conc.), the catalysts (50 μM, final conc.) and 100 mM PBS buffer at pH 7 up to 500 μL. The reaction mixture was vortexed and incubated at 38 °C. After 24 h, the mixture was cooled down and 10 M NaOH (55 μL) was added. The product was extracted with 1 mL ethyl acetate. The organic phase was separated and dried over $MgSO_4$. The dried organic phase was transferred to a clean vial and analysed by HPLC using a Chiracel OD column (1 μL per injection, mobile phase hexane/isopropanol/diethylamine 95/5/0.1, 25 °C, flow rate of 0.85 mL min^{-1}). The concentrations of the imine and amine products were determined by comparison with standard curves and were used to determine the amounts of amine and of imine formed after 24 h. The catalytic results were expressed as the turnover number (TON, defined as μmol of amine produced per μmol of catalyst) and turnover frequency (TOF, min^{-1}, defined as μmol of amine produced per μmol of catalyst per minute). Experiments were run in triplicate and controls were performed with no iridium complexes and no enzymes. A calibration curve was developed under similar conditions with the commercially available substrate and product in a sodium formate and PBS buffer at several concentrations: 0.1 mM, 1 mM, 2 mM, 3 mM and 4 mM.

Author contributions

CMJ and AP developed and conceptualised the research idea and supervised the project, contributed to the planning and design of the experimental methodology, aided in the data interpretation and wrote the manuscript. FLM performed the molecular modelling, designed the methodology with input from CMJ and AP, performed all experimental work and data analysis and contributed to manuscript writing.

Conflicts of interest

There are no conflicts to declare.

Acknowledgements

CMJ gratefully acknowledges the financial support from the Nottingham Advanced Research Fellowship (NRF) scheme for initiating this joint collaboration. The authors further want to thank Prof. Rachel Gomes for her helpful insights in project discussions and Dr Marko Hanževački for support in finalising the manuscript and input into graphical representations. We also gratefully acknowledge access to the University of Nottingham's Augusta HPC service. The authors would also like to thank Dr Gemma Harris for providing access and support for the ITC experiment at the Harwell research complex.

Notes and references

1 S. C. Hammer, A. M. Knight and F. H. Arnold, *Curr. Opin. Green Sustain. Chem.*, 2017, **7**, 23–30.

2 F. Schwizer, Y. Okamoto, T. Heinisch, Y. Gu, M. M. Pellizzoni, V. Lebrun, R. Reuter, V. Köhler, J. C. Lewis and T. R. Ward, *Chem. Rev.*, 2018, **118**, 142–231.

3 B. Large, N. G. Baranska, R. L. Booth, K. S. Wilson and A. K. Duhme-Klair, *Curr. Opin. Green Sustain. Chem.*, 2021, **28**, 1–7.

4 A. S. Klein and C. Zeymer, *Protein Eng., Des. Sel.*, 2021, **34**, gzab003.

5 T. Heinisch and T. R. Ward, *Acc. Chem. Res.*, 2016, **49**, 1711–1721.

6 G. Roelfes, *Acc. Chem. Res.*, 2019, **52**, 545–556.

7 A. D. Liang, J. Serrano-Plana, R. L. Peterson and T. R. Ward, *Acc. Chem. Res.*, 2019, **52**, 585–595.

8 L. Alonso-Cotchico, J. Rodríguez-Guerra, A. Lledós and J.-D. Maréchal, *Acc. Chem. Res.*, 2020, **53**, 896–905.

9 G. B. Akcapinar and O. U. Sezerman, *Biosci. Rep.*, 2017, **37**, BSR20160179.

10 F. Richter, A. Leaver-Fay, S. D. Khare, S. Bjelic and D. Baker, *PLoS One*, 2011, **6**, 1–12.

11 J. K. Leman, B. D. Weitzner, S. M. Lewis, J. Adolf-Bryfogle, N. Alam, R. F. Alford, M. Aprahamian, D. Baker, K. A. Barlow, P. Barth, B. Basanta, B. J. Bender, K. Blacklock, J. Bonet, S. E. Boyken, P. Bradley, C. Bystroff, P. Conway, S. Cooper, B. E. Correia, B. Coventry, R. Das, R. M. De Jong, F. DiMaio, L. Dsilva, R. Dunbrack, A. S. Ford, B. Frenz, D. Y. Fu, C. Geniesse, L. Goldschmidt, R. Gowthaman, J. J. Gray, D. Gront, S. Guffy, S. Horowitz, P. S. Huang, T. Huber, T. M. Jacobs, J. R. Jeliazkov, D. K. Johnson, K. Kappel, J. Karanicolas, H. Khakzad, K. R. Khar, S. D. Khare, F. Khatib, A. Khramushin, I. C. King, R. Kleffner, B. Koepnick, T. Kortemme, G. Kuenze, B. Kuhlman, D. Kuroda, J. W. Labonte, J. K. Lai, G. Lapidoth, A. Leaver-Fay, S. Lindert, T. Linsky, N. London, J. H. Lubin, S. Lyskov, J. Maguire, L. Malmström, E. Marcos, O. Marcu, N. A. Marze, J. Meiler, R. Moretti, V. K. Mulligan, S. Nerli, C. Norn, S. Ó'Conchúir, N. Ollikainen, S. Ovchinnikov, M. S. Pacella,

X. Pan, H. Park, R. E. Pavlovicz, M. Pethe, B. G. Pierce, K. B. Pilla, B. Raveh, P. D. Renfrew, S. S. R. Burman, A. Rubenstein, M. F. Sauer, A. Scheck, W. Schief, O. Schueler-Furman, Y. Sedan, A. M. Sevy, N. G. Sgourakis, L. Shi, J. B. Siegel, D. A. Silva, S. Smith, Y. Song, A. Stein, M. Szegedy, F. D. Teets, S. B. Thyme, R. Y. R. Wang, A. Watkins, L. Zimmerman and R. Bonneau, *Nat. Methods*, 2020, **17**, 665–680.

12 A. Zanghellini, L. Jiang, A. M. Wollacott, G. Cheng, J. Meiler, E. A. Althoff, D. Röthlisberger and D. Baker, *Protein Sci.*, 2006, **15**, 2785–2794.

13 S. Basler, S. Studer, Y. Zou, T. Mori, Y. Ota, A. Camus, H. A. Bunzel, R. C. Helgeson, K. N. Houk, G. Jiménez-Osés and D. Hilvert, *Nat. Chem.*, 2021, **13**, 231–235.

14 S. D. Khare, Y. Kipnis, P. J. Greisen, R. Takeuchi, Y. Ashani, M. Goldsmith, Y. Song, J. L. Gallaher, I. Silman, H. Leader, J. L. Sussman, B. L. Stoddard, D. S. Tawfik and D. Baker, *Nat. Chem. Biol.*, 2012, **8**, 294–300.

15 C. E. Tinberg, S. D. Khare, J. Dou, L. Doyle, J. W. Nelson, A. Schena, W. Jankowski, C. G. Kalodimos, K. Johnsson, B. L. Stoddard and D. Baker, *Nature*, 2013, **501**, 212–216.

16 T. Heinisch, M. Pellizzoni, M. Dürrenberger, C. E. Tinberg, V. Köhler, J. Klehr, D. Häussinger, D. Baker and T. R. Ward, *J. Am. Chem. Soc.*, 2015, **137**, 10414–10419.

17 V. M. Robles, M. Dürrenberger, T. Heinisch, A. Lledós, T. Schirmer, T. R. Ward and J. D. Maréchal, *J. Am. Chem. Soc.*, 2014, **136**, 15676–15683.

18 J. Bos, A. García-Herraiz and G. Roelfes, *Chem. Sci.*, 2013, **4**, 3578–3582.

19 L. Alonso-Cotchico, J. R. G. Pedregal, A. Lledós and J. D. Maréchal, *Front. Chem.*, 2019, **7**, 211.

20 I. Drienovská, L. Alonso-Cotchico, P. Vidossich, A. Lledós, J. D. Maréchal and G. Roelfes, *Chem. Sci.*, 2017, **8**, 7228–7235.

21 M. Basle, H. A. W. Padley, F. L. Martins, G. S. Winkler, C. M. Jäger and A. Pordea, *J. Inorg. Biochem.*, 2021, **220**, 111446.

22 C. Esmieu, M. V. Cherrier, P. Amara, E. Girgenti, C. Marchi-Delapierre, F. Oddon, M. Iannello, A. Jorge-Robin, C. Cavazza and S. Ménage, *Angew. Chem., Int. Ed.*, 2013, **52**, 3922–3925.

23 L. Villarino, K. E. Splan, E. Reddem, L. Alonso-Cotchico, C. Gutiérrez de Souza, A. Lledós, J. D. Maréchal, A. M. W. H. Thunnissen and G. Roelfes, *Angew. Chem., Int. Ed.*, 2018, **57**, 7785–7789.

24 I. A. Guedes, C. S. de Magalhães and L. E. Dardenne, *Biophys. Rev.*, 2014, **6**, 75–87.

25 C. K. Prier and F. H. Arnold, *J. Am. Chem. Soc.*, 2015, **137**, 13992–14006.

26 P. Dydio, H. M. Key, A. Nazarenko, J. Y.-E. Rha, V. Seyedkazemi, D. S. Clark and J. F. Hartwig, *Science*, 2016, **354**, 102–106.

27 S. Morra and A. Pordea, *Chem. Sci.*, 2018, **9**, 7447–7454.

28 I. Kleifeld, O. Shi, S. P. Zarivach, R. Eisenstein and M. Sagi, *Protein Sci.*, 2003, **12**, 468–479.

29 D. J. Raines, J. E. Clarke, E. V. Blagova, E. J. Dodson, K. S. Wilson and A. K. Duhme-Klair, *Nat. Catal.*, 2018, **1**, 680–688.

30 M. Hestericová, T. Heinisch, L. Alonso-Cotchico, J. D. Maréchal, P. Vidossich and T. R. Ward, *Angew. Chem., Int. Ed.*, 2018, **57**, 1863–1868.

31 F. W. Monnard, E. S. Nogueira, T. Heinisch, T. Schirmer and T. R. Ward, *Chem. Sci.*, 2013, **4**, 3269–3274.

32 R. A. Friesner, J. L. Banks, R. B. Murphy, T. A. Halgren, J. J. Klicic, D. T. Mainz, M. P. Repasky, E. H. Knoll, M. Shelley, J. K. Perry, D. E. Shaw, P. Francis and P. S. Shenkin, *J. Med. Chem.*, 2004, **47**, 1739–1749.

33 T. A. Halgren, R. B. Murphy, R. A. Friesner, H. S. Beard, L. L. Frye, W. T. Pollard and J. L. Banks, *J. Med. Chem.*, 2004, **47**, 1750–1759.

34 Y. Korkhin, A. J. Kalb, M. Peretz, O. Bogin, Y. Burstein and F. Frolow, *J. Mol. Biol.*, 1998, **278**, 967–981.

35 J. J. Irwin, T. Sterling, M. M. Mysinger, E. S. Bolstad and R. G. Coleman, *J. Chem. Inf. Model.*, 2012, **52**, 1757–1768.

36 V. Amarnath and K. Amarnath, *J. Org. Chem.*, 1995, **60**, 301–307.

37 L. D. Krasnoslobodskaya and Y. L. Gol'dfarb, *Russ. Chem. Rev.*, 1969, **38**, 389–406.

38 F. Stauffer and R. Neier, *Org. Lett.*, 2000, **2**, 3535–3537.

39 M. Poizat, I. W. C. E. Arends and F. Hollmann, *J. Mol. Catal. B: Enzym.*, 2010, **63**, 149–156.

40 Y. Okamoto, V. Köhler and T. R. Ward, *J. Am. Chem. Soc.*, 2016, **138**, 5781–5784.

41 *Prime, LLC*, New York, NY, 2016, 2002, vol. 320, pp. 597–608.

42 *LigPrep, LLC*, New York, NY, 2016.

43 E. Harder, W. Damm, J. Maple, C. Wu, M. Reboul, J. Y. Xiang, L. Wang, D. Lupyan, M. K. Dahlgren, J. L. Knight, J. W. Kaus, D. S. Cerutti, G. Krilov, W. L. Jorgensen, R. Abel and R. A. Friesner, *J. Chem. Theory Comput.*, 2016, **12**, 281–296.

44 G. Madhavi Sastry, M. Adzhigirey, T. Day, R. Annabhimoju and W. Sherman, *J. Comput.-Aided Mol. Des.*, 2013, **27**, 221–234.

45 M. H. M. Olsson, C. R. SØndergaard, M. Rostkowski and J. H. Jensen, *J. Chem. Theory Comput.*, 2011, **7**, 525–537.

46 R. A. Friesner, R. B. Murphy, M. P. Repasky, L. L. Frye, J. R. Greenwood, T. A. Halgren, P. C. Sanschagrin and D. T. Mainz, *J. Med. Chem.*, 2006, **49**, 6177–6196.

47 J. Li, R. Abel, K. Zhu, Y. Cao, S. Zhao and R. A. Friesner, *Proteins: Struct., Funct., Bioinf.*, 2011, **79**, 2794–2812.

48 K. Krohn, A. Vidal, J. Vitz, B. Westermann, M. Abbas and I. Green, *Tetrahedron: Asymmetry*, 2006, **17**, 3051–3057.

49 Y. L. Song, M. L. Peach, P. P. Roller, S. Qiu, S. Wang and Y. Q. Long, *J. Med. Chem.*, 2006, **49**, 1585–1596.

50 J. G. Kim and D. O. Jang, *Bull. Korean Chem. Soc.*, 2010, **31**, 171–173.

51 Y. K. Tahara, M. Ito, K. S. Kanyiva and T. Shibata, *Chem.–Eur. J.*, 2015, **21**, 11340–11343.

52 J. A. Jiang, W. Bin Huang, J. J. Zhai, H. W. Liu, Q. Cai, L. X. Xu, W. Wang and Y. F. Ji, *Tetrahedron*, 2013, **69**, 627–635.

53 K. Sivagurunathan, S. Raja Mohamed Kamil, S. Syed Shafi, F. Liakth Ali Khan and R. V. Ragavan, *Tetrahedron Lett.*, 2011, **52**, 1205–1207.

54 R. A. Croft, M. A. J. Dubois, A. J. Boddy, C. Denis, A. Lazaridou, A. S. Voisin-chiret, R. Bureau, C. Choi, J. J. Mousseau and J. A. Bull, *Eur. J. Org. Chem.*, 2019, **2019**, 5385–5395.

55 C. Letondor, A. Pordea, N. Humbert, A. Ivanova, S. Mazurek, M. Novic and T. R. Ward, *J. Am. Chem. Soc.*, 2006, **128**, 8320–8328.

56 O. Kleifeld, A. Frenkel, O. Bogin, M. Eisenstein, V. Brumfeld, Y. Burstein and I. Sagi, *Biochemistry*, 2000, **39**, 7702–7711.

Faraday Discussions

ROYAL SOCIETY OF CHEMISTRY

PAPER

Mapping the protonation states of the histidine brace in an AA10 lytic polysaccharide monooxygenase using CW-EPR spectroscopy and DFT calculations†

Peter J. Lindley, Alison Parkin, Gideon J. Davies and Paul H. Walton *

Received 13th October 2021, Accepted 4th November 2021
DOI: 10.1039/d1fd00068c

The active site of the polysaccharide-degrading lytic polysaccharide monooxygenase (LPMO) enzyme features a single copper ion coordinated by a histidine brace. The primary coordination sphere of the copper contains several ligating atoms which are bonded to ionisable protons (*e.g.* OH_2, NH_2), the pK_as of which are unknown. Using a combination of CW-EPR X-band spectroscopy over a range of pH values and DFT calculations, we show that the active site of a chitin-active AA10 LPMO can exist in three different protonation states (p$K_{a1} = 8.7$, p$K_{a2} \sim 11.5$), representing the ionisation of the coordinating groups. The middle pH species (fully formed at pH $\sim$ 10.5) is proposed to be Cu(II)$(His)_2(OH)_2$ (N_2O_2 coordination) with a decoordinated R–NH_3^+ group at the amino terminus. This species also sees a rotation of the SOMO equatorial plane from the canonical histidine brace plane, whereby the nominal Cu d($x^2 - y^2$)-orbital has rotated some 45° along the His–Cu(II)–His axis, driven by the elongation and decoordination of the amino group. The highest pH species (>12) is proposed to exist as a Cu(II)–azanide, in which the NH_2 of the amino terminus has been deprotonated. The high pH means that this species is unlikely to be biologically relevant in the catalytic cycle of AA10 LPMOs.

Introduction

Lytic polysaccharide monooxygenases (LPMOs) are copper-containing enzymes which catalyse the oxidative cleavage of glycosidic bonds within polysaccharides.[1–3] The active site of LPMOs has been shown by crystallography to contain a single copper ion coordinated in a T-shaped configuration by a 'histidine brace', which consists of an N-terminal histidine chelating to the copper *via* its NH_2 and π–N imidazole atoms, along with the τ–N atom of a further histidine.[4]

Department of Chemistry, University of York, Heslington, York, YO10 5DD, UK. E-mail: paul.walton@york.ac.uk

† Electronic supplementary information (ESI) available. See DOI: 10.1039/d1fd00068c

In the copper(II) resting state, the structures further show that one or two exogenous ligands, usually water or hydroxide, occupy the remaining copper coordination sites.

LPMOs are divided into auxiliary activity (AA) families in the Carbohydrate Active enZyme (CAZy) database (www.cazy.org) based on sequence similarity, where so far eight distinct classes have been characterised (AA9–11, AA13–17).[1,2,5–10] The number of exogenous ligands is partly dependent on the genomic class of the LPMO. For instance, structures of AA9-classified enzymes exhibit a single water/hydroxide to give a planar N_3O coordination geometry around the copper, whereas chitin-active AA10-classified enzymes appear to show the copper(II) in a distorted square-pyramidal N_3O_2 geometry, where two exogenous water/hydroxide ligands complete the coordination sphere, as depicted in Fig. 1. In both the AA9 and AA10 cases, EPR spectroscopy has established that the SOMO is dominated by the $d(x^2 - y^2)$ orbital, requiring the ligand field of the copper to have a large sigma-bonding component in the equatorial plane of the metal ion. In the case of AA9 LPMOs, the coordination geometry of the Cu is best described as axially elongated, whereas in the case of AA10 LPMOs, EPR spectroscopy has shown *ca.* 3% of $d(z^2)$ mixing into the largely $d(x^2 - y^2)$ SOMO, commensurate with a distorted square-pyramidal geometry.[11,12]

In chitin-active AA10 LPMOs, the large number of ionisable protons of the ligands, which coordinate to the copper ion in the histidine brace (NH_2, $2 \times H_2O$), opens up the possibility that the copper(II) centre exhibits a range of different protonation states, some of which may be relevant to the enzyme's catalytic cycle. Indeed, this reasoning has extended as far as suggesting that the amino terminus can deprotonate itself to form a copper(II)–azanide bond, the σ-donating power of which is expected to be high and possibly capable of stabilising a high valent copper species.[13] Such a species may be a key intermediate in the catalytic cycle which is capable of abstracting a hydrogen atom from the polysaccharide substrate.[2,14] As such, determining the different degrees of protonation of the active site is germane to any discussion about the catalytic mechanisms of LPMOs.

It is in this context that we aimed to establish the nature of the different protonation states of the active site of an AA10 LPMO from the bacterium *Bacillus*

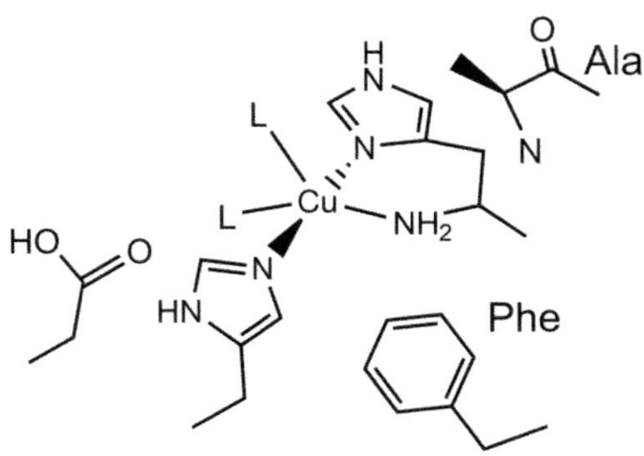

Fig. 1 Active site of a chitin-active AA10 LPMO, depicting the histidine brace exogenous ligands (L) ligating to the Cu(II) ion, and conserved residues.

licheniformis, *Bl*(AA10).[15] To this end, we employed CW-EPR spectroscopy in conjunction with DFT calculations to show that the active site exists in three distinct protonation states across the pH range 6.5 to 12.5. We further show that the highest pH species is most likely to be the copper(II)–azanide species proposed earlier,[2,13] but that it only exists at pH outside the usual biological range, and is not a biologically relevant species for the resting state of the enzyme.

Results and discussion

Preparation and purity of *Bacillus licheniformis* AA10 LPMO

Samples of this enzyme were prepared in *Escherichia coli* (BL21 competent cells) *via* periplasmic expression, previously reported by Courtade *et al.*[16] Crude protein was obtained from the periplasm by osmotic shock and further purified using a 5 mL HiTrap® DEAE FF anion exchanger, eluting with a NaCl gradient from 0–500 mM over 90 column volumes. LPMO-containing fractions were collected and further purified on a Superdex S75 gel filtration column (GE Life Sciences).

The purity of *Bl*(AA10) was established by gel electrophoresis, which displayed a clear and single band at *ca.* 20 kDa (ESI†). For the subsequent EPR experiments, it was important to determine that all copper in the sample was bound stoichiometrically to the active site of the enzyme. On this basis, a single equivalent of Cu was added to the enzyme from a solution of 10 mM $CuSO_4$ prior to size exclusion chromatography, from which any uncoordinated metal ions were separated from the purified enzyme.

CW-EPR X-band spectroscopy

A sample of *Bl*(AA10) (0.4 mM) was prepared in a mixed buffer system of MES, CHES, HEPES and CAPS, each at 5 mM. These buffers were chosen for their known poor-coordination properties toward metal ions and, when used synchronously, buffer a wide range of pHs. The pH of the solution was adjusted up and down by the addition of dilute solutions of NaOH and H_2SO_4, respectively. The pH of the resulting solution at room temperature was determined using a calibrated Fisherbrand™ Accumet™ AE150 Benchtop pH Meter, from which 12 samples from pH 6.5 to 12.5 were prepared in 0.5 pH unit steps. To verify that the pH extremes did not denature the protein, EPR spectra were collected on samples starting at pH 6.5 and then raising to pH 12.5, and then recollected on the same sample at a lower pH, indeed restoring the previous species.

At each pH, the sample was rapidly (<2 s) cooled[17] to 150 K and its CW-EPR X band spectrum collected.‡ The trend of the EPR spectra shows a qualitative change across the pH range, in which it is evident that all species have a spectral envelope which is consistent with a predominantly $d(x^2 - y^2)$ SOMO, showing that the Cu(II) ion at all pHs has a ligand field dominated by sigma donation in its equatorial plane (Fig. 2). Spin quantification by double integration of the spectra

‡ Continuous wave (CW) X-band EPR spectra were obtained using a 100 μL frozen solution of *Bl*AA10 at a concentration of 0.4 mM. Data collection was carried out on a Bruker micro EMX spectrometer using a microwave frequency of *ca.* 9.3 GHz, modulation amplitude of 4G, modulation frequency of 100 kHz and microwave power of 10.02 mW. Spectra were intensity averaged over three scans and simulated using the EasySpin 5.2.28 (ref. 27) open-source toolbox implemented by MATLAB R2020a software on a desktop PC. Raw EPR data are available at DOI: 10.15124/7a485d50-ea4e-4cd1-8fc2-650057c076af.

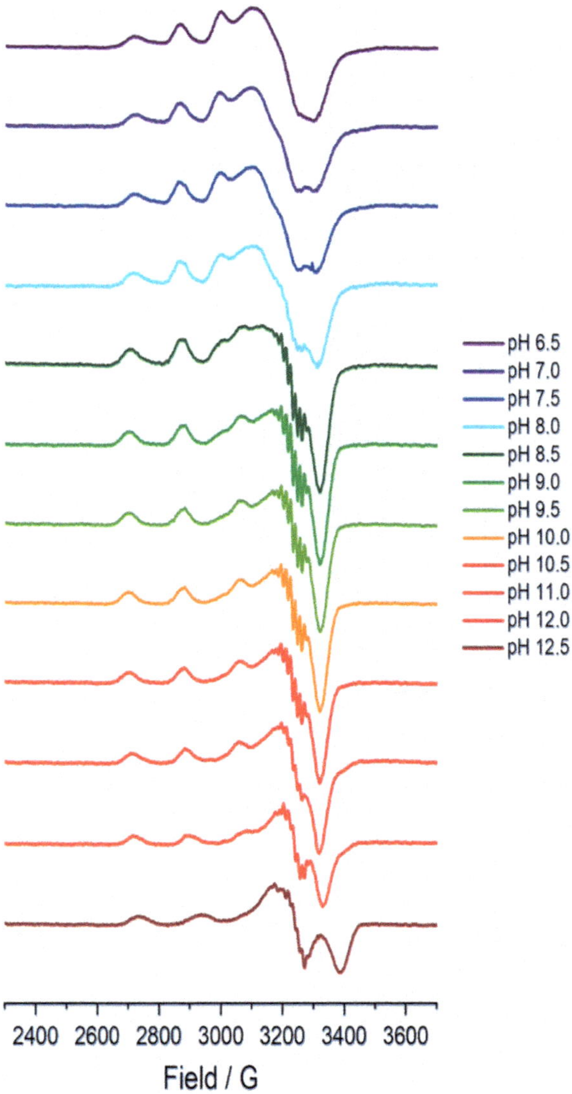

Fig. 2 Frozen solution X-band EPR spectra (150 K) of *Bl*(AA10) (0.4 mM) in a mixed buffer of MES, CHES, HEPES and CAPS (5 mM each) between pH 6.5 and 12.5.

also shows that there are no redox processes that occur at the active site across the pH range (ESI†).

Analysis of EPR spectra and distribution of pH dependent species

Between pHs 6.5 and 8.0, a single species is seen to dominate the spectrum, as evidenced by a single set of Cu hyperfine split peaks at low field ($g_3 = 2.260$, $A_3 = 445$ MHz). Increasing the pH leads to the formation of a new set of peaks in the low field region ($g_3 = 2.233$, $A_3 = 554$ MHz) along with distinctive superhyperfine coupling around $g = 2.07$, commensurate with the formation of a second distinct species, the concentration of which is maximised at pH 10.5. A further increase in pH to 12.5 sees the formation of a further set of peaks in the low field region (g_3 = 2.180, A_3 = 614 MHz), consistent with the formation of a third distinct species. This final species is not fully formed at pH 12.5 but is dominant at that pH. The total number of species across the pH range was assessed using singular value decomposition analysis of the collected spectra (ESI†). From this analysis, three principal components emerged, corroborating the qualitative assessment of the pH dependence of the EPR spectra.

On the basis of the three components (species 1, 2 and 3), we sought to deconvolute the series of spectra. Given the clear separation between the different species across the pH range, it was assumed that the spectrum at pH 6.5 represented 100% of a single complex. Indeed, it was possible to simulate (with EasySpin) the lowest pH species (pH 6.5) with a single set of spin-Hamiltonian parameters (Fig. 3a and Table 1). The same was also true of the spectra collected at pH 10.5 and 12.5 (Fig. 3b and c, respectively, Table 1). In performing the simulation of species 3, we noted the appearance of a significant 'overshoot' feature which appears at high field in the species at pH 12.5. Such a feature has in the past been confused with an anomalously low g_1 value.[18]

Once spin Hamiltonian parameters of each species were obtained, it was possible to simulate the mixed species spectra (pHs 7 to 9.5 and >10.5) to determine the species distributions at each pH, from which the pK_as of species 1 (low pH) and species 2 (middle pH), and the stoichiometry of protons for each ionisation, were determined using standard methods (ESI†). The deprotonation of species 1 to 2 occurs with a pK_a of 8.7 and is a single proton step. The deprotonation of species 2 to 3 occurs with a pK_a of ~11.5 and is also a single proton step.

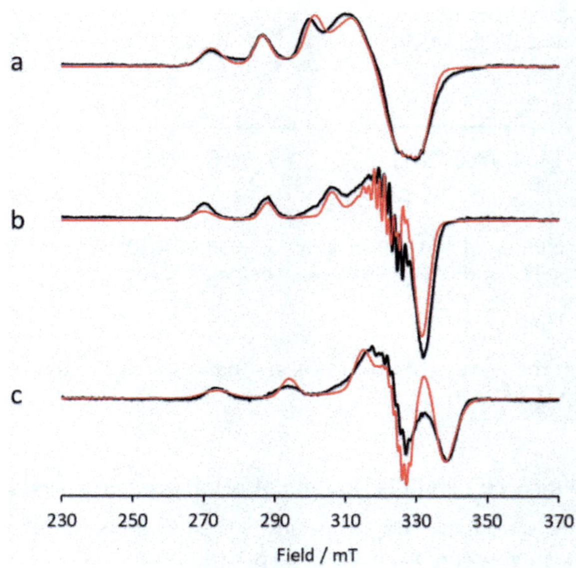

Fig. 3 X-band EPR (a) low pH species 1, with simulation (red), (b) intermediate pH species 2, with simulation (red), (c) high pH species 3, with simulation (red).

Table 1 Spin Hamiltonian parameters of the pH dependent species. It is assumed that the A_3 value is negative in sign, although this is not determined experimentally

	g values			Hyperfine coupling constants/MHz			
Species	g_1	g_2	g_3	$\|A_1\|$	$\|A_2\|$	A_3	b^2/%
1	2.030	2.125	2.260	60	90	−445	3.6
2	2.044	2.069	2.233	60	76	−554	0.8
3	2.038	2.062	2.180	70	90	−614	0.9

Ligand field analyses of spin Hamiltonian parameters

The collected spin Hamiltonian parameters of the three species are shown in Table 1. In all cases, the g values follow the pattern of $g_1 < g_2 \ll g_3$, confirming a SOMO with a high $d(x^2-y^2)$ character. Some $d(z^2)$ mixing is evident for all species from the differences in the values of g_1 and g_2. The degree of mixing, b^2, can be calculated from the standard perturbative EPR formulae as described by Hitchman *et al.*,[19] which give the molecular orbital coefficients, a and b, for the SOMO defined in eqn (1) (Table 1).

$$\psi(\text{SOMO}) = a\psi(x^2 - y^2) + b\psi(z^2) \quad (1)$$

This mixing is low for species 2 and 3, revealing that the ligand field around the Cu in each of these is close to axial four-fold symmetry (note that this does not necessarily translate into a four-fold axial symmetry of the identity of the coordinating atoms, as different coordinating atoms may have similar ligand field parameters). For species 1, more mixing is evident ($b^2 = 3.6\%$), which is in-line with previous EPR studies on AA10 LPMOs[11] and the subsequent structures of these enzymes, which exhibit a distorted square pyramidal geometry around the copper ion (Fig. 1), where the exogenous ligands are two water/hydroxide molecules, one of which forms a Cu–O bond of *ca.* 1.9 A, and the other a more distant bond of 2.2 A.[20]

Taking the value of g_3 as a measure of the relative ligand field strengths in the equatorial plane of the copper, it is evident that this ligand field increases significantly moving from species 1 to 2 to 3, and—moreover, from the values of $d(z^2)$ mixing—that the ligand field in species 2 and 3 is relatively evenly distributed amongst the ligating atoms of the equatorial plane, suggestive of a near four-fold symmetric planar arrangement of the equatorial ligand field. Indeed, corroborating such a ligand field, the increase in A_3 value from species 1 to 2 is also known to be associated with a change in the Fermi coupling as the coordination number of the copper drops from five to four.[15] Without access to the d–d electronic transition energies, it is not possible to be more precise about the nature of the ligand field; nevertheless the trend in increasing ligand field around the equatorial plane is evident from the relative spin-Hamiltonian parameters.

Further information about the ligand field in species 2 can be obtained from the clearly delineated Cu–N superhyperfine coupling (Fig. 3b). This coupling could be accurately simulated with the inclusion of two nitrogen atoms (^{14}N, $I = 1$) in the spin Hamiltonian, in which each N atom couples to the copper with a principal coupling value of 40 MHz. This value is similar to those already known for the Cu–His coupling constants seen in other LPMOs.[17,21] The simulation of species 2 could be adequately performed with the inclusion of two N nuclei, incommensurate with the expected N_3 coordination of the histidine brace. Additionally, the g_3 and A_3 values of species 2 place it squarely within those expected for a CuN_2O_2 coordination geometry on a canonical Peisach–Blumberg plot.[22]

In assigning a putative structure to species 2, it is informative to consider previous EPR studies on the substrate-bound state of chitin-active AA10 enzymes. These species have been well characterised as having a planar CuN_3O coordination geometry with a g_z value of 2.205 and an A_z value of ~620 MHz.[15,23] On this basis, a similar axial CuN_3O (sq. pl) ligand field is unlikely for species 2, given the discrepancies in the g_z and A_z values. This difference in the spin Hamiltonian parameters necessitates an alternative ligand field assignment for species 2. Thus, taking together the Cu spin Hamiltonian values and superhyperfine couplings, a more likely scenario is one in which the NH_2 group of the histidine brace is decoordinated from the copper (see DFT calculations below) and the coordination sphere around the Cu is made up of two histidines and two water/hydroxide ligands (see below for further discussion).

The spin Hamiltonian parameters for species 3 are distinct from those of other known LPMOs, which is particularly evident in the value of g_3 (2.18), which is lower than all known values, including the situation in which a chloride ion has entered the coordination sphere of the copper ($g_3 = 2.23$). Given the low value, we are directed towards the proposal that the ligand field around the copper in species 3 is one that is highly σ-donating, causing a large separation in the $d(x^2 - y^2)$ and $d(xy)$ orbital energies. Such a strong σ donation might be expected of a Cu–azanide interaction, which would arise from the recoordination of the $-NH_3^+$ amino terminus in species 2 which had been doubly deprotonated to give a $-NH^-$ ligating group. Commensurate with this proposal is the simulation of the superhyperfine coupling in species 3, which could only be satisfactorily achieved with the inclusion of three nitrogen atoms in the spin Hamiltonian with principal coupling constants of 35, 35 and 33 MHz (see DFT calculations below).

From the overall ligand field analysis, a hypothesis emerges for the nature of species 1, 2 and 3 and the pH-dependent ionisation processes between them. The hypothesis is depicted in Fig. 4, in which it is shown that species 2 forms *via* the deprotonation of a coordinating water molecule, followed by the decoordination and subsequent protonation of the $-NH_2$ group, driven by an internal proton

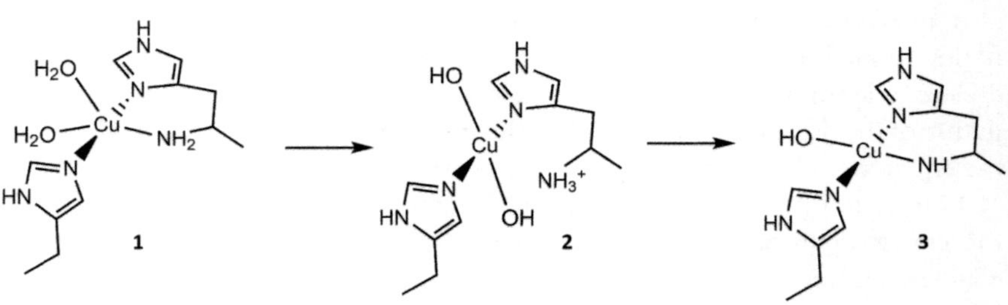

Fig. 4 Hypothetical protonation states and coordination geometries of species 1, 2 and 3.

transfer process from one of two coordinating water molecules in species 1. The formation of species 3 from species 2, on the other hand, sees the reverse proton transfer from the decoordinated $-NH_3^+$ to a coordinated hydroxide with the loss of the resulting water molecule, and the subsequent deprotonation of an N–H proton from the $-NH_2$ and recoordination to the metal as an azanide. In support of such an assignment, the formation of species 3 was accompanied by an unusual colour change from colourless to pale pink (ESI†) due to the formation of a semi-intense absorption band ($\varepsilon \approx 2500\ M^{-1}\ cm^{-1}$) centred at 520 nm. The intensity of this band is in keeping with that of a charge transfer transition – likely arising from a ligand to metal charge transfer (LMCT) from the proposed azanide ($-NH^-$) group to the copper (see TD-DFT calculations, ESI†). Thus, the proposed ligand field changes depicted in Fig. 4 are revealed by the differences in the spin Hamiltonian parameters across the species, in which g_z decreases and A_z increases (going from species 1 to 3), commensurate with an increasing equatorial ligand field.

DFT calculations

Building on the hypothesis depicted in Fig. 4, density functional theory geometry optimisations (uBP86) and EPR calculations (uB3LYP with 38% HF exchange) were performed on the proposed species 1, 2 and 3.

In the absence of a high-resolution crystal structure for the *Bacillus licheniformis* AA10, crystallographic coordinates of the spectroscopically similar enzyme, *Bacillus amyloliquefaciens* AA10 (PDB 5IJU),[20] were used as the basis for these calculations, in keeping with the experiments of Courtade *et al.*[15] Importantly in this regard, the structure of *Ba*AA10 is one of the few LPMO structures in which the copper ion has not been photoreduced from Cu(II) to Cu(I) during the X-ray experiment.[24] Accordingly, key residues from the crystallographic coordinates were used for the basis of the calculations for species 1, retaining the primary coordination sphere and supporting residues (full information and coordinates are given in the ESI†). Following standard practice, the spatial coordinates of certain atoms at the periphery of the model were kept fixed throughout the calculation to prevent the unfettered movement of residues. These atoms are depicted in Fig. S7 (ESI†) and were selected on the basis of earlier DFT studies on a closely related LPMO.[15]

For species 1, the geometry was initially modelled with the inclusion of two ligating water molecules with bond distances based on the positioning of the oxygen atoms in the crystal structure. Notably in this regard, the water situated beneath the histidine brace plane is expected to participate in strong hydrogen bonding with a nearby aspartate residue (glutamate in *Ba*AA10), the potential of which can affect the protonation state and ligating properties of the water molecule. Indeed, upon optimisation, despite the fact that Glu and Asp residues are expected to be deprotonated across the studied pH range owing to the low pK_as of their carboxylate groups ($pK_a = 2–4.5$),[25] one of the protons bound to the water molecule migrated to the neighbouring carboxylate group (Fig. 5) to give a mixed hydroxide/water coordination at the Cu. Confidence in this structural assignment comes from the resulting geometry of the optimised structure, which is a close match to the crystal structure coordinates. Thus, for species 1, the coordination sphere of the Cu is best described as a distorted square-pyramid, where water and hydroxide act as exogenous ligands.

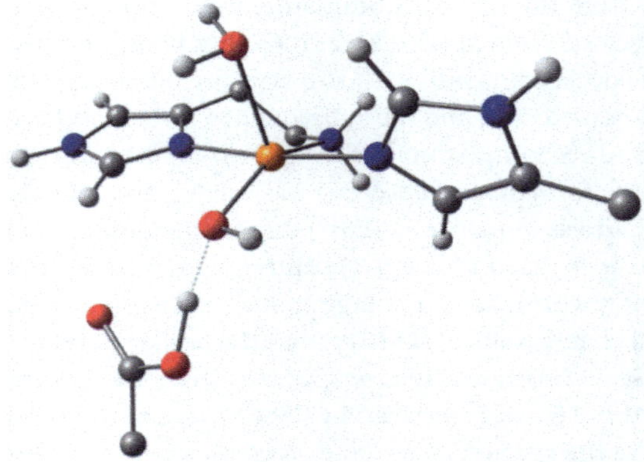

Fig. 5 DFT model of species 1 (truncated for clarity) showing the distorted square-pyramidal geometry and hydroxide/water coordination of the exogenous ligands.

While a good match was found between the optimised and experimental structures, the match between the calculated and experimental EPR parameters (Table 2) was only moderate (*e.g.* calculated $g_3 = 2.216$, experimental $g_3 = 2.26$). Moreover, the calculated Cu hyperfine values, A_1 and A_2, have relative magnitudes which are opposite to those observed experimentally. Notwithstanding the known difficulties in DFT-calculated EPR parameters matching with those from experiments, the relatively large differences seen between the experimental and calculated EPR parameters for species 1 are a recurring theme in DFT calculations on AA10 LPMO active sites in their Cu(II) resting states, even with careful selection of the basis sets and functionals.[26] As such, while the optimised structure of 1 matches well that of the experiment and is likely a good representation of the actual structure, caution must be applied in interpreting the calculated EPR data for species 1, where the reasons for the discrepancies between theoretical and experimental results remain unclear. This situation is in contrast to the calculations of the EPR parameters for species 2 and 3, which give a somewhat closer match to the experimental data (see below).

Following the ligand field analysis above, species 2 was initially modelled by removing the proton shared between the water and glutamate, and by transferring the proton attached to the second water ligand to the $-NH_2$ group (representing an overall removal of a single proton from species 1, as per the experimental

Table 2 Calculated spin Hamiltonian parameters of the pH dependent species. The *x* axis of the calculated *g* tensor lies along the His–Cu–His direction

	g values			Hyperfine coupling constants/MHz		
Species	g_x	g_y	g_z	$\|A_x\|$	$\|A_y\|$	A_z
1	2.049	2.081	2.216	110	29	−598
2	2.070	2.079	2.244	94	3	−582
3	2.037	2.070	2.188	15	140	−496

findings). Upon optimisation, the angle between the hydroxide ligands, ∠O–Cu–O, increased from *ca.* 90° to 157°. The divergence of the hydroxide ligands was also accompanied by the expected decoordination of the now protonated $-NH_3^+$ group from the metal and a rotation of the $d(x^2-y^2)$ SOMO by *ca.* 45° (Fig. 6), such that the equatorial plane was now defined by the two hydroxides and the nitrogen atoms of two histidine ligands, thus completing the CuN_2O_2 coordination geometry expected from the ligand field analysis. The calculated EPR parameters (Table 2) for species 2 reflect the experimental data in that g_x and g_y are closer in values, and that the value of g_z is *ca.* 2.24.

Species 3 was modelled as a Cu(II)–azanide species formed by the expulsion of water from species 2, followed by the further removal of an $-NH_2$ proton, resulting in a short Cu–NH bond of 1.94 A (full computational details can be found in the ESI†). The optimised geometry is best described as CuN_3O planar, albeit distorted (∠HN–Cu–O ~160°). This distortion away from planar has been previously reported for four-coordinate LPMO structures when modelled by DFT.[15] The calculated EPR parameters for species 3 match well with those from experiments,

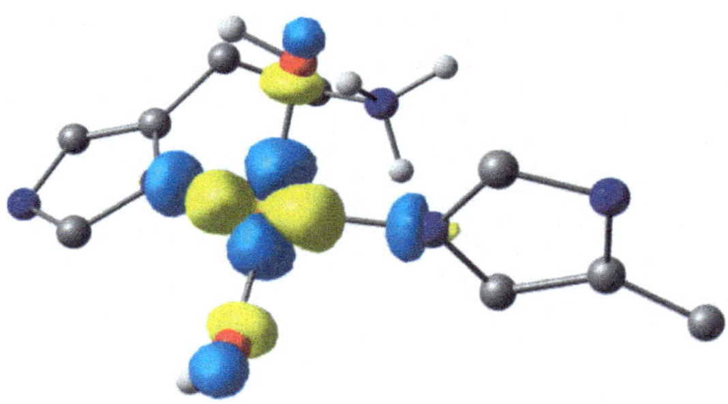

Fig. 6 Hypothetical $d(x^2-y^2)$ dominated SOMO of species 2 showing that the orbital plane has rotated (the colours denote the orbital phases).

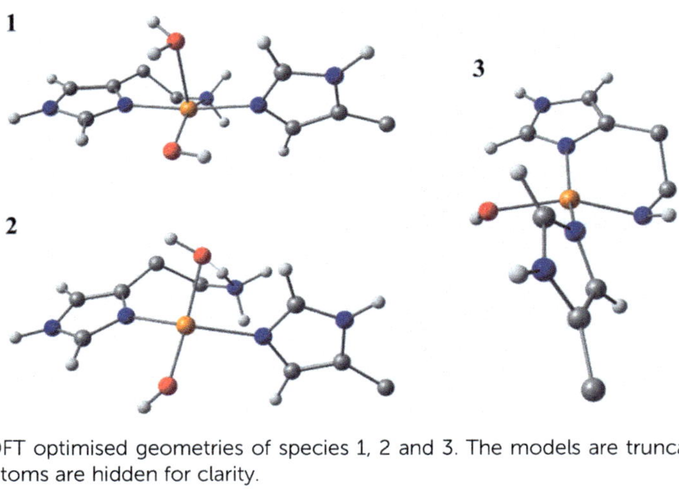

Fig. 7 DFT optimised geometries of species 1, 2 and 3. The models are truncated, and certain atoms are hidden for clarity.

most notably in the $g_{3,z}$ value of 2.18. Also, from a spectroscopic perspective, Species 3 exhibits a semi-intense UV/visible absorption band centred at 520 nm ($\varepsilon \approx 2500\ M^{-1}\ cm^{-1}$). This band is proposed to arise from a LMCT transition from the proposed azanide group to the Cu, which is corroborated with the TD-DFT calculations (ESI†). Both the d–d and LMCT transition energies appear to be significantly blue-shifted with respect to the experimental transitions. Exact transition energies are known to carry significant error when calculated in this way for open shell systems. Nevertheless, these calculations reveal that an intense absorption band is expected for the Cu(II)–azanide complex proposed as species 3 and its energy is predicted to be positioned between the π–π* and the d–d transitions. The three optimised structures for species 1, 2 and 3 are shown in Fig. 7.

Conclusions

Herein, we establish three discrete protonation states for an AA10 LPMO within the pH range of 6.5 to 12.5, two of which are biologically relevant and may be important in the context of the LPMO mechanism. The first species exists at acidic–neutral pH and is believed to be consistent with a 5-coordinate (N_3O_2), in keeping with the current understanding of the chitin-active AA10 coordination sphere. A second species is formed by deprotonating the ligating water molecule (pK_a 8.7); yielding a near-axial EPR spectral envelope ($g_1 = 2.044$, $g_2 = 2.069$ and $g_3 = 2.233$). We hypothesise that this species exists as a N_2O_2, 4-coordinate ligand field with two hydroxide ligands and an elongated/decoordinated –R-NH_3^+ following proton transfer from the coordinating hydroxide to the amino terminus. This gives rise to a rotated SOMO which is no longer defined by the chelating histidine brace. The third species formed at a high pH (p$K_a \sim 11.5$) is proposed to exist as a 4-coordinate, N_3O complex, whereby species 2 has lost a water molecule (from the interaction of an OH^- and NH_3^+) and the –NH_2 group has been deprotonated to form a Cu(II)–NH azanide complex. The high pH at which this species becomes accessible is outside of that which is biologically relevant and, as such, is unlikely to be pertinent in catalysis.

Conflicts of interest

There are no conflicts to declare.

Acknowledgements

We acknowledge the University of York for the support of PJL's PhD studentship. We thank Z. Forsberg, G. Courtade, V. G. H. Eijsink and F. L. Aachmann for providing the clone/cell colonies and purification protocol for the *Bl*AA10 enzyme. GJD thanks the Royal Society for the Ken Murray Research Professorship.

Notes and references

1 G. Vaaje-Kolstad, B. Westereng, S. J. Horn, Z. Liu, H. Zhai, M. Sørlie and V. G. H. Eijsink, *Science*, 2010, **330**, 219–222.
2 R. J. Quinlan, M. D. Sweeney, L. Lo Leggio, H. Otten, J.-C. N. Poulsen, K. S. Johansen, K. B. R. M. Krogh, C. I. Jørgensen, M. Tovborg,

A. Anthonsen, T. Tryfona, C. P. Walter, P. Dupree, F. Xu, G. J. Davies and P. H. Walton, *Proc. Natl. Acad. Sci. U. S. A.*, 2011, **108**, 15079–15084.
3 C. M. Phillips, W. T. Beeson, J. H. Cate and M. A. Marletta, *ACS Chem. Biol.*, 2011, **6**, 1399–1406.
4 L. Ciano, G. J. Davies, W. B. Tolman and P. H. Walton, *Nat. Catal.*, 2018, **1**, 571–577.
5 G. R. Hemsworth, B. Henrissat, G. J. Davies and P. H. Walton, *Nat. Chem. Biol.*, 2014, **10**, 122–126.
6 V. V. Vu, W. T. Beeson, E. A. Span, E. R. Farquhar and M. A. Marletta, *Proc. Natl. Acad. Sci. U. S. A.*, 2014, **111**, 13822–13827.
7 M. Couturier, S. Ladevèze, G. Sulzenbacher, L. Ciano, M. Fanuel, C. Moreau, A. Villares, B. Cathala, F. Chaspoul, K. E. Frandsen, A. Labourel, I. Herpoël-Gimbert, S. Grisel, M. Haon, N. Lenfant, H. Rogniaux, D. Ropartz, G. J. Davies, M.-N. Rosso, P. H. Walton, B. Henrissat and J.-G. Berrin, *Nat. Chem. Biol.*, 2018, **14**, 306.
8 F. Sabbadin, G. R. Hemsworth, L. Ciano, B. Henrissat, P. Dupree, T. Tryfona, R. D. S. Marques, S. T. Sweeney, K. Besser, L. Elias, G. Pesante, Y. Li, A. A. Dowle, R. Bates, L. D. Gomez, R. Simister, G. J. Davies, P. H. Walton, N. C. Bruce and S. J. McQueen-Mason, *Nat. Commun.*, 2018, **9**, 756.
9 C. Filiatrault-Chastel, D. Navarro, M. Haon, S. Grisel, I. Herpoël-Gimbert, D. Chevret, M. Fanuel, B. Henrissat, S. Heiss-Blanquet, A. Margeot and J.-G. Berrin, *Biotechnol. Biofuels*, 2019, **12**, 55.
10 F. Sabbadin, S. Urresti, B. Henrissat, A. O. Avrova, L. R. J. Welsh, P. Lindley, M. Csukai, J. N. Squires, P. H. Walton, G. J. Davies, N. C. Bruce, S. C. Whisson and S. J. McQueen Mason, *Science*, 2021, **373**, 774–779.
11 G. R. Hemsworth, E. J. Taylor, R. Q. Kim, R. C. Gregory, S. J. Lewis, J. P. Turkenburg, A. J. Parkin, G. J. Davies and P. H. Walton, *J. Am. Chem. Soc.*, 2013, **135**, 6069–6077.
12 F. L. Aachmann, M. Sørlie, G. Skjåk-Bræk, V. G. H. Eijsink and G. Vaaje-Kolstad, *Proc. Natl. Acad. Sci. U. S. A.*, 2012, **109**, 18779–18784.
13 N. Gagnon and W. B. Tolman, *Acc. Chem. Res.*, 2015, **48**, 2126–2131.
14 J.-P. Bacik, S. Mekasha, Z. Forsberg, A. Y. Kovalevsky, G. Vaaje-Kolstad, V. G. H. Eijsink, J. C. Nix, L. Coates, M. J. Cuneo, C. J. Unkefer and J. C. H. Chen, *Biochemistry*, 2017, **56**, 2529–2532.
15 G. Courtade, L. Ciano, A. Paradisi, P. J. Lindley, Z. Forsberg, M. Sørlie, R. Wimmer, G. J. Davies, V. G. H. Eijsink, P. H. Walton and F. L. Aachmann, *Proc. Natl. Acad. Sci. U. S. A.*, 2020, **117**, 19178–19189.
16 G. Courtade, S. B. Le, G. I. Sætrom, T. Brautaset and F. L. Aachmann, *Carbohydr. Res.*, 2017, **448**, 212–219.
17 G. R. Hemsworth, L. Ciano, G. J. Davies and P. H. Walton, in *Methods Enzymol.*, ed. F. Armstrong, Academic Press, 2018, vol. 613, pp. 63–90.
18 B. Bissaro, I. Isaksen, G. Vaaje-Kolstad, V. G. H. Eijsink and Å. K. Røhr, *Biochemistry*, 2018, **57**, 1893–1906.
19 M. A. Hitchman and R. L. Belford, *Inorg. Chem.*, 1969, **8**, 958–965.
20 R. C. Gregory, G. R. Hemsworth, J. P. Turkenburg, S. J. Hart, P. H. Walton and G. J. Davies, *Dalton Trans.*, 2016, **45**, 16904–16912.
21 K. E. H. Frandsen, T. J. Simmons, P. Dupree, J.-C. N. Poulsen, G. R. Hemsworth, L. Ciano, E. M. Johnston, M. Tovborg, K. S. Johansen, P. von Freiesleben, L. Marmuse, S. Fort, S. Cottaz, H. Driguez, B. Henrissat,

N. Lenfant, F. Tuna, A. Baldansuren, G. J. Davies, L. Lo Leggio and P. H. Walton, *Nat. Chem. Biol.*, 2016, **12**, 298.
22 J. Peisach and W. E. Blumberg, *Arch. Biochem. Biophys.*, 1974, **165**, 691–708.
23 B. Bissaro, B. Streit, I. Isaksen, V. G. H. Eijsink, G. T. Beckham, J. L. DuBois and Å. K. Røhr, *Proc. Natl. Acad. Sci. U. S. A.*, 2020, **117**, 1504.
24 A. Paradisi, M. J. Steward, P. Lindley, G. J. Davies and P. H. Walton, in *Reference Module in Chemistry, Molecular Sciences and Chemical Engineering*, Elsevier, 2020, DOI: 10.1016/B978-0-12-409547-2.14928-5.
25 M. J. Harms, C. A. Castañeda, J. L. Schlessman, G. R. Sue, D. G. Isom, B. R. Cannon and B. E. García-Moreno, *J. Mol. Biol.*, 2009, **389**, 34–47.
26 Y. A. Theibich, S. P. A. Sauer, L. L. Leggio and E. D. Hedegård, *Comput. Struct. Biotechnol. J.*, 2021, **19**, 555–567.
27 S. Stoll and A. Schweiger, *J. Magn. Reson.*, 2006, **178**, 42–55.

Faraday Discussions

ROYAL SOCIETY OF CHEMISTRY

PAPER

Successes and challenges in multiscale modelling of artificial metalloenzymes: the case study of POP-Rh_2 cyclopropanase†

José-Emilio Sánchez-Aparicio, ‡ Giuseppe Sciortino, ‡§ Eric Mates-Torres, ¶ Agustí Lledós * and Jean-Didier Maréchal *

Received 15th October 2021, Accepted 16th November 2021
DOI: 10.1039/d1fd00069a

Molecular modelling applications in metalloenzyme design are still scarce due to a series of challenges. On top of that, the simulations of metal-mediated binding and the identification of catalytic competent geometries require both large conformational exploration and simulation of fine electronic properties. Here, we demonstrate how the incorporation of new tools in multiscale strategies, namely substrate diffusion exploration, allows taking a step further. As a showcase, the enantioselective profiles of the most outstanding variants of an artificial Rh_2-based cyclopropanase (GSH, HFF and RFY) developed by Lewis and co-workers (*Nat. Commun.*, 2015, **6**, 7789 and *Nat. Chem.*, 2018, **10**, 318–324) have been rationalized. DFT calculations on the free-cofactor-mediated process identify the carbene insertion and the cyclopropanoid formation as crucial events, the latter being the enantiodetermining step, which displays up to 8 competitive orientations easily altered by the protein environment. The key intermediates of the reaction were docked into the protein scaffold showing that some mutated residues have direct interaction with the cofactor and/or the co-substrate. These interactions take the form of a direct coordination of Rh in GSH and HFF and a strong hydrophobic patch with the carbene moiety in RFY. Posterior molecular dynamics sustain that the cofactor induces global re-arrangements of the protein. Finally, massive exploration of substrate diffusion, based on the GPathFinder approach, defines this event as the origin of the enantioselectivity in GSH and RFY. For HFF, fine molecular dockings suggest that it is likely related to local interactions upon diffusion. This work shows how modelling of long-range mutations on the catalytic profiles of

InSiliChem, Department of Chemistry, Universitat Autònoma de Barcelona, 08193 Bellaterra, Spain. E-mail: agusti.lledos@uab.cat; jeandidier.marechal@uab.cat

† Electronic supplementary information (ESI) available: Supplementary scheme, supplementary tables, supplementary figures, details of computational methods, GaudiMM input files, GPathFinder input files, Cartesian coordinates obtained from DFT calculations. See DOI: 10.1039/d1fd00069a

‡ These authors have contributed equally to this work.

§ Current address: Institute of Chemical Research of Catalonia (ICIQ), The Barcelona Institute of Science and Technology, 43007 Tarragona, Spain.

¶ Current address: School of Chemistry, Trinity College Dublin, College Green, Dublin, 2, Ireland.

metalloenzymes may be unavoidable and software simulating substrate diffusion should be applied.

Introduction

Molecular modelling has become a major ally in chemical biology with outstanding contributions in drug design and biocatalysis amongst others.[1–3] It has also reached a prominent position in bioinorganics as demonstrated by a plethora of studies dedicated to studying the catalytic mechanism of metalloenzymes or their inhibitions.[4–11] Nowadays, modelers count on a wide variety of methods whose applicability depends on the system to treat and the molecular variables to analyse. Accurate representations of electronic properties (*e.g.*, for studying catalysis, any formation of bonds or spectroscopic simulations) are resolved by quantum mechanical based approaches (QM). Large conformational explorations (*e.g.*, conformational transitions of a protein) are generally solved based on classical molecular mechanics methods (MM) that are, for systems of similar size, far more rapid than QM approaches. From a biochemical point of view, though, the counterpart of QM accuracy is the limiting of the study to systems of small dimensions. Hybrid methods like QM/MM can partially solve the problem by allowing "full size" modelling although the conformational sampling remains limited.[12] Modelers need to balance between accuracy, sampling and computational power to select the best method for a given study. In some cases, they cannot rely on one unique method but need to combine several. These multilevel pipelines can be necessary when looking for the origin of the enantioselective profiles of metalloenzymes for which there may be a need to identify the best arrangement of protein and substrate for the reaction to occur and simulate the transition paths between reactant and products. Prototypical systems with such multi-scale demands are artificial metalloenzymes (ArMs), a family of new-to-nature biocatalysts obtained by the insertion of homogeneous catalysts into proteic hosts.[13,14]

Artificial metalloenzymes have received increasing attention in the last decades because of several advantages with respect to other *de novo* enzyme designs. These include: (1) a long list of homogeneous catalysts and protein scaffolds available as demonstrated by more than 40 different artificial metalloenzymes reported to date.[15] (2) The homogeneous catalyst brings a stable and well-defined first coordination sphere of the metal and avoids creating from scratch a metal binding site based purely on natural amino acids. (3) The protein can be handled with the same protein engineering techniques as other enzymes (*e.g.*, mutagenesis, directed evolution, *etc.*) to control substrate specificity and enantioselectivity. In many aspects, molecular modelling can be particularly interesting to shed light on ArMs mechanisms and help in their design.[13,14] What makes them more challenging though is the lack of structural information on the orientation and/or location of the cofactor in the protein host and how this relates to other aspects of the catalysis (orientation of the substrate, impact on the first coordination sphere of the metal, *etc.*). In this area, molecular modelling can be particularly informative.

The way our group has been dealing with the computational study of artificial metalloenzymes stands on multi-scale protocols.[16] Those protocols are generally three folded: (**1**) they start with QM studies on models with reduced dimensions of the catalytic centre. They tend to include the first coordination sphere of the metal with a prototypical substrate, the cofactor and eventually few candidate amino acids in the vicinity of the putative active site. The presence of metals in these systems ultimately implies that the quantum level should reach at least density functional theory (DFT). (**2**) Then, key structures extracted from step **1** (*i.e.*, pre-catalytic Michaelis complexes, transition states, reactive intermediates, *etc.*) are incorporated in the protein binding site by protein–ligand dockings. This part has required developments of optimized strategies to predict metal–protein interactions.[17–19] (**3**) The best models from **2** are consequently refined by classical molecular dynamics and/or QM/(MM) approaches to finally identify transition states and reactive paths inside the protein scaffold. Although tedious to set up, these multi-scale approaches have been efficient in decoding the mechanisms and ultimately guide the design of several artificial metalloenzymes.

Molecular modelling has been particularly successful in predicting how mutations of the protein impact on the enantioselective profiles of several ArMs. However, these mutations are generally close to the metal, and impact on its first or second coordination sphere which in turn can alter the electronic properties of the metal or the binding of the substrate.[20] Nonetheless, like in other enzymatic systems, some mutations further from the active site can also affect the catalytic profile. How long-range mutations impact on the enzyme activity is still controversial and can relate to numerous variables like electrostatic patterns, conformational dynamics of pre-catalytic structures, substrate diffusion, and any global motions of the protein that can influence cofactor and/or substrate binding.[21] These aspects are the most challenging to study with today's computational standards. One excellent example to assess how molecular modelling could deal with this problem is the cyclopropanase that Lewis and co-workers developed recently.[22,23]

The cyclopropane core is one of the most important strained rings in chemical intermediates and part of many natural compounds and drug molecules.[24,25] Its synthesis has been intensively investigated with some striking successes obtained by enzyme engineering.[26] The artificial metalloenzymes developed by Lewis and co-workers catalyse the enantioselective cyclopropanation of styrene and result from the inclusion of a dirhodium homogeneous catalyst by covalent linkage (Fig. 1C) into *Pyrococcus furiosus* prolyl oligopeptidase (POP). POP is a protein with a two-domain architecture: a peptidase domain with an α/β-hydrolase fold capped by a seven-bladed β-propeller domain (Fig. 1A).[27] In optimizing this scaffold, Lewis *et al.* played both with cofactor anchoring and amino acid mutations. The variants with the highest enantioselective profiles mutations are called (Fig. 1B) GSH (with mutations Ser301Gly, Tyr326His and Gly99Ser, with 92% ee for the *S*,*R* enantiomer), HFF (with mutations Leu328His, Gly99Phe and Gly594Phe, with 92% ee for the *S*,*R* enantiomer) and RFY (with mutations Gln98Arg, Gly99Phe and Pro239Tyr, with 80% ee for the *R*,*S* enantiomer). In these variants the cofactor is linked to the protein by replacing Ser477 in GSH and HFF, and Phe413 in RFY. In their work, Lewis *et al.* suggest that the global motion of the protein scaffold could drive the changes in the enantioselective profile between the different mutants by altering the dynamics of the opening of the interdomain region.[27–29] Due to its scientific and methodological interest, we decided to apply our multiscale philosophy on this system, also integrating new approaches that can deal with

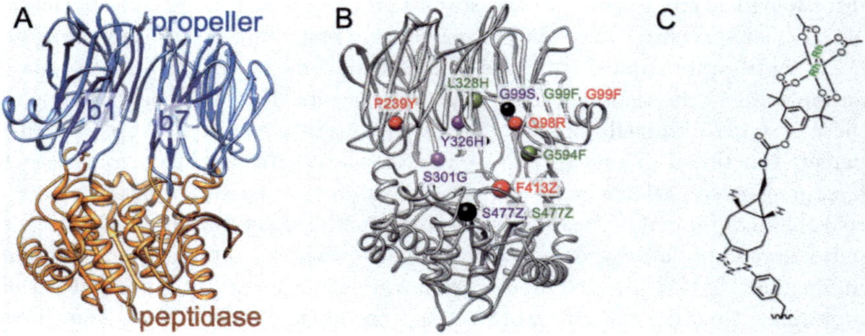

Fig. 1 (A) POP protein scaffold (PDB code 6can (ref. 27)), propeller domain shown in blue and peptidase domain shown in orange. First and last blades of the propeller domain are highlighted in darker blue and labelled as b1 and b7, respectively. (B) Location of the mutations of the three variants studied here: GSH in purple (Ser301Gly, Tyr326His, Gly99Ser, Ser477Z), HFF in green (Leu328His, Gly99Phe, Gly594Phe, Ser477Z) and RFY in red (Gln98Arg, Gly99Phe, Pro239Tyr, Phe413Z); mutations in the pore of the propeller domain are not shown for clarity. (C) Structure of the dirhodium cofactor (named Z in the mutation scheme).

collective motions and substrate diffusion. We believe this study could provide interesting material in the context of this *Faraday Discussion*.

Computational methods

The methodology used in this work is a multi-scale computational pipeline. It is inspired by our previous work on artificial metalloenzymes,[16,30–33] although adapted for the particular problem of the dirhodium cyclopropanase. Up to 5 different methods are used in this workflow. All computational details are presented in the ESI of this manuscript.† Here, we provide the most important information and the general scheme (Fig. 2). The overall workflow consists of:

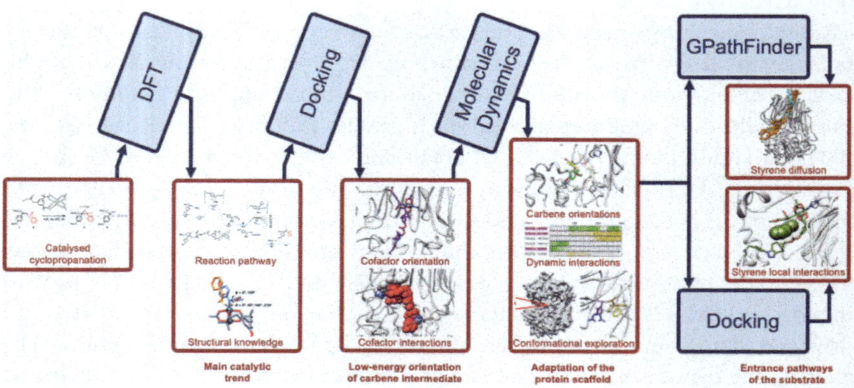

Fig. 2 Computational pipeline employed in the study of the POP-Rh_2 cyclopropanase. Methods are shown in blue, while inputs and outputs of each stage are shown in dark red.

- **Determination of the lowest energetic pathways** for the cyclopropanation of (4-methoxyphenyl)methyldiazoacetate (**1**) and styrene (**2**) mediated by the artificial cofactor of the enzyme Rh_2-$(OAc)_2$(esp-CO_2Ac). Calculations are performed in a water model in absence of any contribution of the protein environment. This model consists of the dirhodium cofactor (Fig. 1C) and the two substrates. Calculations were performed at the DFT level of theory with the Gaussian09 software, using the B3LYP functional combined with Grimme's D3 correction for dispersion and the SMD continuum solvent model.[34–43] This study identifies that the determining steps of the cyclopropanation that could be influenced by the protein scaffold are two intermediate structures (namely **II** and **III**).

- **Docking of the intermediates to POP.** First, we carried out an exploratory docking with the multi-objective genetic algorithm GaudiMM,[17,19] to assess whether the dirhodium cofactor has the possibility to form a coordination bond with an amino acid donor in the vacant site of the Rh cofactor. Dockings were performed on a series of POP structures with wide global conformational changes as extracted from the Protein Data Bank (codes 5t88 and 6can (ref. 27)) and displacements along low-energy normal modes. This allowed us to efficiently explore the large conformational space of the protein.[44–47] Second, a refinement docking was performed on the best-scored structures to evaluate the affinities, relative orientations and interactions of the Rh_2-cofactor into the biohybrid binding pockets. For this stage, we used the Gold5.8 software[48] with our recent version of GoldScore accounting for binding with coordination.[18,49] As a result, the best structures of each variant were selected as initial frames for the further Molecular Dynamics (MD) simulations.

- **Molecular dynamics simulations on the carbene intermediate.** The system was solvated with a cubic box of pre-equilibrated water molecules. The AMBER14SB force field[46] was used for the standard residues, while the GAFF force field was adopted for the remaining atoms. Rh-bonding force constants and equilibrium parameters were obtained through Seminario's method.[34,50–52] The convergence of the trajectories was assessed by RMSD, all-to-all RMSD and cluster counting analyses,[53,54] considering all the atoms of the cofactor and the carbene (Fig. S1–S3†). In light of the analyses, it was decided to extend the MD simulations under periodic boundary conditions[55] up to 300 ns, 400 ns and 200 ns for GSH, HFF and RFY variants, respectively. This step allowed us to obtain a pool of representative orientations of the carbene for each POP variant, as well as analyses[56–59] of the relevant carbene–protein interactions and the POP scaffold dynamics.

- **Simulation of the styrene diffusion pathways.** Calculations to assess the binding route of the styrene and the possibility to arrive either to *Re* or *Si* faces of the carbene were carried out with GPathFinder.[60] Each representative orientation of the carbene (obtained in the previous MD section) was tested for the possibility to bind the styrene at its *Re* and *Si* faces by ubicating the final point of the binding route at the corresponding coordinates. The resulting binding pathways, further validated with Non-Covalent Interaction (NCI) analyses,[61,62] allowed to rationalize the influence of the styrene binding route on the enantioselectivity of the different POP variants.

- **Docking of the styrene.** For the HFF variant, we had to extend the work to evaluate the interactions of the intermediate **III** in the catalytic site. We employed a procedure analogous to that for the intermediate **II** docking. This two-step docking study (exploratory with GaudiMM[17,19,63] and refinement with Gold5.8 (ref. 18, 48 and 49)) allowed to rationalize the enantiomeric preference of the HFF variant.

Results and discussion

Determinant steps of the cyclopropanation

With the aim to define the main features of the catalytic events, a full DFT (B3LYP-D3) analysis of the potential energy surface has been carried out for the isolated Rh_2-cofactor mediated cyclopropanation of (4-methoxyphenyl)methyl-diazoacetate (**1**) and styrene (**2**) in water (Scheme 1).

The general Rh_2-cofactor catalysed cyclopropanation mechanism is the result of two subsequent events: (i) carbene formation through the Rh_2-mediated N_2 extrusion from the co-substrate **1**; and (ii) approach of the styrene to the carbene intermediate with concomitant cyclopropanation of its double bond. The second step, depending on the face with which the styrene approaches the co-substrate, is responsible for the stereo-selectivity of the process. The complete profile is depicted in Scheme 2 and it is in general agreement with the previous work by Autschbach and co-workers.[64]

The carbene formation and concerted N_2 extrusion display an activation barrier of 19.1 kcal mol^{-1} (**TS$_{I-II}$**), which results as the rate determining step of the reaction. The reaction proceeds falling at −15.7 kcal mol^{-1} to intermediate **II** (rhodium carbenoid). The styrene approach requires ~1.0 kcal mol^{-1} forming the supramolecular aggregate **III** (rhodium carbenoid–styrene). It must be highlighted that intermediate **II** shows eight different energetically accessible conformations deriving from the C_2 axis rotation of the carbene and on the methylester flip giving four and two orientations, respectively (Scheme 2, angles ϑ and ϕ). The enantiodetermining and final step is the CC bond formation by cyclopropanation upon approach of the styrene to one of the faces of the plane defined by the (4-methoxyphenyl)methyldiazoacetate. In the case of the isolated achiral cofactor, the symmetric environment prevents any enantiomeric preference. The eight possible transition states display energy barriers for intermediate **II** + separated styrene ranging from 10.1 up to 12.7 kcal mol^{-1} (Scheme 2).

Considering the low yield of the reaction in aqueous media (Scheme 1), we further expanded our analysis to the competitive reaction (carbene insertion into the O–H bond of water), where a water molecule acts as a nucleophile toward the

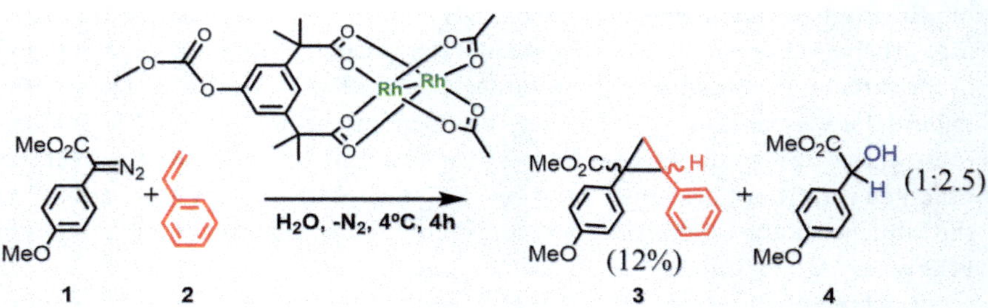

Scheme 1 Rh_2-$(OAc)_2$(esp-CO_2Ac)-mediated cyclopropanation of (4-methoxyphenyl) methyldiazoacetate (1) and styrene (2) in water.[23]

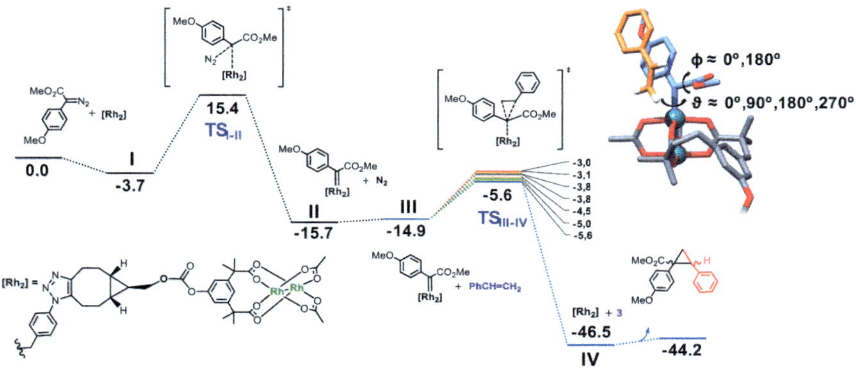

Scheme 2 Gibbs energy profile for the reaction of the complete cofactor in water. All Gibbs energy values are given in kcal mol^{-1}. For intermediates II and III only the lower energy conformations are reported for clarity.

coordinated carbene forming the α-hydroxyester byproduct (**4**). From intermediate **III**′ – the supramolecular adduct analogue of **III** with H_2O instead of styrene – the water nucleophilic attack gives rise to intermediate **IV**′. The following 1,3 proton migration leads to the 1-methoxy-2-phenyl-1,2-ethanediol, evolving into the **4** byproduct throughout a keto–enolic tautomerism (Scheme S1†). The nucleophilic attack was particularly sensitive to microsolvation effects. Increasing the number of explicit H_2O molecules from one to three, the energy barriers decrease by 3.2 kcal mol^{-1}. Although the side reaction barrier is higher by 1.7 kcal mol^{-1} compared to the main process, a full solvated model would be necessary to properly reproduce the experimental yield. Nevertheless, this is out of the scope of this work for which the obtained qualitative trend is enough to define the role of microsolvation: a less microsolvated environment would prevent the formation of the byproduct.

Carbene intermediate inside the POP protein

As seen in the DFT study, the formation of the intermediate **II** (cofactor + carbene adduct) is the rate determining step and slowest part of the reaction. Therefore, most of the reaction time is required for the carbene formation, whereas the cyclopropanation occurs almost instantaneously. With that in mind, we modelled the structure of the POP protein + intermediate **II**. First, by employing docking techniques, we focused on the possibility of finding a coordination bond between a Rh atom of the cofactor and an amino acid of the POP protein. The existence of a Rh–His coordination bond was previously hypothesised in the case of the GSH and HFF variants.[22,23,29] Its presence would confer greater stability to the structure and could be a factor to justify the enantiospecificity of the different variants of the enzyme.

The POP protein is known to explore a wide range of the interdomain angle between the propeller and peptidase domains (named θ),[27] making it unavoidable to take this into consideration in the modelling protocol. Also, the cofactor flexibility and the different intermediate **II** conformations should be considered. To achieve this, we applied a two-step docking protocol. In a first exploratory step with the GaudiMM software,[19] we considered several structures of the POP protein covering the whole range of θ (Fig. 3), together with allowing full flexibility to the intermediate **II**, the candidate sidechains to form a coordination bond and the mutated amino acids of each variant (see Computational methods for further details). This first exploration discarded the possibility of a Rh–POP coordination bond for the RFY variant, whereas suggested the feasibility of His326–Rh for the GSH variant and His328–Rh for the HFF variant (noticeably in both cases are implicated one of the mutated amino acids of the variant). In addition, we observed that coordination in the GSH variant is only possible for intermediate values of θ (29° and 32°, see Fig. 3), whereas for the HFF case the "closed" conformation is also viable. This suggests that the presence of the His–Rh bond constrains the conformations allowed for the protein scaffold.

The best solutions obtained from the exploratory docking were submitted to a further refinement docking procedure using our recently updated version of GoldScore and providing fine evaluation of metal–amino acid interactions.[18,48] While the previous exploratory step allowed to test mainly the geometric feasibility, the main objective in this refinement stage was to obtain the best possible structures in terms of an energy-based score. The His326–Rh and His328–Rh coordination were confirmed for the GSH and HFF variants, whereas interactions between the intermediate **II** and Arg98/Phe99 (two of the mutated amino acids) were found for the RFY case among the best-scored solutions (Fig. 4).

The location and relative orientation of intermediate **II** inside POP could influence the enantiodetermining step of the reaction. We therefore performed first classical molecular dynamics simulations on the three variants loaded with intermediate **II**. The simulations started from the best structures obtained in the previous docking stage (GSH and HFF with Rh–His coordination and RFY without Rh coordination bond). Along the simulations, changes in the relative orientation of the carbene to the cofactor are observed in all three variants. A clustering study taking into consideration the carbene atoms revealed three representative orientations of the carbene for the GSH variant, whereas each HFF and RFY variants account for two representative structures (Fig. 5). The residues interacting with the carbene along the simulations are also dependent on the variant

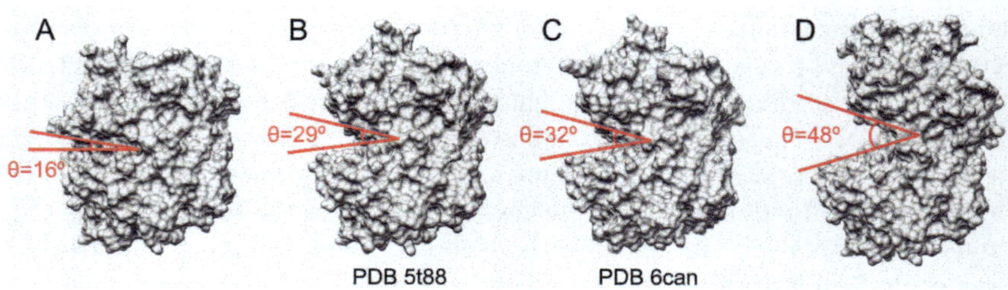

Fig. 3 Different conformations of the POP protein (shown in surface representation) used in the first exploratory docking experiment, allowing coverage of the whole range of POP interdomain angles. Interdomain angle θ (as defined in ref. 27) is indicated in each case. (A) Closed conformation ($\theta = 16°$), generated by displacements along low-energy normal modes on the PDB code 5t88.[27] (B) $\theta = 29°$, PDB code 5t88.[27] (C) $\theta = 32°$, PDB code 6can.[27] (D) Open conformation ($\theta = 48°$), generated by displacements along low-energy normal modes on the PDB code 5t88.[27]

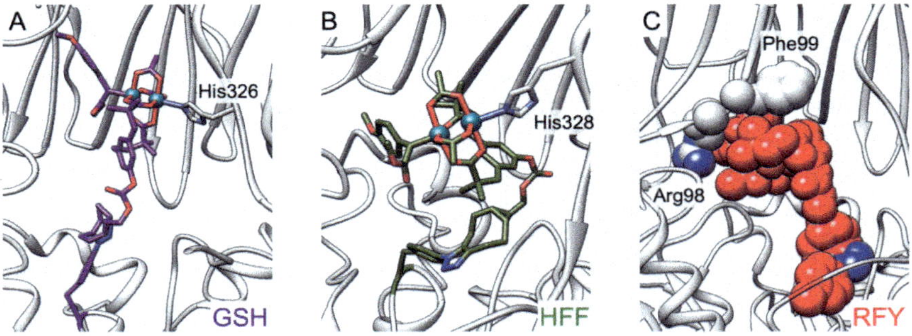

Fig. 4 Best-scored structures obtained from the refinement docking stage. (A) GSH variant. Intermediate II is shown as purple sticks, while His326 is highlighted as grey sticks. (B) HFF variant. Intermediate II is shown as green sticks, while His326 is highlighted as grey sticks. (C) RFY variant. Intermediate II in red spheres, and Phe99 and Arg98 in grey ones.

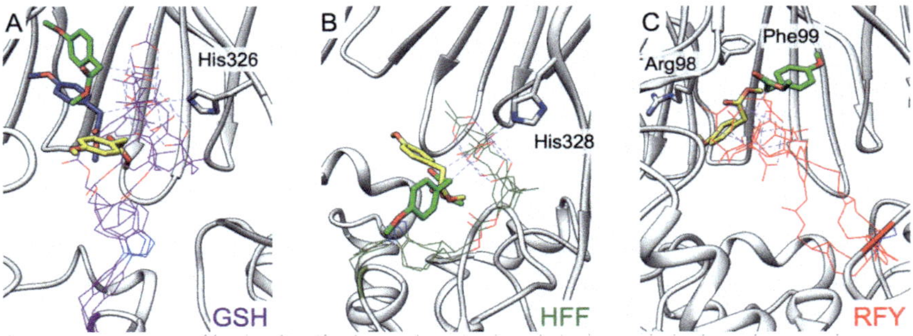

Fig. 5 Representative orientations of the carbene obtained from the MD simulations. The carbene molecule is shown as sticks. The cofactor conformations are shown as wires. Only one conformation of the POP protein backbone is shown as grey ribbon for clarity. (A) GSH variant, three representative carbene orientations, His326 is highlighted as grey sticks. (B) HFF variant, two representative carbene orientations, His328 is highlighted as grey sticks. (C) RFY variant, two representative carbene orientations, Arg98 and Phe99 are highlighted as grey sticks.

(Tables S1–S3†). In the GSH case, the carbene mainly interacts with amino acids ubicated at blades b1, b2 and b3 of the propeller domain, including the mutated Ser99, as well as with some residues of the flexible loop at the peptidase/barrel interface. For the HFF, however, the residues interacting with the carbene are mainly located at the loop linking the third and fourth beta strands of the peptidase domain. Interestingly, in the GSH and HFF variants, the cofactor is anchored at the same residue 477 but the Rh ions coordinate different histidine residues (His326 for GSH and His328 for HFF). This drastically impacts on location of the carbene in the binding site. Although the cofactor is at position 413 in RFY, in this variant, the carbene interacts with residues ubicated at blades b7, b1 and b2 of the propeller domain; in a quite similar manner than for the GSH variant. This shows the importance of the second coordination sphere of the metal to define cofactor and substrate location and orientation. In agreement with the docking results, the interaction of the carbene with Arg98 and Phe99 is maintained, suggesting that these two mutations of the RFY variant are mainly involved in stabilizing the intermediate **II**.

Regarding the analysis of the protein scaffold dynamics, the interdomain angle θ remains stable in the GSH variant (standard deviation of 1.5°), whereas is more flexible for the HFF and RFY variants with standard deviations of 3.7° and 4.3°, respectively (Table 1 and Fig. S4†). In all cases, the range of θ observed in the simulations is narrower than in a previous computational study[27] on the *apo* POP, where a range from ~10° to ~40° was reported. Our findings are supported by a very recent study[29] on a novel enzyme derived from the GSH variant that catalyses diazo cross-coupling. In this study, the MD simulations show a wider range of θ for the *apo* (*i.e.* without cofactor) structure than for the *holo* (*i.e.* with the cofactor) ones (Table 1). Altogether both studies show that the metallic cofactor constrains the POP scaffold and favours tighter conformations of the protein.

Catalytically competent geometries of styrene

Diffusion of the styrene (2) is key in GSH and RFY variants. Once the intermediate **II** is formed, the next step in the reaction pathway is the approach of the styrene to the carbene to form the intermediate **III**. For this step to succeed, the styrene should find a binding route from the solvent to the carbene's site. What faces of the carbene (*Re*, *Si* or both) are accessible to the styrene would ultimately determine the chirality of the product. Two factors may influence the binding routes of the styrene. First, the conformations and orientations of the intermediate **II**, which limit the space available inside the binding pocket of the enzyme. Second, the POP itself, which presents two possible entrance channels,[27] the first one through a pore located at the propeller domain and the other through the interdomain region (Fig. S5†). A preference for the interdomain entrance has been reported for non-modified POPs,[65] but the access through the pore cannot be discarded depending on the substrate or the presence of the cofactor. With the uncertainty about the entrance channel and conformational variability of POP in mind, we opted for an extensive sampling with our ligand pathway simulator GPathFinder.[60] GPathFinder is based on our multi-objective genetic algorithm (GA) GaudiMM and models the entrance of chemicals into proteins using

Table 1 Values of the interdomain angle θ along the MD simulations of this study and the ones reported in ref. 29

Variant	Mean and standard deviation of θ (degrees)	Maximum of θ (degrees)	Minimum of θ (degrees)
GSH with Rh–His	24.6 ± 1.5	30.1	19.8
HFF with Rh–His	23.9 ± 3.7	30.0	11.3
RFY no Rh–His	25.7 ± 4.3	36.3	17.4
5-G apo[29, a]	Not reported	~39	~10
5-G no Rh–His[29, b]	Not reported	~30	~17
5-G with Rh–His[29, c]	Not reported	~25	~13

[a] Enzyme without the dirhodium cofactor. [b] Enzyme with the dirhodium cofactor and without a Rh–His coordination bond. [c] Enzyme with the dirhodium cofactor and a Rh–His coordination bond.

a conformational selection approach. At the start of the simulation, the GA is fed with representative structures of the protein that are extracted from a molecular dynamics run or displacements along collective motions (*e.g.*, from normal mode analysis). During the evaluation process, the best solutions of successive displacements of the ligand are collected and classified to form continuous diffusion paths. Because of the multi-objective philosophy of GPathFinder, different evaluation functions can be used, very wide sampling can be performed, and channelling pathways found without any previous knowledge. It is a very interesting alternative or complementary tool to steered- or meta-dynamics approaches that require previous knowledge about the channel and much higher computational resources.

Here, we sampled the styrene binding routes from the solvent to the *Re* and *Si* faces of the carbene for each orientation of the carbene obtained in the MD simulations of the previous step. GPathFinder was instructed to optimise two objectives along the styrene binding process; the steric clashes and vina score.[66] For every representative orientation of the intermediate **III**, 240 possible binding routes were obtained. Their quality was analysed based on both objectives. The analysis for the clashes was done with standard filtering (threshold of 100 Å)[3,19,60] and post-processing the best pathways with the QM-based method to analyse noncovalent interactions in biomolecules NCIPLOT4 (ref. 62) (Fig. 6). Interestingly, in all three variants, routes for the styrene were found through the inter-domain entrance and the pore channel indicating that both routes are possible. However, striking differences are observed between variants. For the GSH, GPathFinder shows a preference for routes leading to *S*,*R* enantiomer catalytically

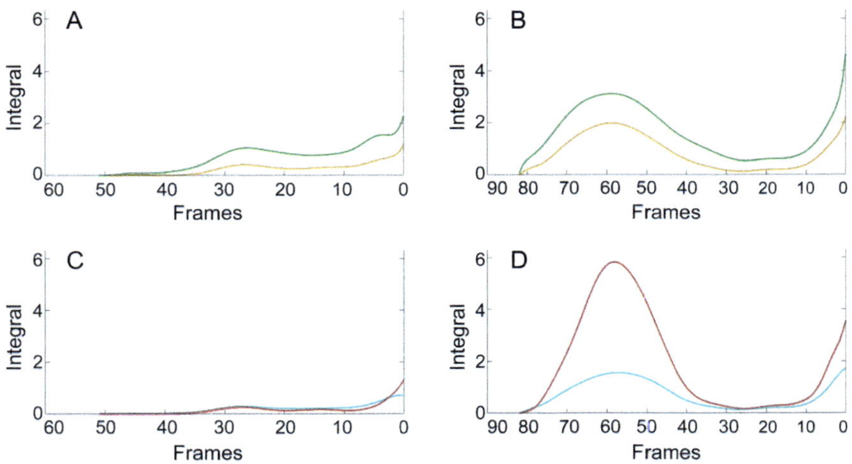

Fig. 6 Example of NCIPLOT analyses over a valid and an invalid binding route for the styrene. All plots show the integral of the interactions for the whole trajectory of the styrene, from the solvent (highest frame number) to the binding site (frame number 0). (A) Example of valid route. The module and sign of van der Waals interactions are shown in green and yellow, respectively. (B) Example of invalid route. The module and sign of van der Waals interactions are shown in green and yellow, respectively. (C) Example of valid route. Attractive and repulsive interactions are shown in blue and red, respectively. (D) Example of invalid route. Attractive and repulsive interactions are shown in blue and red, respectively.

competent geometries (64% for *S*,*R* and 36% for *R*,*S*), in agreement with the experimental observation (Fig. 7A). For the RFY variant, the trend is almost completely inverted with 66% leading to catalytically competent geometries with the *R*,*S* product and 34% for the *S*,*R* enantiomer, also in agreement with experiments (Fig. 7C). However, no such clear tendency was observed for the HFF variant, with much more equilibrated values (54% for *S*,*R* and 46% for *R*,*S*; Fig. 7B). Similar predictions are reached with the analyses of the vina score objective. Considering the maximum vina score of every binding route, which can be understood as a raw estimation of the barrier that the styrene must overcome in its binding process (Table 2), it appears that: (i) in the GSH variant, routes leading to the *S*,*R* enantiomer are favoured by a difference of $\sim$17 kcal mol^{-1}; (ii) in the RFY variant, routes leading to the *R*,*S* enantiomer are favoured by a difference of $\sim$12 kcal mol^{-1}; and (iii) in the HFF variant, the average of the vina-score barriers is similar for the *S*,*R* and *R*,*S* enantiomers (difference of $\sim$1 kcal mol^{-1}). Looking at the number of best-scored routes (those with vina score < 0 kcal mol^{-1} along all the trajectory) instead of the average scores of the whole pool of solutions, the same conclusion is obtained.

Altogether, the modelling of the styrene pathways shows that the GSH and RFY mutants, both coming from a directed evolution approach, apparently gain their enantiospecificity by controlling substrate channelling and consequently the pre-organization of the binding site towards different isomeric catalysis. However, the HFF mutant, which comes from a former series of structure-based optimization, does not produce such differential behaviour. Other molecular reasons have to be responsible for the catalytic preference in this case.

Local interactions in the active site are key in HFF variant. We hypothesized that a more local network of interactions could explain the enantioselective profile of the HFF variant. To check this assumption, we used a two-step docking protocol similar to the procedure employed with the intermediate **II**: a first exploratory docking considering all the flexibility shown in the MD study, followed by GOLD[48] calculations on the most promising combinations to obtain

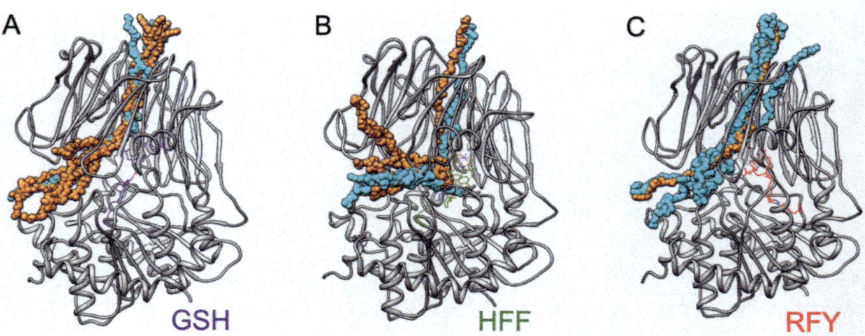

Fig. 7 Schematic representation of the best-scored styrene binding routes. Trajectories favouring the *S*,*R* enantiomer are shown in orange, whereas trajectories favouring the *R*,*S* enantiomer are shown in cyan. Only one conformation of the POP protein backbone is shown as grey ribbon for clarity. (A) GSH variant. Intermediate III and His326 are highlighted as purple wires. (B) HFF variant. Intermediate III and His328 are highlighted as green wires. (C) RFY variant. Intermediate III is highlighted as red wires.

Table 2 Average of vina-score barriers and number of best solutions for the styrene binding routes in the three studied POP variants

Variant	Mean and standard deviation (all solutions[a]) of vina-score barrier for *S*,*R* enantiomer (kcal mol^{-1})	Mean and standard deviation (all solutions[b]) of vina-score barrier for *R*,*S* enantiomer (kcal mol^{-1})	Difference in vina-score barriers (*S*,*R* − *R*,*S* for the mean of all solutions[b]) (kcal mol^{-1})	Number of best-scored solutions[c] for *S*,*R* enantiomer	Number of best-scored solutions[c] for *R*,*S* enantiomer
GSH	24.1 ± 21.7	41.0 ± 25.9	−16.9	32	0
HFF	37.7 ± 18.8	38.8 ± 22.2	−1.1	0	0
RFY	36.4 ± 19.5	24.1 ± 18.5	12.4	0	14

[a] Average and standard deviation calculated for 720 solutions. [b] Average and standard deviation calculated for 480 solutions. [c] Number of solutions with a vina score <0 kcal mol^{-1} along all the route.

reliable energetic scores with GoldScore for metals.[18,49] The results of the docking experiments (Table S4†) show a preference for catalytically competent geometries for the *S*,*R* enantiomer, with a difference of GoldScore of ~4, accounting for the experimental enantiomeric excess. Interactions between the styrene and amino acids of the blades b3 and b4 (Asn143, Phe156, Ser196) were found in the case of best-scored pose leading to the *S*,*R* enantiomer, whereas interactions between the styrene and the cofactor were dominant in the best-scored pose leading to the *R*,*S* enantiomer. As both structures were found almost equally accessible for the styrene from the solvent, the enantiospecificity in the case of the HFF mutant is apparently more related to local interactions in the active site which makes the substrate preferentially adopt a good orientation to produce the *S*,*R* enantiomer.

As a final remark regarding the search of catalytically competent geometries of the styrene, our findings show that the study of interactions away from the catalytic site is sometimes necessary to ascertain key aspects of the reaction (in this case, the determinants of enantioselectivity). An analysis of the substrate binding process could become unavoidable in those cases. However, different methodologies could be used depending on whether the binding channel is known beforehand or not. Indeed, massive conformational exploration is required when no clear route has been identified while more expensive (*ergo* accurate) methods could be used for a specific channel and, in the best-case scenario, lead to energetic magnitudes.

Conclusions

Mutagenic profiles can influence biocatalytic activities in a variety of ways. In metalloenzymes, mutations can impact on the first coordination sphere of the metal and alter, for example, its electronic features. They can also modify substrate binding affinity or how the substrate and cofactors organize for pre-catalytic geometries. The study of the effects of protein mutations on the first and second coordination spheres of the metal is something quite standard nowadays in computational bioinorganics. They can be analysed by the combination of computational methods like QM or classical MD. However, the organization of the pre-catalytic geometries may not be only due to the vicinity of the metal but the result of a cascade of more global events. Such behaviours are still challenging to identify by computational means.

Here, we chose the di-rhodium POP cyclopropanase from Lewis and co-workers as a case study because the most interesting variants in terms of enantioselective profiles, namely GSH (ee 92% *S*,*R*), HFF (ee 92% *S*,*R*) and RFY (ee 80% *R*,*S*) have mutations far from the metallic site. We got a better picture of their mechanisms thanks to the incorporation of a ligand pathway modelling tool into a multi-scale framework and could hypothesize on the molecular role of long-range mutations.

From the start, cluster models studied with DFT calculations show that the carbene and the cyclopropanoid formations have barriers about $\sim$19 and $\sim$10 kcal mol^{-1} respectively. The DFT intermediate structures were docked *a posteriori* into the protein scaffold and refined by classical molecular dynamics. This part of the modelling process shows that some residues mutated in the variants have a direct interaction with the cofactor and/or the substrate either in the form of a coordination bond (His326 and His328 for GSH and HFF, respectively) or a strong hydrophobic patch between Phe99 and Arg98 and the carbene in RFY. Differences between variants also highlight global re-arrangements of the protein scaffold, in particular regarding the interdomain region. Based on the hypothesis that those motions could affect substrate (styrene) binding, we performed GPathFinder calculations on the three variants and analysed their propensity to lead to *S*,*R* and *R*,*S* pre-reactive geometries. This hypothesis appears valid for GSH and RFY, where the enantioselective trends agree with those observed experimentally. For HFF, though, styrene binding pathways are not explicative and additional dockings were necessary to show that the experimental ee may relate to the organization of the second coordination sphere.

Multi-scale modelling applied to ArMs has shown massive success in the last decade although some limitations still exist. These procedures work by sustaining that the best solutions at one level are the most appropriate for the next level. By combining different methods, it is not possible to provide a unique energetic framework for the entire study. It could, moreover, in the worst-case scenario, stay blind when the protein scaffold alters drastically the mechanism of the first coordination sphere of the metal. Another challenge, common to other enzymatic systems, is to understand the impact of certain kinds of mutations, especially those that are not in direct contact with the first coordination sphere of the metal. In this work, in the absence of clear identification of the substrate diffusion paths, we incorporated an atomistic model (GPathFinder) to explore the large conformational space related to ligand channelling. Strikingly, this approach shows that each variant provides better routes for a given enantiomeric pre-catalytic geometry although this may not be a general case (as shown for the HFF variant). Two challenges remain at least. One is to be able to assess more efficiently those systems where substrate channels may be affected by mutations from those who are not. Another relates to the modelling of substrate diffusion. Indeed, when no routes are known beforehand, massive conformational explorations are needed hence requiring methods that calculate rapidly the energy (*i.e.*, those based on docking approaches). It is only with clear knowledge of the location of the diffusion path that one would eventually embark on accurate energetic methods although still expensive by today's standards. Researchers still have room for improvement in balancing between time and accuracy in their studies on how long-range mutations could impact catalytic profiles.

Author contributions

Conceptualization J.-D. M. and A. L.; funding acquisition J.-D. M. and A. L.; supervision J.-D. M. and A. L.; investigation J.-E. S.-A., G. S., E. M.-T., J.-D. M. and A. L.; visualization J.-E. S.-A., G. S. and E. M.-T.; writing J.-E. S.-A., G. S., J.-D. M. and A. L.

Conflicts of interest

There are no conflicts to declare.

Acknowledgements

The authors thank Prof. Jared C. Lewis for helpful discussions. Financial support from Spanish Ministry of Science and Innovation is acknowledged (MICINN project PID-2020-116861GB-I00).

Notes and references

1 M. Aminpour, C. Montemagno and J. A. Tuszynski, *Molecules*, 2019, **24**, 1693.
2 A. Romero-Rivera, M. Garcia-Borràs and S. Osuna, *Chem. Commun.*, 2017, **53**, 284–297.
3 M. Orio and D. A. Pantazis, *Chem. Commun.*, 2021, **57**, 3952–3974.
4 K. D. Vogiatzis, M. V. Polynski, J. K. Kirkland, J. Townsend, A. Hashemi, C. Liu and E. A. Pidko, *Chem. Rev.*, 2019, **119**, 2453–2523.
5 G. Sciortino, J.-D. Maréchal and E. Garribba, *Inorg. Chem. Front.*, 2021, **8**, 1951–1974.
6 W.-J. Wei, H.-X. Qian, W.-J. Wang and R.-Z. Liao, *Front. Chem.*, 2018, **6**, 638.
7 M. Q. E. Mubarak, E. F. Gérard, C. F. Blanford, S. Hay and S. P. de Visser, *ACS Catal.*, 2020, **10**, 14067–14079.
8 X. Sheng, M. Kazemi, F. Planas and F. Himo, *ACS Catal.*, 2020, **10**, 6430–6449.
9 K. Hiraga, P. Mejzlik, M. Marcisin, N. Vostrosablin, A. Gromek, J. Arnold, S. Wiewiora, R. Svarba, D. Prihoda, K. Clarova, O. Klempir, J. Navratil, O. Tupa, A. Vazquez-Otero, M. W. Walas, L. Holy, M. Spale, J. Kotowski, D. Dzamba, G. Temesi, J. H. Russell, N. M. Marshall, G. S. Murphy and D. A. Bitton, *ACS Synth. Biol.*, 2021, **10**, 357–370.
10 M. Kazemi, X. Sheng and F. Himo, *Chem.–Eur. J.*, 2019, **25**, 11945–11954.
11 S. Ahmadi, L. Barrios Herrera, M. Chehelamirani, J. Hostaš, S. Jalife and D. R. Salahub, *Int. J. Quantum Chem.*, 2018, **118**, e25558.
12 H. M. Senn and W. Thiel, *Angew. Chem., Int. Ed.*, 2009, **48**, 1198–1229.
13 V. Muñoz Robles, P. Vidossich, A. Lledós, T. R. Ward and J.-D. Maréchal, *ACS Catal.*, 2014, **4**, 833–842.
14 I. Drienovská, L. Alonso-Cotchico, P. Vidossich, A. Lledós, J.-D. Maréchal and G. Roelfes, *Chem. Sci.*, 2017, **8**, 7228–7235.

15 F. Schwizer, Y. Okamoto, T. Heinisch, Y. Gu, M. M. Pellizzoni, V. Lebrun, R. Reuter, V. Köhler, J. C. Lewis and T. R. Ward, *Chem. Rev.*, 2018, **118**, 142–231.

16 L. Alonso-Cotchico, J. Rodríguez-Guerra, A. Lledós and J.-D. Maréchal, *Acc. Chem. Res.*, 2020, **53**, 896–905.

17 G. Sciortino, E. Garribba, J. Rodríguez-Guerra Pedregal and J.-D. Maréchal, *ACS Omega*, 2019, **4**, 3726–3731.

18 G. Sciortino, J. Rodríguez-Guerra Pedregal, A. Lledós, E. Garribba and J.-D. Maréchal, *J. Comput. Chem.*, 2018, **39**, 42–51.

19 J. Rodríguez-Guerra Pedregal, G. Sciortino, J. Guasp, M. Municoy and J.-D. Maréchal, *J. Comput. Chem.*, 2017, **38**, 2118–2126.

20 V. M. Robles, M. Dürrenberger, T. Heinisch, A. Lledós, T. Schirmer, T. R. Ward and J.-D. Maréchal, *J. Am. Chem. Soc.*, 2014, **136**, 15676–15683.

21 C. G. Acevedo-Rocha, A. Li, L. D'Amore, S. Hoebenreich, J. Sanchis, P. Lubrano, M. P. Ferla, M. Garcia-Borràs, S. Osuna and M. T. Reetz, *Nat. Commun.*, 2021, **12**, 1621.

22 P. Srivastava, H. Yang, K. Ellis-Guardiola and J. C. Lewis, *Nat. Commun.*, 2015, **6**, 7789.

23 H. Yang, A. M. Swartz, H. J. Park, P. Srivastava, K. Ellis-Guardiola, D. M. Upp, G. Lee, K. Belsare, Y. Gu, C. Zhang, R. E. Moellering and J. C. Lewis, *Nat. Chem.*, 2018, **10**, 318–324.

24 D. Y.-K. Chen, R. H. Pouwer and J.-A. Richard, *Chem. Soc. Rev.*, 2012, **41**, 4631.

25 T. T. Talele, *J. Med. Chem.*, 2016, **59**, 8712–8756.

26 P. S. Coelho, E. M. Brustad, A. Kannan and F. H. Arnold, *Science*, 2013, **339**, 307–310.

27 K. Ellis-Guardiola, H. Rui, R. L. Beckner, P. Srivastava, N. Sukumar, B. Roux and J. C. Lewis, *Biochemistry*, 2019, **58**, 1616–1626.

28 J. C. Lewis, *Acc. Chem. Res.*, 2019, **52**, 576–584.

29 D. M. Upp, R. Huang, Y. Li, M. J. Bultman, B. Roux and J. C. Lewis, *Angew. Chem., Int. Ed.*, 2021, **60**, 23672–23677.

30 V. Muñoz Robles, E. Ortega-Carrasco, L. Alonso-Cotchico, J. Rodriguez-Guerra, A. Lledós and J.-D. Maréchal, *ACS Catal.*, 2015, **5**, 2469–2480.

31 L. Villarino, K. E. Splan, E. Reddem, L. Alonso-Cotchico, C. Gutiérrez de Souza, A. Lledós, J.-D. Maréchal, A.-M. W. H. Thunnissen and G. Roelfes, *Angew. Chem., Int. Ed.*, 2018, **57**, 7785–7789.

32 L. Villarino, S. Chordia, L. Alonso-Cotchico, E. Reddem, Z. Zhou, A. M. W. H. Thunnissen, J.-D. Maréchal and G. Roelfes, *ACS Catal.*, 2020, **10**, 11783–11790.

33 F. Christoffel, N. V. Igareta, M. M. Pellizzoni, L. Tiessler-Sala, B. Lozhkin, D. C. Spiess, A. Lledós, J.-D. Maréchal, R. L. Peterson and T. R. Ward, *Nat. Catal.*, 2021, **4**, 643–653.

34 M. J. Frisch, G. W. Trucks, H. B. Schlegel, G. E. Scuseria, M. A. Robb, J. R. Cheeseman, G. Scalmani, V. Barone, B. Mennucci, G. A. Petersson, H. Nakatsuji, M. Caricato, X. Li, H. P. Hratchian, A. F. Izmaylov, J. Bloino, G. Zheng, J. L. Sonnenberg, M. Hada, M. Ehara, K. Toyota, R. Fukuda, J. Hasegawa, M. Ishida, T. Nakajima, Y. Honda, O. Kitao, H. Nakai, T. Vreven, J. A. Montgomery, J. E. Peralta, F. Ogliaro, M. Bearpark, J. J. Heyd, E. Brothers, K. N. Kudin, V. N. Staroverov, R. Kobayashi, J. Normand, K. Raghavachari, A. Rendell, J. C. Burant, S. S. Iyengar, J. Tomasi, M. Cossi,

N. Rega, J. M. Millam, M. Klene, J. E. Knox, J. B. Cross, V. Bakken, C. Adamo, J. Jaramillo, R. Gomperts, R. E. Stratmann, O. Yazyev, A. J. Austin, R. Cammi, C. Pomelli, J. W. Ochterski, R. L. Martin, K. Morokuma, V. G. Zakrzewski, G. A. Voth, P. Salvador, J. J. Dannenberg, S. Dapprich, A. D. Daniels, Ö. Farkas, J. B. Foresman, J. V. Ortiz, J. Cioslowski and D. J. Fox, *Gaussian 09, Revision D.01*, Gaussian Inc., Wallingford CT, 2011.
35 A. V. Marenich, C. J. Cramer and D. G. Truhlar, *J. Phys. Chem. B*, 2009, **113**, 6378–6396.
36 T. Yanai, D. P. Tew and N. C. Handy, *Chem. Phys. Lett.*, 2004, **393**, 51–57.
37 S. Grimme, J. Antony, S. Ehrlich and H. Krieg, *J. Chem. Phys.*, 2010, **132**, 154104.
38 M. M. Francl, W. J. Pietro, W. J. Hehre, J. S. Binkley, M. S. Gordon, D. J. DeFrees and J. A. Pople, *J. Chem. Phys.*, 1982, **77**, 3654–3665.
39 D. Andrae, U. Häußermann, M. Dolg, H. Stoll and H. Preuß, *Theor. Chim. Acta*, 1990, **77**, 123–141.
40 A. W. Ehlers, M. Böhme, S. Dapprich, A. Gobbi, A. Höllwarth, V. Jonas, K. F. Köhler, R. Stegmann, A. Veldkamp and G. Frenking, *Chem. Phys. Lett.*, 1993, **208**, 111–114.
41 F. Weigend and R. Ahlrichs, *Phys. Chem. Chem. Phys.*, 2005, **7**, 3297–3305.
42 V. S. Bryantsev, M. S. Diallo and W. A. Goddard III, *J. Phys. Chem. B*, 2008, **112**, 9709–9719.
43 J. Rodríguez-Guerra Pedregal, P. Gómez-Orellana and J.-D. Maréchal, *J. Chem. Inf. Model.*, 2018, **58**, 561–564.
44 A. Bakan, L. M. Meireles and I. Bahar, *Bioinformatics*, 2011, **27**, 1575–1577.
45 P. Eastman, J. Swails, J. D. Chodera, R. T. McGibbon, Y. Zhao, K. A. Beauchamp, L. P. Wang, A. C. Simmonett, M. P. Harrigan, C. D. Stern, R. P. Wiewiora, B. R. Brooks and V. S. Pande, *PLoS Comput. Biol.*, 2017, **13**, 1–17.
46 V. Hornak, R. Abel, A. Okur, B. Strockbine, A. Roitberg and C. Simmerling, *Proteins: Struct., Funct., Bioinf.*, 2006, **65**, 712–725.
47 E. F. Pettersen, T. D. Goddard, C. C. Huang, G. S. Couch, D. M. Greenblatt, E. C. Meng and T. E. Ferrin, *J. Comput. Chem.*, 2004, **25**, 1605–1612.
48 M. L. Verdonk, J. C. Cole, M. J. Hartshorn, C. W. Murray and R. D. Taylor, *Proteins: Struct., Funct., Bioinf.*, 2003, **52**, 609–623.
49 G. Sciortino, E. Garribba and J.-D. Maréchal, *Inorg. Chem.*, 2019, **58**, 294–306.
50 C. I. Bayly, P. Cieplak, W. D. Cornell and P. A. Kollman, *J. Phys. Chem.*, 1993, **97**, 10269–10280.
51 P. Li and K. M. Merz, *J. Chem. Inf. Model.*, 2016, **56**, 599–604.
52 J. M. Seminario, *Int. J. Quantum Chem.*, 1996, **60**, 1271–1277.
53 G. Sciortino, J.-E. Sánchez-Aparicio, J. Rodríguez-Guerra Pedregal, E. Garribba and J.-D. Maréchal, *Metallomics*, 2019, **11**, 765–773.
54 A. Grossfield and D. M. Zuckerman, *Annu. Rep. Comput. Chem.*, 2009, **5**, 23–48.
55 J. Rodríguez-Guerra Pedregal, L. Alonso-Cotchico, L. Velasco-Carneros and J.-D. Maréchal, *ChemRxiv*, 2018, DOI: 10.26434/chemrxiv.7059263.v1, preprint.
56 I. Y. Ben-Shalom, D. A. Case, S. R. Brozell, D. S. Cerutti, T. E. Cheatham III, V. W. D. Cruzeiro, T. A. Darden, R. E. Duke, D. Ghoreishi, M. K. Gilson, H. Gohlke, A. W. Goetz, D. Greene, R. Harris, N. Homeyer, S. Izadi and A. Kovalenko, *Amber 2018*, University of California, 2018.

57 M. Scheurer, P. Rodenkirch, M. Siggel, R. C. Bernardi, K. Schulten, E. Tajkhorshid and T. Rudack, *Biophys. J.*, 2018, **114**, 577–583.

58 R. González-Alemán, D. Hernández-Castillo, J. Caballero and L. A. Montero-Cabrera, *J. Chem. Inf. Model.*, 2020, **60**, 467–472.

59 Y. Haddad, V. Adam and Z. Heger, *Biophys. J.*, 2019, **116**, 2062–2072.

60 J.-E. Sánchez-Aparicio, G. Sciortino, D. V. Herrmannsdoerfer, P. O. Chueca, J. R.-G. Pedregal and J.-D. Maréchal, *Int. J. Mol. Sci.*, 2019, **20**, 3155.

61 R. A. Boto, F. Peccati, R. Laplaza, C. Quan, A. Carbone, J.-P. Piquemal, Y. Maday and J. Contreras-García, *J. Chem. Theory Comput.*, 2020, **16**, 4150–4158.

62 F. Peccati, *J. Chem. Inf. Model.*, 2020, **60**, 6–10.

63 A. Krammer, P. D. Kirchhoff, X. Jiang, C. M. Venkatachalam and M. Waldman, *J. Mol. Graphics Modell.*, 2005, **23**, 395–407.

64 J. Hansen, J. Autschbach and H. M. L. Davies, *J. Org. Chem.*, 2009, **74**, 6555–6563.

65 M. Kotev, D. Lecina, T. Tarragó, E. Giralt and V. Guallar, *Biophys. J.*, 2015, **108**, 116–125.

66 O. Trott and A. J. Olson, *J. Comput. Chem.*, 2010, **31**, 455–461.

Faraday Discussions

ROYAL SOCIETY OF CHEMISTRY

DISCUSSIONS

Natural and artificial enzymes and medicinal aspects: general discussion

Philip Ash, Akhil R. Chakravarty, Peter Comba, Abhishek Dey, Debabrata Goswami, Christof Martin Jäger, Kenneth Karlin, Subrata Kundu, Salvatore La Gatta, Rocío López Domene, Jean-Didier Maréchal, Shyamalava Mazumdar, Govindasamy Mugesh, Dimitrios Pantazis, Anca Pordea, Peter J. Sadler, Volker Schünemann, Sayam Sen Gupta, Osami Shoji, Edward I. Solomon, Paul Walton and Juliusz A. Wolny

DOI: 10.1039/d2fd90014a

Rocío López Domene opened the discussion of the paper by Govindasamy Mugesh: You developed a novel nanozyme which has dual enzymatic activity. Can you force the reaction in order to have one or the other catalytic activity? Do you have any idea about how to enhance certain activity? Both activities are competing under these reaction conditions.

Govindasamy Mugesh responded: In our study, we preferred to develop a nanozyme with dual enzymatic activity, *i.e.* the nanozyme not only releases NO from *S*-nitrosothiols but also can control the ROS levels by functionally mimicking the antioxidant enzyme glutathione peroxidase (GPx).

Subrata Kundu asked: While the released NO from RSNO has been characterized by $Fe(dtc)_2$ and DAN, how was the released NO quantified? As these NO trapping experiments are diffusion-controlled, how is the dissolved NO in solvent accounted?[1]

1 R. C. Maji, S. K. Barman, S. Roy, S. K. Chatterjee, F. L. Bowles, M. M. Olmstead and A. K. Patra, *Inorg. Chem.*, 2013, **52**, 11084–11095.

Govindasamy Mugesh responded: The amount of NO released during the denitrosylation can be quantified using a DAN based fluorescence assay which was discussed in our previous manuscript.[1] We reported that the amount of NO release correlates well with the decrease in the concentration of *S*-nitrosothiols.

1 S. Ghosh, P. Roy, S. Prasada and G. Mugesh, *Chem. Sci.*, 2019, **10**, 5308–5318.

Akhil R. Chakravarty commented: DMSO solvent is used in fluorescence studies but DMSO is known to be a scavenger so any ROS generated will be scavenged. Does it react with the ROS-based radical species?

Govindasamy Mugesh replied: We do not anticipate any ROS during our studies. The NO produced in the reactions is not affected by DMSO.

Akhil R. Chakravarty added: Anyway, if some ROS is formed it will be scavenged in the DMSO solvent.

Govindasamy Mugesh replied: Yes, you are right.

Shyamalava Mazumdar said: Thank you for the excellent talk. These nanozymes are interesting enzyme mimics. For metalloenzymes we know there are very well-defined active sites. How do you define an active site in a nanozyme? In the case of the nanozyme does it have multiple active sites? If so, are there any cooperative interactions between the active sites?

Govindasamy Mugesh answered: The defects on the surface of nanozymes act as catalytic centres and certainly there are multiple active sites. It is very difficult to study the cooperative interactions between the nanozyme catalytic centres. There may be some cooperativity through alteration in the redox states.

Shyamalava Mazumdar suggested: It might be interesting to look at cooperative effects and allosteric interactions in nanozymes. There might be interesting cross-talk and different reaction pathways at different sites of a nanozyme, giving more versatility to these catalysts.

Govindasamy Mugesh replied: Yes, the nanozymes area is emerging and further studies are required to understand cooperativity and allosteric interactions.

Sayam Sen Gupta asked: You have synthesized substituted vanadates right from Mn all the way to Zn; how does the reactivity change with these different substitutions?

Govindasamy Mugesh replied: The copper substitution worked better, but still the activity was found to be lower than that of vanadium pentoxide nanowires.

Sayam Sen Gupta further queried: What is the resting oxidation state of manganese in the manganese substituted vanadates?

Govindasamy Mugesh answered: The resting oxidation state is +3.

Sayam Sen Gupta commented: Since zinc is redox non-neutral, why doesn't zinc substituted vanadate show some GPx activity?

Govindasamy Mugesh replied: A redox active metal attached to vanadium seems to be important for the partial change in oxidation state of vanadium during the catalytic reduction of hydrogen peroxide in the presence of glutathione.

Akhil R. Chakravarty asked: How is copper bonded to the vanadate lattice in copper vanadate? Is the structure of copper vanadate known?

Govindasamy Mugesh replied: The structure of CuV_2O_6 has been known for a long time.[1]

1 C. Calvo and D. Manolescu, *Acta Crystallogr., Sect. B: Struct. Crystallogr. Cryst. Chem.*, 1973, **29**, 1743–1745.

Akhil R. Chakravarty added: So you can find out which of the coppers is getting reduced? Is it reduced to copper(I)?

Govindasamy Mugesh responded: Yes, the copper undergoes redox reactions in which Cu(II) is reduced to Cu(I) by GSH.

Akhil R. Chakravarty reiterated: So, in copper–vanadate which region on the surface is getting reduced from copper(II) to copper(I)?

Govindasamy Mugesh responded: In the structure of CuV_2O_6, the copper is part of the CuO_6 octahedra, which is reduced from Cu(II) to Cu(I).

Shyamalava Mazumdar opened the discussion of the paper by Osami Shoji: It was an excellent talk. I have one query. You showed the effect of decoy molecules, and showed results on different types of substrates and also used various decoys with a wide range of chain lengths. Can you shed some light on the influence of length of decoy molecule on the reactions? What is the effect of the chain length of the decoy on the reactivity for different substrates?

Osami Shoji answered: Thank you. In the case of P450BSb, heptanoic acid (C7) served as the most effective decoy for peroxidation and preoxygenation. Carboxylic acid with a longer alkyl chain such as decanoic acid has better binding affinity to the enzyme, but its alkyl chain retards access of substrates to the heme pocket.[1]

1 O. Shoji, T. Fujishiro, H. Nakajima, M. Kim, S. Nagano, Y. Shiro and Y. Watanabe, *Angew. Chem., Int. Ed.*, 2007, **46**, 3656–3659.

Shyamalava Mazumdar added: Did you also investigate using larger peroxidase substrates such as ABTS?

Osami Shoji responded: Yes, ABTS can also be the substrate for the peroxidase reaction. P450BSb oxidizes most substrates for one-electron oxidation in the presence of decoy molecules.

Volker Schünemann asked: You report on the basis of UV-Vis spectroscopy that the formation of compound **I** of your enzyme is in the time range of a few seconds. This is very slow for compound **I** formation. High-valent compound **I** porphyrin models decompose into the ferric state very fast, sometimes even in the millisecond range. Can you comment on the timescale you are observing for compound **I** formation?

Osami Shoji answered: The compound **I** formation time range was in the ms order, as shown in Fig. 2A and B in our article (DOI: https://doi.org/10.1039/

d1fd00065a). I'm sorry, I think I misunderstood your question I answered about the rate of reaction in the discussion on the day, so I correct it here.

Volker Schünemann commented: The thermophillic P450 from *Sulfolobus acidocaldarius* displays a very characteristic band at 690 nm after oxidation with *m*-chloroperbenzoic acid as shown by Rittle and Green.[1] This band is regarded as a characteristic UV-Vis marker band for compound **I**. You have shown spectral changes in the 450 nm range. Do you also observe a characteristic band around 690 nm? Are your UV-Vis data comparable with those of Rittle and Green?

1 J. Rittle and M. T. Green, Cytochrome P450 compound I: capture, characterization, and C–H bond activation kinetics, *Science*, 2010, **330**(6006), 933–937.

Osami Shoji responded: Yes, we have observed the UV absorption around 690 nm assigned to the cation radical of porphyrin.

Abhishek Dey said: Questions regarding compound **I:** Do you see an isotope effect in the formation of compound **1**?

Something we showed is that if you dangle a H-bond,[1] not only can you enhance the stability of compound **1** by a couple of seconds, but the absorption spectrum also changes by shifting to around 700 nm – the same red shift you see here.

1 S. Bhunia, A. Rana, S. G. Dey, A. Ivancich and A. Dey, *Chem. Sci.*, 2020, **11**, 2681–2695.

Osami Shoji answered: Thank you for the nice suggestion. We have not yet checked for an isotope effect.

Jean-Didier Maréchal asked: Have you considered using molecular dynamics and docking to try to rationalise the competitions you observe?

Osami Shoji responded: We have carried out a docking simulation of decoy molecules to the enzyme. The docking simulation allowed us to understand the observed stereoselectivity. We also carried out a docking simulation of the second substrate, styrene, to understand enantioselective epoxidation.[1] As for MD, we have not yet tried it in the case of P450BSb and P450SPa. We will report the MD results of P450BM3 soon. Results for the P450BM3-decoy system have been reported.[2–4]

1 T. Fujishiro, O. Shoji, N. Kawakami, T. Watanabe, H. Sugimoto, Y. Shiro and Y. Watanabe, *Chem.–Asian J.*, 2012, 7, 2286–2293.
2 N. Kawakami, O. Shoji and Y. Watanabe, *Angew. Chem., Int. Ed.*, 2011, **50**, 5315–5318.
3 K. Yonemura, S. Ariyasu, J. K. Stanfield, K. Suzuki, H. Onoda, C. Kasai, H. Sugimoto, Y. Aiba, Y. Watanabe and O. Shoji, *ACS Catal.*, 2020, **10**, 9136–9144.
4 J. K. Stanfield, K. Omura, A. Matsumoto, C. Kasai, H. Sugimoto, Y. Shiro, Y. Watanabe and O. Shoji, *Angew. Chem., Int. Ed.*, 2020, **59**, 7611–7618.

Jean-Didier Maréchal remarked: In the modelling you have performed, if I understand correctly, you used a unique X-ray structure so the shape/space of the substrate channel is always the same?

Osami Shoji answered: We speculated a small structural change in P450BSb. We succeeded in the crystal structure analysis of P450SPs with ibuprofen as the decoy molecule.[1]

1 T. Fujishiro, O. Shoji, N. Kawakami, T. Watanabe, H. Sugimoto, Y. Shiro and Y. Watanabe, *Chem.–Asian J.*, 2012, 7, 2286–2293.

Akhil R. Chakravarty opened the discussion of the paper by Peter J. Sadler: How does transfer hydrogenation catalysis work in cells? How is the hydride transfer done?

Peter J. Sadler responded: We have administered half-sandwich Ru(II) and Os(II) complexes to cells in combination with higher non-toxic (millimolar) concentrations of formate as the hydride donor.

In the case of *N*-sulfonyl ethylenediamine as the chelated ligand, we observed a decrease in the NAD^+/NADH ratio, consistent with transfer hydrogenation of NAD^+. This was accompanied by an increase in the cytotoxicity of the Ru(II) complex towards A2780 human ovarian cancer cells.[1] For the chiral half-sandwich Os(II) diphenyl sulfonyl ethylenediamine complex described in our *Faraday Discussions* paper, with formate as the hydride donor, we observed enantioselective reduction of natural pyruvate to lactate, to D- or L-lactate, dependent on the presence of *SS*, or *RR* chiral centres on the chelated ligand. Again there was enhanced activity of the Os(II) complex towards A2780 human ovarian cancer cells. Interestingly enhancement of activity in the presence of formate was not seen for normal ovarian cells.[2]

1 J. J. Soldevila-Barreda, I. Romero-Canelón, A. Habtemariam and P. J. Sadler, *Nat. Commun.*, 2015, **6**, 6582.
2 J. P. C. Coverdale, I. Romero-Canelón, C. Sanchez-Cano, G. J. Clarkson, A. Habtemariam, M. Wills and P. J. Sadler, *Nat. Chem.*, 2018, **10**, 347–354.

Akhil R. Chakravarty remarked: Is NADH a hydride donor for the ATH reactions in the cells?

Peter J. Sadler replied: We have found that NADH is an efficient hydride donor to half-sandwich Cp* Ir(III) complexes, and can occur in cancer cells. These Ir(III) complexes can use NADH to reduce *e.g.* pyruvate to lactate, quinones to semiquinones, and even catalyse efficient H_2 production.

NADH can also donate hydride to half-sandwich Ru(II) arene complexes, and allow conversion of pyruvate to lactate.[1,2]

It is intriguing also that in the presence of an appropriate half-sandwich organometallic complex, formate can donate hydride to NAD^+ to give NADH and create reductive stress in cells, whereas some complexes alone can create oxidative stress by converting NADH to NAD^+. The balance of activity is dependent on the nature of the half-sandwich complex and requires further investigation.

1 S. Betanzos-Lara, Z. Liu, A. Habtemariam, A. M. Pizarro, B. Qamar and P. J. Sadler, *Angew. Chem., Int. Ed.*, 2012, **51**, 3897–3900.
2 Z. Liu, R. J. Deeth, J. S. Butler, A. Habtemariam, M. E. Newton and Peter J. Sadler, *Angew. Chem., Int. Ed.*, 2013, **52**, 4194–4197.

Akhil R. Chakravarty further queried: If stress is generated inside the cell, will that affect cellular organelles like mitochondria or the endoplasmic reticulum? Will the cell undergo apoptosis due to this stress? If that happens, the whole experiment would be void/gone.

Peter J. Sadler replied: It depends on what your overall aim is when attempting to catalyse reactions in cells using synthetic complexes. In our case, if we can induce oxidative (or reductive) stress which kills cancer cells by a novel mechanism at low catalyst doses, then that might be an effective therapeutic strategy.

Akhil R. Chakravarty said: How robust are these asymmetric transfer hydrogenation catalysts, as we have to work at 37 °C because cells in general cannot go beyond this temperature?

Peter J. Sadler replied: Yes, we generally restrict our studies to body temperature, 310 K, or lower, but many half-sandwich organometallic complexes are reasonably stable at higher temperatures too, and it is common for elevated temperatures to be used in synthetic organic chemistry, depending on the solvent.

Akhil R. Chakravarty added: If the catalysts are not robust enough, this means that the concept of using them in cellular system is just at the conceptual stage?

Peter J. Sadler replied: One of the tough challenges is to study the chemistry of the catalysts in intact cancer cells – to understand their distribution amongst the various cellular organelles, and their stability towards redox and ligand exchange reactions.

One way we are doing that is using synchrotron X-ray fluorescence. We can detect *e.g.* Os at physiologically-relevant concentrations in cancer cells, and in one recent experiment we labelled the chelating ligand with Br (not normally detected). We detected significant amounts of both Os and Br in the cytoplasm suggesting that at least this part of the complex has robust enough bonds to stay intact and perhaps retain catalytic activity. Similar distribution data were obtained from determinations of Os and Br in cell fractions by ICP-MS.[1]

1 E. M. Bolitho, J. P. C. Coverdale, H. E. Bridgewater, G. J. Clarkson, P. D. Quinn, C. Sanchez-Cano and P. J. Sadler, *Angew. Chem., Int. Ed.*, 2021, **133**, 6536–6546.

Akhil R. Chakravarty continued: Do you think a macrocyclic cage will be better to study this reaction? Or a 3D cage?

Peter J. Sadler responded: Macrocyclic complexes may offer special features as transfer hydrogenation catalysts for some metal ions, *e.g.* Fe(II).[1]

Cages around metal ions tend to restrict access for substrates, but partial cages, as in tethered half-sandwich complexes can offer extra rigidity and increased enantioselectivity.[2]

1 L. De Luca and A. Mezzetti, *Angew. Chem., Int. Ed.*, 2017, **56**(39), 11949–11953.
2 A. M. Hayes, D. J. Morris, G. J. Clarkson and W. Wills, *J. Am. Chem. Soc.*, 2005, **127**(20), 7318–7319.

Anca Pordea commented: In our work, we tried conjugating Ir(III)Cp* and Rh(III)Cp* complexes to proteins (https://doi.org/10.1039/d1fd00070e). We have been struggling to perform transfer hydrogenation in a more controlled ("clean", *i.e.* protein only) environment because sometimes compounds lose activity on exposure to proteins. In certain cases, we think the Rh(Cp*) or Ir(Cp*) complex loses the ligand (Bipy, Phen, sulfonyl) and binds to the surface side chains on the protein whilst remaining active. Do you think similar events occur in cells? Do you think the bidentate ligands would fall off in the presence of a protein environment in cells, leaving the metal–arene to bind to proteinogenic ligands and perform catalysis in this way?

Peter J. Sadler replied: Yes, these are certainly good questions to ask when we attempt to carry out catalysis in complicated biological media. The behaviour of the catalyst will depend a lot on the exact nature of the ligands, *e.g.* Cp/arene, the chelated diamine, and monodentate ligands. There will be a time dependence – the catalyst may be active for a while inside cells before being deactivated.

With the osmium arene sulfonyldiamine catalysts, our recent X-ray imaging studies suggest that a significant amount of the {Os(sulfonyldiamine} fragment remains intact in the cytoplasm of A549 human lung cancer cells and might be active, although there is loss of some of the Br-labeled diamine ligand, and the Br (but not Os) enters the nucleus: from tracking reactions of asymmetric organo-osmium transfer hydrogenation catalysts in cancer cells.[1]

Our recent studies show that glutathione (GSH), which is abundant in cells, readily reacts with arene Ru(II) sulfonyl-ethylenediamine complexes and displaces the chelated diamine. The effect of cysteine thiols on the catalytic and anticancer activity of Ru(II) sulfonyl-ethylenediamine complexes has also been investigated.[2] Interestingly, thiolate (GS)-bridged dinuclear complexes are formed, and such complexes are known to have anticancer activity.[3]

Interaction of transfer hydrogenation catalysts with substrates can sometimes result in unexpected redox reactions. For example, we found that transfer hydrogenation of quinones using Cp*Ir(III)(phen) complexes leads to a one-electron transfer and semi-quinone formation and involves Ir(III) reduction.[4]

There are competing metal-binding amino acid side-chain ligands in proteins – *e.g.* carboxylates of Asp and Glu, sulfurs of Met and Cys, phenolate oxygen of Tyr, and imidazole ring nitrogen of His – but often they are buried, or partially buried, and not readily accessible. An example is the thiolate sulfur from Cys34 of the blood protein albumin which is in a cleft. The binding of Ru(II) arene complexes for example, can be dependent of the nature of the arene.[5]

1 E. M. Bolitho, J. P. C. Coverdale, H. E. Bridgewater, G. J. Clarkson, P. D. Quinn, C. Sanchez-Cano and P. J. Sadler, *Angew. Chem., Int. Ed.*, 2021, **133**, 6536–6546.
2 F. Chen, I. Romero-Canelón, A. Habtemariam, J.-I. Song, S. Banerjee, G. J. Clarkson, L. Song, I. Prokes and P. J. Sadler, *Dalton Trans.*, 2022, 51, 4447–4457.
3 A.-F. Ibao, M. Gras, B. Therrien, G. Süss-Fink, O. Zava, P. J. Dyson, *Eur. J. Inorg. Chem.*, 2012, **9**, 1531–1535.
4 Z. Liu, R. J. Deeth, J. S. Butler, A. Habtemariam, M. E. Newton and P. J. Sadler, *Angew. Chem., Int. Ed.*, 2013, **52**, 4194–4197.
5 W. Hu, Q. Luo, X. Ma, K. Wu, J. Liu, Y. Chen, S. Xiong, J. Wang, P. J. Sadler and F. Wang, *Chem.–Eur. J.*, 2009, **15**, 6586–6594.

Peter J. Sadler returned a question: Do you think you retain the Cp ligand? It would be good if you can determine by mass spec whether it is still bound.

Anca Pordea responded: This refers to the case of our covalently conjugated rhodium complexes to TbADH.[1] With mutants where the covalent conjugation is sluggish due to an inaccessible cysteine, denaturing ESI-MS in acetonitrile/water with formic acid suggests that the majority of RhCp* is bound unspecifically to the protein, without the phenanthroline ligand;[2] whilst catalytic turnovers remain reasonable.

1 M. Basle, H. A. W. Padley, F. L. Martins, G. S. Winkler, C. M. Jäger and A. Pordea, *J. Inorg. Biochem.*, 2021, **220**, 111446.
2 M. Basle, PhD thesis, University of Nottingham, 2020.

Anca Pordea added: Do you think NADH would bind in a similar way as the acetophenone substrate to the Ru catalyst (*i.e. via* H-bonding to the NH_2 of the ligand)?

Peter J. Sadler replied: We expect the exact nature of the recognition of the hydride donor and the substrate by the catalyst might be different in each case. We have looked at the recognition of NAD^+/formate/NADH by $[(\eta^6\text{-}p\text{-cym})Ru(TsEn(Me,Me))(OH_2)]^+$,[1] in which hydride transfer from formate to Ru(II) and then to NAD^+ has been modelled. But the NADH/pyruvate or acetophenone systems have not been modelled yet.

The DFT calculations suggested an initial substitution of an aqua ligand by formate, hydride transfer to Ru(II) and then to NAD^+. Specific interactions between the aqua complex and both NAD^+ and NADH occur, including T-shaped stacking of adenosine NH_2 with the tosyl ring, see Fig. S8 in the ESI of ref. 1.

1 F. Chen, J. J. Soldevila-Barreda, I. Romero-Canelón, J. P. C. Coverdale, J.-I. Song, G. J. Clarkson, J. Kasparkova, A. Habtemariam, V. Brabec, J. A. Wolny, V. Schünemann and P. J. Sadler, *Dalton Trans.*, 2018, **47**, 7178–7189.

Juliusz A. Wolny added: In principle I wouldn't exclude that. Our recent experience is everything is possible depending on the system that you take. We cannot find the structure with acetophenone bound in a chelating way. If you change ketone to NADH or pyruvate or something else, you may get completely different results. We would have to do very thorough modelling as the picture changes from molecule to molecule. A different system may even interact with a different part of the molecule. What we have been struck by lately is the complexity of this catalyst.

Peter J. Sadler said: It is remarkable that the C–H bonds in some Rh(III)–Cp* complexes are activated towards H/D exchange and Diels–Alder adduct formation with conjugated dienes under mild conditions. The mechanism has been discussed previously.[1,2]

1 S. Banerjee, J. J. Soldevila-Barreda, J. A. Wolny, C. A. Wootton, A. Habtemariam, I. Romero-Canelón, F. Chen, G. J. Clarkson, I. Prokes, L. Song, P. B. O'Connor, V. Schünemann and P. J. Sadler, *Chem. Sci.*, 2018, **9**, 3177–3185.
2 J. Armstrong, S. Banerjee, V. Schünemann, J. A. Wolny, P. J. Sadler, *J. Phys. Chem. Lett.*, 2021, **12**, 658–662.

Juliusz A. Wolny remarked: One sees a system with a modified Cp star, where one methyl group is converted to the fulvene CH_2=, for a rhodium complex. Rh is bound to the OH group and this structure converts to the aqua-complex with a fulvene derivative of Cp*. Then it could react and convert to this ($Rh\_disp\_H_2O$)

system, so the double bond can undergo for example a Diels–Alder reaction and then become something very bulky as shown here. The Sadler group has mass spectral evidence for this structure. We also found the kinetics for it. A proton is between the metal group and oxygen. In a short time in deuterated solvent, all NMR signals of the methyl groups disappear as they are all deuterated.

The reaction profile picture of the Rh_disp_OH structure[1] is shown below as Fig. 1.

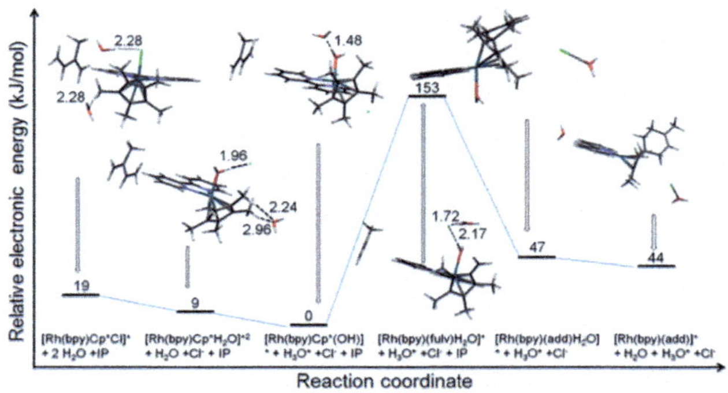

Fig. 1 Electronic energy profile for the reaction of $[(Cp^*)Rh(bpy)Cl]^+$ with isoprene in water at neutral pH (calculated using CAM-B3LYP/CEP-31g, solvent = water, IEFPCM model). Reproduced from ref. 1 with permission from the Royal Society of Chemistry.

1 S. Banerjee, J. J. Soldevila-Barreda, J. A. Wolny, C. A. Wootton, A. Habtemariam, I. Romero-Canelón, F. Chen, G. J. Clarkson, I. Prokes, L. Song, P. B. O'Connor, V. Schünemann and P. J. Sadler, *Chem. Sci.*, 2018, **9**, 3177–3185.

Abhishek Dey commented: The John Hartwig approach for getting a catalyst inside a protein is by over-expressing the apo-protein inside the protein. In these cases the catalyst was inside the protein and not on the surface. While binding to the surface may be convenient, if it is inside a protein it seems to provide both enantio and stereoselectivity. Any thoughts? Putting the catalyst inside a protein cavity could improve selectivity and cell uptake – can you comment on that?

Also, cytoplasm has 100 mM chloride. Several of the intermediates in catalysis have a 3+ oxidation state which should have strong binding to Cl^-. At pH 7 hydroxide concentration is much lower. Any thoughts on that?

Peter J. Sadler responded: Yes, a catalytic metal centre in a crevice in a protein benefits from anchoring to side chains of amino acids as ligands and the often somewhat-fixed spatial arrangement of other surrounding amino acids which can assist the stereoselective recognition of substrates, and act as acids or bases during different stages of the catalytic cycle.

To achieve such stereoselectivity, and especially enantioselectivity, within a catalytic low-molecular-weight metal complex is a challenge, especially as the catalytic metal centre is likely to be more solvent-exposed and susceptible to attack by inhibitors (poisons) in a complicated biological medium.

However, the indications are that this can be achieved to some extent if appropriate sets of ligands are used, as illustrated by our work on chiral half-sandwich Os(II) arene complexes:

Asymmetric transfer hydrogenation by synthetic catalysts in cancer cells.[1] The strategy of conjugating small metal catalysts to proteins to increase selectivity adopted by John Hartwig, Tom Ward, Anca Pordea, and others, certainly offers potential for increased selectivity, and introduces scope for modifying substrate specificity.

As you rightly point out, different biological compartments can have different concentrations of chloride, protons (pH), and other species, and these variations can affect speciation of the metal catalyst. Redox reactions with biological reductants (*e.g.* the millimolar concentrations of glutathione in cells), or oxidants, can also be important. The extent of chloride binding and hydrolysis, as well as the acidity of metal-bound H_2O (OH^-) can depend strongly not only on the metal ion and its oxidation state, but also the on the nature of other ligands.

See for example Fig. 3 in ref. 2.

1 J. P. C. Coverdale, I. Romero-Canelón, C. Sanchez-Cano, G. J. Clarkson, A. Habtemariam, M. Wills and P. J. Sadler, *Nat. Chem.*, 2018, **10**, 347–354.
2 E. J. Anthony, E. M. Bolitho, H. E. Bridgewater, O. W. L. Carter, J. M. Donnelly, C. Imberti, E. C. Lant, F. Lermyte, R. J. Needham, M. Palau, P. J. Sadler, H. Shi, F.-X. Wang, W.-Y. Zhang and Z. Zhang, *Chem. Sci.*, 2020, **11**, 12888–12917.

Peter J. Sadler stated: I think the studies of Mugesh (https://doi.org/10.1039/d1fd00067e) are very nice and they are going to gain increasing medical importance. For example, I have been involved in work in identifying metallic copper and iron nanoparticles, as well as other compounds of these metals, in the brains of Alzheimer's patients.[1] These complexes could be very important in disease processes. Catalytic surfaces can be very important in disease processes, so work like Mugesh's is important.

1 J. Everett, F. Lermyte, J. Brooks, V. Tjendana-Tjhin, G. Plascencia-Villa, I. Hands-Portman, J. Donnelly, K. Billimoria, G. Perry, X. Zhu, P. J. Sadler, P. B. O'Connor, J. F. Collingwood and N. D. Telling, *Sci. Adv.*, 2021, 7, eabf6707.

Akhil R. Chakravarty addressed Peter J. Sadler: How are the nanoparticles formed in the brain?

Peter J. Sadler answered: All we know at the moment, is that nanoscale deposits of Cu(0) and Fe(0), along with Cu(I), Cu(II), Fe(II) and Fe(III) can be characterised by synchrotron X-ray mapping methods within amyloid plaque cores taken from the gray matter of human Alzheimer subjects.[1]

It is perhaps not so difficult to envisage Cu(I) disproportionation into Cu(0) and Cu(II), and there is abundant Cu(I) in the brain, *e.g.* stored in the cysteine-rich protein metallothionein (MT-3), but iron is more difficult to reduce to Fe(0). Perhaps there is reductive stress present under some conditions, and involvement of hydride donors such as NADH, but this all remains to be investigated, as do the surface properties of such nano-metal deposits, which might not only have redox properties, but also catalytic activity. Currently it is not known whether such deposits are present, or have a function, in normal brains too.

1 J. Everett, F. Lermyte, J. Brooks, V. Tjendana-Tjhin, G. Plascencia-Villa, I. Hands-Portman, J. Donnelly, K. Billimoria, G. Perry, X. Zhu, P. J. Sadler, P. B. O'Connor, J. F. Collingwood and N. D. Telling, *Sci. Adv.*, 2021, **7**, eabf6707.

Abhishek Dey added: Regarding your comment on finding Cu nanoparticles in Alzheimer's patient brains, Cu(I) produced at pH 7 would disproportionate to Cu(II) and Cu metal.

Peter J. Sadler replied: A very important question: if formed naturally how are they formed?

The question of how metallic nanoparticles might be formed naturally in the brain and perhaps other tissues, is an important one.

Interestingly, there is thought to be a possible nose-to-brain route that could deliver metal nanoparticles into the brain.[1]

But perhaps they might also form *in situ*.

1 E. F. Craparo, T. Musumeci, A. Bonaccorso, R. Pellitteri, A. Romeo, I. Naletova, L. M. Cucci, G. Cavallaro and C. Satriano, *Pharmaceutics*, 2021, **13**(9), 1508.

Peter J. Sadler added: We don't know enough about transport mechanisms for metals into cells. People attach metal complexes to antibodies for delivery to specific cellular targets.

Some natural transporters appear to handle metallodrugs too, such as the copper transporter Ctr1 for cisplatin and other platinum drugs.[1]

1 I. S. Song, N. Savaraj, Z. H. Siddik, P. Liu, Y. Wei, C. J. Wu, M. T. Kuo, *Mol. Cancer Ther.*, 2004, 3(12), 1543–1549.

Peter J. Sadler continued: The question as to how can we translate our knowledge of the coordination chemistry of metal complexes into understanding their chemistry in complicated biological media, including cells, presents a major challenge.

In one example, we used radio-isotope labelling to study ligand exchange reactions of metallodrugs in cells. ^{131}I labelling of the monodentate iodide ligand in the osmium(II) arene anticancer complex FY26, revealed that although the Os–I bond is strong and relatively inert, it is readily cleaved in cancer cells.[1]

Subsequent experimental and DFT studies suggest that this activation and hydrolysis arises from attack on the azo bond of the N,N-chelated ligand by glutathione:[2]

We are still investigating the range of ketones that might be reduced by transfer hydrogenation in cells. Recently we have shown that [Os^{II}(η^6-arene)(*R*-PhDPEN)] catalysts can reduce duroquinone and menadione to their di-alcohol derivatives using formic acid as the hydride donor.[3]

1 R. J. Needham, C. Sanchez-Cano, X. Zhang, I. Romero-Canelón, A. Habtemariam, M. S. Cooper, L. Meszaros, G. J. Clarkson, P. J. Blower and P. J. Sadler, *Angew. Chem., Int. Ed.*, 2017, **56**, 1017–1020.
2 X. Zhang, F. Ponte, E. Borfecchia, A. Martini, C. Sanchez-Cano, E. Sicilia and P. J. Sadler, *Chem. Commun.*, 2019, **55**, 14602–14605.
3 E. M. Bolitho, N. G. Worby, J. P. C. Coverdale, J. A. Wolny, Schünemann and P. J. Sadler, *Organometallics*, 2021, **40**(17), 3012.

Peter J. Sadler concluded: Metal complexes are often 'pro-drugs'. The oxidation state of the metal and the nature of the bound ligands may change before they reach the target site. Determination of the speciation of metals in intact cells is a major challenge, especially at low physiologically-relevant metallodrug concentrations. We are using synchrotron X-ray methods, which, in favourable situations, allow mapping of the metal distribution and can give information about the oxidation state of the metal and its coordinated ligands.[1–5]

1 C. Sanchez-Cano, I. Romero-Canelón, Y. Yang, I. J. Hands-Portman, S. Bohic, P. Cloetens and P. J. Sadler, *Chem.–Eur. J.*, 2017, **23**, 2512–2516.
2 C. Sanchez-Cano, D. Gianolio, I. Romero-Canelon, R. Tucoulou and P. J. Sadler, *Chem. Commun.*, 2019, **55**, 7065–7068.
3 E. M. Bolitho, J. P. C. Coverdale, H. E. Bridgewater, G. J. Clarkson, P. D. Quinn, C. Sanchez-Cano and P. J. Sadler, *Angew. Chem., Int. Ed.*, 2021, **133**, 6536–6546.
4 E. M. Bolitho, H. E. Bridgewater, R. J. Needham, J. P. C. Coverdale, P. D. Quinn, C. Sanchez-Cano and P. J. Sadler, *Inorg. Chem. Front.*, 2021, **8**, 3675–3685.
5 E. M. Bolitho, C. Sanchez-Cano, H. Shi, P. D. Quinn, M. Harkiolaki, C. Imberti and P. J. Sadler, *J. Am. Chem. Soc.*, 2021, **143**(48), 20224–20240.

Dimitrios Pantazis opened a general discussion of the paper presented by Paul Walton (https://doi.org/10.1039/d1fd00068c): I have a question concerning the EPR simulations in the intermediate pH range. The spectra have a reasonably well defined superhyperfine structure and it is precisely in this region that the simulations for the intermediate pH species seem to agree less well with experiment. Have you tried to fit the spectra with three nitrogen nuclei (probably inequivalent) and can you therefore exclude this possibility based on your simulations?

Paul Walton replied: Thank you. It wasn't possible to fit the middle-pH spectrum easily with 3N superhyperfine couplings. This part of the simulation, especially the number and pattern of SHF peaks, is highly sensitive to the number and size of SHF coupling constants, and the simulation seems like a good fit to the experimental data.

Abhishek Dey remarked: The assignment of the pink CT could benefit from a C/D ratio using CD/MCD and absorption spectroscopy. The mixing from CT can indeed impart strong intensity to a high energy d–d.

Paul Walton replied: Yes agreed, thank you.

Abhishek Dey asked: Could you comment on the nature of the transition at 520 nm? If it is CT, did you try getting Raman from it? Also to probe the exchangeable ligand could you elaborate on any results from labelled ligands and how they changed at different pHs?

Paul Walton replied: Good ideas, thank you. We haven't yet tried resonance Raman spectroscopy on this species.

Abhishek Dey added: In copper nitrate reductases with 3 histidine and –OH there is also an aspartate which participates in the first pK_a event. It ends up tuning the reduction potential substantially.

Paul Walton replied: Copper nitrite reductase. Yes, our DFT calculations suggest that the nearby aspartate in AA10 LPMOs could indeed act as a strong H-bond acceptor for the copper-coordinated water molecule.

Abhishek Dey suggested: It's an open site, right? So it might help to have some water molecules around and see what that does to your calculations.

Paul Walton answered: Agreed, although inclusion of further water molecules creates significant instability in the calculations, such that optimisation becomes problematic.

Abhishek Dey remarked: One may add external ligands like N^{3-} to probe the number of weakly bound ligands or open co-ordination site. We found the addition of azide useful – a yellow, charge transfer, so you can titrate it and see the displacement of ligands. This helps you probe the coordination environment you have with chelating ligands like hydroxyquinoline.

Paul Walton replied: Thank you. Azide does indeed bind to the histidine brace active site, accompanied by the appearance of a band in the visible spectrum, *e.g.* see Fig. S5 in ref. 1.

1 G. R. Hemsworth, E. J. Taylor, R. Q. Kim, R. C. Gregory, S. J. Lewis, J. P. Turkenburg, A. Parkin, G. J. Davies and P. H. Walton, *J. Am. Chem. Soc.*, 2013, **135**(16), 6069–6077.

Philip Ash commented: I am curious about the assertion that an extremely high pK_a means the Cu(II)-azanide species is not pertinent for catalysis at physiological pH. Does an extreme pK_a necessarily preclude catalytic competence in biological systems?

There have been recent reports suggesting the catalytic competence of arginine as a proton transfer site and catalytic base,[1,2] despite the extreme pK_a ($\sim$13.8) of the guanidinium group, including structures of the neutral, deprotonated form in response to the presence of substrate.[3] Arginine need only be transiently deprotonated and/or a fraction of the sites in a sample deprotonated in order to be involved in effective acid–base catalysis, provided the deprotonated portion is catalytically competent. Arginine is often closely associated with carboxylate groups in extensive hydrogen bonding networks, so this proposed catalytic role is presumably not completely accounted for by exclusion of water from the surrounding protein matrix.

1 R. M. Evans, E. J. Brooke, S. A. Wehlin, E. Nomerotskaia, F. Sargent, S. B. Carr, S. E. Phillips and F. A. Armstrong, *Nat. Chem. Biol.*, 2016, **12**, 46–50.
2 M. Lučić, D. A. Svistunenko, M. T. Wilson, A. K. Chaplin, B. Davy, A. Ebrahim, D. Axford, T. Tosha, H. Sugimoto, S. Owada, F. S. N. Dworkowski, I. Tews, R. L. Owen, M. A. Hough and J. A. R. Worrall, *Angew. Chem., Int. Ed.*, 2020, **59**, 21656.
3 H. Kwon, J. Basran, J. M. Devos, R. Suardíaz, M. W. van der Kamp, A. J. Mulholland, T. E. Schrader, A. Ostermann, M. P. Blakely, P. C. E. Moody and E. L. Raven, *Proc. Natl. Acad. Sci. U. S. A.*, 2020, **117**, 6484.

Paul Walton replied: Thank you for this interesting point. In LPMOs the active site is exposed to solvent and I think it's likely that the surrounding environment is one that is not as 'controlled' as would appear in other, more buried, active sites.

Salvatore La Gatta asked: How do you exclude the possibility that the pink colour of the LPMO solution at alkaline pH is not due to a deprotonation of the amide of the backbone?

Paul Walton responded: We cannot specifically exclude any deprotonation event of the protein as the source of the pink colour, although we would note that the relative pK_as of ionisable protons would suggest that deprotonation of the amide backbone (without subsequent coordination to the Cu) is less likely at pH 10 than deprotonation of other amino acid side sides.

Abhishek Dey said: The active site of LMPO and Cu bound to amylin peptide is quite similar. There too the pH dependent changes and the pink color at high pH is observed. Please see ref. 1.

1 M. Seal and S. G. Dey, *Inorg. Chem.*, 2018, **57**(1), 129–138.

Paul Walton responded: Yes, thank you for drawing our attention to this species. There are clear parallels with the pink colour observed in our work. We haven't completed our spectroscopic studies on this pink colour as yet, so cannot comment as to its identity. We would note, however, that there may be several sources of this colour, including tryptophan radicals that may form at high pH.

Salvatore La Gatta commented: Do you think that the structure of the histidine brace in solution is similar to that in the crystal structure? Do you imagine there is equilibrium in solution between unbound terminal amine and bound terminal amine to the copper ion?

Paul Walton answered: Frozen solution state EPR of LPMOs is entirely in accord with the solid state crystal structures. From this it would be reasonable to assume that the solution state structure of the histidine brace is the same as that observed in crystal structures.

Christof Martin Jäger said: It is very interesting to see how selective the protein surface exposed histidine brace copper ion binding appears to be in this example. Have you ever tried exchanging the Cu ion by other transition metal ions, or is there any evidence that this motif binds other ions? If not, what could lead to the Cu selectivity of this seemingly not highly selective motif?

Paul Walton responded: Thanks for the suggestion. We have measured the dissociation constant of the His brace with Cu^{2+} (nM) which shows that it binds strongly. We have thought about other metal ions, but always struggle with any adventitious Cu displacing the other metal ion.

Akhil R. Chakravarty said: You have presented three different copper species at three different pH values. Did you look at the reduction potentials at different pHs?

Paul Walton replied: We didn't. This is a good suggestion.

Akhil R. Chakravarty remarked: Getting a suitable solvent is a problem for electrochemical studies at high pH.

Akhil R. Chakravarty asked: Is Cu(III) generation a possibility at such a high pH electrochemically?

Paul Walton answered: Oxidation to Cu(III) (*cf.* Dale Margerum's work, *e.g.* ref. 1) cannot of course be ruled out. We note however that a suitably positioned protein-backbone amide nitrogen atom that can coordinate to the Cu(III) in a deprotonated form is not evident in the crystal structures of LPMOs without significant distortion of the protein structure.

1 M. R. McDonald, F. C. Fredericks and D. W. Margerum, *Inorg. Chem.*, 1997, **36**(14), 3119–3124.

Jean-Didier Maréchal asked: Have you compared what you observed on the brace motif and what has been done on the Cu-ab1-42?

Paul Walton responded: No, we haven't but thank you for the suggestion. There are clear parallels there.

Jean-Didier Maréchal said: Is the brace motive found in more buried areas in other proteins? And how common is the histidine brace?

Paul Walton answered: All structures of the His brace are at or near an exposed protein surface. In saying that, there are some differences in the nature of the surfaces, with some being notably contoured.

Akhil R. Chakravarty remarked: What kind of similarities are you talking about in histidine brace *vs.* haem?

Paul Walton explained: Enzymes which contain the histidine brace are already known to be oxygenases and peroxygenases (analogous to haem group enzymes), and possibly Cu transport. Other occurrences of the histidine brace have the Cu coordinated not by 3Ns of the brace, but also a further amino acid, essentially making the Cu coordinatively saturated. The role of these latter enzymes is unclear, but there are parallels with cytochromes.

Paul Walton added: The histidine brace is the copper equivalent of haem. Nature has this unit, if no iron is present/available then copper is often doing the same job.

Jean-Didier Maréchal said: Do you think that such complex behavior of the first coordination sphere of the copper in the brace motive would also happen if it is inserted deep inside a protein?

Paul Walton replied: It's hard to tell (but see also my answer to the follow-up question below). My view is that the first coordination sphere is largely driven by the energetic demands of the Cu and the immediate hydrogen-bonds which are formed between the coordinating atoms and other residues. If the latter of these were to change from enclosing the site within the protein matrix, then I would expect the primary coordination sphere also to change.

Jean-Didier Maréchal continued: What is your feeling regarding the possible modification of the protonation state of the brace system when it's in a protein instead of exposed to solvent. I would think there will be a change when exposed or buried. What do you think?

Paul Walton replied: We haven't tried changing the electronic properties of the active site environment in our calculations, so I am not able to give a definitive answer. However, I can imagine that the relative stability of the different complexes we see in our paper may well be altered by a different protein environment, with the 'less charged' species (*i.e.* complexes **2** and **3**) becoming more stable.

Paul Walton said: Is the inside of a protein more hydrophobic than the outside/in the solvent?

Jean-Didier Maréchal answered: Buried environments tend to present higher hydrophobicity than exposed parts or regions even partially exposed to the solvent. For that reason, it is frequent to simulate QM cluster models using a continuum model with low dielectric constant values from 2 (buried) to 8 (partially exposed surface) sites.

Jean-Didier Maréchal asked: Did you consider solvent models in your DFT calculations?

Paul Walton answered: No we didn't, so I am afraid I cannot provide any more detail.

Abhishek Dey opened a general discussion of the paper by Christof Martin Jäger: The protonation states of arginine and phosphate have rather high pK_a's and are rather sensitive to hydrogen bonding.

Christof Martin Jäger replied: Yes, this is certainly the case and also a key feature defining many enzymes utilizing NAD(P) as co-factors. Their importance is also demonstrated by examples of single polar residue mutations that can determine NAD/NADP selectivity. Logically, that meant we were looking for hydrogen bonding anchoring motifs for the design of our catalyst, that create binding affinity and bring the catalyst in the correct positioning.

Abhishek Dey commented: Did you do MM (if FF are available) to try to guess the residency time of the normal *vs.* inverted orientations?

Also, how are you assigning the protonation states of the ionizable groups?

Christof Martin Jäger answered: That is a very good question highlighting the importance of assessing the thermodynamic equilibrium of different binding poses and orientations to assess binding poses and affinities accurately. In the current approach, we have not yet implemented molecular-dynamics (MD) simulations to evaluate the dynamic stability of different binding poses more thoroughly. In regards to assessing accurate binding free energies it should be noted that doing so *via* simulating ligand residency times *via* MD of the different poses would need an extremely high number of simulations and long simulation times to retrieve statistically meaningful data. Other methods to assess accurate relative and absolute binding affinities would include free energy perturbation (FEP) simulations. Those, although also still computationally expensive, are more and more often used in similar workflows, and would be targeted for our workflows in future as well. An approximate solution to binding free energies, however, is already tested in our approach where we compare standard docking scoring (Glide SP and XP) with more accurate end-point molecular-mechanics general born surface area (GBSA) calculations giving a better estimate for (de-)solvation and entropic effects on binding affinities of different molecular conformations and binding poses. In regards to the protonation states of ionizible groups our protocol uses,[1] we generate all possible protonation and ionisation state combinations in solution at pH of 7.0 ± 2.0 before docking all those structures. In the case of Glide XP subsequently different desolvation penalties are applied to polar and charged groups in the final docking score.

1 J. C. Shelley, A. Cholleti, L. L. Frye, J. R. Greenwood, M. R. Timlin and M. Uchimaya, *J. Comput.-Aided Mol. Des.*, 2007, **21**, 681–691.

Abhishek Dey said: Lack of reactivity of the polyhydroxy compound, I wonder about solvation energy? If your case you are trying to coax the well solvated compound in to an active site – there is probably a very large barrier to reactivity? Desolvation of solvated substrate or polyhydroxy compounds could be an issue.

Christof Martin Jäger responded: It is correct that desolvation penalties for the ligands need to be taken into account, in particular in the case of polyhydroxy compounds like our best working ligand. Our methodologies include accounting for desolvation penalties in the Glide XP method and the MM-GBSA refinement. The observation that the polyhydroxy ligand still shows good binding affinity in computation and experiment lies in the more hydrophilic nature of the co-factor binding pocket in comparison to the hydrophobic substrate binding pocket. Furthermore, the observation that the more hydrophobic furan based ligand (and catalyst) also shows comparable affinity, but no catalytic activity, and also fits this assessment, as it is likely that this ligand binds in the hydrophobic substrate pocket hindering substrate binding and thus explaining the lack of catalysis.

Jean-Didier Maréchal asked: Why did you choose Glide for performing your protein–ligand docking calculations?

Christof Martin Jäger answered: The start of this project was formulated around workflows to be used with the Schrödinger software package, which also allows straight forward inclusion of MM-GBSA and free energy perturbation (FEP) methods, as well as covalent docking workflows we have used in related projects in our group (see *e.g.* ref. 1). Schrödinger, exquisitely works with Glide, which has demonstrated good performance in comparison to other formulations particularly those applied in drug discovery and high throughput virtual screening (see ref. 2), which stands at the start of our workflow. In a current follow up, we also test other docking software such as GOLD or AutoDock for comparison and to test which packages are best to implement adequate and flexible models of transition metal complex 'substrates' for supramolecular and covalent docking.

1 M. Basle, H. A. W. Padley, F. L. Martins, G. S. Winkler, C. M. Jäger and A. Pordea, *J. Inorg. Biochem.*, 2021, **220**, 111446.
2 Z. Zhou, A. K. Felts, R. A. Friesner and R. M. Levy, *J. Chem. Inf. Model.*, 2007, **47**(4), 1599–1608.

Jean-Didier Maréchal added: We use a combination of Gold and GaudiMM, the latest is a software we develop in our group.

Akhil R. Chakravarty opened the discussion of the paper by Jean-Didier Maréchal: Why have you chosen this dirhodium complex instead of the mononuclear structure? This cage is well known to bind at the open axial sites.

Jean-Didier Maréchal responded: The dirhodium complex has been chosen by the experimental group of Prof. Lewis and we worked on the modelling on the systems that have already been characterized experimentally. The binding in axial sites does not necessarily represent a problem in the course of the design.

Anca Pordea asked: Is the Rh complex flexible inside the protein, does the docking step take this into account, and does this influence the styrene diffusion pathway?

Jean-Didier Maréchal answered: Indeed, the Rh complex is flexible inside the protein site and the docking takes into account this aspect. However, some orientations are preferential. Where the Rh stands indeed it drives the diffusion pathway.

Anca Pordea further asked: About the GPathFinder – does this method also allow us to find tunnels for larger substrates?

Jean-Didier Maréchal answered: GPathFinder[1] has been benchmarked for a series of systems with known channels for ligands (substrates) that are small and, not so small. It performed very well in all cases. From our research, the only limitation that could occur is when the sampling of the protein dynamics (generally provided by normal modes or molecular dynamics) is not sufficient to satisfy the adaptation of the ligand binding. In those cases, we tend to find the correct pathways although some steps along the trajectory could present energies that are not optimal.

1 J.-E. Sánchez-Aparicio, G. Sciortino, D. V. Herrmannsdoerfer, P. O. Chueca, J. R.-G. Pedregal and J.-D. Maréchal, *Int. J. Mol. Sci.*, 2019, **20**, 3155.

Anca Pordea asked: Related to my previous question, is the use of GPathFinder more problematic with longer, flexible structures such as alkane chains (or non aromatic structures)?

Jean-Didier Maréchal answered: Like any computational methods, the more you increase the number of degrees of freedom, the more conformational exploration you need and time for calculation. So in principle, longer and flexible ligands require more intensive calculations than small and conformationally restricted ones (*e.g.* aromatic). It does not necessarily mean it is more "problematic" but yes, depending on the system, it could be. However, one of the strengths of GPathFinder is that it is based on a multi objective genetic algorithm. If you have very flexible molecules, you could decrease the level of the representation of the energy and, for example, switch from molecular mechanics force field or scoring functions to simple calculations of good and bad contacts (simple van der Waals measures). That would give you faster ways to make intensive sampling.

Debabrata Goswami addressed all: Addressed to all working on DFT. Most of the studies here concern excited states, and I am worried about predictions based on DFT for excited states. The question is, how much is DFT applicable to most of the studies being presented that involve an excited state? How to justify that?

Paul Walton answered: Our principal analysis is one of ligand field analysis, DFT is secondary, for the reasons you have mentioned. You cannot base deep analysis on spin-Hamiltonian analysis from DFT. The power of ligand field theory is in obtaining an experimentally informed answer rather than just a theoretical answer – you have to get the balance right.

Akhil R. Chakravarty said: For beginners, it is best to work at DFT level to get preliminary information. A lot of synthetic chemists use DFT without really understanding the theoretical methodologies they are using in detail.

Debabrata Goswami prompted: Is DFT applicable to excited states?

Akhil R. Chakravarty replied: DFT in general is a very useful tool. If it is not fitting/reaching expectations, then one can ask an expert to resolve the deficiencies.

Edward I. Solomon commented: Prof. Debabrata Goswami asked the question: "Is DFT applicable to excited states?" I do not agree with some of the responses. Ligand field theory focuses on d to d transitions and is quite good when a metal site is not very covalent, as in high spin Fe(II) in non-heme ligand fields. However, a good example of its significant limitations when a site is quite covalent is provided by my comments in the discussion (DOI: https://doi.org/10.1039/d2fd90013k) of Peter Comba's paper on the incorrect assignment of a ligand field transition in the Fe(IV)=O TMC complex by an angular overlap ligand field model (DOI: https://doi.org/10.1039/d1fd00073j). Of course experimental spectroscopy (absorption (polarized), CD, MCD, resonance Raman) provides detailed insight into the nature of an excited state. As I mentioned in the Introductory lecture (DOI: https://doi.org/10.1039/d2fd00001f) different functionals and hybrids give quantitative differences in their descriptions of the electronic structure. But the correlation to excited state spectroscopic properties by TD DFT provides a strong basis for experimentally evaluating the specific DFT calculation's description of the electronic structure, thus locking in on the description supported by experiment and then evaluating key frontier molecular orbitals and electronic structure contributions to reactivity.

Christof Martin Jäger added: Witnessing the growing importance of accurate computational predictions for artificial metalloenzyme design and a lack of accurate transition metal complex models for molecular modelling at the molecular mechanics level, I believe it is important to think about creating a shared *in silico* (community) resource database for the benefit of the whole community. The question is how to do this, maybe in alignment with other efforts in that direction like ioChem-BD (a computational chemistry results repository)?

Jean-Didier Maréchal responded: I agree with this statement. I also think that the ioChem-BD model would be one of the best options to follow.

We have not tried to generate a data sharing platform in our community. It might not be too difficult to build and could very helpful. Still we are a small community of computational chemists devoted to designing metalloenzymes. That is a complicated field, but highly gratifying.

Edward I. Solomon opened the discussion of the Concluding remarks by Kenneth Karlin (DOI: https://doi.org/10.1039/d2fd00073c): Great overview on the evolution of the field of bioinorganic chemistry including your recent results on ligand hydroxylation by a Cu(I) H_2O_2 system modelling the LPMO reaction. In the early part of your talk you emphasized the potential for the development of new materials for catalysis and electrocatalysis. I wanted to point out that one area of catalytic material that has strong overlap with bioinorganic systems is metallozeolites. These have Cu and Fe active sites that turn methane to methanol, with the Fe at room temp. Many of the concepts of bioinorganic chemistry, for example the entatic state and the local cage preventing radical escape, directly transfer to metallozeolite reactivity. Also the same spectroscopic methods are needed to define these sites, but in an active site selective way. These parallels between metallozeolites and metalloenzymes are developed in our recent work.[1]

1 H. M. Rhoda, A. J. Heyer, B. E. R. Snyder, D. Plessers, M. L. Bols, R. A. Schoonheydt, B. F. Sels and E. I. Solomon, *Chem. Rev.*, DOI: https://doi.org/10.1021/acs.chemrev.1c00915.

Paul Walton added: LPMOs appear to have means of 'extinguishing' any reactive species that are created in the absence of substrate (to prevent self-oxidation). Do you think that this is the next evolution of synthetic catalysts – to couple the generation of reactive species in the presence of substrate and have means of dealing with reactive intermediates in the absence of substrate?

Kenneth Karlin answered: The proposal to couple the generation of a highly reactive species but only in the presence of a substrate is indeed a near future step or goal in the development of synthetic catalysts, the purpose is preventing self-oxidation, as appears to occur in LPMOs and of course is a well established phenomenon in cytochrome P450s. Perhaps this is not "the" next evolution of synthetic catalysts, but certainly it is "a" soon-to-come evolution. Your question comes from your expertise in LPMOs, and perhaps also in part from some of the new research results I presented in my lecture, pertaining to a peroxygenase type mechanism where a reduced copper(I) center reacts with hydrogen peroxide to produce a reactive hydroxyl radical. In our model system this is a single-turnover reaction, where one ligand-copper(I) center reacts with hydrogen peroxide to give 1 mol-equiv ·OH, which attacks the ligand (as the substrate). At this time, I don't know how one would or could modify the synthetic system in order to obtain catalysis, *i.e.*, turnover. Perhaps this is future research.

Debabrata Goswami said: I like the idea of looking for model systems that can mimic the overall concept of a complicated reaction. That's perhaps one of the goals of many physical chemists – they want to break it down into more straightforward problems so they can solve what they are looking for. I like the approach you mentioned. However, is there any case where this approach completely fails?

Kenneth Karlin responded: Well, sure, if the model system is not well designed, well thought out, then some new chemistry may be completed (which always has general value), but specific questions may not be well answered, *i.e.*, the factors in question for the natural system, to hopefully be (partially) answered by the model study, may not be answered, or at least not completely. To carry out a good model study, the scientist must have studied the natural system very well, and thoroughly understand it, so as to know what questions need answering *via* the model approach.

Conflicts of interest

There are no conflicts to declare.

Faraday Discussions

ROYAL SOCIETY OF CHEMISTRY

PAPER

Concluding remarks: discussion on natural and artificial enzymes including synthetic models

Kenneth D. Karlin, * Pradip K. Hota and Bohee Kim

Received 7th April 2022, Accepted 19th April 2022
DOI: 10.1039/d2fd00073c

This paper overviews the final remarks lecture delivered (by K. D. K.) at the end of this bioinorganic chemistry *Faraday Discussion*, held online for a worldwide audience from January 31 – February 3, 2022. This paper provides discussion in six sections: (1) the Introductory lecture, from Ed Solomon, emphasized past and present uses of advanced spectroscopic methods and theoretical approaches to elucidate metalloenzyme active site structure, physical properties and function. (2) The discussion topics are divided into groups having similar research themes, as seen from this author's perspective. Emphasis is given to the non-heme iron group of articles with dioxygen activation research. (3) Small molecule activation (*e.g.*, N_2, CO_2 and O_2 reduction; CH_4 or H_2O oxidation) is widely covered in this discussion; this authors' view of the important reactions in bioinorganic chemistry is discussed. (4) We discuss current practice and vision for employing materials chemistry to widely apply to electrocatalytic methods to effect small molecule activation (as above) to fulfill societal energy demands. (5) A discussion is given on the topic of synthetic models and the approach utilized therein. (6) New research on the authors' synthetic modeling is presented; preliminary results are given in the area of copper mediated peroxide activation.

1. Introduction

It was a great pleasure and honor to be invited to participate in this 2022 *Faraday Discussion*. This important scientific gathering was conceived of and organized (*via* the Royal Society of Chemistry) by a long-time participant in this field, from the synthetic modelling side of research, Professor Rabindranath (Rabi) Mukherjee of the Indian Institute of Technology (IIT) Kanpur, India. The broader field is these days widely referred to as *Bioinorganic Chemistry*, or *Metal Ions in Biology*. It is worth noting that the last *Faraday Discussion* devoted to this field, with closely related or even the same specific topics (*e.g.*, oxygen activation, artificial enzymes, hydrogenases) occurred in 2010, with the discussion title "Spectroscopy, Theory and Mechanism in *Bioinorganic Chemistry*"; for that,

Department of Chemistry, Johns Hopkins University, Baltimore, Maryland 21218, USA. E-mail: karlin@jhu.edu

a themed collection of articles was published as *Faraday Discuss.*, 2011, **148**, pages 1–460.

As outlined on the RSC Web page for this discussion, at https://www.rsc.org/events/detail/46673/natural-and-artificial-metalloenzymes-faraday-discussion, the main themes defined were:

(1) Small molecule activation and synthetic analogues.

(2) Techniques for studies of the kinetics and thermodynamics of metal–ligand exchange reactions.

(3) Electron transfer, spectroscopy and theory.

(4) Natural & artificial enzymes and medicinal aspects.

In fact, on the basis of the 20 short lectures presented, which were accompanied by considerable discussion (on-line, and recorded) and full manuscripts published in *Faraday Discuss.* (*vide supra*), topics 1, 3 and 4 were very well addressed, but topic 2 much less so. Bioinorganic chemistry is a very exciting field and it is much too broad in scope to cover in any one gathering of scientists in a single symposium format.

This *Faraday Discussion* included speakers, session chairpersons poster presenters and attendees from all over the globe, providing a challenge for an on-line gathering. A good number of the presentations and discussions occurred at times which made it very difficult to attend (*e.g.*, 4:00 to 7:00 am for K. D. K.). Thus, participants often had to rely on the published papers from speakers (which were made available prior to the actual on-line event) or recordings of the presentations and discussion to stay on top of the meeting.

The first scientific speaker, presenting the "Introductory lecture", was Edward Solomon of Stanford University, who provided a lecture entitled "Spiers Memorial Lecture: activating metal sites for biological electron transfer" (https://doi.org/10.1039/D2FD00001F). This talk emphasized the spectroscopic methods utilized in bioinorganic chemistry research, themed around the elucidation of the structure and electronic structure (*i.e.*, bonding and spectroscopic properties) of metalloenzyme active sites (Fig. 1). The article emphasized the finding that these often display distinctive spectroscopic characteristics which reflect the

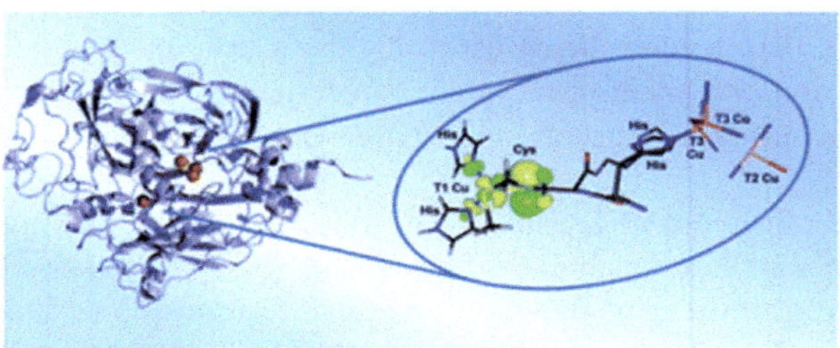

Fig. 1 Depiction of a multicopper oxidase (MCO) emphasizing the zoomed out active-site, the Type-1 (T1) His_2–Cys copper site which draws electrons from an external substrate and passes them on to the trinuclear Type-2/Type 3 (Cu_2) center. The latter binds molecular oxygen and effects its reduction–protonation, giving $2H_2O$. The point in the drawn out expanded view is to depict the ground state wave-functions, emphasizing the T1 copper–cysteine sulfur atom bonding which is key to the electron-transfer process. Reproduced from Solomon and coworkers (https://doi.org/10.1039/D2FD00001F) with permission from the Royal Society of Chemistry.

coordination geometry and electronic structure (*i.e.*, bonding) imposed by the protein matrix (and the specific ligation) that are critical to the enzyme's reactivity.

Solomon began his lecture with the very basics, the structure and bonding (using computational methods), spectroscopy (EPR, electronic absorption and charge-transfer) and the spectroscopy of square-planar $[Cu^{II}Cl_4]^{2-}$. This led to the advanced subject matter, electron-transfer sites in bioinorganic chemistry, especially the paradigmatic "blue" copper sites whose unique spectroscopic properties and Cu^{II}–cysteinate bonding were elucidated by Solomon and coworkers. However, the real emphasis of Solomon's lecture and paper was the delineation and description of the numerous physical and theoretical methods now available to interrogate metalloenzyme active sites. These include Cu L-edge or sulfur-ligand K-edge X-ray absorption spectroscopy (XAS), low-temperature magnetic circular dichroism (MCD) spectroscopy, resonant inelastic X-ray scattering (RIXS) and ultrafast XFEL spectroscopy. Solomon emphasized the critical need to accompany all of these studies with theoretical–computational methods, such as density functional theory (DFT). The material discussed in the Solomon lecture and paper included discussion of perturbed or "entatic state" active site structures, resulting in distinctive spectroscopic properties contributing to the unique reactivity. These concepts, accompanied by the utilization of theoretical–computational methods, overlap in one way or another with nearly all efforts in bioinorganic chemistry research, including the topics discussed by the other contributors to this *Faraday Discussion*, even including those whose studies/research involve synthetic chemistry.

2. Groupings and survey of the lectures/papers presented

To a reasonable extent, one could group the lectures presented into various subtopics, perhaps useful in order to have a view of the subject matter focused upon in this discussion, given the incredibly broad scope of bioinorganic chemistry. The choice of groupings and the extent of the authors' discussion of these lectures/papers is to a great extent dictated by the authors' particular research interests and biases, those being especially the activation of molecular oxygen with copper and/or heme-iron coordination complexes. These, of course, lead to biases in the authors' particular perspectives and preferences for discussion in this article.

These groupings can be labelled in the following manner: (a) non-heme iron coordination complexes, elaborated upon below. (b) Copper bioinorganic chemistry, including synthetic modeling chemistry in relationship to dioxygen activation (Stack and coworkers, https://doi.org/10.1039/d1fd00071c, Schindler and coworkers, https://doi.org/10.1039/d1fd00062d, and our research (*vide infra*)). Walton and coworkers (https://doi.org/10.1039/d1fd00068c) described aspects of the chemistry within a lytic polysaccharide monooxygenase. The contribution by Mugesh and coworkers (https://doi.org/10.1039/d1fd00067e) was rather different, and focused on a new material which can deliver NO (g) as a drug, so it possibly also comes under the label of medicinal chemistry. (c) Organometallic chemistry, with synthetic modeling of hydrogenase chemistry by Artero, Duboc and coworkers (https://doi.org/10.1039/d1fd00085c), a DFT study of Ru or Os asymmetric transfer hydrogenation catalysis (Wolny, Sadler and coworkers, https://doi.org/10.1039/d1fd00075f) and the application of a new/novel analytical physical technique (X-ray absorption spectroelectrochemistry) to hydrogenases (Rüdiger, DeBeer and coworkers, https://doi.org/10.1039/d1fd00079a). (d) Heme (bio)chemistry, with contributions by Dey and coworkers (https://doi.org/10.1039/d1fd00076d) on O_2-reduction and Shoji and coworkers (https://doi.org/10.1039/d1fd00065a) on insights into the previously unknown functions of a CYP fatty acid peroxygenase. (e) The topic of artificial enzymes, which consisted of very exciting computational contributions. The paper by Lledós, Maréchal and coworkers (https://doi.org/10.1039/d1fd00069a) reported a multiscale model of an artificial enzyme, a POP-Rh_2 cyclopropanase, and Jäger and coworkers (https://doi.org/10.1039/d1fd00070e) presented a computational perspective on iridium based alcohol dehydrogenases. There is considerable research activity on artificial enzymes; it is a very hot area and promising results for practical applications are apparent.

Two excellent discussion presentations and accompanying papers which did not fall into the above categories, but which certainly belong in this bioinorganic discussion, are (i) the work by Parker and coworkers (https://doi.org/10.1039/d1fd00059d) and (ii) that by Lakshmi, Brudvig, and coworkers (https://doi.org/10.1039/d1fd00094b). The former study centered around lanthanide luminescence, since this is involved in bioanalytical applications. Here, the scope and mechanisms involved in the excited state quenching of 9-coordinate terbium and europium(III) were delineated. The latter paper involved the application of an advanced pulsed electron paramagnetic resonance (EPR) spectroscopic method, two-dimensional (2D) hyperfine sublevel correlation (HYSCORE) spectroscopy, to the study of isotopically labelled MeOH (as a probe molecule) binding to the Mn_4Ca-oxo cluster in the S_2 state of the oxygen evolving complex of a cyanobacterial photosystem II (PSII).

With so many papers covering somewhat disparate subjects, we choose here to elaborate on the subtopic having the most speaker contributors, and possessing a number of highly exciting new results. This we label as the non-heme iron coordination complex area; five speakers gave lectures/papers in this area, with four emphasizing synthetic systems and reactivity studies. Sen Gupta and coworkers (https://doi.org/10.1039/d1fd00074h) reported on the use of a later generation bTAML derived iron complex (TAML = tetra-amido macrocyclic ligand, developed and championed by T. J. Collins)[1] to carry out the efficient electrochemical oxidation of alcohols to aldehydes/ketones. The researchers were able to obtain considerable mechanistic information. They identified a high-valent $Fe^V{=}O$ species as a reactive intermediate formed by the PCET of a starting TAML aqua complex, which affects the rate-determining (second-order) substrate C–H abstraction in a process involving overall net hydride transfer. The use of electrochemistry is notable, as mentioned below.

Kumar, de Visser, Sastri and coworkers (https://doi.org/10.1039/d1fd00064k) reported on the much less common use of high-valent non-heme iron-oxo species. The $Fe^{IV}{=}O$ complex with a quinolyl and pyridyl containing pentadentate ligand is able to oxidatively degrade halo-substituted phenols, which are nasty pollutants. Mechanistic studies revealed the occurrence of a phenol H-atom abstraction reaction and the authors suggested that a new kind of species is involved in the overall reaction; here, a phenoxyl radical is proposed to bind to an iron(III)-hydroxide complex. Interestingly, *o*-quinones were found to be the major products following oxidative dehalogenation.

Comba and coworkers (https://doi.org/10.1039/d1fd00073j) reported on the extraordinary reactivity, in fact record reactivity, of a new intermediate-spin ($S = 1$) $Fe^{IV}{=}O$ complex employing a pentadentate bispidine ligand. For example, C–H abstraction on cyclohexane as a substrate occurs (in propionitrile solvent at -90 °C) at a rate comparable to those of non-heme iron enzymes. The authors complemented their work using ligand field theory to compare the properties of their bispidine complex with those of the well-known $Fe^{IV}{=}O$ complex with tetramethylcyclam, toward providing information on the complex spin-state gap size and its correlation to oxidative reactivity.

A most exciting advance was discussed and published by Que and coworkers as part of this *Faraday Discussion* (https://doi.org/10.1039/d1fd00066g), a breakthrough in the field of di-iron chemistry relevant to oxygen activating proteins possessing a μ-oxo di-iron active site such as in the hydroxylase component of soluble methane monooxygenase (sMMOH). A powerful oxidant is generated, derived from molecular oxygen, going through a diferric-peroxo intermediate P, which undergoes O–O cleavage to give a CH_4-oxidizing intermediate called Q, which most spectroscopic studies have suggested to have a $Fe_2^{IV}(\mu\text{-}O)_2$ "diamond core" structure. While the Que group was some time ago able to characterize a related and relevant high-valent $(\mu\text{-}O)_2Fe^{IV}Fe^{III}$ complex, they only now (in 2021–2022) could describe the successful synthesis of a model compound with a diiron(IV) "diamond core", and its X-ray structure (Fig. 2).

The *Faraday Discussion* publication accompanying the short lecture delivered by Que is a very important paper, one which will be very highly cited over the next few years. It is very detailed and describes the synthetic approaches/methods, and it tabulates a variety of important structural (*e.g.*, Fe···Fe distance) and

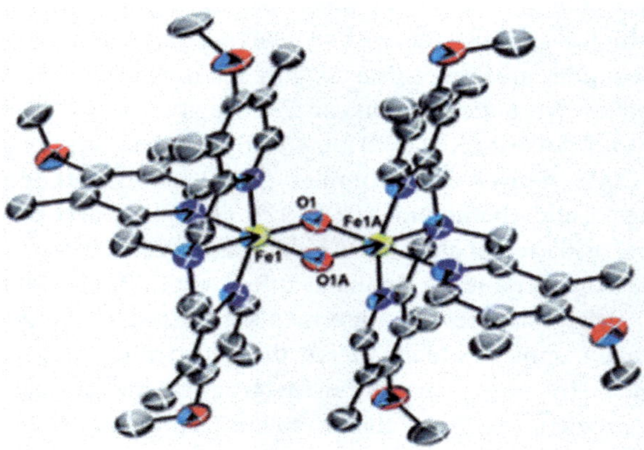

Fig. 2 Representation of the X-ray structure of the $[(TPA^*)Fe_2^{IV}(\mu\text{-}O)_2]^{4+}$ (TPA* = tris(3,5-dimethyl-4-methoxypyridyl-2-methyl)amine; crystallized with perchlorate as the counter-anion). The Fe···Fe distance is 2.711(4) Å, which can be compared to other relevant diiron complexes and proteins. Reproduced from Que and coworkers (https://doi.org/10.1039/d1fd00066g) with permission from the Royal Society of Chemistry.

spectroscopic data (*e.g.*, UV-vis, resonance Raman, Mössbauer and EXAFS) for the precursor complexes, the di-iron(IV) product (Fig. 2), related high-valent or other diiron complexes, as well as diiron proteins. Furthermore, this paper gives reactivity comparisons which very interestingly show that this new complex $[(TPA^*)Fe_2^{IV}(\mu\text{-}O)_2]^{4+}$ is amongst the least reactive of a related series of compounds, based on comparisons of the oxidative reactivity with the substrate dihydroanthracene. The more highly reactive complexes, by many orders of magnitude, have a partially or fully open core structure possessing a mono μ-oxo bridge. Many questions still remain, such as how to explain the reactivity variations observed in terms of the structure and iron ion (and complex) spin-state, and there are still differing interpretations of structural data or theoretical analyses of the actual enzyme Q intermediate. As already suggested, this work represents a significant advance and seminal contribution in bioinorganic oxidative chemistry.

The fifth paper in this group of non-heme iron papers is a strictly computational work on first-row metal ion high-valent oxo complexes by Sen and Rajaraman (https://doi.org/10.1039/d1fd00072a). They use DFT and an *ab initio* method to explore the electronic structure of high-valent oxo complexes of cobalt in different geometries or spin states, asking when do $Co^{IV}{=}O$ *versus* $Co^{III}–O^{•}$ descriptions apply, the former violating the oxo-wall concept long ago described by Ballhausen and Gray[2,3] for metal-oxo complexes, but only in tetragonal geometries. The authors present calculations indicating that $Fe^{IV}{=}O$ and $Mn^{IV}{=}O$ are the best descriptions for these species; they do exist and are pre-oxo-wall complexes. For cobalt, the authors studied different possible ligand architectures and varying cobalt ion spin states, and find that a mixing of the $Co^{IV}{=}O \Leftrightarrow Co^{III}–O^{•}$ electromer description is considerable in most geometries and this also depends on the cobalt spin state. They conclude that discussion of the percentage of $Co^{III}–O^{•}$ character for certain complexes can be reasonable.

3. Small molecule activation and key reactions in (bioinorganic) chemistry

As mentioned, the organizers of the present discussion enumerated small molecule activation as being one of the major themes, and they stated that abundant and mostly inert species include O_2, N_2, CO_2 and CH_4. Of course, molecular oxygen is not inert and lectures and papers on transition metal mediated oxygen activation are strongly represented in this discussion. However, there are many other entities, neutral molecules or ions, which are biologically important and whose chemistry includes transition metal manipulation. These include CO and H_2O, along with a good many nitrogen oxides, hydroxylamine (NH_2OH), N_2O, NO (nitrogen monoxide; nitric oxide), HNO (nitroxyl), nitrite (NO_2^-) and others, including sulfur containing species. Clearly, the research represented by the presentations in this *Faraday Discussion* covers only a small fraction of possible bioinorganic chemistry involving small molecules.

We believe it is useful to consider a listing of perhaps the most important reactions in chemistry, those that are critically important in biological processes and which impact society in terms of energy conservation, energy needs and/or environmental demands. In fact, most of the listing below comes *via* the senior author of this article; however, the ideas here come from his listening to lectures (*e.g.*, at American Chemical Society national meetings) by one of the giants in inorganic and bioinorganic chemistry, Harry B. Gray, of Caltech:

$$O_2 \rightarrow H_2O \text{ (fuel cell reaction)} \quad \text{(a)}$$

$$H_2O \rightarrow O_2 \text{ (water oxidation)} \quad \text{(b)}$$

$$H_2O \rightarrow H_2 + O_2 \text{ (water splitting)} \quad \text{(c)}$$

$$CO_2 \text{ reduction} \rightarrow CO\text{, formate, } CH_4\text{, } C_2O_4^{2-} \quad \text{(d)}$$

$$2H^+ \text{ (aq)} \rightarrow H_2 \text{ (g) (cheap fuel production)} \quad \text{(e)}$$

$$N_2 \rightarrow 2NH_3 \text{ (nitrogen fixation) (recent: } NH_3 \text{ as fuel)} \quad \text{(f)}$$

$$CH_4 \rightarrow MeOH \text{ (hydrocarbon functionalization)} \quad \text{(g)}$$

Six of the reactions listed do occur in nature and require mediation or catalysis by metal ions, such as in metalloenzyme active sites, enhancing the broad interest in bioinorganic chemistry. The reduction of dioxygen to water (a) is an oxidase reaction and is coupled to some other process, such as membrane proton translocation (which is then linked to the bio-generation of adenosine triphosphate (ATP) from ADP) in cytochrome c oxidases, containing heme and copper ion centers.[4] Multicopper oxidases (4 coppers per functional subunit)[5] also efficiently reduce O_2 to two waters, coupling this reaction to substrate dehydrogenation (oxidation) chemistry.

Water oxidation is of course the reaction in photosynthesis, so it is mediated by the OEC Mn_4Ca-oxo cluster.[6,7] There is considerable current interest and activity by (bio)inorganic researchers to find and define catalysts for practical water oxidation (*i.e.*, the oxygen evolution reaction; OER) (b); most of the research community employs electrocatalytic methods. Water splitting, akin to water electrolysis, to give $2H_2$ (g) + O_2 (g) (c) as a single process, is not a reaction carried out by *nature*, but there is interest in carrying out such chemistry *via* photocatalytic methods.

Carbon dioxide reduction to carbon monoxide (d) is carried out by protein CO dehydrogenases (CODHs).[8] Aerobic bacteria carry out this biochemistry using copper–molybdenum flavoenzymes with a Mo-[2Fe–2S]-FAD active site,[9] while anaerobes employ a Ni–Fe catalyst where the active site possesses a [Ni–3Fe–4S] cluster and another distal iron center.[10] Because of our societal environmental excess of CO_2, there are considerable efforts to develop electrocatalysts which can selectively carry out a reduction directly to methane or C_2 products such as ethylene (d). Treated copper-based material electrocatalysts seem to be particularly effective in facilitating CO_2 reduction with H_2 (g).[11,12]

Two reduction reactions of great biological and societal importance are (e), the production of H_2 as a fuel from water, the hydrogen evolution reaction (HER) and (f), dinitrogen reduction to give ammonia as a useful precursor (*e.g.*, as a fertilizer) for the nitrogen needed for protein and nucleic acid biosynthesis. Hydrogenases efficiently interconvert protons and dihydrogen[13,14] and, of course, the latter is important as a fuel in energy applications. The key sulfur rich FeMo-cluster in nitrogenase enzymes found in cyanobacteria and rhizobacteria possesses a Fe_7MoS_9C stoichiometry; it has a somewhat complicated structure, where a central carbon atom connects to six (6) surrounding iron ions, where the latter are bridged by either two or three sulfide (S^{2-}) ions. In a highly energy demanding (*via* ATP) reaction, dinitrogen molecules react as follows:

$$N_2 + 8H^+ + 8e^- + 16MgATP \rightarrow 2NH_3 + H_2 + 16MgADP + 16P_i$$

Atmospheric dinitrogen conversion into ammonia gives us our primary source of bioavailable nitrogen. Ammonia oxidation is also a critical step in global nitrogen cycling, and research on ammonia oxidation has attracted great interest in clean energy and environmental pollution control.[15–18] NH_3 can be readily liquified and its oxidation (chemical or electrochemical) may yield dinitrogen plus H_2 (g), thus NH_3 can be thought of as a hydrogen carrying fuel (f). Again illustrating the extremely broad based subject matter of bioinorganic chemistry, none of the lectures/papers in this *Faraday Discussion* in any way pertained to nitrogenase or N_2-reduction.

Reaction (g) is the non-heme iron or copper catalyzed O_2 oxygenation of methane to give methanol, which is a well-known bioinorganic chemistry topic, falling under the oxygen-activation label. Aerobic methanotrophic organisms are Gram-negative bacteria that occur in a variety of environments and utilize methane as their sole carbon source. There is a soluble cytoplasmic enzyme form, sMMO, which possesses an active-site non-heme di-iron center. In this *Faraday Discussion*, the lecture and paper from Que and coworkers (https://doi.org/10.1039/d1fd00066g) detail aspects of both the protein and model chemistry efforts, (see above). When the environmental copper ion is present, the predominant form of methane monooxygenase is the particulate membrane-bound form, pMMO. The nature of the copper-containing active-site(s) has been controversial, and it is still not completely clear if the active site is comprised of a mono- or di-copper center.[19,20] More importantly, future studies must focus on the mechanistic aspects of these enzymes and/or appropriate model compounds. Certainly, for copper ion mediated methane/hydrocarbon oxygenation chemistry, it is quite uncertain as to what sort of O_2-derived ligand–copper chemistry leading to reactive intermediates is required to hydroxylate methane, other hydrocarbons, or polysaccharides (*vide infra*).

4. Small molecule activation and energy concerns *via* electrocatalysis and materials/solid-state chemistry

As a practicing member of the synthetic bioinorganic chemistry community, of course I have a constant eye on the chemical literature reporting on the fast moving and considerable advances in the activation/manipulation of small molecules, in our case especially molecular oxygen. And it was with pleasure and admiration that I listened to the lecture presentations and read the accompanying papers and saw that these fit right in to the contemporary landscape of the field. However, in my normal perusing of prominent journals in chemistry, I have not been able to avoid noticing the huge number of high quality papers that involve small molecule manipulation but approach the issues and chemistry using materials/solid-state chemistry and even an engineering approach. Design or manipulation, including systematic variation, including the use of theoretical–computational methods of evaluation, lead to new materials as "reagents" or catalysts for carrying out the relevant chemistry. Prominent among the approaches utilized include electrochemistry and electrocatalysis, where materials/solid-state chemists supply one or the other of the electrode materials.

The prominent use and discussions of the likely future employment of electrocatalysis led back to a key publication in *Science* magazine by Seh and coworkers,[21] entitled "Combining theory and experiment in electrocatalysis: insights into materials design" which was noted in our Concluding remarks presentation (Fig. 3). Here, Seh *et al.*[21] reviewed advancements in electrocatalyst development, which also necessitates complementary theoretical–computational research, which may enhance the electrocatalysis of the water-splitting reaction (*vide supra*), that being the reverse of the reaction that underlies fuel cell chemistry. Related dioxygen, dinitrogen, and carbon dioxide reductions are also emphasized (Fig. 3). The authors presented a theoretical framework which highlighted the need for catalyst design strategies that may stabilize distinctive reaction intermediates in a selective manner, relative to each other.

Seh *et al.*[21] also described how electrocatalysis research and implementation of such electrocatalytic processes can fit into the need for safekeeping our energy future, given a rising global population, accompanying increased energy demands and impending climate change. Fig. 4 depicts the conceivable sustainable production and transformation of fuels and chemicals, such as hydrogen, hydrocarbons, oxygenates and ammonia. Water, carbon dioxide and dinitrogen are plentiful and available from the Earth's atmosphere and may be transformed into the products mentioned above *via* electrocatalytic processes coupled to renewable energy (Fig. 4) if electrocatalysts with the requisite properties (*e.g.*, efficiency and reaction selectivity) can be advanced.

Seh *et al.*[21] has been heavily cited, especially including papers utilizing materials chemistry approaches to generate electrode materials for electrocatalysis, for example (i) promoting the use of "high-entropy materials" in catalysts,[22] (ii) the

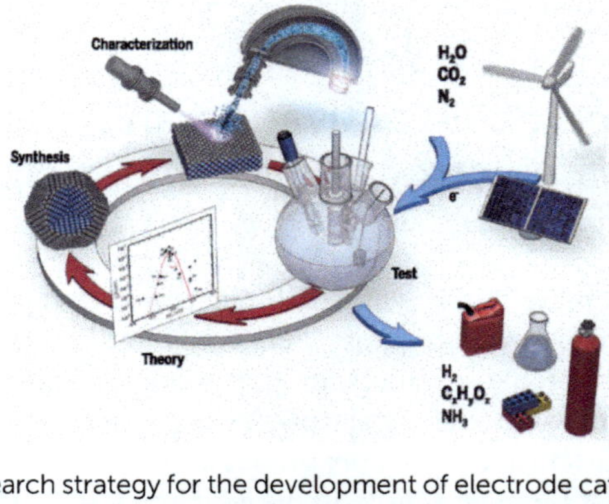

Fig. 3 General research strategy for the development of electrode catalysts, *via* materials/solid-state chemical synthesis, to be utilized in electrocatalytic processes involving important small molecules, such as H_2O, CO_2 and N_2. Reprinted from ref. 21 with permission from The American Association for the Advancement of Science.

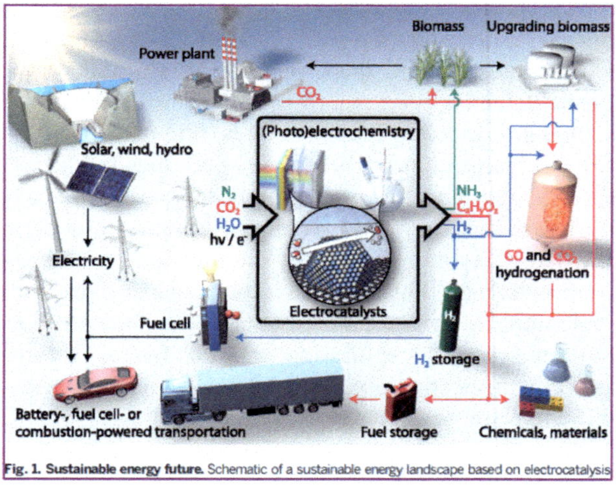

Fig. 4 Conceivable sustainable pathways for the production of essential fuels and chemicals. See the text. Reprinted from ref. 21 with permission from The American Association for the Advancement of Science.

implementation of specialized proton shuttles to aid the electrocatalytic reduction of N_2 to NH_3,[23] and (iii) the introduction of a nanoscale conductive copper-based metal–organic framework (MOF) layer supported over iron [$Fe(OH)_x$] nanoboxes to catalyze the HER.[24] Yet another recent publication emphasizes the use of "nature's blueprint to expand catalysis with Earth-abundant metals". It is notable that the discussion in this paper and the composition of the coauthor pool is dominated by bioinorganic chemistry and bioinorganic research chemists.[25]

A number of bioinorganic chemists participating in this *Faraday Discussion* employed electrocatalysis, notably (i) Artero, Duboc and coworkers (https://doi.org/10.1039/d1fd00085c), describing FeFe, FeNi or CoFe binuclear complexes which may or may not mediate the HER in an organic solvent and (ii) Sen Gupta and coworkers, in utilizing a non-heme iron complex to effect electrocatalytic alcohol oxidation. In fact, within the broader bioinorganic chemistry community, electrocatalysis is now a very popular research tool, especially for the HER, O_2-reduction, CO_2-reduction and water oxidation. A very new publication[26] presents a striking result, where a heme possessing synthetically derived tethered urea functions as a highly efficient photocatalyst for CO_2 reduction to CO; mechanistic insights suggest (i) the involvement of the urea functional groups to aid docking of the CO_2 substrate and (ii) that the transformation of an Fe^{I}–CO_2 adduct is involved in the rate determining reaction step.

5. Synthetic models in bioinorganic chemistry & synergy of physical methods–spectroscopy–theory

As per my Introduction, Solomon and Jose (https://doi.org/10.1039/D2FD00001F) emphasized the elucidation of enzyme active sites by utilizing a variety of approaches: spectroscopy, X-ray crystallography, enzymology, *molecular biology*, and, as per my own expertise and research activity, the synergy arising from the study of carefully designed and then well-defined synthetic models (also see below). Professors Solomon and Richard H. Holm from Harvard University have been instrumental in expressing these ideas, starting in 1996 with their conception and editorial organization of three thematic issues published in *Chemical Reviews* (ACS) (1996, 2004 and 2014). These publications featured a huge range of outstanding review articles from key researchers active in bioinorganic chemistry, presenting the latest results from many bioinorganic sub-topics, including "Bioinorganic Enzymology" (Fig. 5). It should be kept in mind the this *Faraday Discussion*'s Solomon/Jose presentation & publication (https://doi.org/10.1039/D2FD00001F) emphasized the utilization of a good many recently developed advanced physical and spectroscopic methods, now critical to the advancement of bioinorganic chemistry research.

5.1 Models in bioinorganic chemistry and in science

As biased by our own research interests, we feel that it is important to provide some discussion concerning bioinorganic synthetic models. The building and consideration of models is critical in complementing research on the system of interest, *e.g.*, the enzyme and its active site. "Models are the building blocks of science".[28] "The model is the most basic element of the scientific method. Everything done in science is done with models. A model is any simplification, substitute or stand-in for what you are actually studying or trying to predict. Models are used because they are convenient substitutes".[28] "A model is a substitute, but it is also similar to what it represents."[28] "Models are used in most if not all subfields of chemistry". With respect to biological research and the

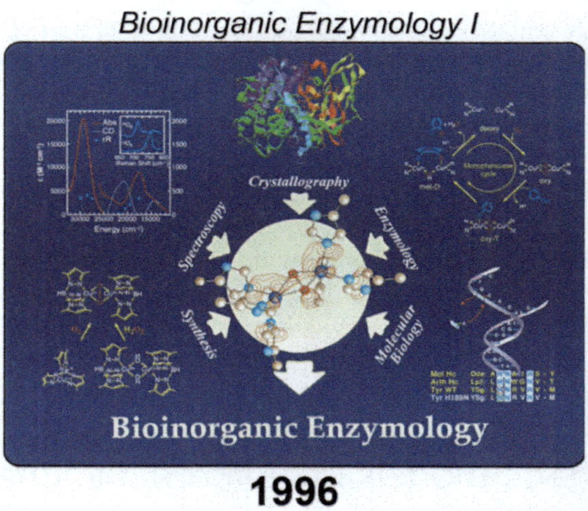

Fig. 5 Subfield research contributing to the elucidation of a metalloenzyme's active site structure and electronic structure, leading to an understanding of function. Results from computational–theoretical chemistry studies should be added as an important contributor. Reprinted (adapted) with permission from ref. 27. Copyright 2014 American Chemical Society.

understanding of protein function, "Virtually everything that is understood about biological systems is derived from work with model systems".[29] And models are useful in many diverse fields, such as membrane fusion[30] or to understand "lizard tail autotomy".[31]

Thus, as they pertain to synthetic models, biomimetic studies need not necessarily lead to the duplication of the chemical or physical characteristics of the enzyme, but rather serve to sharpen or focus the scientific questions asked. By careful design and systematic variations (*e.g.*, in the nature of the ligand employed), one might help identify those factors (*e.g.*, donor atom type, metal redox potential, coordination number and/or geometry) contributing to the observed enzyme structure, spectroscopy and chemical mechanism of action. Chemical models can provide reference compounds for structural and coordination aspects, or provide insights into the plausibility of biochemically proposed 'active' intermediates and their reaction competency. Bioinorganic models may also lead to compounds which mimic enzyme function and provide new reagents or catalysts for practical application.[32]

The present authors would like to express some frustration and mention some comments from reviewers/referees received over the years, illustrating how many in the field just do not understand concepts in bioinorganic chemistry (or any) modeling. (1) Reviewer: "the studies were not performed in aqueous, physiologically relevant conditions, comparable to the natural reaction in the protein." This is a ridiculous comment; most metalloenzyme active sites are not hydrophilic, and may be highly hydrophobic. (2) Reviewer: "unfortunately, the results do not provide proof of the actual mechanism of action in the protein." By definition, the mechanism of action of any enzyme can only be absolutely proven on studies of the enzyme itself. Models have other purposes.

A classic example of a synthetic model having a great impact in the subfield of dioxygen activation by copper and Cu-metalloproteins came in 1989 (Fig. 6).[33] This was a breakthrough study. The binuclear dioxygen complex generated and characterized possessed structural features (*i.e.*, a μ–η^2:η^2- (*i.e.*, side-on) peroxo ligand binding, Cu···Cu distance and spectroscopic properties (UV-vis, resonance Raman, *etc.*)) which closely matched those known for oxy-hemocyanin (an arthropodal and molluscan O_2-carrier) and oxy-tyrosinase (monooxygenase). The matching of properties occurred in spite of the fact that a highly non-biomimetic ligand was employed, a sterically hindered trispyrazolylborate anionic species (Fig. 6). This study in fact was published several years prior to the confirmation of the oxy-hemocyanin active site structure determined by X-ray crystallography.[34]

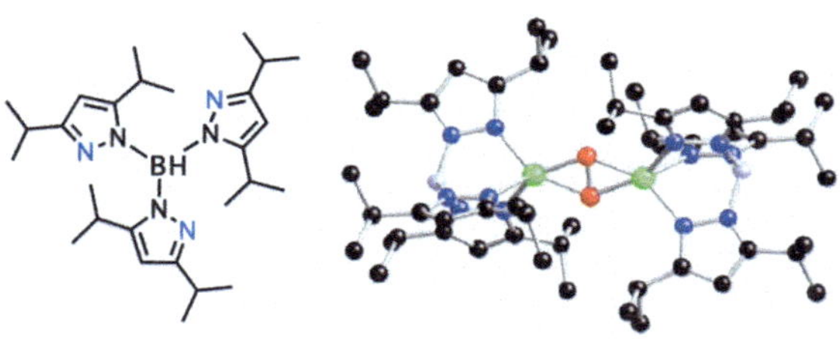

Fig. 6 A classically known synthetic model from Fujisawa, Kitajima and coworkers. See the text for further discussion.

The approach and results illustrate the synthetic model approach and how critically important results can come from such a study. Here, in this 2022 *Faraday Discussion*, the successful generation and characterization of the $[(TPA^*)Fe_2^{IV}(\mu\text{-}O)_2]^{4+}$ model compound (*vide supra*) (https://doi.org/10.1039/d1fd00066g) closely compares in importance and likely impact.

6. New synthetic bioinorganic chemistry relevant to LPMOs

As this paper's authors are actively involved in synthetic bioinorganic chemistry relevant to dioxygen activation by copper, the senior author here chose to present new research during his Concluding remarks lecture. The class of copper metalloproteins of interest/relevance are lytic polysaccharide monooxygenases (LPMOs; Fig. 7).[35–42] The active site has a single copper ion with three protein derived ligands, a terminal histidine forming a chelate *via* coordination of the amino terminus and an imidazole nitrogen atom (the "histidine brace"), along with a second histidine derived imidazole donor (Fig. 7). Recent research indicates that LPMOs possess both monooxygenase and peroxygenase activity, but the latter occurs at considerably faster rates. Thus, research aimed at providing insight into the peroxygenase reaction, *via* new Cu/H_2O_2 model chemistry, may provide interesting insights.

It has come to light that the classic Fenton reaction, the process where a reduced metal ion (Fe(II) or Cu(I)) reacts with hydrogen peroxide, is not as simple as producing an oxidized metal ion (Fe(III) or Cu(II)) plus hydroxide plus a hydroxyl radical ($^{\bullet}OH$). For iron, depending on the ligand environment for Fe(II) and other reaction conditions (such as acidity), the product is usually a high-valent $Fe^{IV}{=}O$ species.[43] For the copper ion, much less is known, and the reaction of a (ligand)–Cu^{I} complex and H_2O_2 could lead to the following outcomes:[44]

$$LCu^{I} + H_2O_2 \rightarrow LCu^{II}\text{–}O^{\bullet} + H_2O \quad (1)$$

$$LCu^{I} + H_2O_2 \rightarrow LCu^{II}\text{–}OH + {}^{\bullet}OH \quad (2)$$

$$LCu^{I} + H_2O_2 \rightarrow LCu^{III}\text{–}(OH)_2 \quad (3)$$

A combination of reactions might occur, such as reaction (2) being followed by the Cu^{II}-bound hydroxide losing $^{\bullet}H$ to the hydroxyl radical *via* HAT, giving the products in eqn (1). In any case, a substrate R–H (with C–H bond) could undergo

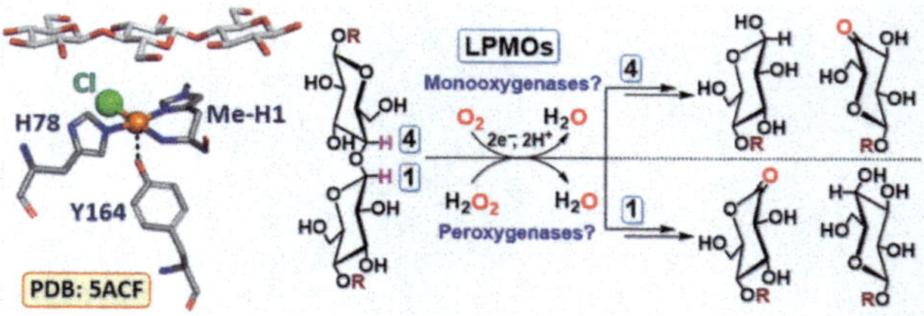

Fig. 7 Depiction of one derivative form of the active site and polysaccharide oxidation reactions carried out by LPMOs. See the text for further discussion.

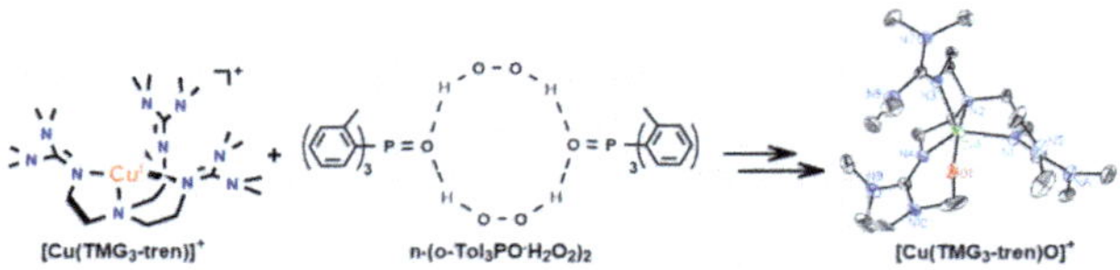

Fig. 8 The chemical system examined in this research, wherein a known ligand–copper(I) complex is exposed to a 'dry' hydrogen peroxide source under cryogenic conditions in organic solvents. With excess H_2O_2, the reaction results in the hydroxylation of one of the exterior ligand methyl groups, which can lead to the copper(II)–alkoxide structure shown. See the text for further discussion.

hydrogen-atom abstraction (HAT) by the cupryl species in eqn (1), or by the freed hydroxyl radical in eqn (2), or by the copper(III) species in eqn (3), where a bound hydroxide could be a proton acceptor. In all cases, a substrate carbon radical would form but rebound of a copper bound $^{\bullet}$OH (formally) group would lead to hydroxylated product R–OH, leaving behind copper(I).

The chemical system we have chosen to study is shown in Fig. 8. Here, the copper(I) complex $[Cu^I(TMG_3\text{-tren})]^+$, which otherwise has been shown to reversibly bind molecular oxygen,[45,46] can also under certain conditions be oxidized to give an alkoxide–copper(II) complex $[Cu^{II}(TMG_3\text{-tren})O]^+$, which has been crystallographically characterized.[47] In the present ongoing work, we have been using low-temperature UV-vis spectroscopic monitoring in organic solvents of the reaction of $[Cu^I(TMG_3\text{-tren})]^+$ with a dry source of hydrogen peroxide, *n*-(*o*-$Tol_3\cdot POH_2O_2)_2$, shown in Fig. 8.

Some of the experimental observations are: (i) the use of an excess of the H_2O_2 reagent gives good yields of $[Cu^{II}(TMG_3\text{-tren})O]^+$, (ii) when the H_2O_2 reagent is added but the reaction is quenched prior to alkoxide complex formation, the stripping of copper and analysis of organics reveals that good yields of the hydroxylated ligand are obtained; one of the ligand methyl groups has been converted to a $-CH_2OH$ group, giving L–OH, (iii) when excess copper(I) complex is reacted with the H_2O_2 reagent, L–OH is still generated, in amounts correlating with the quantity of H_2O_2 added. These results are consistent with the nature of the reaction occurring as being $[Cu^I(TMG_3\text{-tren})]^+ + H_2O_2 \rightarrow [Cu^I(L\text{–}OH)]^+$, suggesting that one of the reactions (1), (2) or (3) (*vide supra*) took place, *i.e.*, a ligand methyl group was attacked and $^{\bullet}$OH rebound may have ensued, leaving copper(I). It may be that we can show that LCu^I reacts with hydrogen peroxide to produce (L–OH)Cu^I plus H_2O, in line with literature proposals for catalytic peroxygenation. Many more experiments need to be carried out, including utilizing trapping reagents which possibly could show that $^{\bullet}$OH, for example, is produced as an intermediate.

7. Conclusions

It is clear that bioinorganic chemistry is alive and well and the 2022 *Faraday Discussion* on natural and artificial enzymes, including synthetic models, has provided contemporary and exciting contributions for a range of subtopics in the field. As mentioned, bioinorganic chemistry is a hugely broad discipline and it includes the study of metal ions in biology, which possesses contributions from spectroscopy (and, as noted, many very new advanced techniques and applications), X-ray crystallography, enzymology, molecular biology, computational/theoretical studies and synthetic modeling. The expectation is that bioinorganic chemistry will remain a very vibrant field and there will be much more to come.

Conflicts of interest

There are no conflicts to declare.

Acknowledgements

The authors acknowledge the generous research support of the National Institutes of Health of the USA.

Notes and references

1 T. J. Collins, *Acc. Chem. Res.*, 1994, **27**, 279–285.
2 C. J. Ballhausen and H. B. Gray, *Inorg. Chem.*, 1962, **1**, 111–122.
3 J. R. Winkler and H. B. Gray, *Struct. Bonding*, 2012, **142**, 17–28.
4 S. M. Adam, G. B. Wijeratne, P. J. Rogler, D. E. Diaz, D. A. Quist, J. J. Liu and K. D. Karlin, *Chem. Rev.*, 2018, **118**, 10840–11022.
5 E. I. Solomon, D. E. Heppner, E. M. Johnston, J. W. Ginsbach, J. Cirera, M. Qayyum, M. T. Kieber-Emmons, C. H. Kjaergaard, R. G. Hadt and L. Tian, *Chem. Rev.*, 2014, **114**, 3659–3853.
6 Y. Umena, K. Kawakami, J.-R. Shen and N. Kamiya, *Nature*, 2011, **473**, 55–60.
7 J. P. McEvoy and G. W. Brudvig, *Chem. Rev.*, 2006, **106**, 4455–4483.
8 A. M. Appel, J. E. Bercaw, A. B. Bocarsly, H. Dobbek, D. L. DuBois, M. Dupuis, J. G. Ferry, E. Fujita, R. Hille, P. J. A. Kenis, C. A. Kerfeld, R. H. Morris, C. H. F. Peden, A. R. Portis, S. W. Ragsdale, T. B. Rauchfuss, J. N. H. Reek, L. C. Seefeldt, R. K. Thauer and G. L. Waldrop, *Chem. Rev.*, 2013, **113**, 6621–6658.
9 H. Dobbek, L. Gremer, R. Kiefersauer, R. Huber and O. Meyer, *Proc. Natl. Acad. Sci. U. S. A.*, 2002, **99**, 15971–15976.
10 J.-H. Jeoung and H. Dobbek, *Science*, 2007, **318**, 1461–1464.
11 M. B. Ross, P. De Luna, Y. Li, C.-T. Dinh, D. Kim, P. Yang and E. H. Sargent, *Nat. Catal.*, 2019, **2**, 648–658.
12 Y. Zheng, A. Vasileff, X. Zhou, Y. Jiao, M. Jaroniec and S.-Z. Qiao, *J. Am. Chem. Soc.*, 2019, **141**, 7646–7659.
13 D. Schilter, J. M. Camara, M. T. Huynh, S. Hammes-Schiffer and T. B. Rauchfuss, *Chem. Rev.*, 2016, **116**, 8693–8749.
14 W. Lubitz, H. Ogata, O. Rüdiger and E. Reijerse, *Chem. Rev.*, 2014, **114**, 4081–4148.
15 Y. Liu and T. J. Meyer, *Proc. Natl. Acad. Sci. U. S. A.*, 2019, **116**, 2794.
16 O. Elishav, B. Mosevitzky Lis, E. M. Miller, D. J. Arent, A. Valera-Medina, A. Grinberg Dana, G. E. Shter and G. S. Grader, *Chem. Rev.*, 2020, **120**, 5352–5436.

17 M. Raghibi Boroujeni, C. Greene, J. A. Bertke and T. H. Warren, *ChemRxiv*, 2019, DOI: 10.26434/chemrxiv.9729635.v9729631, Cambridge: Cambridge Open Engage. This content is a preprint and has not been peer-reviewed.

18 M. D. Zott and J. C. Peters, *J. Am. Chem. Soc.*, 2021, **143**, 7612–7616.

19 G. E. Cutsail, M. O. Ross, A. C. Rosenzweig and S. DeBeer, *Chem. Sci.*, 2021, **12**, 6194–6209.

20 C. W. Koo, F. J. Tucci, Y. He and A. C. Rosenzweig, *Science*, 2022, **375**, 1287–1291.

21 Z. W. Seh, J. Kibsgaard, C. F. Dickens, I. Chorkendorff, J. K. Nørskov and T. F. Jaramillo, *Science*, 2017, **355**, eaad4998.

22 Y. Sun and S. Dai, *Sci. Adv.*, 2021, **7**, eabg1600.

23 B. H. R. Suryanto, K. Matuszek, J. Choi, R. Y. Hodgetts, H.-L. Du, J. M. Bakker, C. S. M. Kang, P. V. Cherepanov, A. N. Simonov and D. R. MacFarlane, *Science*, 2021, **372**, 1187–1191.

24 W. Cheng, H. Zhang, D. Luan and W. Lou Xiong, *Sci. Adv.*, 2021, **7**, eabg2580.

25 R. M. Bullock, J. G. Chen, L. Gagliardi, P. J. Chirik, O. K. Farha, C. H. Hendon, C. W. Jones, J. A. Keith, J. Klosin, S. D. Minteer, R. H. Morris, A. T. Radosevich, T. B. Rauchfuss, N. A. Strotman, A. Vojvodic, T. R. Ward, J. Y. Yang and Y. Surendranath, *Science*, 2020, **369**, eabc3183.

26 E. Pugliese, P. Gotico, I. Wehrung, B. Boitrel, A. Quaranta, M.-H. Ha-Thi, T. Pino, M. Sircoglou, W. Leibl, Z. Halime and A. Aukauloo, *Angew. Chem., Int. Ed.*, 2022, **61**, e202117530.

27 R. H. Holm and E. I. Solomon, *Chem. Rev.*, 2014, **114**, 3367–3368.

28 https://utw10426.utweb.utexas.edu/Topics/Models/Text.html.

29 S. C. Zimmerman and A. D. Hamilton, *Curr. Opin. Chem. Biol.*, 1999, **3**, 711–713.

30 H. R. Marsden, I. Tomatsu and A. Kros, *Chem. Soc. Rev.*, 2011, **40**, 1572–1585.

31 N. S. Baban, A. Orozaliev, S. Kirchhof, C. J. Stubbs and Y.-A. Song, *Science*, 2022, **375**, 770–774.

32 Y. Lee and K. D. Karlin, in *Concepts and Models in Bioinorganic Chemistry*, ed. N. Metzler-Nolte and H.-B. Kraatz, Wiley-VCH, New York, 2006, pp. 363–395.

33 N. Kitajima, K. Fujisawa, Y. Morooka and K. Toriumi, *J. Am. Chem. Soc.*, 1989, **111**, 8975–8976.

34 K. A. Magnus, H. Ton-That and J. E. Carpenter, *Chem. Rev.*, 1994, **94**, 727–735.

35 J. Ø. Ipsen, M. Hallas-Møller, S. Brander, L. Lo Leggio and K. S. Johansen, *Biochem. Soc. Trans.*, 2021, **49**, 531–540.

36 L. Rieder, A. A. Stepnov, M. Sørlie and V. G. H. Eijsink, *Biochemistry*, 2021, **60**, 3633–3643.

37 S. M. Jones, W. J. Transue, K. K. Meier, B. Kelemen and E. I. Solomon, *Proc. Natl. Acad. Sci. U. S. A.*, 2020, **117**, 11916.

38 B. Bissaro, B. Streit, I. Isaksen, V. G. H. Eijsink, G. T. Beckham, J. L. DuBois and Å. K. Røhr, *Proc. Natl. Acad. Sci. U. S. A.*, 2020, **117**, 1504, DOI: 10.1073/pnas.1904889117.

39 B. Wang, Z. Wang, G. J. Davies, P. H. Walton and C. Rovira, *ACS Catal.*, 2020, **10**, 12760–12769.

40 K. K. Meier, S. M. Jones, T. Kaper, H. Hansson, M. J. Koetsier, S. Karkehabadi, E. I. Solomon, M. Sandgren and B. Kelemen, *Chem. Rev.*, 2018, **118**, 2593–2635.

41 L. Ciano, G. J. Davies, W. B. Tolman and P. H. Walton, *Nat. Catal.*, 2018, **1**, 571–577.

42 J. A. Hangasky, T. C. Detomasi and M. A. Marletta, *Trends Chem.*, 2019, **1**, 198–209.
43 E. Illés, A. Mizrahi, V. Marks and D. Meyerstein, *Free Radical Biol. Med.*, 2019, **131**, 1–6.
44 R. Davydov, A. E. Herzog, R. J. Jodts, K. D. Karlin and B. M. Hoffman, *J. Am. Chem. Soc.*, 2022, **144**, 377–389.
45 M. Bhadra, W. J. Transue, H. Lim, R. E. Cowley, J. Y. C. Lee, M. A. Siegler, P. Josephs, G. Henkel, M. Lerch, S. Schindler, A. Neuba, K. O. Hodgson, B. Hedman, E. I. Solomon and K. D. Karlin, *J. Am. Chem. Soc.*, 2021, **143**, 3707–3713.
46 C. Würtele, E. Gaoutchenova, K. Harms, M. C. Holthausen, J. Sundermeyer and S. Schindler, *Angew. Chem., Int. Ed.*, 2006, **45**, 3867–3869.
47 D. Maiti, D.-H. Lee, K. Gaoutchenova, C. Würtele, M. C. Holthausen, A. A. N. Sarjeant, J. Sundermeyer, S. Schindler and K. D. Karlin, *Angew. Chem., Int. Ed.*, 2008, **47**, 82–85.

Poster titles

Understanding the effect of hydrotrope towards stability and activity of cytochrome c, **P. Bharadwaj, S. M. Shet, S. K. Nataraj and D. Mondal**, *Centre for Nano & Material Sciences, JAIN (Deemed to be University), Bangalore-562112, India*

Single-atom catalyst boost enzyme-like activity for biosensing, **S. Ding, Z. Lyu, D. Du and Y. Lin**, *School of Mechanical and Materials Engineering, Washington State University, USA*

The activation of glycyl radical enzymes catalysed by the iron–sulfur containing proteins, **M. Hanževački, A. Croft and C. Jäger**, *Department of Chemical and Environmental Engineering, University of Nottingham, UK*

Copper complexes as antimicrobial resistance breakers, **A. Mishra and J. W. Walton**, *Chemistry Department, Durham University, UK*

Structural and spectroscopic evidences for a side-on Fe(III)-superoxo complex featuring discrete O–O bond distances, **H.-R. Pan, H.-J. Chen, Z.-H. Wu, P. Ge, S. Ye, G.-H. Lee and H.-F. Hsu**, *Department of Chemistry, National Cheng Kung University, Chinese Taipei*

Engineering cytochrome c with quantum dots and ionic liquids: a win-win strategy for protein packaging against multiple stresses, **S. M. Shet, P. Bharadwaj, S. K. Nataraj and D. Mondal**, *Centre for Nano & Material Sciences, JAIN (Deemed to be University) Bangalore-562112, India*

Benchmarking density functional theory for metalloenzyme reactions, **D. Wappett and L. Goerigk**, *Melbourne Centre for Theoretical and Computational Chemistry, School of Chemistry, University of Melbourne, Australia*

Understanding the oxygen sensitivity of [FeFe]-hydrogenase: a reactive study of active site mimics and functional enzyme, **S. Yadav, S. T. Stripp and U.-P. Apfel**, *Inorganic Chemistry I, Ruhr-Universität Bochum, Germany*

Engineered metalloproteins as artificial biocatalysts, **R. L. Domene, A. Beloqui and A. L. Cortajarena**, *POLYMAT and Applied Chemistry Department, Faculty of Chemistry, University of the Basque Country UPV/EHU, Spain; CIC biomaGUNE, Spain*

The affiliation of the presenting author is listed above.

The Faraday Division Poster Prize for the best poster was awarded to Hung-Ruei Pan of National Cheng Kung University, Chinese Taipei for a poster on "Structural and spectroscopic evidences for a side-on Fe(III)-superoxo complex featuring discrete O–O bond distances".

List of participants

Dr Mohsen Ahmadi, *Leibniz Institute for Plasma Science and Technology (INP Greifswald), Germany*
Dr Leigh Aldous, *King's College London*, *United Kingdom*
Miss Anjali Anilkumar, *Technical University Braunschweig*, *Germany*
Dr Philip Ash, *University of Leicester*, *United Kingdom*
Mr Pranav Bharadwaj, *JAIN(Deemed-to-be University)*, *India*
Professor Dr Akhil Chakravarty, *Indian Institute of Science*, *India*
Dr Sudipta Chatterjee, *King Abdullah University of Science and Technology*, *Saudi Arabia*
Ms Yunzhou Chen, *Hong Kong Baptist University*, *Hong Kong*
Dr Luisa Ciano, *University of Nottingham*, *United Kingdom*
Miss Eleanor Clifford, *Royal Society of Chemistry*, *United Kingdom*
Professor Dr Peter Comba, *Universität Heidelberg*, *Germany*
Dr James Coverdale, *Univeristy of Birmingham*, *United Kingdom*
Professor Dr Serena DeBeer, *Max Planck Institute for Chemical Energy Conversion*, *Germany*
Professor Abhishek Dey, *Indian Association for the Cultivation of Science*, *India*
Mr Shichao Ding, *Washington State University*, *USA*
Mr Zachary Douglas, *University of Leicester*, *United Kingdom*
Dr Apparao Draksharapu, *Indian Institute of Technology Kanpur*, *India*
Mrs Carole Duboc, *Université Grenoble Alpes, CNRS*, *France*
Dr Ouissam El Bakouri, *Universitat de Girona*, *Spain*
Mr Andrey Fedulin, *University of Hamburg*, *Germany*
Dr Stefano Fornasaro, *University of Trieste*, *Italy*
Mr Rufus Frew, *University of Leicester*, *United Kingdom*
Miss Zahra Ghafari, *Kharazmi University*, *Iran*
Professor Debabrata Goswami, *Indian Institute of Technology Kanpur*, *India*
Dr Sheetal Handa, *bp International Ltd*, *United Kingdom*
Dr Marko Hanzevacki, *University of Nottingham*, *United Kingdom*
Dr Katy Hattingh, *Illumina*, *United Kingdom*
Dr Jeannine Hess, *The Francis Crick Institute, King's College London*, *United Kingdom*
Professor Hua-Fen Hsu, *National Chang Kung University*, *Chinese Taipei*
Professor Shinobu Itoh, *Osaka University*, *Japan*
Dr Christof Jaeger, *University of Nottingham*, *United Kingdom*
Dr Amanda Jarvis, *University of Edinburgh*, *United Kingdom*
Professor Kenneth Karlin, *Johns Hopkins University*, *USA*
Mr Desmond Ato Koomson, *King's College London*, *United Kingdom*
Dr Subrata Kundu, *Indian Institute of Science Education and Research Thiruvananthapuram*, *India*
Dr Salvatore La Gatta, *University of Naples Federico II*, *Italy*
Professor K. V. Lakshmi, *Rensselaer Polytechnic Institute*, *USA*
Miss Rocío López Domene, *CIC biomaGUNE*, *Spain*

Professor Dr Jean-Didier Maréchal, *Universitat Autònoma de Barcelona*, *Spain*
Professor Shyamalava Mazumdar, *Tata Institute of Fundamental Research*, *India*
Dr Aidan McDonald, *Trinity College Dublin, the University of Dublin*, *Ireland*
Miss Atreyee Mishra, *Durham University*, *United Kingdom*
Dr Kinga Mlekodaj, *J. Heyrovsky Institute of Physical Chemistry of the ASCR, v. v. i.*, *Czech Republic*
Professor Govindasamy Mugesh, *Indian Institute of Science*, *India*
Professor Rabindranath Mukherjee, *Indian Institute of Technology (Indian School of Mines) Dhanbad*, *India*
Mr George Nunn, *University of York*, *United Kingdom*
Mr Sean O'Halloran, *Dublin City University*, *Ireland*
Dr Philip Palacios, *Carnegie Mellon University*, *USA*
Mr Hung-Ruei Pan, *National Chang Kung University*, *Chinese Taipei*
Dr Dimitrios Pantazis, *Max-Planck-Institut für Kohlenforschung*, *Germany*
Professor David Parker, *Durham University*, *United Kingdom*
Dr Anca Pordea, *University of Nottingham*, *United Kingdom*
Dr Vikki Pritchard, *Royal Society of Chemistry*, *United Kingdom*
Professor Dr Lawrence Que, *University of Minnesota*, *USA*
Professor Gopalan Rajaraman, *Indian Institute of Technology Bombay*, *India*
Professor Dr Sankar Rath, *Indian Institute of Technology Kanpur*, *India*
Professor Peter Sadler, *University of Warwick*, *United Kingdom*
Ms Paramita Saha, *Indian Association for the Cultivation of Science*, *India*
Professor Chivukula Sastri, *Indian Institute of Technology Guwahati*, *India*
Mr Frederik Scherz, *Universität Heidelberg*, *Germany*
Professor Dr Siegfried Schindler, *Justus-Liebig-Universität*, *Germany*
Professor Dr Volker Schünemann, *University of Kaiserslautern*, *Germany*
Mr Dmitry Sedush, *University of Leicester*, *United Kingdom*
Dr Sayam Sen Gupta, *Indian Institute of Science Education and Research Kolkata*, *India*
Dr Kushal Sengupta, *Max Planck Institute for Chemical Energy Conversion*, *Germany*
Mr Sachin Shet, *JAIN(Deemed-to-be-university) Centre for Nano and Material Science*, *India*
Professor Dr Osami Shoji, *Nagoya University*, *Japan*
Dr Donna Smith, *Royal Society of Chemistry*, *United Kingdom*
Professor Edward I. Solomon, *Stanford University*, *USA*
Professor Dan Stack, *Stanford University*, *USA*
Professor Dr Marcel Swart, *Catalan Institution for Research and Advanced Studies and Universitat de Girona*, *Spain*
Professor Dr Edyta Tabor, *J. Heyrovsky Institute of Physical Chemistry of the ASCR, v. v. i.*, *Czech Republic*
Professor Paul Walton, *University of York*, *United Kingdom*
Ms Dominique Wappett, *The University of Melbourne*, *Australia*
Dr Juliusz A. Wolny, *Technische Universität Kaiserslautern*, *Germany*
Mr Kin Long Wong, *University of Oxford*, *United Kingdom*
Miss Shanika Yadav, *Ruhr-Universität Bochum*, *Germany*
Ms Anna Ziemele, *Dublin City University*, *Ireland*